Springer Handbook of Auditory Research

Series Editors: Richard R. Fay and Arthur N. Popper

Springer
New York
Berlin
Heidelberg
Hong Kong
London
Milan
Paris
Tokyo

SPRINGER HANDBOOK OF AUDITORY RESEARCH

Volume 1: The Mammalian Auditory Pathway: Neuroanatomy
Edited by Douglas B. Webster, Arthur N. Popper, and Richard R. Fay

Volume 2: The Mammalian Auditory Pathway: Neurophysiology
Edited by Arthur N. Popper and Richard R. Fay

Volume 3: Human Psychophysics
Edited by William Yost, Arthur N. Popper, and Richard R. Fay

Volume 4: Comparative Hearing: Mammals
Edited by Richard R. Fay and Arthur N. Popper

Volume 5: Hearing by Bats
Edited by Arthur N. Popper and Richard R. Fay

Volume 6: Auditory Computation
Edited by Harold L. Hawkins, Teresa A. McMullen, Arthur N. Popper, and Richard R. Fay

Volume 7: Clinical Aspects of Hearing
Edited by Thomas R. Van De Water, Arthur N. Popper, and Richard R. Fay

Volume 8: The Cochlea
Edited by Peter Dallos, Arthur N. Popper, and Richard R. Fay

Volume 9: Development of the Auditory System
Edited by Edwin W Rubel, Arthur N. Popper, and Richard R. Fay

Volume 10: Comparative Hearing: Insects
Edited by Ronald Hoy, Arthur N. Popper, and Richard R. Fay

Volume 11: Comparative Hearing: Fish and Amphibians
Edited by Richard R. Fay and Arthur N. Popper

Volume 12: Hearing by Whales and Dolphins
Edited by Whitlow W.*L. Au, Arthur N. Popper, and Richard R. Fay*

Volume 13: Comparative Hearing: Birds and Reptiles
Edited by Robert Dooling, Arthur N. Popper, and Richard R. Fay

Volume 14: Genetics and Auditory Disorders
Edited by Bronya J.B. Keats, Arthur N. Popper, and Richard R. Fay

Volume 15: Integrative Functions in the Mammalian Auditory Pathway
Edited by Donata Oertel, Richard R. Fay, and Arthur N. Popper

Volume 16: Acoustic Communication
Edited by Andrea Simmons, Arthur N. Popper, and Richard R. Fay

Volume 17: Compression: From Cochlea to Cochlear Implants
Edited by Sid P. Bacon, Richard R. Fay, and Arthur N. Popper

Volume 18: Speech Processing in the Auditory System
Edited by Steven Greenberg, William Ainsworth, Arthur N. Popper, and Richard R. Fay

Volume 19: The Vestibular System
Edited by Stephen M. Highstein, Richard R. Fay, and Arthur N. Popper

(continued after index)

Steven Greenberg William A. Ainsworth
Arthur N. Popper Richard R. Fay
Editors

Speech Processing in the Auditory System

With 83 Illustrations

Springer

Steven Greenberg
The Speech Institute
Berkeley, CA 94704, USA

Arthur N. Popper
Department of Biology and
Neuroscience and Cognitive Science
Program and
Center for Comparative and
Evolutionary Biology of Hearing
University of Maryland
College Park, MD 20742-4415, USA

William A. Ainsworth (deceased)
Department of Communication and
Neuroscience
Keele University
Keele, Staffordshire ST5 3BG, UK

Richard R. Fay
Department of Psychology and
Parmly Hearing Institute
Loyola University of Chicago
Chicago, IL 60626 USA

Series Editors: Richard R. Fay and Arthur N. Popper

Cover illustration: Details from Figs. 5.8: Effects of reverberation on speech spectrogram (p. 270) and 8.4: Temporospatial pattern of action potentials in a group of nerve fibers (p. 429).

Library of Congress Cataloging-in-Publication Data
Speech processing in the auditory system / editors, Steven Greenberg . . . [et al.].
p. cm.—(Springer handbook of auditory research ; v. 18)
Includes bibliographical references and index.
ISBN 0-387-00590-0 (hbk. : alk. paper)
1. Audiometry–Handbooks, manuals, etc. 2. Auditory pathways–Handbooks, manuals, etc. 3. Speech perception–Handbooks, manuals, etc. 4. Speech processing systems–Handbooks, manuals, etc. 5. Hearing–Handbooks, manuals, etc.
I. Greenberg, Steven. II. Series.
RF291.S664 2203
617.8′075—dc21 2003042432

ISBN 0-387-00590-0 Printed on acid-free paper.

Printed in the United States of America. (EB)

9 8 7 6 5 4 3 2 1 SPIN 10915684

Springer-Verlag is a part of *Springer Science+Business Media*

springeronline.com

In Memoriam

William A. Ainsworth

1941–2002

This book is dedicated to the memory of Bill Ainsworth, who unexpectedly passed away shortly before this book's completion. He was an extraordinarily gifted scientist who pioneered many areas of speech research relating to perception, production, recognition, and synthesis. Bill was also an exceptionally warm and friendly colleague who touched the lives of many in the speech community. He will be sorely missed.

Series Preface

The *Springer Handbook of Auditory Research* presents a series of comprehensive and synthetic reviews of the fundamental topics in modern auditory research. The volumes are aimed at all individuals with interests in hearing research including advanced graduate students, post-doctoral researchers, and clinical investigators. The volumes are intended to introduce new investigators to important aspects of hearing science and to help established investigators to better understand the fundamental theories and data in fields of hearing that they may not normally follow closely.

Each volume presents a particular topic comprehensively, and each chapter serves as a synthetic overview and guide to the literature. As such, the chapters present neither exhaustive data reviews nor original research that has not yet appeared in peer-reviewed journals. The volumes focus on topics that have developed a solid data and conceptual foundation rather than on those for which a literature is only beginning to develop. New research areas will be covered on a timely basis in the series as they begin to mature.

Each volume in the series consists of a few substantial chapters on a particular topic. In some cases, the topics will be ones of traditional interest for which there is a substantial body of data and theory, such as auditory neuroanatomy (Vol. 1) and neurophysiology (Vol. 2). Other volumes in the series will deal with topics that have begun to mature more recently, such as development, plasticity, and computational models of neural processing. In many cases, the series editors will be joined by a co-editor having special expertise in the topic of the volume.

Richard R. Fay, Chicago, Illinois
Arthur N. Popper, College Park, Maryland

Preface

Although our sense of hearing is exploited for many ends, its communicative function stands paramount in our daily lives. Humans are, by nature, a vocal species and it is perhaps not too much of an exaggeration to state that what makes us unique in the animal kingdom is our ability to communicate via the spoken word. Virtually all of our social nature is predicated on verbal interaction, and it is likely that this capability has been largely responsible for the rapid evolution of humans. Our verbal capability is often taken for granted; so seamlessly does it function under virtually all conditions encountered. The intensity of the acoustic background hardly matters—from the hubbub of a cocktail party to the roar of waterfall's descent, humans maintain their ability to interact verbally in a remarkably diverse range of acoustic environments. Only when our sense of hearing falters does the auditory system's masterful role become truly apparent.

This volume of the *Springer Handbook of Auditory Research* examines speech communication and the processing of speech sounds by the nervous system. As such, it is a natural companion to many of the volumes in the series that ask more fundamental questions about hearing and processing of sound. In the first chapter, Greenberg and the late Bill Ainsworth provide an important overview on the processing of speech sounds and consider a number of the theories pertaining to detection and processing of communication signals.

In Chapter 2, Avendaño, Deng, Hermansky, and Gold discuss the analysis and representation of speech in the brain, while in Chapter 3, Diehl and Lindblom deal with specific features and phonemes of speech. The physiological representations of speech at various levels of the nervous system are considered by Palmer and Shamma in Chapter 4. One of the most important aspects of speech perception is that speech can be understood under adverse acoustic conditions, and this is the theme of Chapter 5 by Assmann and Summerfield. The growing interest in speech recognition and attempts to automate this process are discussed by Morgan, Bourlard, and Hermansky in Chapter 6. Finally, the very significant issues related to hearing impairment and ways to mitigate these issues are considered first by Edwards (Chapter 7) with regard to hearing aids and then by Clark (Chapter 8) for cochlear implants and speech processing.

Clearly, while previous volumes in the series have not dealt with speech processing per se, chapters in a number of volumes provide background and related topics from a more basic perspective. For example, chapters in *The Mammalian Auditory Pathway: Neurophysiology* (Vol. 2) and in *Integrative Functions in the Mammalian Auditory Pathway* (Vol. 15) help provide an understanding of central processing of sounds in mammals. Various chapters in *Human Psychophysics* (Vol. 3) deal with sound perception and processing by humans, while chapters in *Auditory Computation* (Vol. 6) discuss computational models related to speech detection and processing.

The editors would like to thank the chapter authors for their hard work and diligence in preparing the material that appears in this book. Steven Greenberg expresses his gratitude to the series editors, Arthur Popper and Richard Fay, for their encouragement and patience throughout this volume's lengthy gestation period.

Steven Greenberg, Berkeley, California
Arthur N. Popper, College Park, Maryland
Richard R. Fay, Chicago, Illinois

Contents

Contributors

William A. Ainsworth†
Department of Communication & Neuroscience, Keele University, Keele, Staffordshire ST5 3BG, UK

Peter Assmann
School of Human Development, University of Texas–Dallas, Richardson, TX 75083-0688, USA

Carlos Avendaño
Creative Advanced Technology Center, Scotts Valley, CA 95067, USA

Hervé Bourlard
Dalle Molle Institute for Perceptual Artificial Intelligence, CH-1920 Martigny, Switzerland

Graeme Clark
Centre for Hearing Communication Research and Co-operative Research Center for Cochlear Implant Speech and Hearing Center, Melbourne, Australia

Li Deng
Microsoft Corporation, Redmond, WA 98052, USA

Randy Diehl
Psychology Department, University of Texas, Austin, TX 78712, USA

Brent Edwards
Sound ID, Palo Alto, CA 94303, USA

Ben Gold
MIT Lincoln Laboratory, Lexington, MA 02173, USA

† Deceased

STEVEN GREENBERG
The Speech Institute, Berkeley, CA 94704, USA

HYNEK HERMANSKY
Dalle Molle Institute for Perceptual Artificial Intelligence, CH-1920, Martigny, Switzerland

BJÖRN LINDBLOM
Department of Linguistics, Stockholm University, S-10691 Stockholm, Sweden

NELSON MORGAN
International Computer Science Institute, Berkeley, CA 94704, USA

ALAN PALMER
MRC Institute of Hearing Research, University Park, Nottingham NG7 2RD, UK

SHIHAB SHAMMA
Department of Electrical and Computer Engineering, University of Maryland, College Park, MD 20742, USA

QUENTIN SUMMERFIELD
MRC Institute of Hearing Research, University Park, Nottingham NG7 2RD, UK

1
Speech Processing in the Auditory System: An Overview

Steven Greenberg and William A. Ainsworth

1. Introduction

Although our sense of hearing is exploited for many ends, its communicative function stands paramount in our daily lives. Humans are, by nature, a vocal species, and it is perhaps not too much of an exaggeration to state that what makes us unique in the animal kingdom is our ability to communicate via the spoken word (Hauser et al. 2002). Virtually all of our social nature is predicated on verbal interaction, and it is likely that this capability has been largely responsible for *Homo sapiens'* rapid evolution over the millennia (Lieberman 1990; Wang 1998). So intricately bound to our nature is language that those who lack it are often treated as less than human (Shattuck 1980).

Our verbal capability is often taken for granted, so seamlessly does it function under virtually all conditions encountered. The intensity of the acoustic background hardly matters—from the hubbub of a cocktail party to the roar of waterfall's descent, humans maintain their ability to verbally interact in a remarkably diverse range of acoustic environments. Only when our sense of hearing falters does the auditory system's masterful role become truly apparent (cf. Edwards, Chapter 7; Clark, Chapter 8). For under such circumstances the ability to communicate becomes manifestly difficult, if not impossible. Words "blur," merging with other sounds in the background, and it becomes increasingly difficult to keep a specific speaker's voice in focus, particularly in noise or reverberation (cf. Assmann and Summerfield, Chapter 5). Like a machine that suddenly grinds to a halt by dint of a faulty gear, the auditory system's capability of processing speech depends on the integrity of most (if not all) of its working elements.

Clearly, the auditory system performs a remarkable job in converting physical pressure variation into a sequence of meaningful elements composing language. And yet, the process by which this transformation occur is poorly understood despite decades of intensive investigation.

The role of the auditory system has traditionally been viewed as a frequency analyzer (Ohm 1843; Helmholtz 1863), albeit of limited precision

(Plomp 1964), providing a faithful representation of the spectro-temporal properties of the acoustic waveform for higher-level processing. According to Fourier theory, any waveform can be decomposed into a series of sinusoidal constituents, which mathematically describe the acoustic waveform (cf. Proakis and Manolakis 1996; Lynn and Fuerst 1998). By this analytical technique it is possible to describe all speech sounds in terms of an energy distribution across frequency and time. Thus, the Fourier spectrum of a typical vowel is composed of a series of sinusoidal components whose frequencies are integral multiples of a common (fundamental) frequency (f_0), and whose amplitudes vary in accordance with the resonance pattern of the associated vocal-tract configuration (cf. Fant 1960; Pickett 1980). The vocal-tract transfer function modifies the glottal spectrum by selectively amplifying energy in certain regions of the spectrum (Fant 1960). These regions of energy maxima are commonly referred to as "formants" (cf. Fant 1960; Stevens 1998). The spectra of nonvocalic sounds, such as stop consonants, affricates, and fricatives, differ from vowels in a number of ways potentially significant for the manner in which they are encoded in the auditory periphery. These segments typically exhibit formant patterns in which the energy peaks are considerably reduced in magnitude relative to those of vowels. In certain articulatory components, such as the stop release and frication, the energy distribution is rather diffuse, with only a crude delineation of the underlying formant pattern. In addition, many of these segments are voiceless, their waveforms lacking a clear periodic quality that would otherwise reflect the vibration of the vocal folds of the larynx. The amplitude of such consonantal segments is typically 30 to 50 dB sound pressure level (SPL), up to 40 dB less intense than adjacent vocalic segments (Stevens 1998). In addition, the rate of spectral change is generally greater for consonants, and they are usually of brief duration compared to vocalic segments (Avendaño et al., Chapter 2; Diehl and Lindblom, Chapter 3). These differences have significant consequences for the manner in which consonants and vowels are encoded in the auditory system.

Within this traditional framework each word spoken is decomposed into constituent sounds, known as phones (or phonetic segments), each with its own distinctive spectral signature. The auditory system need only encode the spectrum, time frame by time frame, to provide a complete representation of the speech signal for conversion into meaning by higher cognitive centers. Within this formulation (known as articulation theory), speech processing is a matter of frequency analysis and little else (e.g., French and Steinberg 1947; Fletcher and Gault 1950; Pavlovic et al. 1986; Allen 1994). Disruption of the spectral representation, by whatever means, results in phonetic degradation and therefore interferes with the extraction of meaning. This "spectrum-über-alles" framework has been particularly influential in the design of automatic speech recognition systems (cf. Morgan et al., Chapter 6), as well as in the development of algorithms for the prosthetic amelioration of sensorineural hearing loss (cf. Edwards, Chapter 7; Clark, Chapter 8).

However, this view of the ear as a mere frequency analyzer is inadequate for describing the auditory system's ability to process speech. Under many conditions its frequency-selective properties bear only a tangential relationship to its ability to convey important information concerning the speech signal, relying rather on the operation of integrative mechanisms to isolate information-laden elements of the speech stream and provide a continuous event stream from which to extract the underlying message. Hence, cocktail party devotees can attest to the fact that far more is involved in decoding the speech signal than merely computing a running spectrum (Bronkhorst 2000). In noisy environments a truly faithful representation of the spectrum could actually serve to hinder the ability to understand due to the presence of background noise or competing speech. It is likely that the auditory system uses very specific strategies to focus on those elements of speech most likely to extract the meaningful components of the acoustic signal (cf. Brown and Cooke 1994; Cooke and Ellis 2001). Computing a running spectrum of the speech signal is a singularly inefficient means to accomplish this objective, as much of the acoustics is extraneous to the message. Instead, the ear has developed the means to extract the information-rich components of the speech signal (and other sounds of biological significance) that may resemble the Fourier spectral representation only in passing.

As the chapters in this volume attest, far more is involved in speech processing than mere frequency analysis. For example, the spectra of speech sounds change over time, sometimes slowly, but often quickly (Liberman et al. 1956; Pols and van Son 1993; Kewley-Port 1983; van Wieringen and Pols 1994, 1998; Kewley-Port and Neel 2003). These dynamic properties provide information essential for distinguishing among phones. Segments with a rapidly changing spectrum sound very different from those whose spectra modulate much more slowly (e.g., van Wieringen and Pols 1998, 2003).

Thus, the concept of "time" is also important for understanding how speech is processed in the auditory system (Fig. 1.1). It is not only the spectrum that changes with time, but also the energy. Certain sounds (typically vowels) are far more intense than others (usually consonants). Moreover, it is unusual for a segment's amplitude to remain constant, even over a short interval of time. Such modulation of energy is probably as important as spectral variation (cf. Van Tassell 1987; Drullman et al. 1994a,b; Kollmeier and Koch 1994; Drullman 2003; Shannon et al. 1995), for it provides information crucial for segmentation of the speech signal, particularly at the syllabic level (Greenberg 1996b; Shastri et al. 1999).

Segmentation is a topic rarely discussed in audition, yet is of profound importance for speech processing. The transition from one syllable to the next is marked by appreciable variation in energy across the acoustic spectrum. Such changes in amplitude serve to delimit one linguistic unit from

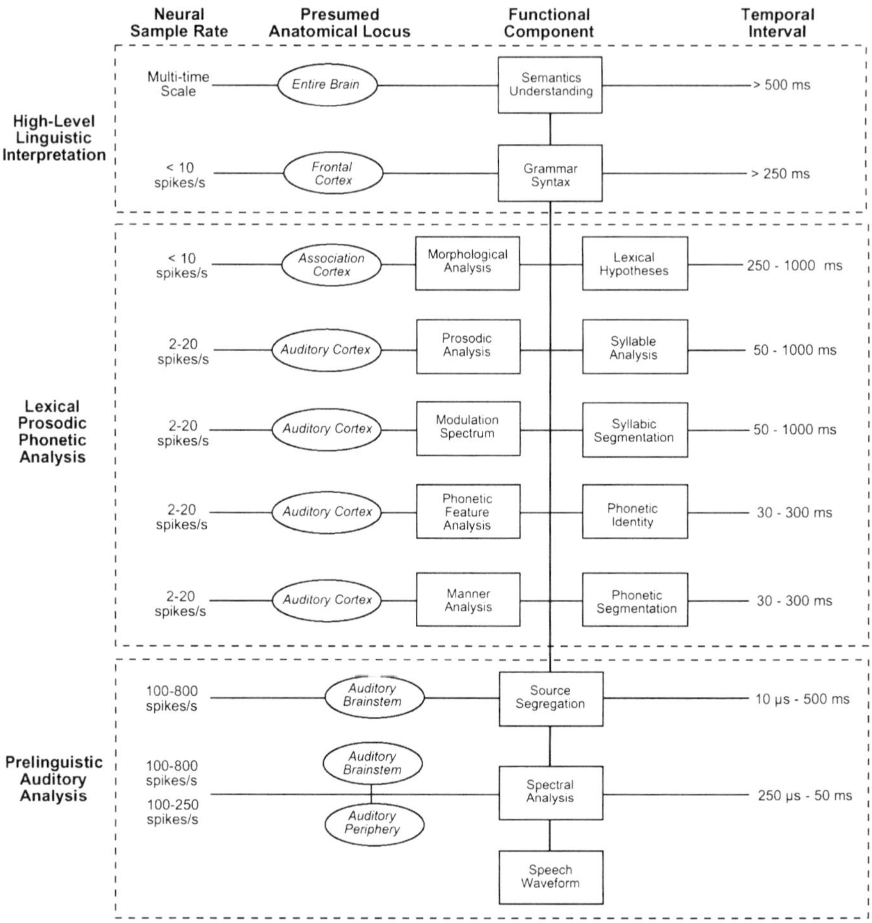

FIGURE 1.1. A temporal perspective of speech processing in the auditory system. The time scale associated with each component of auditory and linguistic analysis is shown, along with the presumed anatomical locus of processing. The auditory periphery and brain stem is presumed to engage solely in prelinguistic analysis relevant for spectral analysis, noise robustness, and source segregation. The neural firing rates at this level of the auditory pathway are relatively high (100–800 spikes/s). Phonetic and prosodic analyses are probably the product of auditory cortical processing, given the relatively long time intervals required for evaluation and interpretation at this linguistic level. Lexical processing probably occurs beyond the level of the auditory cortex, and involves both memory and learning. The higher-level analyses germane to syntax and semantics (i.e., meaning) is probably a product of many different regions of the brain and requires hundreds to thousands of milliseconds to complete.

the next, irrespective of spectral properties. Smearing segmentation cues has a profound impact on the ability to understand speech (Drullman et al. 1994a,b; Arai and Greenberg 1998; Greenberg and Arai 1998), far more so than most forms of spectral distortion (Licklider 1951; Miller 1951; Blesser 1972). Thus, the auditory processes involved in coding syllable-length fluctuations in energy are likely to play a key role in speech processing (Plomp 1983; Drullman et al. 1994a; Grant and Walden 1996a; Greenberg 1996b).

Accompanying modulation of amplitude and spectrum is a variation in fundamental frequency that often spans hundreds, or even thousands, of milliseconds (e.g., Ainsworth 1986; Ainsworth and Lindsay 1986; Lehiste 1996). Such f_0 cues are usually associated with prosodic properties such as intonation and stress (Lehiste 1996), but are also relevant to emotion and semantic nuance embedded in an utterance (Williams and Stevens 1972; Lehiste 1996). In addition, such fluctuations in fundamental frequency (and its perceptual correlate, pitch) may be important for distinguishing one speaker from another (e.g., Weber et al. 2002), as well as locking onto to a specific speaker in a crowded environment (e.g., Brokx and Nooteboom 1982; Cooke and Ellis 2001). Moreover, in many languages (e.g., Chinese and Thai), pitch (referred to as "tone") is also used to distinguish among words (Wang 1972), providing yet another context in which the auditory system plays a key role in the processing of speech.

Perhaps the most remarkable quality of speech is its multiplicity. Not only are its spectrum, pitch, and amplitude constantly changing, but the variation in these properties occurs, to a certain degree, independently of each other, and is decoded by the auditory system in such seamless fashion that we are rarely conscious of the "machinery" underneath the "hood." This multitasking capability is perhaps the auditory system's most important capability, the one enabling a rich stream of information to be securely transmitted to the higher cognitive centers of the brain.

Despite the obvious importance of audition for speech communication, the neurophysiological mechanisms responsible for decoding the acoustic signal are not well understood, either in the periphery or in the more central stations of the auditory pathway (cf. Palmer and Shamma, Chapter 4). The enormous diversity of neuronal response properties in the auditory brainstem, thalamus, and cortex (cf. Irvine 1986; Popper and Fay 1992; Oertel et al. 2002) is of obvious relevance to the encoding of speech and other communicative signals, but the relationship between any specific neuronal response pattern an information contained in the speech signal has not been precisely delineated.

Several factors limit our ability to generalize from brain physiology to speech perception. First, it is not yet possible to record from single neuronal elements in the auditory pathway of humans due to the invasive nature of the recording technology. For this reason, current knowledge concerning the physiology of hearing is largely limited to studies on nonhuman species lacking linguistic capability. Moreover, most of these physiological studies have been performed on anesthesized, nonbehaving animals, rendering the neuronal responses recorded of uncertain relevance to the awake preparation, particularly with respect to the dorsal cochlear nucleus (Rhode and Kettner 1987) and higher auditory stations.

Second, it is inherently difficult to associate the neuronal activity recorded in any single part of the auditory pathway with a specific behavior given the complex nature of decoding spoken language. It is likely that many different regions of the auditory system participate in the analysis and interpretation of the sound patterns associated with speech, and therefore the conclusions that can be made via recordings from any single neuronal site are limited.

Ultimately, sophisticated brain-imaging technology using such methods as functional magnetic resonance imaging (e.g., Buchsbaum et al. 2001) and magnetoencephalography (e.g., Poeppel et al. 1996) is likely to provide the sort of neurological data capable of answering specific questions concerning the relation between speech decoding and brain mechanisms. Until the maturation of such technology much of our knowledge will necessarily rely on more indirect methods such as perceptual experiments and modeling studies.

One reason why the relationship between speech and auditory function has not been delineated with precision is that, historically, hearing has been largely neglected as an explanatory framework for understanding the structure and function of the speech signal itself. Traditionally, the acoustic properties of speech have been ascribed largely to biomechanical constraints imposed by the vocal apparatus (e.g., Ohala 1983; Lieberman 1984). According to this logic, the tongue, lips, and jaw can move only so fast and so far in a given period of time, while the size and shape of the oral cavity set inherent limits on the range of achievable vocal-tract configurations (e.g., Ladefoged 1971; Lindblom 1983; Lieberman 1984).

Although articulatory properties doubtless impose important constraints, it is unlikely that such factors, in and of themselves, can account for the full constellation of spectro-temporal properties of speech. For there are sounds that the vocal apparatus can produce, such as coughing and spitting, that do not occur in any language's phonetic inventory. And while the vocal tract is capable of chaining long sequences composed exclusively of vowels or consonants together in succession, no language relies on either segmental form alone, nor does speech contain long sequences of acoustically similar elements. And although speech can be readily whispered, it is only occasionally done.

Clearly, factors other than those pertaining to the vocal tract per se are primarily responsible for the specific properties of the speech signal. One important clue as to the nature of these factors comes from studies of the evolution of the human vocal tract, which anatomically has changed dramatically over the course of the past several hundred thousand years (Lieberman 1984, 1990, 1998). No ape is capable of spoken language, and the vocal repertoire of our closest phylogenetic cousins, the chimpanzees and gorillas, is impoverished relative to that of humans[1] (Lieberman 1984). The implication is that changes in vocal anatomy and physiology observed over the course of human evolution are linked to the dramatic expansion of the brain (cf. Wang 1998), which in turn suggests that a primary selection factor shaping vocal-tract function (Carré and Mrayati 1995) is the capability of transmitting large amounts of information quickly and reliably.

However, this dramatic increase in information transmission has been accompanied by relatively small changes in the anatomy and physiology of the human auditory system. Whereas a quantal leap occurred in vocal capability from ape to human, auditory function has not changed all that much over the same evolutionary period. Given the conservative design of the auditory system across mammalian species (cf. Fay and Popper 1994), it seems likely that the evolutionary innovations responsible for the phylogenetic development of speech were shaped to a significant degree by anatomical, physiological, and functional constraints imposed by the auditory nervous system in its role as transmission route for acoustic information to the higher cortical centers of the brain (cf. Ainsworth 1976; Greenberg 1995, 1996b, 1997a; Greenberg and Ainsworth 2003).

2. How Does the Brain Proceed from Sound to Meaning?

Speech communication involves the transmission of ideas (as well as desires and emotions) from the mind of the speaker to that of the listener via an acoustic (often supplemented by a visual) signal produced by the vocal apparatus of the speaker. The message is generally formulated as a sequence of words chosen from a large but finite set known to both the speaker and the listener. Each word contains one or more syllables, which are themselves composed of sequences of phonetic elements reflecting the manner in which the constituent sounds are produced. Each phone has a number of distinctive attributes, or features, which encode the manner of production and place of articulation. These features form the acoustic pattern that the listener decodes to understand the message.

The process by which the brain proceeds from sound to meaning is not well understood. Traditionally, models of speech perception have assumed that the speech signal is decoded phone by phone, analogous to the manner in which words are represented on the printed page as a sequence of

[1] However, it is unlikely that speech evolved de novo, but rather represents an elaboration of a more primitive form of acoustic communication utilized by our primate forebears (cf. Hauser 1996). Many of the selection pressures shaping these nonhuman communication systems, such as robust transmission under uncertain acoustic conditions (cf. Assmann and Summerfield, Chapter 5), apply to speech as well.

discrete orthographic characters (Klatt 1979; Pisoni and Luce 1987; Goldinger et al. 1996). The sequence of phones thus decoded enables the listener to match the acoustic input to an abstract phone-sequence representation stored in the brain's mental lexicon. According to this perspective the process of decoding is a straightforward one in which the auditory system performs a spectral analysis over time that is ultimately associated with an abstract phonetic unit known as the phoneme.

Such sequential models assume that each phone is acoustically realized in comparable fashion from one instance of a word to the next, and that the surrounding context does not affect the manner in which a specific phone is produced. A cursory inspection of a speech signal (e.g., Fig. 2.5 in Avendaño et al., Chapter 2) belies this simplistic notion. Thus, the position of a phone within the syllable has a noticeable influence on its acoustic properties. For example, a consonant at the end (coda) of a syllable tends to be shorter than its counterpart in the onset. Moreover, the specific articulatory attributes associated with a phone also vary as a function of its position within the syllable and the word. A consonant at syllable onset is often articulated differently from its segmental counterpart in the coda. For example, voiceless, stop consonants, such as [p], [t], and [k] are usually produced with a complete articulatory constriction ("closure") followed by an abrupt release of oral pressure, whose acoustic signature is a brief (ca. 5–10ms) transient of broadband energy spanning several octaves (the "release"). However, stop consonants in coda position rarely exhibit such a release. Thus, a [p] at syllable onset often differs substantially from one in the coda (although they share certain features in common, and their differences are largely predictable from context).

The acoustic properties of vocalic segments also vary greatly as a function of segmental context. The vowel [ɑ] (as in the word "hot") varies dramatically, depending on the identity of the preceding and/or following consonant, particularly with reference to the so-called formant transitions leading into and out of the vocalic nucleus (cf. Avendaño et al., Chapter 2; Diehl and Lindblom, Chapter 3). Warren (2003) likens the syllable to a "temporal compound" in which the identity of the individual constituent segments is not easily resolvable into independent elements; rather, the segments garner their functional specificity through combination within a larger, holistic entity.

Such context-dependent variability in the acoustics raises a key issue: Precisely "where" in the signal does the information associated with a specific phone reside? And is the phone the most appropriate unit with which to decode the speech signal? Or do the "invariant" cues reside at some other level (or levels) of representation?

The perceptual invariance associated with a highly variable acoustic signal has intrigued scientists for many years and remains a topic of intense controversy to this day. The issue of invariance is complicated by other sources of variability in the acoustics, either of environmental origin (e.g., reverberation and background noise), or those associated with differences in speaking style and dialect (e.g., pronunciation variation). There are dozens of different ways in which many common words are pronounced (Greenberg 1999), and yet listeners rarely have difficulty understanding the spoken message. And in many environments acoustic reflections can significantly alter the speech signal in such a manner that the canonical cues for many phonetic properties are changed beyond recognition (cf. Fig. 5.1 in Assmann and Summerfield, Chapter 5). Given such variability in the acoustic signal, how do listeners actually proceed from sound to meaning?

The auditory system may well hold the key for understanding many of the fundamental properties of speech and answer the following age-old questions:

1. *What* is the information conveyed in the acoustic signal?
2. *Where* is it located in time and frequency?
3. *How* is this information encoded in the auditory pathway and other parts of the brain?
4. *What* are the mechanisms for *protecting* this information from the potentially deleterious effects of the acoustic background to ensure reliable and accurate transmission?
5. *What* are the *consequences* of such mechanisms and the structure of the speech signal for higher-level properties of spoken language?

Based on this information-centric perspective, we can generalize from such queries to formulate several additional questions:

1. To what extent can general auditory processes account for the major properties of speech perception? Can a comprehensive account of spoken language be derived from a purely auditory-centric perspective, or must speech-specific mechanisms (presumably localized in higher cortical centers) be invoked to fully account for what is known about human speech processing (e.g., Liberman and Mattingly 1989)?

2. How do the structure and function of the auditory system shape the spectrotemporal properties of the speech signal?

3. How can we use knowledge concerning the auditory foundations of spoken language to benefit humankind?

We shall address these questions in this chapter as a means of providing the background for the remainder of volume.

3. Static versus Dynamic Approaches to Decoding the Speech Signal

As described earlier in this chapter, the traditional approach to spoken language assumes a relatively *static* relationship between segmental identity and the acoustic spectrum. Hence, the spectral cues for the vowel [i^y] ("heat") differ in specific ways from the vowel [ae] ("hat") (cf. Avendaño et al., Chapter 2); the anti-resonance (i.e., spectral zero) associated with an

[m] is lower in frequency than that of an [n], and so on. This approach is most successfully applied to a subset of segments such as fricatives, nasals, and certain vowels that can be adequately characterized in terms of relatively steady-state spectral properties. However, many segmental classes (such as the stops and diphthongs) are not so easily characterizable in terms of a static spectral profile. Moreover, the situation is complicated by the fact that certain spectral properties associated with a variety of different segments are often vitally dependent on the nature of speech sounds preceding and/or following (referred to as "coarticulation").

3.1 The Motor Theory of Speech Perception

An alternative approach is a dynamic one in which the core information associated with phonetic identity is bound to the movement of the spectrum over time. Such spectral dynamics reflect the movement of the tongue, lips, and jaw over time (cf. Aveñdano et al., Chapter 2). Perhaps the invariant cues in speech are contained in the underlying articulatory gestures associated with the spectrum? If so, then all that would be required is for the brain to back-compute from the acoustics to the original articulatory gestures. This is the essential idea underlying the motor theory of speech perception (Liberman et al. 1967; Liberman and Mattingly 1985), which tries to account for the brain's ability to reliably decode the speech signal despite the enormous variability in the acoustics. Although the theory elegantly accounts for a wide range of articulatory and acoustic phenomena (Liberman et al. 1967), it is not entirely clear precisely how the brain proceeds from sound to (articulatory) gesture (but cf. Ivry and Justus 2001; Studdert-Kennedy 2002) on this basis alone. The theory implies (among other things) that those with a speaking disorder should experience difficulty understanding spoken language, which is rarely the case (Lenneberg 1962; Fourcin 1975). Moreover, the theory assumes that articulatory gestures are relatively stable and easily characterizable. However, there is almost as much variability in the production as there is in the acoustics, for there are many different ways of pronouncing words, and even gestures associated with a specific phonetic segment can vary from instance to instance and context to context. Ohala (1994), among others, has criticized production-based perception theories on several grounds: (1) the phonological systems of languages (i.e., their segment inventories and phonotactic patterns) appear to optimize sounds, rather than articulations (cf. Liljencrants and Lindblom 1971; Lindblom 1990); (2) infants and certain nonhuman species can discriminate among certain sound contrasts in human speech even though there is no reason to believe they know how to produce these sounds; and (3) humans can differentiate many complex nonspeech sounds such as those associated with music and machines, as well as bird and monkey vocalizations, even though humans are unable to recover the mechanisms producing the sounds.

Ultimately, the motor theory deals with the issue of invariance by displacing the issues concerned with linguistic representation from the acoustics to production without any true resolution of the problem (Kleunder and Greenberg 1989).

3.2 The Locus Equation Model

An approach related to motor theory but more firmly grounded in acoustics is known as the "locus equation" model (Sussman et al. 1991). Its basic premise is as follows: although the trajectories of formant patterns vary widely as a function of context, they generally "point" to a locus of energy in the spectrum ranging between 500 and 3000 Hz (at least for stop consonants). According to this perspective, it is not the trajectory itself that encodes information but rather the frequency region thus implied. The locus model assumes some form of auditory extrapolation mechanism capable of discerning end points of trajectories in the absence of complete acoustic information (cf. Kleunder and Jenison 1992). While such an assumption falls within the realm of biological plausibility, detailed support for such a mechanism is currently lacking in mammals.

3.3 Quantal Theory

Stevens (1972, 1989) has observed that there is a nonlinear relationship between vocal tract configuration and the acoustic output in speech. The oral cavity can undergo considerable change over certain parts of its range without significant alteration in the acoustic signal, while over other parts of the range even small vocal tract changes result in large differences. Stevens suggests that speech perception takes advantage of this quantal character by categorizing the vocal tract shapes into a number of discrete states for each of several articulatory dimensions (such as voicing, manner, and place of articulation), thereby achieving a degree of representational invariance.

4. Amplitude Modulation Patterns

Complementary to the spectral approach is one based on modulation of energy over time. Such modulation occurs in the speech signal at rates ranging between 2 and 6000 Hz. Those of most relevance to speech perception and coding lie between 2 and 2500 Hz.

4.1 Low-Frequency Modulation

At the coarsest level, slow variation in energy reflects articulatory gestures associated with the syllable (Greenberg 1997b, 1999) and possibly the phrase. These low-frequency (2–20 Hz) modulations encode not only information pertaining to syllables but also phonetic segments and articulatory features (Jakobson et al. 1952), by virtue of variation in the modulation pattern across the acoustic spectrum. In this sense the modulation approach is complementary to the spectral perspective. The latter emphasizes energy variation as a function of frequency, while the former focuses on such fluctuations over time.

In the 1930s Dudley (1939) applied this basic insight to develop a reasonably successful method for simulating speech using a Vocoder. The basic idea is to partition the acoustic spectrum into a relatively small number (20 or fewer) of channels and to capture the amplitude fluctuation patterns in an efficient manner via low-pass filtering of the signal waveform (cf. Avendaño et al., Chapter 2). Dudley was able to demonstrate that the essential information in speech is encapsulated in modulation patterns lower than 25 Hz distributed over as few as 10 discrete spectral channels. The Vocoder thus demonstrates that much of the detail contained in the speech signal is largely "window dressing" with respect to information required to decode the message contained in the acoustic signal.

Houtgast and Steeneken (1973, 1985) took Dudley's insight one step further by demonstrating that modulation patterns over a restricted range, between 2 and 10 Hz, can be used as an objective measure of intelligibility (the speech transmission index, STI) for quantitative assessment of speech transmission quality over a wide range of acoustic environments. Plomp and associates (e.g., Plomp 1983; Humes et al. 1986; cf. Edwards, Chapter 7) extended application of the STI to clinical assessment of the hearing impaired.

More recently, Drullman and colleagues (1994a,b) have demonstrated a direct relationship between the pattern of amplitude variation and the ability to understand spoken language through systematic low-pass filtering of the modulation spectrum in spoken material.

The modulation approach is an interesting one from an auditory perspective, as certain types of neurons in the auditory cortex have been shown to respond most effectively to amplitude-modulation rates comparable to those observed in speech (Schreiner and Urbas 1988). Such studies suggest a direct relation between syllable-length units in speech and neural response patterns in the auditory cortex (Greenberg 1996b; Wong and Schreiner 2003). Moreover, human listeners appear to be most sensitive to modulation within this range (Viemeister 1979, 1988). Thus, the rate at which speech is spoken may reflect not merely biomechanical constraints (cf. Boubana and Maeda 1998) but also an inherent limitation in the capacity of the auditory system to encode information at the cortical level (Greenberg 1996b).

4.2 Fundamental-Frequency Modulation

The vocal folds in the larynx vibrate during speech at rates between 75 and 500 Hz, and this phonation pattern is referred to as "voicing." The lower portion of the voicing range (75–175 Hz) is characteristic of adult male speakers, while the upper part of the range (300–500 Hz) is typical of infants and young children. The midrange (175–300 Hz) is associated with the voice pitch of adult female speakers.

As a function of time, approximately 80% of the speech signal is voiced, with a quasi-periodic, harmonic structure. Among the segments, vowels, liquids ([l], [r]), glides ([y], [w]), and nasals ([m], [n], [ng]) ("sonorants") are almost always voiced (certain languages manifest voiceless liquids, nasals, or vowels in certain restricted phonological contexts), while most of the consonantal forms (i.e., stops, fricatives, affricates) can be manifest as either voiced or not (i.e., unvoiced). In such consonantal segments, voicing often serves as a phonologically contrastive feature distinguishing among otherwise similarly produced segments (e.g., [p] vs. [b], [s] vs. [z], cf. Diehl and Lindblom, Chapter 3).

In addition to serving as a form of phonological contrast, voice pitch also provides important information about the speaker's gender, age, and emotional stage. Moreover, much of the prosody in the signal is conveyed by pitch, particularly in terms of fundamental frequency variation over the phrase and utterance (Halliday 1967). Emotional content is also transmitted in this manner (Mozziconacci 1995), as is grammatical and syntactic information (Bolinger 1986, 1989).

Voice pitch also serves to "bind" the signal into a coherent entity by virtue of common periodicity across the spectrum (Bregman 1990; Langner 1992; Cooke and Ellis 2001). Without this temporal coherence various parts of the spectrum could perceptually fission into separate streams, a situation potentially detrimental to speech communication in noisy environments (cf. Cooke and Ellis 2001; Assmann and Summerfield, Chapter 5).

Voicing also serves to shield much of the spectral information contained in the speech signal from the potentially harmful effects of background noise (see Assmann and Summerfield, Chapter 5). This protective function is afforded by intricate neural mechanisms in the auditory periphery and brain stem synchronized to the fundamental frequency (cf. section 9). This "phase-locked" response increases the effective signal-to-noise ratio of the neural response by 10 to 15 dB (Rose et al. 1967; Greenberg 1988), and thereby serves to diminish potential masking effects exerted by background noise.

4.3 Periodicity Associated with Phonetic Timbre and Segmental Identity

The primary vocal-tract resonances of speech range between 225 and 3200 Hz (cf. Avendaño et al., Chapter 2). Although there are additional resonances in the higher frequencies, it is common practice to ignore those above the third formant, as they are generally unimportant from a perceptual perspective, particularly for vowels (Pols et al. 1969; Carlson and Granström 1982; Klatt 1982; Chistovich 1985; Lyon and Shamma 1996). The first formant varies between 225 Hz (the vowel [iʸ] and 800 Hz ([ɑ]). The second formant ranges between 600 Hz ([ʊ]) and 2500 ([iʸ]), while the third formant usually lies in the range of 2500 to 3200 Hz for most vowels (and many consonantal segments).

Strictly speaking, formants are associated exclusively with the vocal-tract resonance pattern and are of equal magnitude. It is difficult to measure formant patterns directly (but cf. Fujimura and Lundqvist 1971); therefore, speech scientists rely on computational methods and heuristics to estimate the formant pattern from the acoustic signal (cf. Avendaño et al., Chapter 2; Flanagan 1972). The procedure is complicated by the fact that spectral maxima reflect resonances only indirectly (but are referred to as "formants" in the speech literature). This is because the phonation produced by glottal vibration has its own spectral roll-off characteristic (ca. −12 dB/octave) that has to be convolved with that of the vocal tract. Moreover, the radiation property of speech, upon exiting the oral cavity, has a +6 dB/octave characteristic that also has to be taken into account. To simplify what is otherwise a very complicated situation, speech scientists generally combine the glottal spectral roll-off with the radiation characteristic, producing a −6 dB/octave roll-off term that is itself convolved with the transfer function of the vocal tract. This means that the amplitude of a spectral peak associated with a formant is essentially determined by its frequency (Fant 1960). Lower-frequency formants are therefore of considerably higher amplitude in the acoustic spectrum than their higher-frequency counterparts. The specific disparity in amplitude can be computed using the −6 dB/octave roll-off approximation described above. There can be as much as a 20-dB difference in sound pressure level between the first and second formants (as in the vowel [iʸ]).

5. Auditory Scene Analysis and Speech

The auditory system possesses a remarkable ability to distinguish and segregate sounds emanating from a variety of different sources, such as talkers or musical instruments. This capability to filter out extraneous sounds underlies the so-called cocktail-party phenomenon in which a listener filters out background conversation and nonlinguistic sounds to focus on a single speaker's message (cf. von Marlsburg and Schneider 1986). This feat is of particular importance in understanding the auditory foundations of speech processing. Auditory scene analysis refers to the process by which the brain reconstructs the external world through intelligent analysis of acoustic cues and information (cf. Bregman 1990; Cooke and Ellis 2001).

It is difficult to imagine how the ensemble of frequencies associated with a complex acoustic event, such as a speech utterance, could be encoded in the auditory pathway purely on the basis of (tonotopically organized) spectral place cues; there are just too many frequency components to track through time. In a manner yet poorly understood, the auditory system utilizes efficient parsing strategies not only to encode information pertaining to a sound's spectrum, but also to track that signal's acoustic trajectory through time and space, grouping neural activity into singular acoustic events attached to specific sound sources (e.g., Darwin 1981; Cooke 1993).

There is an increasing body of evidence suggesting that neural temporal mechanisms play an important role. Neural discharge synchronized to specific properties of the acoustic signal, such as the glottal periodicity of the waveform (which is typically correlated with the signal's fundamental frequency) as well as onsets (Bregman 1990; Cooke and Ellis 2001), can function to mark activity as coming from the same source. The operational assumption is that the auditory system, like other sensory systems, has evolved to focus on acoustic events rather than merely performing a frequency analysis of the incoming sound stream. Such relevant signatures of biologically relevant events include common onsets and offsets, coherent modulation, and spectral trajectories (Bregman 1990). In other words, the auditory system performs intelligent processing on the incoming sound stream to re-create as best it can the physical scenario from which the sound emanates.

This ecological acoustical approach to auditory function stems from the pioneering work of Gibson (1966, 1979), who considered the senses as intelligent computational resources designed to re-create as much of the external physical world as possible. The Gibsonian perspective emphasizes the deductive capabilities of the senses to infer the conditions behind the sound, utilizing whatever cues are at hand. The limits of hearing capability are ascribed to functional properties interacting with the environment. Sensory systems need not be any more sensitive or discriminating than they need to be in the natural world. Evolutionary processes have assured that the auditory system works sufficiently well under most conditions. The direct realism approach espoused by Fowler (1986, 1996) represents a contemporary version of the ecological approach to speech. We shall return to this issue of intelligent processing in section 11.

6. Auditory Representations

6.1 Rate-Place Coding of Spectral Peaks

In the auditory periphery the coding of speech and other complex sounds is based on the activity of thousands of auditory-nerve fibers (ANFs) whose tuning characteristics span a broad range in terms of sensitivity, frequency selectivity, and threshold. The excitation pattern associated with speech signals is inferred through recording the discharge activity from hundreds of individual fibers to the same stimulus. In such a "population" study the characteristic (i.e., most sensitive) frequency (CF) and spontaneous activity of the fibers recorded are broadly distributed in a tonotopic manner thought to be representative of the overall tuning properties of the auditory nerve. Through such studies it is possible to infer how much information is contained in the distribution of neural activity across the auditory nerve pertinent to the speech spectrum (cf. Young and Sachs 1979; Palmer and Shamma, Chapter 4).

At low sound pressure levels (<40 dB), the peaks in the vocalic spectrum are well resolved in the population response, with the discharge rate roughly proportional to the cochlear-filtered energy level. Increasing the sound pressure level by 20 dB alters the distribution of discharge activity such that the spectral peaks are no longer so prominently resolved in the tonotopic place-rate profile. This is a consequence of the fact that the discharge of fibers with CFs near the formant peaks has saturated relative to those with CFs corresponding to the spectral troughs. As the stimulus intensity is raised still further, to a level typical of conversational speech, the ability to resolve the spectral peaks on the basis of place-rate information is compromised even further.

On the basis of such population profiles, it is difficult to envision how the spectral profile of vowels and other speech sounds could be accurately and reliably encoded on the basis of place-rate information at any but the lowest stimulus intensities. However, a small proportion of ANFs (15%), with spontaneous (background) rates (SRs) less than 0.5 spikes/s, may be capable of encoding the spectral envelope on the basis of rate-place information, even at the highest stimulus levels (Sachs et al. 1988; Blackburn and Sachs 1990). Such low-SR fibers exhibit extended dynamic response ranges and are more sensitive to the mechanical suppression behavior of the basilar membrane than their higher SR counterparts (Schalk and Sachs 1980; Sokolowski et al. 1989). Thus, the discharge rate of low-SR fibers, with CFs close to the formant peaks, will continue to grow at high sound pressure levels, and the activity of low-SR fibers responsive to the spectral troughs should, in principle, be suppressed by energy associated with the formants. However, such rate suppression also reduces the response to the second and third formants (Sachs and Young 1980), thereby decreasing the resolution of the spectral peaks in the rate-place profile at higher sound pressure levels. For this reason it is not entirely clear that lateral suppression, by itself, actually functions to provide an adequate rate-place representation of speech and other spectrally complex signals in the auditory nerve.

The case for a rate-place code for vocalic stimuli is therefore equivocal at the level of the auditory nerve. The discharge activity of a large majority of fibers is saturated at these levels in response to vocalic stimuli. Only a small proportion of ANFs resolve the spectral peaks across the entire dynamic range of speech. And the representation provided by these low-SR units is less than ideal, particularly at conversational intensity levels (i.e., 75 dB SPL).

The rate-place representation of the spectrum may be enhanced in the cochlear nucleus and higher auditory stations relative to that observed in the auditory nerve. Such enhancement could be a consequence of preferential projection of fibers or through the operation of lateral inhibitory networks that sharpen still further the contrast between excitatory and background neural activity (Shamma 1985b; Palmer and Shamma, Chapter 4).

Many chopper units in the anteroventral cochlear nucleus (AVCN) respond to steady-state vocalic stimuli in a manner similar to that of low-SR ANFs (Blackburn and Sachs 1990). The rate-place profile of these choppers exhibit clearly delineated peaks at CFs corresponding to the lower formant frequencies, even at 75 dB SPL (Blackburn and Sachs 1990). In principle, a spectral peak would act to suppress the activity of choppers with CFs corresponding to less intense energy, thereby enhancing the neural contrast between spectral maxima and minima. Blackburn and Sachs have proposed that such lateral inhibitory mechanisms may underlie the ability of AVCN choppers to encode the spectral envelope of vocalic stimuli at sound pressure levels well above those at which the average rate of the majority of ANFs saturate. Palmer and Shamma discuss such issues in greater detail in Chapter 4.

The evidence is stronger for a rate-place representation of certain consonantal segments. The amplitude of most voiceless consonants is sufficiently low (<50 dB SPL) as to evade the rate saturation attendant in the coding of vocalic signals. The spectra of plosive bursts, for example, is generally broadband, with several local maxima. Such spectral information is not likely to be temporally encoded due to its brief duration and the lack of sharply defined peaks. Physiological studies have shown that such segments are adequately represented in the rate-place profile of all spontaneous rate groups of ANFs across the tonotopic axis (e.g., Miller and Sachs 1983; Delgutte and Kiang 1984).

Certain phonetic parameters, such as voice-onset time, are signaled through absolute and relative timing of specific acoustic cues. Such cues are observable in the tonotopic distribution of ANF responses to the initial portion of these segments (Miller and Sachs 1983; Delgutte and Kiang 1984). For example, the articulatory release associated with stop consonants has a broadband spectrum and a rather abrupt onset, which evokes a marked flurry of activity across a wide CF range of fibers. Another burst of activity occurs at the onset of voicing. Because the dynamic range of ANF discharge is much larger during the initial rapid adaptation phase (0–10 ms) of the response, there is relatively little or no saturation of discharge rate during this interval at high sound pressure levels (Sachs et al. 1988; Sinex and Geisler 1983). In consequence, the onset spectra serving to distinguish the stop consonants (Stevens and Blumstein 1978, 1981) are adequately represented in the distribution of rate-place activity across the auditory nerve (Delgutte and Kiang 1984) over the narrow time window associated with articulatory release.

This form of rate information differs from the more traditional "average" rate metric. The underlying parameter governing neural magnitude at onset is actually the probability of discharge over a very short time interval. This probability is usually converted into effective discharge rate normalized to units of spikes per second. If the analysis window (i.e., bin width) is sufficiently short (e.g., 100 μs), the apparent rate can be exceedingly high (up to 10,000 spikes/s). Such high onset rates reflect two properties of the neural discharge: the high probability of firing correlated with stimulus onset, and the small degree of variance associated with this first-spike latency. This measure of onset response magnitude is one form of instantaneous discharge rate. "Instantaneous," in this context, refers to the spike rate measured over an interval corresponding to the analysis bin width, which generally ranges between 10 and 1000 μs. This is in contrast to average rate, which reflects the magnitude of activity occurring over the entire stimulus duration. Average rate is essentially an integrative measure of activity that counts spikes over relatively long periods of time and weights each point in time equally. Instantaneous rate emphasizes the clustering of spikes over small time windows and is effectively a correlational measure of neural response. Activity that is highly correlated in time, upon repeated presentations will, over certain time intervals, have very high instantaneous rates of discharge. Conversely, poorly correlated response patterns will show much lower peak instantaneous rates whose magnitudes are close to that of the average rate. The distinction between integrative and correlational measures of neural activity is of critical importance for understanding how information in the auditory nerve is ultimately processed by neurons in the higher stations of the auditory pathway.

Place-rate models of spectral coding do not function well in intense background noise. Because the frequency parameter is coded though the spatial position of active neural elements, the representation of complex spectra is particularly vulnerable to extraneous interference (Greenberg 1988). Intense noise or background sounds with significant energy in spectral regions containing primary information about the speech signal possess the capability of compromising the auditory representation of the speech spectrum. This vulnerability of place representations is particularly acute when the neural information is represented in the form of average rate. This vulnerability is a consequence of there being no neural marker other than tonotopic affiliation with which to convey information pertaining to the frequency of the driving signal. In instances where both fore- and background signals are sufficiently intense, it will be exceedingly difficult to distinguish that portion of the place representation driven by the target signal from that driven by interfering sounds. Hence, there is no systematic way of separating the neural activity associated with each source *purely* on the basis of rate-place–encoded information. We shall return to the issue of information coding robustness in section 9.

The perceptual implications of a strictly rate-place model are counter-intuitive, for it is implied that the intelligibility of speech should decline with increasing sound pressure level above 40 dB. Above this level the rate-place representation of the vocalic spectrum for most AN fibers becomes much less well defined, and only the low-SR fiber population continues to encode the spectral envelope with any degree of precision. In actuality, speech intelligibility *improves* above this intensity level at a point where the rate-place representation is not nearly so well delineated.

6.2 Latency-Phase Representations

In a linear system the phase characteristics of a filter are highly correlated with its amplitude response. On the skirts of the filter, where the amplitude response diminishes quickly, the phase of the output signal also changes rapidly. The phase response, by itself, can thus be used in such a system to infer the properties of the filter (cf. Huggins 1952). For a nonlinear system, such as pertains to signal transduction in the cochlea, phase and latency (group delay) information may provide a more accurate estimate of the underlying filter characteristics than average discharge rate because they are not as sensitive to such cochlear nonlinearities as discharge-rate compression and saturation, which typically occur above 40 dB SPL.

Several studies suggest that such phase and latency cues are exhibited in the auditory nerve across a very broad range of intensities. A large phase transition is observed in the neural response distributed across ANFs whose CFs span the lower tonotopic boundary of a dominant frequency component (Anderson et al. 1971), indicating that the high-frequency skirt of the cochlear filters is sharply tuned across intensity. A latency shift of the neural response is observed over a small range of fiber CFs. The magnitude of he shift can be appreciable, as much as half a cycle of the driving frequency (Anderson et al. 1971; Kitzes et al. 1978). For a 500-Hz signal, this latency change would be on the order of 1 ms. Because this phase transition may not be subject to the same nonlinearities that result in discharge-rate saturation, fibers with CFs just apical to the place of maximal response can potentially encode a spectral peak in terms of the onset phase across a wide range of intensities.

Interesting variants of this response-latency model have been proposed by Shamma (1985a,b, 1988) and Deng et al. (1988). The phase transition for low-frequency signals should, in principle, occur throughout the entire response, not just at the beginning, as a result of ANFs' phase-locking properties. Such ongoing phase disparities could be registered by some form of neural circuitry presumably located in the cochlear nucleus. The output of such networks would magnify activity in those tonotopic regions over which the phase and/or latency changes rapidly through some form of cross-frequency–channel correlation. In the Shamma model, the correlation is performed through the operation of a lateral inhibitory network, which subtracts the auditory nerve (AN) output of adjacent channels. The effect of this cross-channel subtraction is to null out activity for channels with similar phase and latency characteristics, leaving only that portion of the activity pattern where rapid phase transitions occur. The Deng model uses cross-channel correlation (i.e., multiplication) instead of subtraction to locate the response boundaries. Correlation magnifies the activity of channels with similar response patterns and reduces the output of dissimilar adjacent channels. Whether the cross-channel comparison is performed through subtraction, multiplication, or some other operation, the consequence of such neural computation is to provide "pointers" to those tonotopic regions where a boundary occurs that might otherwise be hidden if analyzed solely on the basis of average rate. These pointers, in principle, could act in a manner analogous to peaks in the excitation pattern but with the advantage of being preserved across a broad range of sound pressure levels.

6.3 Synchrony-Place Information

Place and temporal models of frequency coding are generally discussed as if they are diametrically opposed perspectives. Traditionally, temporal models have de-emphasized tonotopic organization in favor of the fine-temporal structure of the neural response. However, place and temporal coding need not be mutually exclusive. The concept of the central spectrum (Goldstein and Srulovicz 1977; Srulovicz and Goldstein 1983) attempts to reconcile the two approaches through their combination within a single framework for frequency coding. In this model, both place and temporal information are used to construct the peripheral representation of the spectrum. Timing information, as reflected in the interval histogram of ANFs, is used to estimate the driving frequency. The model assumes that temporal activity is keyed to the tonotopic frequency representation. In some unspecified way, the system "knows" what sort of temporal activity corresponds to each tonotopic location, analogous to a matched filter.

The central spectrum model is the intellectual antecedent of the peripheral representational model of speech proposed by Young and Sachs (1979), whose model is based on the auditory-nerve population response study discussed in section 6.1. As with place schemes in general, spectral frequency is mapped onto tonotopic place (i.e., ANF characteristic frequency), while the amplitude of each frequency is associated with the magnitude of the neural response synchronized to that component by nerve fibers whose CFs lay within close proximity ($^{1}/_{4}$ octave). The resulting average localized synchronized rate (ALSR) is a parsimonious representation of the stimulus signal spectrum (cf. Figs. 4.5 and 4.7 in Palmer and Shamma, Chapter 4). The ALSR is a computational procedure for estimating the magnitude of neural response in a given frequency channel based on the product of firing rate and temporal correlation with a predefined frequency band. The spectral peaks associated with the three lower formants (F_1, F_2, F_3) are clearly delineated in the ALSR representation, in marked contrast to the rate-place representation.

The mechanism underlying the ALSR representation is referred to as "synchrony suppression" or "synchrony capture." At low sound pressure levels, temporal activity synchronized to a single low-frequency (<4 kHz) spectral component is generally restricted to a circumscribed tonotopic region close to that frequency. Increasing the sound pressure level results in a spread of the synchronized activity, particularly toward the region of high-CF fibers. In this instance, the spread of temporal activity occurs in roughly tandem relationship with the activation of fibers in terms of average discharge rate. At high sound pressure levels (ca. 70–80 dB), a large majority of ANFs with CFs below 10 kHz are phase-locked to low-frequency components of the spectrum. This upward spread of excitation into the high-frequency portion of the auditory nerve is a consequence of the unique filter characteristics of high-CF mammalian nerve fibers. Although the filter function for such units is sharply bandpass within 20 to 30 dB of rate threshold, it becomes broadly tuned and low pass at high sound pressure levels. This tail component of the high-CF fiber frequency-threshold curve (FTC) renders such fibers extremely responsive to low-frequency signals at sound pressure levels typical of conversational speech. The consequence of this low-frequency sensitivity, in concert with the diminished selectivity of low-CF fibers, is the orderly basal recruitment (toward the high-frequency end of the auditory nerve) of ANFs as a function of increasing sound pressure level.

Synchrony suppression is intricately related to the frequency selectivity of ANFs. At low sound pressure levels, most low-CF nerve fibers are phase-locked to components in the vicinity of their CF. At this sound pressure level the magnitude of a fiber's response, measured in terms of either synchronized or average rate, is approximately proportional to the signal energy at the unit CF, resulting in rate-place and synchrony-place profiles relatively isomorphic to the input stimulus spectrum. At higher sound pressure levels, the average-rate response saturates across the tonotopic array of nerve fibers, resulting in significant degradation of the rate-place representation of the formant pattern, as described above. The distribution of temporal activity also changes, but in a somewhat different manner. The activity of fibers with CFs near the spectral peaks remains phase-locked to the formant frequencies. Fibers whose CFs lie in the spectral valleys, particularly between F_1 and F_2, become synchronized to a different frequency, most typically F_1.

The basis for this suppression of synchrony may be as follows: the amplitude of components in the formant region (particularly F_1) are typically 20 to 40 dB greater than that of harmonics in the valleys. When the amplitude of the formant becomes sufficiently intense, its energy "spills" over into neighboring frequency channels as a consequence of the broad tuning of low-frequency fibers referred to above. Because of the large amplitude disparity between spectral peak and valley, there is now more formant-related energy passing through the fiber's filter than energy derived from components in the CF region of the spectrum. Suppression of the original timing pattern actually begins when the amount of formant-related energy equals that of the original signal. Virtually complete suppression of the less intense signal results when the amplitude disparity is greater than 15 dB (Greenberg et al. 1986). In this sense, encoding frequency in terms of neural phase-locking acts to enhance the peaks of the spectrum at the expense of less intense components.

The result of this synchrony suppression is to reduce the amount of activity phase-locked to frequencies other than the formants. At higher sound pressure levels, the activity of fibers with CFs in the spectral valleys are indeed phase-locked, but to frequencies distant from their CFs. In the ALSR model the response of these units contributes to the auditory representation of the signal spectrum only in an indirect fashion, since the magnitude of temporal activity is measured only for frequencies near the fiber CF. In this model, only a small subset of ANFs, with CFs near the formant peaks, directly contribute to the auditory representation of the speech spectrum in the model.

6.4 Cortical Representations of the Speech Signal

Neurons do not appear to phase-lock to frequencies higher than 200 to 300 Hz above the level of inferior colliculus, implying that spectral information based on timing information in the peripheral and brain stem regions of the auditory pathway is transformed into some other representation in the auditory cortex. Moreover, most auditory cortical neurons respond at very low discharge rates, typically less than 10 spikes/s. It is not uncommon for units at this level of the auditory pathway to respond only once per acoustic event, with the spike associated with stimulus onset.

Shamma and colleagues describe recent work from their laboratory in Chapter 4 that potentially resolves some of the issues discussed earlier in this section. Most of the responsiveness observed at this level of the auditory pathway appears to be associated with low-frequency properties of the spectrally filtered waveform envelope, suggesting a neural basis for the perceptual and synthesis studies described in section 4. In this sense, the cortex appears to be concerned primarily with events occurring over much longer time spans than those of the brain stem and periphery.

7. Functional Properties of Hearing Relevant to Speech

For many applications, such as speech analysis, syntheses, and coding, it is useful to know the perceptual limits pertaining to speech sounds. For example, how accurately do we need to specify the frequency or amplitude of a formant for such applications? Such functional limits can be estimated using psychophysical techniques.

7.1 Audibility and Dynamic Range Sensitivity

The human ear responds to frequencies between 30 and 20,000 Hz, and is most sensitive between 2.5 and 5 kHz (Wiener and Ross 1946). The upper limit of 20 kHz is an average for young adults with normal hearing. As individuals age, sensitivity to high-frequencies diminishes, so much so that by the age of 60 it is unusual for a listener to hear frequencies above 12 kHz. Below 400 Hz, sensitivity decreases dramatically. The threshold of detectability at 100 Hz is about 30 dB higher (i.e., less sensitive) than at 1 kHz. Above 5 kHz, sensitivity declines steeply as well. Most of the energy in the speech signal lies below 2 kHz (Fig. 5.1 in Assmann and Summerfield, Chapter 5). The peak in the *average* speech spectrum is about 500 Hz, falling off at about 6 dB/octave thereafter (Fig. 5.1 in Assmann and Summerfield, Chapter 5). There is relatively little energy of informational relevance above 10 kHz in the speech signal. Thus, there is a relatively good match between the spectral energy profile in speech and human audibility. Formant peaks in the very low frequencies are high in magnitude, largely compensating for the decreased sensitivity in this portion of the spectrum. Higher-frequency formants are of lower amplitude but occur in the most sensitive part of the hearing range. Thus, the shape of the speech spectrum is remarkably well adapted to the human audibility curve.

Normal-hearing listeners can generally detect sounds as low as –10 dB SPL in the most sensitive part of the spectrum (ca. 4 kHz) and are capable of withstanding sound pressure levels of 110 dB without experiencing pain. Thus, the human ear is capable of transducing about 120-dB (1 : 1,000,000) dynamic range of sound pressure under normal-hearing conditions. The SPL of the most intense speech sounds (usually vowels) generally lies between 70 and 85 dB, while the SPL of certain consonants (e.g., fricatives) can be as low as 35 dB. The dynamic range of speech sounds is therefore about 50 dB. (This estimate of SPL applies to the entire segment. Prior to initiation of a speech gesture, there is little or no energy produced, so the *true* dynamic range of the speech signal from instant to instant is probably about 90 dB.)

Within this enormous range the ability to discriminate fluctuations in intensity (ΔI) varies. At low sound pressure levels (<40 dB) the difference limen (DL) lies between 1 and 2 dB (Riesz 1928; Jesteadt et al. 1977; Viemeister 1988). Above this limit, the DL can decline appreciably (i.e., discriminability improves) to about half of this value (Greenwood 1994). Thus, within the core range of the speech spectrum, listeners are exceedingly sensitive to variation in intensity. Flanagan (1957) estimated that ΔI for formants in the speech signal to be about 2 dB.

7.2 Frequency Discrimination and Speech

Human listeners can distinguish exceedingly fine differences in frequency for sinusoids and other narrow-band signals. At 1 kHz the frequency DL (Δf) for such signals can be as small as 1 to 2 Hz (i.e., 0.1–0.2%) (Wier et al. 1977). However, Δf varies as a function of frequency, sound pressure level, and duration. Frequency discriminability is most acute in the range between 500 and 1000 Hz, and falls dramatically at high frequencies (>4 kHz), particularly when the signal-to-noise ratio is held constant (Dye and Hafter 1980). Thus, discriminability is finest for those parts of the spectrum in which most of the information in the speech spectrum resides. With respect to duration, frequency discriminability is most acute for signals longer than 80 to 100 ms (at any frequency), and signals greater than 40 dB SPL are generally more finely discriminated in terms of frequency than those of lower intensity (Wier et at. 1977).

The discriminability of broadband signals, such as formants in a speech signal, is not nearly as fine as for narrowhand stimuli. In an early study, Flanagan (1955) found that Δf ranged between 3% and 5% of the formant frequency for steady-state stimuli. More recent studies indicate that Δf can be as low as 1% when listeners are highly trained (Kewley-Port and Watson 1994). Still, the DL for frequency appears to be an order of magnitude greater for formants than for sinusoidal signals.

Of potentially greater relevance for speech perception is discriminability of non–steady-state formants, which possess certain properties analogous to formant transitions interposed between consonants and vowels. Mermelstein (1978) estimated that the DL for formant transitions ranges between 49 and 70 Hz for F_1 and between 171 and 199 Hz for F_2. A more recent study by van Wieringen and Pols (1994) found that the DL is sensitive to the rate and duration of the transition. For example, the DL is about 70 Hz for F_1 when the transition is 20 ms, but decreases (i.e., improves) to 58 Hz when transition duration is increased to 50 ms.

Clearly, the ability to distinguish fine gradations in frequency is much poorer for complex signals, such as speech formants, relative to spectrally simple signals, such as sinusoids. At first glance such a relation may appear puzzling, as complex signals provide more opportunities for comparing details of the signal than simple ones. However, from an information-theoretic perspective, this diminution of frequency discriminability could be of utility for a system that generalizes from signal input to a finite set of classes through a process of learned association, a topic that is discussed further in section 11.

8. The Relation Between Spectrotemporal Detail and Channel Capacity

It is important for any information-rich system that the information carrier be efficiently and reliably encoded. For this reason a considerable amount of research has been performed over the past century on efficient methods of coding speech (cf. Avendaño et al., Chapter 2). This issue was of particular concern for analog telephone systems in which channel capacity was severely limited (in the era of digital communications, channel capacity is much less of a concern for voice transmission, except for wireless communication, e.g., cell phones). Pioneering studies by Harvey Fletcher and associates at Bell Laboratories,[2] starting in the 1910s, systematically investigated the factors limiting intelligibility as a means of determining how to reduce the bandwidth of the speech signal without compromising the ability to communicate using the telephone (cf. Fletcher 1953).

In essence, Fletcher's studies were directed toward determining the information-laden regions of the spectrum. Although information theory had yet to be mathematically formulated (Shannon's paper on the mathematical foundation of information theory was originally published in the Bell System Technical Journal, and was issued in book form the following year—Shannon and Weaver 1949), it was clear to Fletcher that the ability to decode the speech signal into constituent sounds could be used as a quantitative means of estimating the amount of information contained. Over a period of 20 years various band-limiting experiments were performed in an effort to ascertain the frequency limits of information contained in speech (Miller 1951; Fletcher 1953; Allen 1994). The results of these studies were used to define the bandwidth of the telephone (300–3400 Hz), a standard still in use today. Although there is information in the frequency spectrum residing outside these limits, Fletcher's studies revealed that its absence did not significantly impair verbal interaction and could therefore be tolerated over the telephone.

More recent work has focused on delineating the location of information contained in both frequency and time. Spectral maxima associated with the three lowest formants are known to carry much of the timbre information associated with vowels and other phonetic classes (e.g., Ladefoged 1967, 2001; Pols et al. 1969). However, studies using "sine-wave" speech suggest that spectral maxima, in and of themselves, are not the ultimate carriers of information in the signal. The speech spectrum can be reduced to a series of three sinusoids, each associated with the center frequency of a formant

[2]Fletcher began his speech research at Western Electric, which manufactured telephone equipment for AT&T and other telephone companies. In 1925, Western Electric was merged with AT&T, and Bell Laboratories was established. Fletcher directed the acoustics research division at Bell Labs for many years before his retirement from AT&T in 1951.

(Remez et at. 1981, 1994). When played, this stimulus sounds extremely unnatural and is difficult to understand without prior knowledge of the words spoken.[3] In fact, Kakusho and colleagues (1971) demonstrated many years ago that for such a sparse spectral representation to sound speech-like and be identified reliably, each spectral component in this sparse representation must be coherently amplitude-modulated at a rate within the voice-pitch range. This finding is consistent with the notion that the auditory system requires complex spectra, preferably with glottal periodicity, to associate the signal with information relevant to speech. (Whispered speech lacks a glottal excitation source, yet is comprehensible. However, such speech is extremely fragile, vulnerable to any sort of background noise, and is rarely used except in circumstances where secrecy is of paramount concern or vocal pathology has intervened.)

Less radical attempts to reduce the spectrum have proven highly successful. For example, smoothing the spectral envelope to minimize fine detail in the spectrum is a common technique used in digital coding of speech (cf. Avendaño et al., Chapter 2), a result consistent with the notion that *some* property associated with spectral maxima is important, even if it is not the absolute peak by itself (cf. Assmann and Summerfield, Chapter 5). Such spectral envelope smoothing has been successfully applied to automatic speech recognition as a means of reducing extraneous detail for enhanced acoustic-phonetic pattern classification (cf. Davis and Mermelstein 1980; Ainsworth 1988; Hermansky 1990; Morgan et al., Chapter 6). And perceptual studies, in which the depth and detail of the spectral envelope is systematically manipulated, have demonstrated the importance of such information for speech intelligibility both in normal and hearing-impaired individuals (ter Keurs et al. 1992, 1993; Baer and Moore 1993).

Intelligibility can remain high even when much of the spectrum is eliminated in such a manner as to discard many of the spectral peaks in the signal. As few as four band-limited ($^1/_3$ octave) channels distributed across the spectrum, *irrespective of the location of spectral maxima*, can provide nearly perfect intelligibility of spoken sentences (Greenberg et al. 1998). Perhaps the spectral peaks, in and of themselves, are not as important as functional *contrast* across frequency and over time (cf. Lippmann 1996; Müsch and Buus 2001b).

How is such information extracted from the speech signal? Everything we know about speech suggests that the mechanisms responsible for decoding the signal must operate over relatively long intervals of time, between 50 and 1000 ms (if not longer), which are characteristic of cortical rather than brain stem or peripheral processing (Greenberg 1996b). At the corti-

[3] Remez and associates would disagree with this statement, claiming in their paper and in subsequent publications and presentations that sine-wave speech is indeed intelligible. The authors of this chapter (and many others in the speech community) respectfully disagree with their assertion.

cal level, auditory neurons respond relatively infrequently, and this response is usually associated with the onset of discrete events (cf. section 6.4; Palmer and Shamma, Chapter 4). It is as if cortical neurons respond primarily to truly informative features in the signal and otherwise remain silent. A potential analog of cortical speech processing is the highly complex response patterns observed in the auditory cortex of certain echo-locating bats in response to target-ranging or Doppler-shifted signals (Suga et al. 1995; Suga 2003). Many auditory cortical neurons in such species as *Pteronotus parnellii* require specific combinations of spectral components distributed over frequency and/or time in order to fire (Suga et al. 1983). Perhaps comparable "combination-sensitive" neurons function in human auditory cortex (Suga 2003).

If it is mainly at the level of the cortex that information relevant to speech features is extracted, what role is played by more peripheral stations in the auditory pathway?

9. Protecting Information Contained in the Speech Signal

Under many conditions speech (and other communication signals) is transmitted in the presence of background noise and/or reverberation. The sound pressure level of this background can be considerable and thus poses a considerable challenge to any receiver intent on decoding the message contained in the foreground signal. The problem for the receiver, then, is not just to decode the message, but also to do so in the presence of variable and often unpredictable acoustic environments. To accomplish this objective, highly sophisticated mechanisms must reside in the brain that effectively shield the message in the signal.

This informational shielding is largely performed in the auditory periphery and central brain stem regions. In the periphery are mechanisms that serve to enhance spectral peaks, both in quiet and in noise. Such mechanisms rely on automatic gain control (AGC), as well as mechanical and neural suppression of those portions of spectrum distinct from the peaks (cf. Rhode and Greenberg 1994; Palmer and Shamma, Chapter 4). The functional consequence of such spectral-peak enhancement is the capability of preserving the general shape of the spectrum over a wide range of background conditions and signal-to-noise ratios (SNRs).

In the cochlea are several mechanisms operating to preserve the shape of the spectrum. Mechanical suppression observed in the basilar membrane response to complex signals at high sound pressure levels serves to limit the impact of those portions of the spectrum significantly below he peaks, effectively acting as a peak clipper. This form of suppression appears to be enhanced under noisy conditions (Rhode and Greenberg 1994), and is potentially mediated through the olivocochlear bundle (Liberman 1988;

Warr 1992; Reiter and Liberman 1995) passing from the brain stem down into the cochlea itself.

A second means with which to encode and preserve the shape of spectrum is through the spatial frequency analysis performed in the cochlea (cf. Greenberg 1996a; Palmer and Shamma, Chapter 4; section 6 of this chapter). As a consequence of the stiffness gradient of the basilar membrane, its basal portion is most sensitive to high frequencies (>10 kHz), while the apical end is most responsive to frequencies below 500 Hz. Frequencies in between are localized to intermediate positions in the cochlea in a roughly logarithmic manner (for frequencies greater than 1 kHz). In the human cochlea approximately 50% of the 35-mm length of the basilar membrane is devoted to frequencies below 2000 kHz (Greenwood 1961, 1990), suggesting that the spectrum of the speech signal has been tailored, at least in part, to take advantage of the considerable amount of neural "real estate" devoted to low-frequency signals.

The frequency analysis performed by the cochlea appears to be quantized with a resolution of approximately $^1/_4$ octave. Within this "critical band" (Fletcher 1953; Zwicker et al. 1957) energy is quasi-linearly integrated with respect to loudness summation and masking capability (Scharf 1970). In many ways the frequency analysis performed in the cochlea behaves as if the spectrum is decomposed into separate (and partially independent) channels. This sort of spectral decomposition provides an effective means of protecting the most intense portions of the spectrum from background noise under many conditions.

A third mechanism preserving spectral shape is based on neural phase-locking, whose origins arise in the cochlea. The release of neurotransmitter in inner hair cells (IHCs) is temporally modulated by the stimulating (cochlear) waveform and results in a temporal patterning of ANF responses that is "phase-locked" to certain properties of the stimulus. The effectiveness of this response modulation depends on the ratio of the alternating current (AC) to the direct current (DC) components of the IHC receptor potential, which begins to diminish for signals greater than 800 Hz. Above 3 kHz, the AC/DC ratio is sufficiently low that the magnitude of phase-locking is negligible (cf. Greenberg 1996a for further details). Phase-locking is thus capable of providing an effective means of temporally coding information pertaining to the first, second, and third formants of the speech signal (Young and Sachs 1979). But there is more to phase-locking than mere frequency coding.

Auditory-nerve fibers generally phase-lock to the portion of the local spectrum of greatest magnitude through a combination of AGC (Geisler and Greenberg 1986; Greenberg et al. 1986) and a limited dynamic range of about 15 dB (Greenberg et al. 1986; Greenberg 1988). Because ANFs phase-lock poorly (if at all) to noise, signals with a coherent temporal structure (e.g., harmonics) are relatively immune to moderate amounts of background noise. The temporal patterning of the signal ensures that peaks in the foreground signal rise well above the average noise level at all but the lowest SNRs. Phase-locking to those peaks riding above the background effectively suppresses the noise (cf. Greenberg 1996a).

Moreover, such phase-locking enhances the effective SNR of the spectral peaks through a separate mechanism that distributes the temporal information across many neural elements. The ANF response is effectively "labeled" with the stimulating frequency by virtue of the temporal properties of the neural discharge. At moderate-to-high sound pressure levels (40–80 dB), the number of ANFs phase-locked to the first formant grows rapidly, so that it is not just fibers most sensitive to the first formant that respond. Fibers with characteristic (i.e., most sensitive) frequencies as high as several octaves above F_1 may also phase-lock to this frequency region (cf. Young and Sachs 1979; Jenison et al. 1991). In this sense, the auditory periphery is exploiting redundancy in the neural timing pattern distributed across the cochlear partition to robustly encode information associated with spectral peaks. Such a distributed representation renders the information far less vulnerable to background noise (Ghitza 1988; Greenberg 1988), and provides an indirect measure of peak magnitude via determining the number of auditory channels that are coherently phase-locked to that frequency (cf. Ghitza 1988).

This phase-locked information is preserved to a large degree in the cochlear nucleus and medial superior olive. However, at the level of the inferior colliculus it is rare for neurons to phase-lock to frequencies above 1000 Hz. At this level the temporal information has probably been recoded, perhaps in the form of spatial modulation maps (Langner and Schreiner 1988; Langner 1992).

Phase-locking provides yet a separate means of protecting spectral peak information through binaural cross-correlation. The phase-locked input from each ear meets in the medial superior olive, where it is likely that some form of cross-correlational analysis is computed. Additional correlational analyses are performed in the inferior colliculus (and possibly the lateral lemniscus). Such binaural processing provides a separate means of increasing the effective SNR, by weighting that portion of the spectrum that is binaurally coherent across the two ears (cf. Stern and Trahiotis 1995; Blauert 1996).

Yet a separate means of shielding information in speech is through temporal coding of the signal's fundamental frequency (f_0). Neurons in the auditory periphery and brain stem nuclei can phase-lock to the signal's f_0 under many conditions, thus serving to bind the discharge patterns associated with different regions of the spectrum into a coherent entity, as well as enhance the SNR via phase-locking mechanisms described above. Moreover, fundamental-frequency variation can serve, under appropriate circumstances, as a parsing cue, both at the syllabic and phrasal levels (Brokx and Nooteboom 1982; Ainsworth 1986; Bregman 1990; Darwin and Carlyon 1995; Assmann and Summerfield, Chapter 5). Thus, pitch cues can serve to guide the segmentation of the speech signal, even under relatively low SNRs.

10. When Hearing Fails

The elaborate physiological and biochemical machinery associated with acoustic transduction in the auditory periphery may fail, thus providing a natural experiment with which to ascertain the specific role played by various cochlear structures in the encoding of speech. Hearing impairment also provides a method with which to estimate the relative contributions made by bottom-up and top-down processing for speech understanding (Grant and Walden 1995; Grant and Seitz 1998; Grant et al. 1998).

There are two primary forms of hearing impairment—conductive hearing loss and sensorineural loss—that affect the ability to decode the speech signal. Conductive hearing loss is usually the result of a mechanical problem in the middle ear, with attendant (and relatively uniform) loss of sensitivity across much of the frequency spectrum. This form of conductive impairment can often be ameliorated through surgical intervention.

Sensorineural loss originates in the cochlea and has far more serious consequences for speech communication. The problem lies primarily in the outer hair cells (OHCs), which can be permanently damaged as a result of excessive exposure to intense sound (cf. Bohne and Harding 2000; Patuzzi 2002). Outer hair cell stereocilia indirectly affect the sensitivity and tuning of IHCs via their articulation with the underside of the tectorial membrane (TM). Their mode of contact directly affects the TM's angle of orientation with respect to the IHC stereocilia and hence can reduce the ability to induce excitation in the IHCs via deflection of their stereocilia (probably through fluid coupling rather than direct physical contact). After exposure to excessive levels of sound, the cross-linkages of actin in OHC stereocilia are broken or otherwise damaged, resulting in ciliary floppiness that reduces OHC sensitivity substantially and thereby also reduces sensitivity in the IHCs (cf. Gummer et al. 1996, 2002). In severe trauma the stereocilia of the IHCs are also affected. Over time both the OHCs and IHCs of the affected frequency region are likely to degenerate, making it impossible to stimulate ANFs innervating this portion of the cochlea. Eventually, the ANFs themselves lose their functional capacity and whither, which in turn can result in degeneration of neurons further upstream in the central brain stem pathway and cortex (cf. Gravel and Ruben 1996).

When the degree of sensorineural impairment is modest, it is possible to partially compensate for the damage through the use of a hearing aid (Edwards, Chapter 7). The basic premise of a hearing aid is that audibility has been compromised in selected frequency regions, thus requiring some form of amplification to raise the level of the signal to audible levels (Steinberg and Gardner 1937). However, it is clear from recent studies of the hearing impaired that audibility is not the only problem. Such individuals also manifest under many (but not all) circumstances a significant reduction in frequency and temporal resolving power (cf. Edwards, Chapter 7).

A separate but related problem concerns a drastic decrease in dynamic range of intensity coding. Because the threshold of neural response is significantly elevated, without an attendant increase in the upper limit of sound pressure transduction, the effective range between the softest and most intense signals is severely compressed. This reduction in dynamic range means that the auditory system is no longer capable of using energy modulation for reliable segmentation in the affected regions of the spectrum, and therefore makes the task of parsing the speech signal far more difficult.

Modern hearing aids attempt to compensate for this dynamic-range reduction through frequency-selective compression. Using sophisticated signal-processing techniques, a 50-dB range in the signal's intensity can be "squeezed" into a 20-dB range as a means of simulating the full dynamic range associated with the speech signal. However, such compression only partially compensates for the hearing impairment, and does not fully restore the patient's ability to understand speech in noisy and reverberant environments (cf. Edwards, Chapter 7).

What other factors may be involved in the hearing-impaired's inability to reliably decode the speech signal? One potential clue is encapsulated in the central paradox of sensorineural hearing loss. Although most of the energy (and information) in the speech signal lies *below* 2 kHz, most of the impairment in the clinical population is *above* 2 kHz. In quiet, the hearing impaired rarely experience difficulty understanding speech. However, in noisy and reverberant conditions, the ability to comprehend speech completely falls apart (without some form of hearing aid or speech-reading cues).

This situation suggests that there is information in the mid- and high-frequency regions of the spectrum that is of the utmost importance under acoustic-interference conditions. In quiet, the speech spectrum below 2 kHz can provide sufficient cues to adequately decode the signal. In noise and reverberation, the situation changes drastically, since most of the energy produced by such interference is also in the low-frequency range. Thus, the effective SNR in the portion of the spectrum where hearing function is relatively normal is reduced to the point where information from other regions of the spectrum are required to supplement and disambiguate the speech cues associated with the low-frequency spectrum.

There is some evidence to suggest that normal-hearing individuals do indeed utilize a spectrally adaptive process for decoding speech. Temporal scrambling of the spectrum via desynchonization of narrowband ($^1/_3$ octave) channels distributed over the speech range simulates certain properties of reverberation. When the channels are desynchronized by modest amounts, the intelligibility of spoken sentences remains relatively high. As the amount of asynchrony across channels increases, intelligibility falls. The rate at which intelligibility decreases is consistent with the hypothesis that for small degrees of cross-spectral asynchrony (i.e., weak reverberation), the lower parts of the spectrum (<1500 Hz) are responsible for most of the intelligibility performance, while for large amounts of asynchrony (i.e., strong reverberation) it is channels above 1500 Hz that are most highly correlated with intelligibility performance (Arai and Greenberg 1998; Greenberg and Arai 1998). This result is consistent with the finding that the best single psychoacoustic (nonspeech) predictor of speech intelligibility capability in quiet is the pure-tone threshold *below* 2 kHz, while the best predictor of speech intelligibility in noise is the pure-tone threshold *above* 2 kHz (Smoorenburg 1992; but cf. Festen and Plomp 1981 for an alternative perspective).

What sort of information is contained in the high-frequency portion of the spectrum that could account for this otherwise paradoxical result? There are two likely possibilities. The first pertains to articulatory place of articulation, information that distinguishes, for example, a [p] from [t] and [k]. The locus of maximum articulatory constriction produces an acoustic "signature" that requires reliable decoding of the entire spectrum between 500 and 3500 Hz (Stevens and Blumstein 1978, 1981). Place-of-articulation cues are particularly vulnerable to background noise (Miller and Nicely 1955; Wang and Bilger 1973), and removal of any significant portion of the spectrum is likely to degrade the ability to identify consonants on this articulatory dimension. Place of articulation is perhaps the single most important acoustic feature dimension for distinguishing among words, particularly at word onset (Rabinowitz et al. 1992; Greenberg and Chang 2000). It is therefore not surprising that much of the problem the hearing impaired manifest with respect to speech decoding pertains to place-of-articulation cues (Dubno and Dirks 1989; Dubno and Schaefer 1995).

A second property of speech associated with the mid- and high-frequency channels is prosodic in nature. Grant and Walden (1996) have shown the portion of the spectrum above 3 kHz provides the most reliable information concerning the number of syllables in an utterance. It is also likely that these high-frequency channels provide reliable information pertaining to syllable boundaries (Shastri et al. 1999). To the extent that this sort of knowledge is important for decoding the speech signal, the high-frequency channels can provide information that supplements that of the low-frequency spectrum. Clearly, additional research is required to more fully understand the contribution made by each part of the spectrum to the speech-decoding process.

Grant and colleagues (Grant and Walden 1995; Grant and Seitz 1998; Grant et al. 1998) estimate that about two thirds of the information required to decode spoken material (in this instance sentences) is bottom-up in nature, derived from detailed phonetic and prosodic cues. Top-down information concerned with semantic and grammatical context accounts for perhaps a third of the processing involved. The relative importance of the spectro-temporal detail for understanding spoken language is certainly consistent with the communication handicap experienced by the hearing impaired.

In cases where there is little hearing function left in any portion of the spectrum, a hearing aid is of little use to the patient. Under such circumstances a more drastic solution is required, namely implantation into the cochlea of an electrode array capable of direct stimulation of the auditory nerve (Clark 2003; Clark, Chapter 8). Over the past 25 years the technology associated with cochlear implants has progressed dramatically. Whereas in the early 1980s such implants were rarely capable of providing more than modest amelioration of the communication handicap associated with profound deafness, today there are many thousands who communicate at near normal levels, both in face-to-face interaction and (in the most successful cases) over the telephone (i.e., unaided by visible speech-reading cues) using such technology. The technology has been particularly effective for young children who have been able to grow up using spoken language to a degree that would have been unimaginable 20 years ago.

The conceptual basis of cochlear-implant technology is simple (although the surgical and technical implementation is dauntingly difficult to properly execute). An array of about 24 electrodes is threaded into the scala media of the cochlea. Generally, the end point of the array reaches into the basal third of the partition, perhaps as far as the 800 to 1000 Hz portion of the cochlea. Because there is often some residual hearing in the lowest frequencies, this technical limitation is not as serious as it may appear. The 24 electrodes generally span a spectral range between about 800 and 6000 Hz. Not all of the electrodes are active. Rather, the intent is to chose between four and eight electrodes that effectively sample the spectral range. The speech signal is spectrally partitioned so that lower frequencies stimulate the most apical electrodes and the higher frequencies are processed through the more basal ones, in a frequency-graded manner. Thus, the implant performs a crude form of spatial frequency analysis, analogous to that performed by the normal-functioning cochlea.

Complementing the cochlear place cues imparted by the stimulating electrode array is low-frequency periodicity information associated with the waveform's fundamental frequency. This voice-pitch information is signaled through the periodic nature of the stimulating pulses emanating from each electrode. In addition, coarse amplitude information is transmitted by the overall pulse rate.

Although the representation of the speech signal provided by the implant is a crude one, it enables most patients to verbally interact effectively. Shannon and colleagues (1995) have explored the nature of this coarse representation in normal-hearing individuals, demonstrating that only four spectrally discrete channels are required (under ideal listening conditions) to transmit intelligible speech using a noise-like phonation source. Thus, it would appear that the success of cochlear implants relies, to a certain extent, on the relatively coarse spectro-temporal representation of information in the speech signal (cf. section 4.1).

11. The Influence of Learning on Auditory Processing of Speech

Language represents the culmination of the human penchant for communicating vocally and appears to be unique in the animal kingdom (Hauser 1996). Much has been made of the creative aspect of language that enables the communication of ideas virtually without limit (cf. Chomsky 1965; Hauser et al. 2002; Studdert-Kennedy and Goldstein 2003). Chomsky (2000) refers to this singular property as "discrete infinity."

The limitless potential of language is grounded, however, in a vocabulary with limits. There are 415,000 word forms listed in the unabridged edition of the *Oxford English Dictionary*, the gold standard of English lexicography. Estimates of the average individual's *working* vocabulary range from 10,000 to 100,000. But a statistical analysis of spontaneous dialogues (American English) reveals an interesting fact—90% of the words used in casual discussions can be covered by less than 1000 distinctive lexical items (Greenberg 1999). The 100 most frequent words from the corpus Switchboard account for two thirds of the lexical tokens, and the 10 most common words account for nearly 25% of the lexical usage (Greenberg 1999). Comparable statistics were compiled by French and colleagues (1930). Thus, while a speaker may possess the potential for producing tens of thousands of different words, in daily conversation this capacity is rarely exercised. Most speakers get by with only a few thousand words most of the time.

This finite property of spoken language is an important one, for it provides a means for the important elements to be learned effectively (if not fully mastered) to facilitate rapid and reliable communication. While "discrete infinity" is attractive in principle, it is unlikely to serve as an accurate description of spoken language in the real world, where speakers are rarely creative or original. Most utterances are composed of common words sequenced in conventional ways, as observed by Skinner (1957) long ago. This stereotypy is characteristic of an overlearned system designed for rapid and reliable communication. With respect to spontaneous speech, Skinner is probably closer to the mark than Chomsky.

11.1 Auditory Processing with an Interpretative Linguistic Framework

Such constraints on lexical usage are important for understanding the role of auditory processing in linguistic communication. Auditory patterns, as processed by the brain, bear no significance except as they are interpretable with respect to the real world. In terms of language, this means that the sounds spoken must be associated with specific events, ideas, and objects. And given the very large number of prospective situations to describe, some form of structure is required so that acoustic patterns can be readily associated with meaningful elements.

Such structure is readily discernible in the syntax and grammar of any language, which constrain the order in which words occur relative to each other. On a more basic level, germane to hearing are the constraints imposed on the sound shapes of words and syllables, which enable the auditory system to efficiently decode complex acoustic patterns within a meaningful linguistic framework. The examples that follow illustrate the importance of structure (and constraints implied) for efficiently decoding the speech signal.

The 100 most frequent words in English (accounting for 67% of the lexical instances) tend to contain but a single syllable, and the exceptions contain only two (Greenberg 1999). This subset of spoken English generally consists of the "function" words such as pronouns, articles, and locatives, and is generally of Germanic origin.

Moreover, most of these common words have a simple syllable structure, containing either a consonant followed by a vowel (CV), a consonant followed by a vowel, followed by another consonant (CVC), a vowel followed by a consonant (VC), or just a vowel by itself (V). Together, these three syllable forms account for more than fourth fifths of the syllables encountered (Greenberg 1999).

In contrast to function words are the "content" lexemes that provide the specific referential material enabling listeners to decode the message with precision and confidence. Such content words occur less frequently than their function-word counterparts, often contain three or more syllables, and are generally nouns, adjectives, or adverbs. Moreover, their linguistic origin is often non-Germanic—Latin and Norman French being the most common sources of this lexicon. When the words are of Germanic origin, their syllable structure is often complex (i.e., consonant clusters in either the onset or coda, or both). Listeners appear to be aware of such statistical correlations, however loose they may be.

The point reinforced by these statistical patterns is that spoken forms in language are far from arbitrary, and are highly constrained in their structure. Some of these structural constraints are specific to a language, but many appear to be characteristic of all languages (i.e., universal). Thus, all utterances are composed of syllables, and every syllable contains a nucleus, which is virtually always a vowel. Moreover, syllables can begin with a consonant, and most of them do. And while a syllable can also end with a consonant, this is much less likely to happen. Thus, the structural nature of the syllable is asymmetric. The question arises as to why.

Syllables can begin and end with more than a single consonant in many (but not all) languages. For example, in English, a word can conform to the syllable structure CCCVCCC ("strengths"), but rarely does so. When consonants do occur in sequence within a syllable, their order is nonrandom, but conforms to certain phonotactic rules. These rules are far from arbitrary, but conform to what is known as the "sonority hierarchy" (Clements 1990; Zec 1995), but which is really a cover term for sequencing segments in a quasi-continuous "energy arc" over the syllable.

Syllables begin with gradually increasing energy over time that rises to a crescendo in the nucleus before descending in the coda (or the terminal portion of the nucleus in the absence of a coda segment). This statement is an accurate description only for energy integrated over 25-ms time windows. Certain segments, principally the stops and affricates, begin with a substantial amount of energy that is sustained over a brief (ca. 10-ms) interval of time, which is followed by a more gradual buildup of energy over the following 40 to 100 ms. Vowels are the most energetic (i.e., intense) of segments, followed by the liquids, and glides (often referred to as "semivowels") and nasals. The least intense segments are the fricatives (particularly of the voiceless variety), the affricates, and the stops. It is a relatively straightforward matter of predicting the order of consonant types in onset and coda from the energy-arc principle. More intense segments do not precede less intense ones in the syllable onset building up to the nucleus. Conversely, less intense segments do not precede more intense ones in the coda. If the manner (mode) of production is correlated with energy level, adjacent segments within the syllable should rarely (if ever) be of the same manner class, which is the case in spontaneous American English (Greenberg et al. 2002).

Moreover, the entropy associated with the syllable onset appears to be considerably greater than in the coda or nucleus. Pronunciation patterns are largely canonical (i.e., of the standard dictionary form) at onset, with a full range of consonant segments represented. In coda position, three segments—[t], [d], and [n]—account for over 70% of the consonantal forms (Greenberg et al. 2002).

Such constraints serve to reduce the perplexity of constituents within a syllable, thus making "infinity" more finite (and hence more learnable) than would otherwise be the case. More importantly, they provide an auditory-based framework with which to interpret auditory patterns within a linguistic framework, reducing the effective entropy associated with many parts of the speech signal to manageable proportions (i.e., much of the entropy is located in the syllable onset, which is more likely to evoke neural discharge in the auditory cortex). In the absence of such an interpretive framework auditory patterns could potentially lose all meaning and merely register as sound.

11.2 Visual Information Facilitates Auditory Interpretation

Most verbal interaction occurs face to face, thus providing visual cues with which to supplement and interpret the acoustic component of the speech signal. Normally, visual cues are unconsciously combined with the acoustic signal and are largely taken for granted. However, in noisy environments, such "speech-reading" information provides a powerful assist in decoding speech, particularly for the hearing impaired (Sumby and Pollack 1954; Breeuer and Plomp 1984; Massaro 1987; Summerfield 1992; Grant and Walden 1996b; Grant et al. 1998; Assmann and Summerfield, Chapter 5).

Because speech can be decoded without visual input much of the time (e.g., over the telephone), the significance of speech reading is seldom fully appreciated. And yet there is substantial evidence that such cues often provide the extra margin of information enabling the hearing impaired to communicate effectively with others. Grant and Walden (1995) have suggested that the benefit provided by speech reading is comparable to, or even exceeds, that of a hearing aid for many of the hearing impaired.

How are such cues combined with the auditory representation of speech? Relatively little is known about the specific mechanisms. Speech-reading cues appear to be primarily associated with place-of-articulation information (Grant et al. 1998), while voicing and manner information are derived almost entirely from the acoustic signal.

The importance of the visual modality for place-of-articulation information can be demonstrated through presentation of two different syllables, one using the auditory modality, the other played via the visual channel. If the consonant in the acoustic signal is [p] and in the visual signal is [k] (all other phonetic properties of the signals being played equal), listeners often report "hearing" [t], which represents a blend of the audiovisual streams with respect to place of articulation (McGurk and McDonald 1976). Although this "McGurk effect" has been studied intensively (cf. Summerfield 1992), the underlying neurological mechanisms remain obscure. Whatever its genesis in the brain, the mechanisms responsible for combining auditory and visual information must lie at a fairly abstract level of representation. It is possible for the visual stream to precede the audio by as much as 120 to 200 ms without an appreciable affect on intelligibility (Grant and Greenberg 2001). However, if the audio precedes the video, intelligibility falls dramatically for leads as small as 50 to 100 ms. The basis of this sensory asymmetry in stream asynchrony is the subject of ongoing research. Regardless of the specific nature of the neurological mechanisms underlying auditory-visual speech processing, it serves as a powerful example of how the brain is able to interpret auditory processing within a larger context.

11.3 Informational Constraints on Auditory Speech Processing

It is well known that the ability to recognize speech depends on the size of the response set—the smaller the number of linguistic categories involved, the easier it is for listeners to correctly identify words and phonetic segments (Pollack 1959) for any given SNR. In this sense, the amount of inherent information [often referred to as (negative) "entropy"] associated with a recognition or identification task has a direct impact on performance (cf. Assmann and Summerfield, Chapter 5), accounting to a certain degree for variation in performance using different kinds of speech material. Thus, at an SNR of 0 dB, spoken digits are likely to be recognized with 100% accuracy, while for words of a much larger response set (in the hundreds or thousands) the recognition score will be 50% or less under comparable conditions.

However, if these words were presented at the same SNR in a connected sentence, the recognition score would rise to about 80%. Presentation of spoken material within a grammatical and semantic framework clearly improves the ability to identify words.

The articulation index was originally developed using nonsense syllables devoid of semantic context, on the assumption that the auditory processes involved in this task are comparable to those operating in a more realistic linguistic context. Hence, a problem decoding the phonetic properties of nonsense material should, in principle, also be manifest in continuous speech. This is the basic premise underlying extensions of the articulation index to meaningful material (e.g., Boothroyd and Nittrouer 1988; cf. Assmann and Summerfield, Chapter 5). However, this assumption has never been fully verified, and therefore the relationship between phonetic-segment identification and decoding continuous speech remains to be clarified.

11.4 Categorical Perception

The importance of learning and generalization in speech decoding is amply illustrated in studies on categorical perception (cf. Rosen and Howell 1987). In a typical experiment, a listener is asked to denote a speech segment as an exemplar of either class A or B. Unbeknownst to the subject, a specific acoustic parameter has been adjusted in fine increments along a continuum. At one end of the continuum virtually all listeners identify the sounds as A, while at the other end, all of the sounds are classified as B. In the middle responses are roughly equally divided between the two. The key test is one in which discrimination functions between two members of the continuum are produced. In instances where one stimulus has been clearly identified as A and the other as B, these signals are accurately distinguished and labeled as "different." In true categorical perception, listeners are able to reliably discriminate only between signals identified as different phones. Stimuli from within the same labeled class, even though they differ along a specific acoustic dimension, are not reliably distinguished (cf. Liberman et al. 1957).

A number of specific acoustic dimensions have been shown to conform to categorical perception, among them voice onset time (VOT; cf. Lisker and Abramson 1964) and place of articulation. VOT refers to the interval of time separating the articulatory release from glottal vibration (cf. Arendaño et al., Chapter 2; Diehl and Lindblom, Chapter 3). For a segment, such as [b], VOT is short, typically less than 20 ms, while for its voiceless counterpart, [p], the interval is generally 40 ms or greater. Using synthetic stimuli, it is possible to parametrically vary VOT between 0 and 60 ms, keeping other properties of the signal constant. Stimuli with a VOT between 0 and 20 ms are usually classified as [b], while those with a VOT between 40 and 60 ms are generally labeled as [p]. Stimuli with VOTs between 20 and 40 ms often sound ambiguous, eliciting [p] and [b] responses in varying proportions. The VOT boundary is defined as that interval for which [p] and [b] responses occur in roughly equal proportion. Analogous experiments have been performed for other stop consonants, as well as for segments associated with different manner-of-articulation classes (for reviews, see Liberman et al. 1967; Liberman and Mattingly 1985).

Categorical perception provides an illustration of the interaction between auditory perception and speech identification using a highly stylized signal. In this instance listeners are given only two response classes and are forced to choose between them. The inherent entropy associated with the task is low (essentially a single bit of information, given the binary nature of the classification task), unlike speech processing in more natural conditions where the range of choices at any given instant is considerably larger. However, the basic lesson of categorical perception is still valid—that perception can be guided by an abstraction based on a learned system, rather than by specific details of the acoustic signal. Consistent with this perspective are studies in which it is shown that the listener's native language has a marked influence on the location of the category boundary (e.g., Miyawaki et al. 1975).

However, certain studies suggest that categorical perception may not reflect linguistic processing per se, but rather is the product of more general auditory mechanisms. For example, it is possible to shift the VOT boundary by selective adaptation methods, in which the listener is exposed to repeated presentation of the same stimulus (usually an exemplar of one end of the continuum) prior to classification of a test stimulus. Under such conditions the boundary shifts away (usually by 5 to 10 ms) from the exemplar (Eimas and Corbit 1973; Ganong 1980). The standard interpretation of this result is that VOT detectors in the auditory system have been "fatigued" by the exemplar.

Categorical perception also has been used to investigate the ontogeny of speech processing in the maturing brain. Infants as young as 1 month are able to discriminate, as measured by recovery from satiation, two stimuli from different acoustic categories more reliably than signals with comparable acoustic distinctions from the same phonetic category (Eimas et al. 1971). Such a result implies that the basic capability for phonetic-feature detection may be "hard-wired" into the brain, although exposure to language-specific patterns appears to play an important role as well (Strange and Dittman 1983; Kuhl et al. 1997).

The specific relation between categorical perception and language remains controversial. A number of studies have shown that nonhuman species, such as chinchilla (Kuhl and Miller 1978), macaque (Kuhl and Padden 1982), and quail (Kluender 1991), all exhibit behavior comparable in certain respects to categorical perception in humans. Such results suggest that at least some properties of categorical perception are not strictly language-bound but rather reflect the capability of consistent generalization between classes regardless of their linguistic significance (Kluender et al. 2003).

12. Technology, Speech, and the Auditory System

Technology can serve as an effective proving ground for ideas generated during the course of scientific research (Greenberg 2003). Algorithms based on models of the auditory system's processing of speech, in principle, can be used in auditory prostheses, as well as for automatic speech recognition systems and other speech applications. To the extent that these auditory-inspired algorithms improve performance of the technology, some degree of confidence is gained that the underlying ideas are based on something more than wishful thinking or mathematical elegance. Moreover, careful analysis of the problems encountered in adapting scientific models to real-world applications can provide insight into the limitations of such models as a description of the processes and mechanisms involved (Greenberg 2003).

12.1 Automatic Speech Recognition (Front-End Features)

The historical evolution of automatic speech recognition (ASR) can be interpreted as a gradually increasing awareness of the specific problems required to be solved (Ainsworth 1988). For example, an early, rather primitive system developed by Davis and colleagues (1952) achieved a word-recognition score of 98% correct for digits spoken by a single speaker. However, the recognition score dropped to about 50% when the system was tested on other speakers. This particular system measured the zero-crossing rate of the speech signal's pressure waveform after it had been filtered into two discrete frequency channels roughly corresponding to the range associated with the first and second formants. The resulting outputs were cross-correlated with a set of stored templates associated with representative exemplars for each digit. The digit template associated with the highest correlation score was chosen as the recognized word.

This early system's structure—some form of frequency analysis followed by a pattern matcher—persists in contemporary systems, although the nature of the analyses and pattern recognition techniques used in contemporary systems has markedly improved in recent years. Early recognition systems used pattern-matching methods to compare a sequence of incoming feature vectors derived from the speech signal with a set of stored word templates. Recognition error rates for *speaker-dependent* recognizers dropped appreciably when dynamic-time-warping (DTW) techniques were introduced as a means of counteracting durational variability (Velichko and Zagoruyko 1970; Sakoe and Chiba 1978). However the problem associated with *speaker-independent* recognition remained until statistical methods were introduced in the late 1970s.

Over the past 25 years, statistical approaches have replaced the correlational and DTW approaches of the early ASR systems and are embedded within a mathematical framework known as hidden Markov models (HMMs) (e.g., Jelinek 1976, 1977), which are used to represent each word and sub-word (usually phoneme) unit involved in the recognition task. Associated with each HMM state is a probability score associated with the likelihood of a particular unit occurring in that specific context given the training data used to develop the system.

One of the key problems that a speech recognizer must address is how to efficiently reduce the amount of data representing the speech signal without compromising recognition performance. Can principles of auditory function be used to achieve this objective as well as to enhance ASR performance, particularly in background noise?

Speech technology provides an interesting opportunity to test many of the assumptions that underlie contemporary theories of hearing (cf. Hermansky 1998; Morgan et al., Chapter 6). For example, the principles underlying the spectral representation used in ASR systems are directly based on perceptual studies of speech and other acoustic signals. In contrast to Fourier analysis, which samples the frequency spectrum linearly (in terms of Hz units), modern approaches (Mel frequency cepstral coefficients—Davis and Mermelstein 1980; perceptual linear prediction—Hermansky 1990) warp the spectral representation, giving greater weight to frequencies below 2 kHz. The spatial-frequency mapping is logarithmic above 800 Hz (Avendaño et al., Chapter 2; Morgan et al., Chapter 6), in a manner comparable to what has been observed in both perceptual and physiological studies. Moreover, the granularity of the spectral representation is much coarser than the fast Fourier transform (FFT), and is comparable to the critical-band analysis performed in the cochlea (section 9). The representation of the spectrum is highly smoothed, simulating integrative processes in both the periphery and central regions of the auditory pathway. In addition, the representation of spectral magnitude is not in terms of decibels (a physical measure), but rather in units analogous to sones, a perceptual measure of loudness rooted in the compressive nature of transduction in the cochlea and beyond (cf. Zwicker 1975; Moore 1997). This sort of transformation has the effect of compressing the variation in peak magnitude across the spectrum, thereby providing a parsimonious and effective method of preserving the shape of the spectral envelope across a wide variety of environmental conditions.

RASTA is yet another example of auditory-inspired signal processing that has proven useful in ASR systems. Its conceptual roots lie in the sensory and neural adaptation observed in the cochlea and other parts of the auditory pathway. Auditory neurons adapt their response level to the acoustic context in such a manner that a continuous signal evokes a lower level of activity during most of its duration than at stimulus onset (Smith 1977). This reduction in responsiveness may last for hundreds or even thousands of milliseconds after cessation of the signal, and can produce an auditory "negative afterimage" in which a phantom pitch is "heard" in the region of the spectrum close to that of the original signal (Zwicker 1964). Summerfield et al. (1987) demonstrated that such an afterimage could be generated using a steady-state vowel in a background of noise. Once the original vowel was turned off, subjects faintly perceived a second vowel whose spectral properties were the inverse of the first.

This type of phenomenon implies that the auditory system should be most responsive to signals whose spectral properties evade the depressive consequences of adaptation through constant movement at rates that lie outside the time constants characteristic of sensorineural adaptation. The formant transitions in the speech signal move at such rates over much of their time course, and are therefore likely to evoke a relatively high level of neural discharge across a tonotopically organized population of auditory neurons. The rate of this formant movement can be modeled as a temporal filter with a specific time constant (ca. 160 ms), and used to process the speech signal in such a manner as to provide a representation that weights the spectrally dynamic portions of the signal much more highly than the steady-state components. This is the essence of RASTA, a technique that has been used to shield the speech spectrum against the potential distortion associated with microphones and other sources of extraneous acoustic energy (Hermansky and Morgan 1994; Morgan et al., Chapter 6).

12.2 Speech Synthesis

Computational simulation of the speech-production process, known as speech synthesis, affords yet a separate opportunity to evaluate the efficacy of auditory-inspired algorithms. Synthesis techniques have focused on three broad issues: (1) intelligibility, (2) quality (i.e., naturalness), and (3) computational efficiency. Simulating the human voice in a realistic manner requires knowledge of the speech production process, as well as insight into how the auditory system interprets the acoustic signal.

Over the years two basic approaches have been used, one modeling the vocal production of speech, the other focusing on spectro-temporal manipulation of the acoustic signal. The vocal-tract method was extensively investigated by Flanagan (1972) at Bell Labs and by Klatt (1987) at the Massachusetts Institute of Technology (MIT). The entire speech production process is simulated, from the flow of the air stream through the glottis into the oral cavity and out of the mouth, to the movement of the tongue, lips, velum, and jaw. These serve as control parameters governing the acoustic resonance patterns and mode of vocal excitation. The advantage of this method is representational parsimony—a production-based model that generally contains between 30 and 50 parameters updated 100 times per second. Because many of the control states do not change from frame to frame, it is possible to specify an utterance with perhaps a thousand different parameters (or less) per second. In principle, any utterance, from any language, can be generated from such a model, as long as the relationship between the control parameters and the linguistic input is known. Although such vocal tract synthesizers are generally intelligible, they are typically judged as sounding unnatural by human listeners. The voice quality has a metallic edge to it, and the durational properties of the signal are not quite what a human would produce.

The alternative approach to synthesis starts with a recording of the human voice. In an early version of this method, as exemplified by the Vocoder (section 4.1), the granularity of the speech signal was substantially reduced both in frequency and in time, thereby compressing the amount of information required to produce intelligible speech. This synthesis technique is essentially a form of recoding the signal, as it requires a recording of the utterance to be made in advance. It does not provide a principled method for extrapolating from the recording to novel utterances.

Concantenative synthesis attempts to fill this gap by generating continuous speech from several hours of prerecorded material. Instead of simulating the vocal production process, it assumes that the elements of any and all utterances that might ever be spoken are contained in a finite sample of recorded speech. Thus, it is a matter of splicing the appropriate intervals of speech together in the correct order. The "art" involved in this technique pertains to the algorithms used to determine the length of the spliced segments and the precise context from which they come. At its best, concantenative synthesis sounds remarkably natural and is highly understandable. For these reasons, most contemporary commercial text-to-speech applications are based on this technology. However, there are two significant limitations. First, synthesis requires many hours of material to be recorded from each speaker used in the system. The technology does not provide a principled method of generating voices other than those previously recorded. Second, the concantenative approach does not, in fact, handle all instances of vocal stitching well. Every so often such systems produce unintelligible utterances in circumstances where the material to be spoken lies outside the range of verbal contexts recorded.

A new form of synthesis, known as "STRAIGHT," has the potential to rectify the problems associated with production-based models and concantenative approaches. STRAIGHT is essentially a highly granular Vocoder, melded with sophisticated signal-processing algorithms that enable flexible and realistic alteration of the formant patterns and fundamental frequency contours of the speech signal (Kawahara et al. 1999). Although the synthesis method uses prerecorded material, it is capable of altering the voice quality in almost unlimited ways, thereby circumventing the most serious limitation of concantenative synthesis. Moreover, it can adapt the spectrotemporal properties of the speech waveform to any specifiable target. STRAIGHT requires about 1000 separate channels to fully capture the natural quality of the human voice, 100 times as many channels as used by the original Vocoder of the 1930s. Such a dense sampling of the spectrum is consistent with the innervation density of the human cochlea—3000 IHCs projecting to 30,000 ANFs—and suggests that undersampling of spectral information may be a major factor in the shortcomings of current-generation hearing aids in rendering sound to the ear.

12.3 Auditory Prostheses

Hearing-aid technology stands to benefit enormously from insights into the auditory processing of speech and other communication signals. A certain amount of knowledge, pertaining to spectral resolution and loudness compression, has already been incorporated into many aids (e.g., Villchur 1987; cf. Edwards, Chapter 7). However, such aids do not entirely compensate for the functional deficit associated with sensorineural damage (cf. section 10). The most sophisticated hearing aids incorporate up to 64 channels of quasi-independent processing, with four to eight different compression settings specifiable over the audio range. Given the spectral-granularity capability of the normal ear (cf. sections 7 and 9), it is conceivable that hearing aids would need to provide a much finer-grained spectral representation of the speech signal in order to provide the sort of natural quality characteristic of the human voice. On the other hand, it is not entirely clear whether the damaged ear would be capable of exploiting such fine spectral detail.

One of the most significant problems with current-generation hearing aids is the difficulty encountered processing speech in noisy backgrounds. Because the basic premise underlying hearing-aid technology is amplification ("power to the ear!"), boosting the signal level per se also increases the noise background. The key is to enhance the speech signal and other foreground signals while suppressing the background. To date, hearing-aid technology has not been able to solve this problem despite some promising innovations. One method, called the "voice activity detector," adjusts the compression parameters in the presence (or absence) of speech, based on algorithms similar in spirit of RASTA. Modulation of energy at rates between 3 and 10 Hz are interpreted as speech, with attendant adjustment of the compression parameters. Unfortunately, this form of quasi-dynamic range adjustment is not sufficient to ameliorate the acoustic interference problem. Other techniques, based on deeper insight into auditory processes, will be required (cf. section 13).

12.4 Automatic Speech Recognition (Lexical Decoding)

There is far more to decoding speech than mere extraction of relevant information from the acoustic signal. It is for this reason that ASR systems focus much of their computational power on associating spectro-temporal features gleaned from the "front end" with meaningful linguistic units such as phones, syllables, and words.

Most current-generation ASR systems use the phoneme as the basic decoding unit (cf. Morgan et al., Chapter 6). Words are represented as linear sequences of phonemic elements, which are associated with spectro-temporal cues in the acoustic signal via acoustic models trained on context-dependent phone models. The problem with this approach is the enormous amount of pronunciation variation characteristic of speech spoken in the real world. Much of this variation is inherent to the speaking process and reflects dialectal, gender, emotional, socioeconomic, and stylistic factors. The phonetic properties can vary significantly from one context to the next, even for the same speaker (section 2).

Speech recognition systems currently do well only in circumstances where they have been trained on extensive amounts of data representative of the task domain and where the words spoken (and their order) are known in advance. For this reason, ASR systems tend to perform best on prompted speech, where there is a limited set of lexical alternatives (e.g., an airline reservation system), or where the spoken material is read in a careful manner (and hence the amount of pronunciation variation is limited). Thus, current ASR systems function essentially as sophisticated decoders rather than as true open-set recognition devices. For this reason, automatic speech recognition is expensive, time-consuming technology to develop and is not easily adaptable to novel task domains.

13. Funture Trends in Auditory Research Germane to Speech

Spoken language is based on processes of enormous complexity, involving many different regions of the brain, including those responsible for hearing, seeing, remembering, and interpreting. This volume focuses on just one of these systems, hearing, and attempts to relate specific properties of the auditory system to the structure and function of speech. In coming years our knowledge of auditory function is likely to increase substantially and in ways potentially capable of having a direct impact on our understanding of the speech decoding process.

It is becoming increasing clear that the auditory pathway interacts either directly or indirectly with many other parts of the brain. For example, visual input can directly affect the response properties of neurons in the auditory cortex (Sharma et al. 2000), and there are instances where even somatosensory input can affect auditory processing (Gao and Suga 2000). It is thus becoming increasingly evident that auditory function cannot be entirely understood without taking such cross-modal interactions into consideration. Moreover, the auditory system functions as part of an integrated behavioral system where, in many circumstances, it may provide only a small part of the information required to perform a task. Many properties of hearing can only be fully appreciated within such a holistic framework. Spoken language is perhaps the most elaborate manifestation of such integrated behavior and thus provides a fertile framework in which to investigate the interaction among various brain regions involved in the execution of complex behavioral tasks.

Future research pertaining to auditory function and speech is likely to focus on several broad areas. Human brain-imaging technology has improved significantly over the past decade, so that is it now possible to visualize neural activation associated with specific behavioral tasks with a degree of spatial and temporal resolution undreamt of in the recent past. Such techniques as functional magnetic resonance imaging (fMRI) and magnetoencephalography (MEG) will ultimately provide (at least in principle) the capability of answering many of the "where" and "when" questions posed in this chapter. Dramatic discoveries are likely to be made using such imaging methods over the next decade, particularly with respect to delineating the interaction and synergy among various neurological systems involved in processing spoken language.

Language is a highly learned behavior, and it is increasing clear that learning plays an important role in auditory function (Recanzone et al. 1993; Wright et al. 1997) germane to speech processing. How does the auditory system adapt to experience with specific forms of acoustic input? Do sensory maps of fundamental auditory features change over time in response to such acoustic experience (as has been demonstrated in the barn owl, cf. Knudsen 2002)? What is the role of attentional processes in the development of auditory representations and the ability to reliably extract behaviorally relevant information? Do human listeners process sounds differently depending on exposure to specific acoustic signals? Are certain language-related disorders the result of a poor connection between the auditory and learning systems? These and related issues are likely to form the basis of much hearing-based research over the next 20 years.

Technology has historically served as a "forcing function," driving the pace of innovation in many fields of scientific endeavor. This technology-driven research paradigm is likely to play an ever-increasing role in the domains of speech and auditory function.

For example, hearing aids do not currently provide a truly effective means of shielding speech information in background noise, nor are automatic speech recognition systems fully capable of decoding speech under even moderately noisy conditions. For either technology to evolve, the noise-robustness problem needs to be solved, both from an engineering and (more importantly) a scientific perspective. And because of this issue's strategic importance for speech technology, it is likely that a considerable amount of research will focus on this topic over the next decade.

Intelligibility remains perhaps the single most important issue for auditory prostheses. The hearing impaired wish to communicate easily with others, and the auditory modality provides the most effective means to do so. To date, conventional hearing aids have not radically improved the ability to understand spoken language except in terms of enhanced audibility. Digital compression aids provide some degree of improvement with respect to noise robustness and comfort, but a true breakthrough in terms of speech comprehension awaits advances in the technology. One of the obstacles to achieving such an advance is our limited knowledge of the primary cues in the speech signal required for a high degree of intelligibility (cf. Greenberg et al. 1998; Greenberg and Arai 2001; Müsch and Buus 2001a,b). Without such insight it is difficult to design algorithms capable of significantly enhancing speech understanding. Thus, it is likely that a more concerted effort will be made over the next few years to develop accurate speech intelligibility metrics germane to a broad range of acoustic-environment conditions representative of the real world, and which are more accurate than the articulation index (AI) and STI.

Related to this effort will be advances in cochlear implant design that provide a more natural-sounding input to the auditory pathway than current devices afford. Such devices are likely to incorporate a more fine-grained representation of the speech spectrum than is currently provided, as well as using frequency-modulation techniques in tandem with those based on amplitude modulation to simulate much of the speech signal's spectro-temporal detail.

The fine detail of the speech signal is also important for speech synthesis applications, where a natural-sounding voice is often of paramount importance. Currently, the only practical means of imparting a natural quality to the speech is by prerecording the materials with a human speaker. However, this method ("concantenative synthesis") limits voice quality and speaking styles to the range recorded. In the future, new synthesis techniques (such a s STRAIGHT, cf. Kawahara et al. 1999), will enable life-like voices to be created, speaking in virtually any style and tone imaginable (and for a wide range of languages). Moreover, the acoustic signal will be melded with a visual display of a talking avatar simulating the look and feel of a human speaker. Achieving such an ambitious objective will require far more detailed knowledge of the auditory (and visual) processing of the speech stream, as well as keen insight into the functional significance of the spectro-temporal detail embedded in the speech signal.

Automatic speech recognition is gaining increasing commercial acceptance and is now commonly deployed for limited verbal interactions over the telephone. Airplane flight and arrival information, credit card and telephone account information, stock quotations, and the like are now often mediated by speaker-independent, constrained-vocabulary ASR systems in various locations in North America, Europe and Asia. This trend is likely to continue, as companies learn how to exploit such technology (often combined with speech synthesis) to simulate many of the functions previously performed by human operators.

However, much of ASR's true potential lies beyond the limits of current technology. Currently, ASR systems perform well only in highly constrained, linguistically prompted contexts, where very specific information is elicited through the use of pinpoint questions(e.g., Gorin et al. 1997). This form of interaction is highly unnatural and customers quickly tire of its repetitive, tedious nature. Truly robust ASR would be capable of providing the illusion of speaking to a real human operator, an objective that lies many years in the future. The knowledge required to accomplish this objective is immense and highly variegated. Detailed information about spoken language structure and its encoding in the auditory system is also required before speech recognition systems achieve the level of sophistication required to successfully simulate human dialogue.

Advances in speech recognition and synthesis technology may ultimately advance the state of auditory prostheses. The hearing aid and cochlear implant of the future are likely to utilize such technology as a means of providing a more intelligible and life-like signal to the brain. Adapting the auditory information provided, depending on the nature of the interaction context (e.g., the presence of speech-reading cues and/or background noise) will be commonplace.

Language learning is yet another sector likely to advance as a consequence of increasing knowledge of spoken language and the auditory system. Current methods of teaching pronunciation of foreign languages are often unsuccessful, focusing on the articulation of phonetic segments in isolation, rather than as an integrated whole organized prosodically. Methods for providing accurate, production-based feedback based on sophisticated phonetic and prosodic classifiers could significantly improve pronunciation skills of the language student. Moreover, such technology could also be used in remedial training regimes for children with specific articulation disorders.

Language is what makes humans unique in the animal kingdom. Our ability to communicate via the spoken word is likely to be associated with the enormous expansion of the frontal regions of the human cortex over the course of recent evolutionary history and probably laid the behavioral groundwork for development of complex societies and their attendant cultural achievements. A richer knowledge of this crucial behavioral trait depends in large part on deeper insight into the auditory foundations of speech communication.

List of Abbreviations

AC	alternating current
AGC	automatic gain control
AI	articulation index
ALSR	average localized synchronized rate
AN	auditory nerve
ANF	auditory nerve fiber
ASR	automatic speech recognition
AVCN	anteroventral cochlear nucleus
CF	characteristic frequency
CV	consonant-vowel
CVC	consonant-vowel-consonant
Δf	frequency DL
ΔI	intensity DL
DC	direct current
DL	difference limen
DTW	dynamic time warping
F_1	first formant
F_2	second formant
F_3	third formant
FFT	fast Fourier transform
fMRI	functional magnetic resonance imaging
f_0	fundamental frequency
FTC	frequency threshold curve
HMM	hidden Markov model
IHC	inner hair cell
MEG	magnetoencephalography
OHC	outer hair cell
PLP	perceptual linear prediction
SNR	signal-to-noise ratio
SPL	sound pressure level
SR	spontaneous rate
STI	speech transmission index
TM	tectorial membrane
V	vowel
VC	vowel-consonant
VOT	voice onset time

References

Ainsworth WA (1976) Mechanisms of Speech Recognition. Oxford: Pergamon Press.
Ainsworth WA (1986) Pitch change as a cue to syllabification. J Phonetics 14:257–264.
Ainsworth WA (1988) Speech Recognition by Machine. Stevenage, UK: Peter Peregrinus.
Ainsworth WA, Lindsay D (1986) Perception of pitch movements on tonic syllables in British English. J Acoust Soc Am 79:472–480.
Allen JB (1994) How do humans process and recognize speech? IEEE Trans Speech Audio Proc 2:567–577.
Anderson DJ, Rose JE, Brugge JF (1971) Temporal position of discharges in single auditory nerve fibers within the cycle of a sine-wave stimulus: frequency and intensity effects. J Acoust Soc Am 49:1131–1139.
Arai T, Greenberg S (1988) Speech intelligibility in the presence of cross-channel spectral asynchrony. Proc IEEE Int Conf Acoust Speech Sig Proc (ICASSP-98), pp. 933–936.
Baer T, Moore BCJ (1993) Effects of spectral smearing on the intelligibility of sentences in noise. J Acoust Soc Am 94:1229–1241.
Blackburn CC, Sachs MB (1990) The representation of the steady-state vowel sound [ε] in the discharge patterns of cat anteroventral cochlear nucleus neurons. J Neurophysiol 63:1191–1212.
Blauert J (1996) Spatial Hearing: The Psychophysics of Human Sound Localization, 2nd ed. Cambridge, MA: MIT Press.
Blesser B (1972) Speech perception under conditions of spectral transformation. I. Phonetic characteristics. J Speech Hear Res 15:5–41.
Bohne BA, Harding GW (2000) Degeneration in the cochlea after noise damage: primary versus secondary events. Am J Otol 21:505–509.
Bolinger D (1986) Intonation and Its Parts: Melody in Spoken English. Stanford: Stanford University Press.
Bolinger D (1989) Intonation and Its Uses: Melody in Grammar and Discourse. Stanford: Stanford University Press.
Boothroyd A, Nittrouer S (1988) Mathematical treatment of context effects in phoneme and word recognition. J Acoust Soc Am 84:101–114.
Boubana S, Maeda S (1998) Multi-pulse LPC modeling of articulatory movements. Speech Comm 24:227–248.
Breeuer M, Plomp R (1984) Speechreading supplemented with frequency-selective sound-pressure information. J Acoust Soc Am 76:686–691.
Bregman AS (1990) Auditory Scene Analysis. Cambridge, MA: MIT Press.
Brokx JPL, Nooteboom SG (1982) Intonation and the perceptual separation of simultaneous voices. J Phonetics 10:23–36.
Bronkhorst AW (2000) The cocktail party phenomenon: a review of research on speech intelligibility in multiple-talker conditions. Acustica 86:117–128.
Brown GJ, Cooke MP (1994) Computational auditory scene analysis. Comp Speech Lang 8:297–336.
Buchsbaum BR, Hickok G, Humphries C (2001) Role of left posterior superior temporal gyrus in phonological processing for speech perception and production. Cognitive Sci 25:663–678.

Carlson R, Granström B (eds) (1982) The Representation of Speech in the Peripheral Auditory System. Amsterdam: Elsevier.

Carré R, Mrayati M (1995) Vowel transitions, vowel systems and the distinctive region model. In: Sorin C, Méloni H, Schoentingen J (eds) Levels in Speech Communication: Relations and Interactions. Amsterdam: Elsevier, pp. 73–89.

Chistovich LA (1985) Central auditory processing of peripheral vowel spectra. J Acoust Soc Am 77:789–805.

Chomsky N (1965) Aspects of the Theory of Syntax. Cambridge, MA: MIT Press.

Chomsky N (2000) New Horizons in the Study of Language and Mind. Cambridge: Cambridge University Press.

Clark GM (2003) Cochlear Implants: Fundamentals and Applications. New York: Springer-Verlag.

Clements GN (1990) The role of the sonority cycle in core syllabification. In: Kingston J, Beckman M (eds) Papers in Laboratory Phonology I: Between the Grammar and Physics of Speech. Cambridge: Cambridge University Press, pp. 283–325.

Cooke MP (1993) Modelling Auditory Processing and Organisation. Cambridge: Cambridge University Press.

Cooke M, Ellis DPW (2001) The auditory organization of speech and other sources in listeners and computational models. Speech Comm 35:141–177.

Darwin CJ (1981) Perceptual grouping of speech components different in fundamental frequency and onset-time. Q J Exp Psychol 3(A):185–207.

Darwin CJ, Carlyon RP (1995) Auditory grouping. In: Moore BCJ (ed) The Handbook of Perception and Cognition, Vol. 6, Hearing. London: Academic Press, pp. 387–424.

Davis K, Biddulph R, Balashek S (1952) Automatic recognition of spoken digits. J Acoust Soc Am 24:637–642.

Davis SB, Mermelstein P (1980) Comparison of parametric representation for monosyllabic word representation in continuously spoken sentences. IEEE Trans Acoust Speech Sig Proc 28:357–366.

Delgutte B, Kiang NY-S (1984) Speech coding in the auditory nerve: IV. Sounds with consonant-like dynamic characteristics. J Acoust Soc Am 75:897–907.

Deng L, Geisler CD, Greenbery S (1988) A composite model of the auditory periphery for the processing of speech. J Phonetics 16:93–108.

Drullman R (2003) The significance of temporal modulation frequencies for speech intelligibility. In: Greenberg S, Ainsworth WA (eds) Listening to Speech: An Auditory Perspective. Hillsdale, NJ: Erlbaum.

Drullman R, Festen JM, Plomp R (1994a) Effect of temporal envelope smearing on speech reception. J Acoust Soc Am 95:1053–1064.

Drullman R, Festen JM, Plomp R (1994b) Effect of reducing slow temporal modulations on speech reception. J Acoust Soc Am 95:2670–2680.

Dubno JR, Dirks DD (1989) Auditory filter characteristics and consonant recognition for hearing-impaired listeners. J Acoust Soc Am 85:1666–1675.

Dubno JR, Schaefer AB (1995) Frequency selectivity and consonant recognition for hearing-impaired and normal-hearing listeners with equivalent masked thresholds. J Acoust Soc Am 97:1165–1174.

Dudley H (1939) Remaking speech. J Acoust Soc Am 11:169–177.

Dye RH, Hafter ER (1980) Just-noticeable differences of frequency for masked tones. J Acoust Soc Am 67:1746–1753.

Eimas PD, Corbit JD (1973) Selective adaptation of linguistic feature detectors. Cognitive Psychol 4:99–109.
Eimas PD, Siqueland ER, Jusczyk P, Vigorito J (1971) Speech perception in infants. Science 171:303–306.
Fant G (1960) Acoustic Theory of Speech Production. The Hague: Mouton.
Fay RR, Popper AN (1994) Comparative Hearing: Mammals. New York: Springer-Verlag.
Festen JM, Plomp R (1981) Relations between auditory functions in normal hearing. J Acoust Soc Am 70:356–369.
Flanagan JL (1955) A difference limen for vowel formant frequency. J Acoust Soc Am 27:613–617.
Flanagan JL (1957) Estimates of the maximum precision necessary in quantizing certain "dimensions" of vowel sounds. J Acoust Soc Am 29:533–534.
Flanagan JL (1972) Speech Analysis, Synthesis and Perception, 2nd ed. Berlin: Springer-Verlag.
Fletcher H (1953) Speech and Hearing in Communication. New York: Van Nostrand.
Fletcher H, Gault RH (1950) The perception of speech and its relation to telephony. J Acoust Soc Am 22:89–150.
Fourcin AJ (1975) Language development in the absence of expressive speech. In: Lenneberg EH, Lenneberg E (eds) Foundations of Language Development, Vol. 2. New York: Academic Press, pp. 263–268.
Fowler C (1986) An event approach to the study of speech perception from a direct-realist perspective. J Phonetics 14:3–28.
Fowler CA (1996) Listeners do hear sounds, not tongues. J Acoust Soc Am 99:1730–1741.
French NR, Steinberg JC (1947) Factors governing the intelligibility of speech sounds. J Acoust Soc Am 19:90–119.
French NR, Carter CW, Koenig W (1930) The words and sounds of telephone conversations. Bell System Tech J 9:290–324.
Fujimura O, Lindqvist J (1971) Sweep-tone measurements of vocal tract characteristics. J Acoust Soc Am 49:541–558.
Ganong WF (1980) Phonetic categorization in auditory word recognition. J Exp Psych (HPPP) 6:110–125.
Gao E, Suga N (2000) Experience-dependent plasticity in the auditory cortex and the inferior colliculus of bats: role of the corticofugal system. Proc Natl Acad Sci USA 97:8081–8085.
Geisler CD, Greenberg S (1986) A two-stage automatic gain control model predicts the temporal responses to two-tone signals. J Acoust Soc Am 80:1359–1363.
Ghitza O (1988) Temporal non-place information in the auditory-nerve firing patterns as a front-end for speech recognition in a noisy environment. J Phonetics 16:109–123.
Gibson JJ (1966) The Senses Considered as Perceptual Systems. Boston: Houghton Miflin.
Gibson JJ (1979) The Ecological Approach to Visual Perception. Boston: Houghton Miflin.
Goldinger SD, Pisoni DB, Luce P (1996) Speech perception and spoken word recognition: research and theory. In: Lass N (ed) Principles of Experimental Phonetics. St. Louis: Mosby, pp. 277–327.

Goldstein JL, Srulovicz P (1977) Auditory nerve spike intervals as an adequate basis for aural spectrum analysis. In: Evans EF, Wilson JP (eds) Psychophysics and Physiology of Hearing. London: Academic Press, pp. 337–346.

Gorin AL, Riccardi G, Wright JH (1997) How may I help you? Speech Comm 23:113–127.

Grant K, Greenberg S (2001) Speech intelligibility derived from asynchronous processing of auditory-visual information. Proc Workshop Audio-Visual Speech Proc (AVSP-2001), pp. 132–137.

Grant KW, Seitz PF (1998) Measures of auditory-visual integration in nonsense syllables and sentences. J Acoust Soc Am 104:2438–2450.

Grant KW, Walden BE (1995) Predicting auditory-visual speech recognition in hearing-impaired listeners. Proc XIIIth Int Cong Phon Sci, Vol. 3, pp. 122–125.

Grant KW, Walden BE (1996a) Spectral distribution of prosodic information. J Speech Hearing Res 39:228–238.

Grant KW, Walden BE (1996b) Evaluating the articulation index for auditory-visual consonant recognition. J Acoust Soc Am 100:2415–2424.

Grant KW, Walden BE, Seitz PF (1998) Auditory-visual speech recognition by hearing-impaired subjects: consonant recognition, sentence recognition, and auditory-visual integration. J Acoust Soc Am 103:2677–2690.

Gravel JS, Ruben RJ (1996) Auditory deprivation and its consequences: from animal models to humans. In: Van De Water TR, Popper AN, Fay RR (eds) Clinical Aspects of Hearing. New York: Springer-Verlag, pp. 86–115.

Greenberg S (1988) The ear as a speech analyzer. J Phonetics 16:139–150.

Greenberg S (1995) The ears have it: the auditory basis of speech perception. Proc 13th Int Cong Phon Sci, Vol. 3, pp. 34–41.

Greenberg S (1996a) Auditory processing of speech. In: Lass N (ed) Principles of Experimental Phonetics. St. Louis: Mosby, pp. 362–407.

Greenberg S (1996b) Understanding speech understanding—towards a unified theory of speech perception. Proc ESCA Tutorial and Advanced Research Workshop on the Auditory Basis of Speech Perception, pp. 1–8.

Greenberg S (1997a) Auditory function. In: Crocker M (ed) Encyclopedia of Acoustics. New York: John Wiley, pp. 1301–1323.

Greenberg S (1997b) On the origins of speech intelligibility in the real world. Proc ESCA Workshop on Robust Speech Recognition in Unknown Communication Channels, pp. 23–32.

Greenberg S (1999) Speaking in shorthand—a syllable-centric perspective for understanding pronunciation variation. Speech Comm 29:159–176.

Greenberg S (2003) From here to utility—melding phonetic insight with speech technology. In: Barry W, Domelen W (eds) Integrating Phonetic Knowledge with Speech Technology, Dordrecht: Kluwer.

Greenberg S, Ainsworth WA (2003) Listening to Speech: An Auditory Perspective. Hillsdale, NJ: Erlbaum.

Greenberg S, Arai T (1998) Speech intelligibility is highly tolerant of cross-channel spectral asynochrony. Proc Joint Meeting Acoust Soc Am and Int Cong Acoust, pp. 2677–2678.

Greenberg S, Arai T (2001) The relation between speech intelligibility and the complex modulation spectrum. Proc 7th European Conf Speech Comm Tech (Eurospeech-2001), pp. 473–476.

Greenberg S, Arai T, Silipo R (1998) Speech intelligibility derived from exceedingly sparse spectral information, Proc 5th Int Conf Spoken Lang Proc, pp. 74–77.
Greenberg S, Chang S (2000) Linguistic dissection of switchboard-corpus automatic speech recognition systems. Proc ISCA Workshop on Automatic Speech Recognition: Challenges for the New Millennium, pp. 195–202.
Greenberg S, Geisler CD, Deng L (1986) Frequency selectivity of single cochlear nerve fibers based on the temporal response patterns to two-tone signals. J Acoust Soc Am 79:1010–1019.
Greenberg S, Carvey HM, Hitchcock L, Chang S (2002) Beyond the phoneme—a juncture-accent model for spoken language. Proc Human Language Technology Conference, pp. 36–44.
Greenwood DD (1961) Critical bandwidth and the frequency coordinates of the basilar membrane. J Acoust Soc Am 33:1344–1356.
Greenwood DD (1990) A cochlear frequency-position function for several species—29 years later. J Acoust Soc Am 87:2592–2650.
Greenwood DD (1994) The intensitive DL of tones: dependence of signal/masker ratio on tone level and spectrum of added noise. Hearing Res 65:1–39.
Gummer AW, Hemmert W, Zenner HP (1996) Resonant tectorial membrane motion in the inner ear: its crucial role in frequency tuning. Proc Natl Acad Sci USA 93:8727–8732.
Gummer AW, Meyer J, Frank G, Scherer MP, Preyer S (2002) Mechanical transduction in outer hair cells. Audiol Neurootol 7:13–16.
Halliday MAK (1967) Intonation and Grammar in British English. The Hague: Mouton.
Hauser MD (1996) The Evolution of Communication. Cambridge, MA: MIT Press.
Hauser MD, Chomsky N, Fitch H (2002) The faculty of language: What is it, who has it, and how did it evolve? Science 298:1569–1579.
Helmholtz HLF von (1863) Die Lehre von Tonemfindungen als Physiologie Grundlage dur die Theorie der Musik. Braunschweige: F. Vieweg und Sohn. [On the Sensations of Tone as a Physiological Basis for the Theory of Music (4th ed., 1897), trans. by A J. Ellis. New York: Dover (reprint of 1897 edition).]
Hermansky H (1990) Perceptual linear predictive (PLP) analysis of speech. J Acoust Soc Am 87:1738–1752.
Hermansky H (1998) Should recognizers have ears? Speech Comm 25:3–27.
Hermansky H, Morgan N (1994) RASTA processing of speech. IEEE Trans and Audio 2:578–589.
Houtgast T, Steeneken HJM (1973) The modulation transfer function in room acoustics as a predictor of speech intelligibility. Acustica 28:66–73.
Houtgast T, Steeneken H (1985) A review of the MTF concept in room acoustics and its use for estimating speech intelligibility in auditoria. J Acoust Soc Am 77:1069–1077.
Huggins WH (1952) A phase principle for complex-frequency analysis and its implications in auditory theory. J Acoust Soc Am 24:582–589.
Humes LE, Dirks DD, Bell TS, Ahlstrom C, Kincaid GE (1986) Application of the Articulation Index and the Speech Transmission Index to the recognition of speech by normal-hearing and hearing-impaired listeners. J Speech Hear Res 29:447–462.
Irvine DRF (1986) The Auditory Brainstem. Berlin: Springer-Verlag.

Ivry RB, Justus TC (2001) A neural instantiation of the motor theory of speech perception. Trends Neurosci 24:513–515.

Jakobson R, Fant G, Halle M (1952) Preliminaries to Speech Analysis. Tech Rep 13. Cambridge, MA: Massachusetts Institute of Technology [reprinted by MIT Press, 1963].

Jelinek F (1976) Continuous speech recognition by statistical methods. Proc IEEE 64:532–556.

Jelinek F (1997) Statistical Methods for Speech Recognition. Cambridge, MA: MIT Press.

Jenison R, Greenberg S, Kluender K, Rhode WS (1991) A composite model of the auditory periphery for the processing of speech based on the filter response functions of single auditory-nerve fibers. J Acoust Soc Am 90:773–786.

Jesteadt W, Wier C, Green D (1977) Intensity discrimination as a function of frequency and sensation level. J Acoust Soc Am 61:169–177.

Kakusho O, Hirato H, Kato K, Kobayashi T (1971) Some experiments of vowel perception by harmonic synthesizer. Acustica 24:179–190.

Kawahara H, Masuda-Katsuse I, de Cheveigné A (1999) Restructuring speech representations using a pitch-adaptive time-frequency smoothing and an instantaneous-frequency-based f_0 extraction: possible role of a repetitive structure in sounds. Speech Comm 27:187–207.

Kewley-Port D (1983) Time-varying features as correlates of place of articulation in stop consonants. J Acoust Soc Am 73:322–335.

Kewley-Port D, Neel A (2003) Perception of dynamic properties of speech: peripheral and central processes. In: Greenberg S, Ainsworth WA (eds) Listening to Speech: An Auditory Perspective. Hillsdale, NJ: Erlbaum.

Kewley-Port D, Watson CS (1994) Formant-frequency discrimination for isolated English vowels. J Acoust Soc Am 95:485–496.

Kitzes LM, Gibson MM, Rose JE, Hind JE (1978) Initial discharge latency and threshold considerations for some neurons in cochlear nucleus complex of the cat. J Neurophysiol 41:1165–1182.

Klatt DH (1979) Speech perception: a model of acoustic-phonetic analysis and lexical access. J Phonetics 7:279–312.

Klatt DH (1982) Speech processing strategies based on auditory models. In: Carlson R, Granstrom B (eds) The Representation of Speech in the Peripheral Auditory System. Amsterdam: Elsevier.

Klatt D (1987) Review of text-to-speech conversion for English. J Acoust Soc Am 82:737–793.

Kluender KR (1991) Effects of first formant onset properties on voicing judgments result from processes not specific to humans. J Acoust Soc Am 90:83–96.

Kluender KK, Greenberg S (1989) A specialization for speech perception? Science 244:1530(L).

Kluender KR, Jenison RL (1992) Effects of glide slope, noise intensity, and noise duration on the extrapolation of FM glides though noise. Percept Psychophys 51:231–238.

Kluender KR, Lotto AJ, Holt LL (2003) Contributions of nonhuman animal models to understanding human speech perception. In: Greenberg S, Ainsworth WA (eds) Listening to Speech: An Auditory Perspective. Hillsdale, NJ: Erlbaum.

Knudsen EI (2002) Instructed learning in the auditory localization pathway of the barn owl. Nature 417:322–328.

Kollmeier B, Koch R (1994) Speech enhancement based on physiological and psychoacoustical models of modulation perception and binaural interaction. J Acoust Soc Am 95:1593–1602.

Kuhl PK, Miller JD (1978) Speech perception by the chinchilla: Identification functions for synthetic VOT stimuli. J Acoust Soc Am 63:905–917.

Kuhl PK, Padden DM (1982) Enhanced discriminability at the phonetic boundaries for the voicing feature in Macaques. Percept Psychophys 32:542–550.

Kuhl PK, Andruski JE, Chistovich IA, Chistovich LA, et al. (1997) Cross-language analysis of phonetic units in language addressed to infants. Science 277:684–686.

Ladefoged P (1967) Three Areas of Experimental Phonetics. Oxford: Oxford University Press.

Ladefoged P (1971) Preliminaries to Linguistic Phonetics. Chicage: University of Chicago Press.

Ladefoged P (2001) A Course in Phonetics, 4th ed. New York: Harcourt.

Ladefoged P, Maddieson I (1996) The Sounds of the World's Languages. Oxford: Blackwell.

Langner G (1992) Periodicity coding in the auditory system. Hearing Res 60:115–142.

Langner G, Schreiner CE (1988) Periodicity coding in the inferior colliculus of the cat. I. Neuronal mechanisms. J Neurophys 60:1799–1822.

Lehiste I (1996) Suprasegmental features of speech. In: Lass N (ed) Principles of Experimental Phonetics. St. Louis: Mosby, pp. 226–244.

Lenneberg EH (1962) Understanding language without ability to speak: A case report. J Abnormal Soc Psychol 65:419–425.

Liberman AM, Mattingly IG (1985) The motor theory of speech perception revised. Cognition 21:1–36.

Liberman AM, Mattingly IG (1989) A specialization for speech perception. Science 243:489–494.

Liberman AM, Delattre PC, Gerstman LJ, Cooper FS (1956) Tempo of frequency change as a cue for distinguishing classes of speech sounds. J Exp Psychol 52:127–137.

Liberman AM, Harris KS, Hoffman HS, Griffith BC (1957) The discrimination of speech sounds within and across phoneme boundaries. J Exp Psychol 53:358–368.

Liberman AM, Cooper FS, Shankweiler DS, Studdert-Kennedy M (1967) Perception of the speech code. Psychol Rev 74:431–461.

Liberman MC (1988) Response properties of cochlear efferent neurons: Monaural vs. binaural stimulation and the effects of noise. J Neurophys 60:1779–1798.

Licklider JCR (1951) A duplex theory of pitch perception. Experientia 7:128–133.

Lieberman P (1984) The Biology and Evolution of Language. Cambridge, MA: Harvard University Press.

Lieberman P (1990) Uniquely Human: The Evolution of Speech, Thought and Selfless Behavior. Cambridge, MA: Harvard University Press.

Lieberman P (1998) Eve Spoke: Human Language and Human Evolution. New York: Norton.

Liljencrants J, Lindblom B (1972) Numerical simulation of vowel quality systems: The role of perceptual contrast. Language 48:839–862.

Lindblom B (1983) Economy of speech gestures. In: MacNeilage PF (ed) Speech Production. New York: Springer-Verlag, pp. 217–245.

Lindblom B (1990) Explaining phonetic variation: A sketch of the H & H theory. In: Hardcastle W, Marchal A (eds) Speech Production and Speech Modeling. Dordrecht: Kluwer, pp. 403–439.

Lippmann RP (1996) Accurate consonant perception without mid-frequency speech energy. IEEE Trans Speech Audio Proc 4:66–69.

Lisker L, Abramson A (1964) A cross-language study of voicing in initial stops: Acoustical measurements. Word 20:384–422.

Lynn PA, Fuerst W (1998) Introductory Digital Signal Processing with Computer Applications, 2nd ed. New York: John Wiley.

Lyon R, Shamma SA (1996) Auditory representations of timbre and pitch. In: Hawkins H, Popper AN, Fay RR (eds) Auditory Computation. New York: Springer-Verlag, pp. 221–270.

Massaro DM (1987) Speech Perception by Ear and by Eye. Hillsdale, NJ: Erlbaum.

McGurk H, MacDonald J (1976) Hearing lips and seeing voices. Nature 264:746–778.

Mermelstein P (1978) Difference limens for formant frequencies of steady-state and consonant-bound vowels. J Acoust Soc Am 63:572–580.

Miller GA (1951) Language and Communication. New York: McGraw-Hill.

Miller GA, Nicely PE (1955) An analysis of perceptual confusions among some English consonants. J Acoust Soc Am 27:338–352.

Miller MI, Sachs MB (1983) Representation of stop consonants in the discharge patterns of auditory-nerve fibers. J Acoust Soc Am 74:502–517.

Miyawaki K, Strange W, Verbrugge R, Liberman AM, Jenkins JJ, Fujimura O (1975) An effect of linguistic experience: the discrimination of [r] and [l] of Japanese and English. Percept Psychophys 18:331–340.

Moore BCJ (1997) An Introduction to the Psychology of Hearing, 4th ed. London: Academic Press.

Mozziconacci SJL (1995) Pitch variations and emotions in speech. Proc 13th Intern Cong Phon Sci Vol. 1, pp. 178–181.

Müsch H, Buus S (2001a) Using statistical decision theory to predict speech intelligibility. I. Model structure. J Acoust Soc Am 109:2896–2909.

Müsch H, Buss S (2001b) Using statistical decision theory to predict speech intelligibility. II. Measurement and prediction of consonant-discrimination performance. J Acoust Soc Am 109:2910–2920.

Oertel D, Popper AN, Fay RR (2002) Integrative Functions in the Mammalian Auditory System. New York: Springer-Verlag.

Ohala JJ (1983) The origin of sound patterns in vocal tract constraints. In: MacNeilage P (ed) The Production of Speech. New York: Springer-Verlag, pp. 189–216.

Ohala JJ (1994) Speech perception is hearing sounds, not tongues. J Acoust Soc Am 99:1718–1725.

Ohm GS (1843) Über die definition des Tones, nebst daran geknupfter Theorie der Sirene und ahnlicher Tonbildener Vorrichtungen. Ann D Phys 59:497–565.

Patuzzi R (2002) Non-linear aspects of outer hair cell transduction and the temporary threshold shifts after acoustic trauma. Audiol Neurootol 7:17–20.

Pavlovic CV, Studebaker GA, Sherbecoe RL (1986) An articulation index based procedure for predicting the speech recognition performance of hearing-impaired individuals. J Acoust Soc Am 80:50–57.

Picker JM (1980) The Sounds of Speech Communication. Baltimore: University Park Press.

Pisoni DB, Luce PA (1987) Acoustic-phonetic representations in word recognition. In: Frauenfelder UH, Tyler LK (eds) Spoken Word Recognition. Cambridge, MA: MIT Press, pp. 21–52.

Plomp R (1964) The ear as a frequency analyzer. J Acoust Soc Am 36:1628–1636.

Plomp R (1983) The role of modulation in hearing. In: Klinke R (ed) Hearing: Physiological Bases and Psychophysics. Heidelberg: Springer-Verlag, pp. 270–275.

Poeppel D, Yellin E, Phillips C, Roberts TPL, et al. (1996) Task-induced asymmetry of the auditory evoked M100 neuromagnetic field elicited by speech sounds. Cognitive Brain Res 4:231–242.

Pollack I (1959) Message uncertainty and message reception. J Acoust Soc Am 31:1500–1508.

Pols LCW, van Son RJJH (1993) Acoustics and perception of dynamic vowel segments. Speech Comm 13:135–147.

Pols LCW, van der Kamp LJT, Plomp R (1969) Perceptual and physical space of vowel sounds. J Acoust Soc Am 46:458–467.

Popper AN, Fay RR (1992) The Mammalian Auditory Pathway: Neurophysiology. New York: Springer-Verlag.

Proakis JG, Manolakis DG (1996) Digital Signal Processing: Principles, Algorithms and Applications. New York: Macmillan.

Rabinowitz WM, Eddington DK, Delhorne LA, Cuneo PA (1992) Relations among different measures of speech reception in subjects using a cochlear implant. J Acoust Soc Am 92:1869–1881.

Recanzone GH, Schreiner CE, Merzenich MM (1993) Plasticity of frequency representation in the primary auditory cortex following discrimination training in adult owl monkeys. J Neurosci 13:87–103.

Reiter ER, Liberman MC (1995) Efferent-mediated protection from acoustic overexposure: Relation to slow effects of olivocochlear stimulation. J Neurophysiol 73:506–514.

Remez RE, Rubin PE, Pisoni DB, Carrell TD (1981) Speech perception without traditional speech cues. Science 212:947–950.

Remez RE, Rubin PE, Berns SM, Pardo JS, Lang JM (1994) On the perceptual organization of speech. Psychol Rev 101:129–156.

Rhode WS, Greenberg S (1994) Lateral suppression and inhibition in the cochlear nucleus of the cat. J Neurophys 71:493–519.

Rhode WS, Kettner RE (1987) Physiological study of neurons in the dorsal and posteroventral cochlear nucleus of the unanesthesized cat. J Neurophysiol 57: 414–442.

Riesz RR (1928) Differential intensity sensitivity of the ear for pure tones. Phys Rev 31:867–875.

Rose JE, Brugge JF, Anderson DJ, Hind JE (1967) Phase-locked response to low-frequency tones in single auditory nerve fibers of the squirrel monkey. J Neurophysiol 30:769–793.

Rosen S, Howell P (1987) Auditory, articulatory, and learning explanations of categorical perception in speech. In: Harnad S (ed) Categorical Perception. Cambridge: Cambridge University Press, pp. 113–160.

Sachs MB, Young ED (1980) Effects of nonlinearities on speech encoding in the auditory nerve. J Acoust Soc Am 68:858–875.

Sachs MB, Blackbum CC, Young ED (1988) Rate-place and temporal-place representations of vowels in the auditory nerve and anteroventral cochlear nucleus. J Phonetics 16:37–53.

Sakoe H, Chiba S (1978) Dynamic programming algorithms optimization for spoken word recognition. IEEE Trans Acoust Speech Sig Proc 26:43–49.

Schalk TB, Sachs MB (1980) Nonlinearities in auditory-nerve responses to band-limited noise. J Acoust Soc Am 67:903–913.

Scharf B (1970) Critical bands. In: Tobias JV (ed) Foundations of Modern Auditory Theory, Vol. 1. New York: Academic Press, pp. 157–202.

Schreiner CE, Urbas JV (1988) Representation of amplitude modulation in the auditory cortex of the cat. I. The anterior auditory field (AAF). Hearing Res 21:227–241.

Shamma SA (1985a) Speech processing in the auditory system I: the representation of speech sounds in the responses of the auditory nerve. J Acoust Soc Am 78: 1612–1621.

Shamma SA (1985b) Speech processing in the auditory system II: Lateral inhibition and central processing of speech evoked activity in the auditory nerve. J Acoust Soc Am 78:1622–1632.

Shamma SA (1988) The acoustic features of speech sounds in a model of auditory processing: Vowels and voiceless fricatives. J Phonetics 16:77–91.

Shannon CE, Weaver W (1949) A Mathematical Theory of Communication. Urbana: University of Illinois Press.

Shannon RV, Zeng FG, Kamath V, Wygonski J (1995) Speech recognition with primarily temporal cues. Science 270:303–304.

Sharma J, Angelucci A, Sur M (2000) Induction of visual orientation modules in auditory cortex. Nature 404:841–847.

Shastri L, Chang S, Greenberg S (1999) Syllable detection and segmentation using temporal flow neural networks. Proc 14th Int Cong Phon Sci, pp. 1721–1724.

Shattuck R (1980) The Forbidden Experiment: The Story of the Wild Boy of Aveyron. New York: Farrar Straus Giroux.

Sinex DG, Geisler CD (1983) Responses of auditory-nerve fibers to consonant-vowel syllables. J Acoust Soc Am 73:602–615.

Skinner BF (1957) Verbal behavior. New York: Appleton-Century-Crofts.

Smith RL (1977) Short-term adaptation in single auditory nerve fibers: some post-stimulatory effects. J Neurophys 40:1098–1111.

Smoorenburg GF (1992) Speech reception in quiet and in noisy conditions by individuals with noise-induced hearing loss in relation to their tone audiogram. J Acoust Soc Am 91:421–437.

Sokolowski BHA, Sachs MB, Goldstein JL (1989) Auditory nerve rate-level functions for two-tone stimuli: possible relation to basilar membrane nonlinearity. Hearing Res 41:115–124.

Srulovicz P, Goldstein JL (1983) A central spectrum model: a synthesis of auditory-nerve timing and place cues in monaural communication of frequency spectrum. J Acoust Soc Am 73:1266–1275.

Steinberg JC, Gardner MB (1937) The dependence of hearing impairment on sound intensity. J Acoust Soc Am 9:11–23.

Stern RM, Trahiotis C (1995) Models of binaural interaction. In: Moore BCJ (ed) Hearing: Handbook of Perception and Cognition. San Diego: Academic Press, pp. 347–386.

Stevens KN (1972) The quantal nature of speech: evidence from articulatory-acoustic data. In: David EE, Denes PB (eds) Human Communication: A Unified View. New York: McGraw-Hill, pp. 51–66.

Stevens KN (1989) On the quantal nature of speech. J Phonetics 17:3–45.

Stevens KN (1998) Acoustic Phonetics. Cambridge, MA: MIT Press.

Stevens KN, Blumstein SE (1978) Invariant cues for place of articulation in stop consonants. J Acoust Soc Am 64:1358–1368.

Stevens KN, Blumstein SE (1981) The search for invariant acoustic correlates of phonetic features. In: Eimas PD, Miller JL (eds) Perspectives on the Study of Speech. Hillsdale, NJ: Erlbaum, pp. 1–38.

Strange W, Dittman S (1984) Effects of discrimination training on the perception of /r-1/ by Japanese adults learning English. Percept Psychophys 36:131–145.

Studdert-Kennedy M (2002) Mirror neurons, vocal imitation, and the evolution of particulate speech. In: Stamenov M, Gallese V (eds) Mirror Neurons and the Evolution of Brain and Language. Amsterdam: Benjamins John Publishing.

Studdert-Kennedy M, Goldstein L (2003) Launching language: The gestural origin of discrete infinity. In: Christiansen M, Kirby S (eds) Language Evolution: The States of the Art. Oxford: Oxford University Press.

Suga N (2003) Basic acoustic patterns and neural mechanisms shared by humans and animals for auditory perception. In: Greenberg S, Ainsworth WA (eds) Listening to Speech: An Auditory Perspective. Hillsdale, NJ: Erlbaum.

Suga N, O'Neill WE, Kujirai K, Manabe T (1983) Specificity of combination-sensitive neurons for processing of complex biosonar signals in the auditory cortex of the mustached bat. J Neurophysiol 49:1573–1626.

Suga N, Butman JA, Teng H, Yan J, Olsen JF (1995) Neural processing of target-distance information in the mustached bat. In: Flock A, Ottoson D, Ulfendahl E (eds) Active Hearing. Oxford: Pergamon Press, pp. 13–30.

Sumby WH, Pollack I (1954) Visual contribution to speech intelligibility in noise. J Acoust Soc Am 26:212–215.

Summerfield Q (1992) Lipreading and audio-visual speech perception. In: Bruce V, Cowey A, Ellis AW, Perrett DI (eds) Processing the Facial Image. Oxford: Oxford University Press, pp. 71–78.

Summerfield AQ, Sidwell A, Nelson T (1987) Auditory enhancement of changes in spectral amplitude. J Acoust Soc Am 81:700–708.

Sussman HM, McCaffrey HAL, Matthews SA (1991) An investigation of locus equations as a source of relational invariance for stop place categorization. J Acoust Soc Am 90:1309–1325.

ter Keurs M, Festen JM, Plomp R (1992) Effect of spectral envelope smearing on speech reception. I. J Acoust Soc Am 91:2872–2880.

ter Keurs M, Festen JM, Plomp R (1993) Effect of spectral envelope smearing on speech reception. II. J Acoust Soc Am 93:1547–1552.

Van Tassell DJ, Soli SD, Kirby VM, Widin GP (1987) Speech waveform envelope cues for consonant recognition. J Acoust Soc Am 82:1152–1161.

van Wieringen A, Pols LCW (1994) Frequency and duration discrimination of short first-formant speech-like transitions. J Acoust Soc Am 95:502–511.

van Wieringen A, Pols LCW (1998) Discrimination of short and rapid speechlike transitions. Acta Acustica 84:520–528.

van Wieringen A, Pols LCW (2003) Perception of highly dynamic properties of speech. In: Greenberg S, Ainsworth WA (eds) Listening to Speech: An Auditory Perspective. Hillsdale, NJ: Erlbaum.

Velichko VM, Zagoruyko NG (1970) Automatic recognition of 200 words. Int J Man-Machine Studies 2:223–234.

Viemeister NF (1979) Temporal modulation transfer functions based upon modulation thresholds. J Acoust Soc Am 66:1364–1380.

Viemeister NF (1988) Psychophysical aspects of auditory intensity coding. In: Edelman G, Gall W, Cowan W (eds) Auditory Function. New York: Wiley, pp. 213–241.

Villchur E (1987) Multichannel compression for profound deafness. J Rehabil Res Dev 24:135–148.

von Marlsburg C, Schneider W (1986) A neural cocktail-party processor. Biol Cybern 54:29–40.

Wang MD, Bilger RC (1973) Consonant confusions in noise: a study of perceptual features. J Acoust Soc Am 54:1248–1266.

Wang WS-Y (1972) The many uses of f_0. In: Valdman A (ed) Papers in Linguistics and Phonetics Dedicated to the Memory of Pierre Delattre. The Hague: Mouton, pp. 487–503.

Wang WS-Y (1998) Language and the evolution of modern humans. In: Omoto K, Tobias PV (eds) The Origins and Past of Modern Humans. Singapore: World Scientific, pp. 267–282.

Warr WB (1992) Organization of olivocochlear efferent systems in mammals. In: Webster DB, Popper AN, Fay RR (eds) The Mammalian Auditory Pathway: Neuroanatomy. New York: Springer-Verlag, pp. 410–448.

Warren RM (2003) The relation of speech perception to the perception of non-verbal auditory patterns. In: Greenberg S, Ainsworth WA (eds) Listening to Speech: An Auditory Perspective. Hillsdale, NJ: Erlbaum.

Weber F, Manganaro L, Peskin B, Shriberg E (2002) Using prosodic and lexical information for speaker identification. Proc IEEE Int Conf Audio Speech Sig Proc, pp. 949–952.

Wiener FM, Ross DA (1946) The pressure distribution in the auditory canal in a progressive sound field. J Acoust Soc Am 18:401–408.

Wier CC, Jestaedt W, Green DM (1977) Frequency discrimination as a function of frequency and sensation level. J Acoust Soc Am 61:178–184.

Williams CE, Stevens KN (1972) Emotions and speech: Some acoustical factors. J Acoust Soc Am 52:1238–1250.

Wong S, Schreiner CE (2003) Representation of stop-consonants in cat primary auditory cortex: intensity dependence. Speech Comm 41:93–106.

Wright BA, Buonomano DV, Mahncke HW, Merzenich MM (1997) Learning and generalization of auditory temporal-interval discrimination in humans. J Neurosci 17:3956–3963.

Young ED, Sachs MB (1979) Representation of steady-state vowels in the temporal aspects of the discharge patterns of auditory-nerve fibers. J Acoust Soc Am 66:1381–1403.

Zec D (1995) Sonority constraints on syllable structure. Phonology 12:85–129.

Zwicker E (1964) "Negative afterimage" in hearing. J Acoust Soc Am 36:2413–2415.

Zwicker E (1975) Scaling. In: Keidel W, Neff WD Handbook of Sensory Physiology V. Hearing. Heidelberg: Springer-Verlag, pp. 401–448.

Zwicker E, Flottorp G, Stevens SS (1957) Critical bandwidth in loudness summation. J Acoust Soc Am 29:548–557.

2
The Analysis and Representation of Speech

Carlos Avendaño, Li Deng, Hynek Hermansky, and Ben Gold

1. Introduction

The goal of this chapter is to introduce the reader to the acoustic and articulatory properties of the speech signal, as well as some of the methods used for its analysis. Presented, in some detail, are the mechanisms of speech production, aiming to provide the reader with the background necessary to understand the different components of the speech signal. We then briefly discuss the history of the development of some of the early speech analysis techniques in different engineering applications. Finally, we describe some of the most commonly used speech analysis techniques.

2. The Speech Signal

Speech, as a physical phenomenon, consists of local changes in acoustic pressure resulting from the actions of the human vocal apparatus. It is produced mainly for the purpose of verbal communication. The pressure changes generate acoustic waves that propagate through the communication medium (generally air). At the receiving end, speech is processed by the auditory system and higher cortical regions of the brain.

A transducer (microphone) in the acoustic field "follows" the speech signal, which can be analyzed numerically. In the case of a microphone, the speech signal is electrical in nature and describes the acoustic pressure changes as voltage variations with respect to time (Fig. 2.1).

The speech signal contains information not only about just what has been said (the linguistic message), but also about who has said it (speaker-dependent information), in which environment it was said (e.g., noise or reverberation), over which communication channel it was transmitted (e.g., microphone, recording equipment, transmission line, etc.), the health of the speaker, and so on. Not all of the information sources are of interest for any given application. For instance, in some automatic speech recognition applications the goal is to recover only the linguistic message regardless of

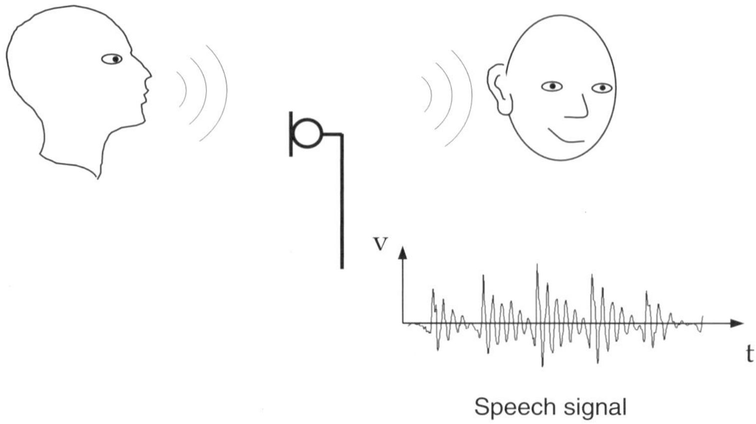

Figure 2.1. The speech communication chain. A microphone placed in the acoustic field captures the speech signal. The signal is represented as voltage (V) variations with respect to time (t).

the identity of the speaker, the acoustic environment, or the transmission channel. In fact, the presence of additional information sources may be detrimental to the decoding process.

3. Speech Production and Phonetic-Phonological Processes

Many speech analysis and coding (as well as speech recognition) techniques have been based on some form of speech production model. To provide readers with a solid background in understanding these techniques, this section describes several modern models of speech production. A discussion of speech production models will also help in understanding the relevance of production-based speech analysis methods to speech perception. A possible link between speech production and speech perception has been eloquently addressed by Dennis Klatt (1992).

3.1 Anatomy, Physiology, and Functions of the Speech Organs

The lungs are the major source of exhalation and thus serve as the primary power supply required to produce speech. They are situated in the chest cavity (thorax). The diaphragm, situated at the bottom of the thorax, contracts and expands. During expansion, the lungs exhale air, which is forced up into the trachea and into the larynx, where it passes between the vocal folds.

In spoken English roughly three types of laryngeal gestures are possible. First, if the vocal folds are far apart, the air passes through the pharynx and mouth relatively easily. This occurs during breathing and during the aspirated segments of speech (e.g., the [h] sound and after release of a voiceless stop consonant). Second, if the vocal folds are still apart but some constriction(s) is (are) made, the air from the lungs will not pass through as easily (as occurs in voiceless speech segments, e.g., voiceless fricatives, [s]). Third, when the vocal folds are adjusted such that only a narrow opening is created between the vocal folds, the air pushed out of the lungs will set them into a mode of a quasi-periodic vibration, as occurs in voiced segments (e.g., vowels, nasals, glides, and liquids). In this last instance, the quasi-periodic opening and closing of the glottis produces a quasi-periodic pressure wave that serves as an excitation source, located at the glottis, during normal production of voiced sounds.

The air passages anterior to the larynx are referred to as the vocal tract, which in turn can be divided into oral and nasal compartments. The former consists of the pharynx and the mouth, while the latter contains the nasal cavities. The vocal tract can be viewed as an acoustic tube extending from the larynx to the lips. It is the main source of the resonances responsible for shaping the spectral envelope of the speech signal.

The shape of the vocal tract (or more accurately its area function) at any point in time is the most important determinant of the resonant frequencies of the cavity. The articulators are the movable components of the vocal tract and determine its shape and, hence, its resonance pattern. The principal articulators are the jaw, lips, tongue, and soft palate (or velum). The pharynx is also an articulator, but its role in shaping the speech sounds of English is relatively minor. Although the soft palate acts relatively independently of the other articulators, the movements of the jaw, lips, and tongue is highly coordinated during speech production. This kind of articulatory movement is sometimes referred to as compensatory articulation. Its function has been hypothesized as a way of using multiple articulatory degrees of freedom to realize, or to enhance, specific acoustic goals (Perkell 1969, 1980; Perkell et al. 1995) or given tasks in the vocal tract constriction (Saltzman and Munhall 1989; Browman and Goldstein 1989, 1992).

Of all the articulators, the tongue is perhaps the most important for determining the resonance pattern of the vocal tract. The tip and blade of the tongue are highly movable; their actions or gestures, sometimes called articulatory features in the literature (e.g., Deng and Sun 1994), determine a large number of consonantal phonetic segments in the world's languages. Behind the tongue blade is the tongue dorsum, whose movement is relatively slower. A large number of different articulatory gestures formed by the tongue dorsum determines almost all of the variety of vocalic and consonantal segments observed in the world's languages (Ladefoged and Maddieson 1990).

The velum is involved in producing nasal sounds. Lowering of the velum opens up the nasal cavity, which can be thought of as an additional

TABLE 2.1. Place and manner of articulation for the consonants, glides, and liquids of American English

Place of articulation	Glides	Liquids	Nasals	Stops voiced	Stop unvoiced	Fricatives voiced	Fricatives unvoiced	Affricates voiced	Affricates unvoiced
Bilabial	w		m	b	p				
Labiodental						v	f		
Apicodental		l				ð	θ		
Alveolar			n	d	t	z	s	ʒ	ʧ
Palatal	y	r					ʃ		
Velar		l	ŋ	g	k				
Glottal					ʔ		h		

acoustic tube coupled to the oral cavity during production of nasal segments. There are two basic types of nasal sounds in speech. One involves sound radiating from both the mouth and nostrils of the speaker (e.g., vowel nasalization in English and nasal vowels in French), while the other involves sounds radiating only from the nostrils (e.g., nasal murmurs or nasal stops).

Table 2.1 lists the major consonantal segments of English, along with their most typical place and manner of articulation. A more detailed description of this material can be found in Ladefoged (1993).

3.2 Phonetic Processes of Speech

Early on, Dudley (1940) described in detail what has become known as the "carrier nature of speech." The relatively broad bandwidth of speech (approximately 10 kHz) is caused by the sudden closure of the glottis as well as by the turbulence created by vocal tract constriction. The relatively narrow bandwidth of the spectral envelope modulations is created by the relatively slow motion of the vocal tract during speech. In this view, the "message" (the signal containing information of vocal tract motions with a narrow bandwidth) modulates the "carrier" signal (high frequency) analogous to the amplitude modulation (AM) used in radio communications. Over 50 years of research has largely confirmed the view that the major linguistically significant information in speech is contained in the details of this low-frequency vocal tract motion. This perspective deviates somewhat from contemporary phonetic theory, which posits that the glottal and turbulence excitations (the so-called carrier) also carry some phonologically significant information, rather than serving only as a medium with which to convey the message signal.

The study of phonetic processes of speech can be classified into three broad categories: (1) Articulatory phonetics addresses the issue of what the components of speech-generation mechanisms are and how these mechanisms are used in speech production. (2) Acoustic phonetics addresses what acoustic characteristics are associated with the various speech sounds generated by the articulatory system. (3) Auditory phonetics addresses the issue of how a listener derives a perceptual impression of speech sounds based on properties of the auditory system.

3.3 Coarticulation and Acoustic Transitions in Speech

One important characteristic of speech production is "coarticulation," the overlapping of distinctive articulatory gestures, and its manifestation in the acoustic domain is often called context dependency.

Speech production involves a sequence of articulatory gestures overlapped in time so that the vocal tract shape and its movement are strongly dependent on the phonetic contexts. The result of this overlapping is the simultaneous adjustment of articulators and a coordinated articulatory structure for the production of speech. The need for gestural overlap can be appreciated by looking at how fast is the act of speaking. In ordinary conversation a speaker can easily produce 150 to 200 words per minute, or roughly 10 to 12 phones per second (Greenberg et al. 1996).

Coarticulation is closely related to the concept of a target in speech production, which forms the basis for speech motor control in certain goal-oriented speech production theories. The mechanism underlying speech production can be viewed as a target or goal-oriented system. The articulators have a specific inertia, which does not allow the articulators to move instantaneously from one configuration to another. Thus, with rapid rates of production, articulators often only move toward specific targets, or following target trajectories (diphthongs and glides, for example). Articulatory motions resulting from the target-oriented mechanism produce similar kinds of trajectories in speech acoustics, such as formant movements. These dynamic properties, either in articulatory or in acoustic domains, are perceptually as important as the actual attainment of targets in each of the domains.

Speech is produced by simultaneous gestures loosely synchronized with one another to approach appropriate targets. The process can be regarded as a sequence of events that occur in moving from one target to another during the act of speaking. This process of adjustment is often referred to as a transition. While its origin lies in articulation, such a transition is also capable of evoking a specific auditory phonetic percept. In speech production, especially for fast or casual speech, the ideal targets or target regions are often not reached. Acoustic transitions themselves already provide sufficient cues about the targets intended for when the speech is slowly and carefully spoken.

A common speech phenomenon closely related to coarticulation and to target transitions is vocalic reduction. Vowels are significantly shortened when reduction occurs, and their articulatory positions, as well as their formant patterns, tend to centralize to a neutral vowel or to assimilate to adjacent phones. Viewing speech production as a dynamic and coarticulated process, we can treat any speech utterance as a sequence of vocalic (usually the syllabic nucleus) gestures occurring in parallel with consonantal gestures (typically syllable onset or coda), where there is partial temporal overlap between the two streams.

The speaking act proceeds by making ubiquitous transitions (in both the articulatory and acoustic domains) from one target region to the next. The principal transitions are from one syllabic nucleus to another, effected mainly by the movement of the tongue body and the jaw. Shortening of such principal transitions due to an increase in speaking rate, or other factors, produces such reduction. Other articulators (the tongue blade, lips, velum, and glottis) often move concurrently with the tongue body and jaw, superimposing their consonantal motion on the principal vocalic gestures. The addition of consonantal gestures locally perturbs the principal acoustic transitions and creates acoustic turbulences (as well as closures or short acoustic pauses), which provide the listener with added information for identifying fricatives and stops. However, the major cues for identifying these consonants appear to be the nature of the perturbation in the acoustic transitions caused by vocal tract constriction.

3.4 Control of Speech Production

An important property of the motor control of speech is that the effects of motor commands derived from phonetic instructions are self-monitored and -regulated. Auditory feedback allows the speaker to measure the success in achieving short-term communicative goals while the listener receives the spoken message, as well as establishing long-term, stable goals of phonetic control. Perhaps more importantly, self-regulating mechanisms involve use of taction and proprioception (i.e., internal tension of musculature) to provide immediate control and compensation of articulatory movements (Perkell 1980). With ready access to this feedback information, the control system is able to use such information about the existing state of the articulatory apparatus and act intelligently to achieve the phonetic goals.

Detailed mechanisms and functions of speech motor control have received intensive study over the past 30 years, resulting in a number of specific models and theories such as the location programming model, mass-spring theory, the auditory distinctive-feature target model, orosensory goal and intrinsic timing theory, the model-reference (or internal control) model, and the coordinative structure (or task dynamic) model (Levelt 1989). There is a great deal of divergence among these models in terms of the nature of the phonetic goals, of the nature of the motor commands, and of the precise motor execution mechanisms. However, one common view that seems to be shared, directly or indirectly, by most models is the importance of the syllable as a unit of speech motor execution. Because the significance of the syllable has been raised again in the auditory speech perception community (Greenberg and Kingsbury 1997), we briefly address here the issues related to the syllable-based speech motor control.

Intuitively, the syllable seems to be a natural unit for articulatory control. Since consonants and vowels often involve separate articulators (with the exception of a few velar and palatal consonants), the consonantal cues can be relatively reliably separated from the core of syllabic nucleus (typically a vowel). This significantly reduces otherwise more random effects of coarticulation and hence constrains the temporal dynamics in both the articulatory and acoustic domains.

The global articulatory motion is relatively slow, with frequencies in the range of 2 to 16 Hz (e.g., Smith et al. 1993; Boubana and Maeda 1998) due mainly to the large mass of the jaw and tongue body driven by the slow action of the extrinsic muscles. Locally, where consonantal gestures are intruding, the short-term articulatory motion can proceed somewhat faster due to the small mass of the articulators involved, and the more slowly acting intrinsic muscles on the tongue body. These two sets of articulatory motions (a locally fast one superimposed on a globally slow one) are transformed to acoustic energy during speech production, largely maintaining their intrinsic properties. The slow motion of the articulators is reflected in the speech signal. Houtgast and Steeneken (1985) analyzed the speech signal and found that, on average, the modulations present in the speech envelope have higher values at modulation frequencies of around 2 to 16 Hz, with a dominant peak at 4 Hz. This dominant peak corresponds to the average syllabic rate of spoken English, and the distribution of energy across this spectral range corresponds to the distribution of syllabic durations (Greenberg et al. 1996).

The separation of control for the production of vowels, and consonants (in terms of the specific muscle groups involved, the extrinsic muscles control the tongue body and are more involved in the production of vowels, while the intrinsic muscles play a more important role in many consonantal segments) allows movements of the respective articulators to be more or less free of interference. In this way, we can view speech articulation as the production of a sequence of slowly changing syllable nuclei, which are perturbed by consonantal gestures. The main complicating factor is that aerodynamic effects and spectral zeros associated with most consonants regularly interrupt the otherwise continuous acoustic dynamic pattern. Nevertheless, because of the syllabic structure of speech, the aerodynamic effects (high-frequency frication, very fast transient stop release, closures, etc.) are largely localized at or near the locus of articulatory perturbations, interfering minimally with the more global low-frequency temporal dynamics of articulation. In this sense, the global temporal dynamics reflecting vocalic production, or syllabic peak movement, can be viewed as the carrier waveform for the articulation of consonants.

The discussion above argues for the syllable as a highly desirable unit for speech motor control and a production unit for optimal coarticulation.

Recently, the syllable has also been proposed as a desirable and biologically plausible unit for speech perception. An intriguing question is asked in the article by Greenberg (1996) as to whether the brain is able to back-compute the temporal dynamics that underlie both the production and perception of speech. A separate question concerns whether such global dynamic information can be recovered from the appropriate auditory representation of the acoustic signal. A great deal of research is needed to answer the above questions, which certainly have important implications for both the phonetic theory of speech perception and for automatic speech recognition.

3.5 Acoustic Theory of Speech Production

The sections above have dealt largely with how articulatory motions are generated from phonological and phonetic specifications of the intended spoken messages and with the general properties of these motions. This section describes how the articulatory motions are transformed into an acoustic signal.

The speech production process can be described by a set of partial differential equations pertaining to the physical principles of acoustic wave propagation. The following factors determine the final output of the partial differential equations:

1. Time-varying area functions, which can be obtained from geometric consideration of the vocal tract pattern of the articulatory movements
2. Nasal cavity coupling to the vocal tract
3. The effects of the soft tissue along the vocal-tract walls
4. Losses due to viscous friction and heat conduction in the vocal tract walls
5. Losses due to vocal-tract wall vibration
6. Source excitation and location in the vocal tract

Approximate solutions to the partial differential equations can be obtained by modeling the continuously variable tube with a series of uniform circular tubes of different lengths and cross sections. Such structures can be simulated with digital, wave-guide models. Standard textbooks (e.g., Rabiner and Schafer 1978), usually begin discussing the acoustic theory of speech production by using a uniform tube of fixed length. The sound generated by such a tube can be described by a wave equation. The solution of the wave equation leads to a traveling or standing wave. Reflections at the boundaries between adjacent sections can be determined as a function of the tube dimensions.

Building up from this simple uniform tube model, one can create increasingly complicated multiple-tube models for the vocal tract shape associated with different vowels and other sounds. The vocal tract shape can be specified as a function of time, and the solution to the time-varying partial differential equations will yield dynamic speech sounds. Alternatively, just from the time-varying vocal tract shape or the area function one can compute the time-varying transfer functions corresponding to the speech sounds generated from the partial differential equations.

3.5.1 The Linear Model of Speech Production

In this section we establish a link between the physiology and functions of the speech organs studied in section 3.1 and a linear model of speech production.

In the traditional acoustic theory of speech production, which views speech production as a linear system (Fant 1960), factors 1 to 5, listed in the previous section, pertain to the series as transfer function (also known as the filter) of the system. Factor 6, known as the "source," is considered as the input to system. This traditional acoustic theory of speech production is referred to as the source-filter (linear) model of speech production.

Generally, the source can be classified into two types of components. One is quasi-periodic, related to the third laryngeal gesture described in section 3.1, and which is responsible mainly for the production of vocalic sounds including vowels and glides. It is also partly responsible for the production of voiced consonants (fricatives, nasals, and stops). The location of this quasi-periodic source is the glottis.

The other type of source is due to aerodynamic processes that generate sustained or transient frication, and is related to the first two types of laryngeal gestures described in section 3.1. The sustained frication noise-like source is responsible for generating voiceless fricatives (constriction located above the glottis) and aspirated sounds (e.g., /h/, constriction located *at* the glottis). The transient noise source is responsible for generating stop consonants (constriction located *above* the glottis). Mixing the two types of sources gives rise to affricates and stops.

The speech production model above suggests separating the articulatory system into two independent subsystems. The transfer function related to the vocal tract can be modeled as a linear filter. The input to the filter is the source signal, which is modeled as a train of pulses, in the case of a quasi-periodic component, or as a random signal in the case of the noise-like component. The output of the filter yields the speech signal.

Figure 2.2 illustrates this model, where $u(t)$ is the source, $h(t)$ is the filter, and $s(t)$ is a segment of the speech signal. The magnitude spectrum associated with each of these components for a voiced segment is also shown. The voice source corresponds to the fine structure of the spectrum, while the filter corresponds to the spectral envelope. The peaks of the spectral envelope $H(\omega)$ represent the formants of the vowel. Notice that the spectra of the quasi-periodic signal has a roll- off of approximately –6 dB/oct. This is due to the combined effect of the glottal pulse shape (–12 dB/oct) and lip radiation effects (6 dB/oct).

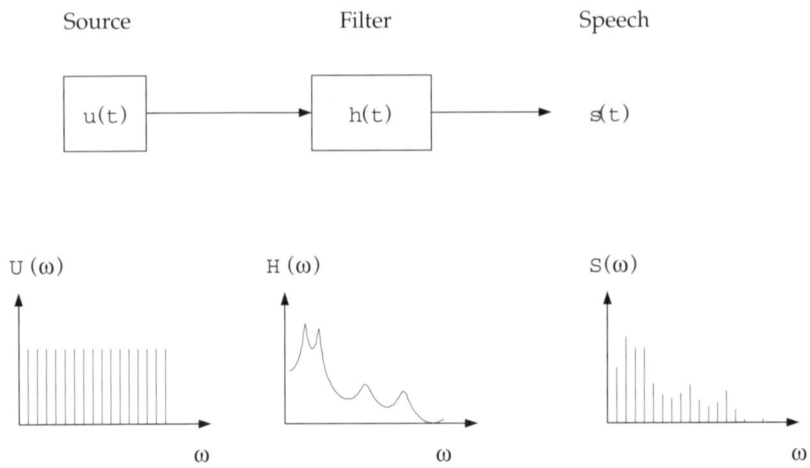

FIGURE 2.2. A linear model of speech production. A segment of a voiced signal is generated by passing a quasi-periodic pulse train $u(t)$ through the spectral shaping filter $h(t)$. The spectra of the source $U(\omega)$, filter $H(\omega)$, and speech output $S(\omega)$ are shown below.

In fluent speech, the characteristics of the filter and source change over time, and the formant peaks of speech are continuously changing in frequency. Consonantal segments often interrupt the otherwise continuously moving formant trajectories. This does not mean that the formants are absent during these consonantal segments. Vocal tract resonances and their associated formants are present for all speech segments including consonants. These slowly time-varying resonant characteristics constitute one aspect of the global speech dynamics discussed in section 3.2. Formants trajectories are interrupted by the consonantal segments only because their spectral zeros cancel out the poles in the acoustic domain.

4. Speech Analysis

In its most elementary form, speech analysis attempts to break the speech signal into its constituent frequency components (signal-based analysis). On a higher level, it may attempt to derive the parameters of a speech production model (production-based analysis), or to simulate the effect that the speech signal has on the speech perception system (perception-based analysis). In section 5 we discuss each of these analyses in more detail.

The method for specific analysis is determined by the purpose of the analysis. For example, if accurate reconstruction (resynthesis) of the speech signal after analysis is required, then signal-based techniques, such as perfect-reconstruction filter banks could be used (Vaidyanathan 1993). In contrast, compression applications, such as low-bit rate coding or speech recognition would benefit by having knowledge provided by production or perception-based analysis techniques.

4.1 History and Basic Principles of Speech Analysis for Engineering Applications

Engineering applications of speech analysis have a long history that goes back as far as the late 17th century. In the following sections we provide a brief history of some of the discoveries and pioneering efforts that laid the foundations for today's speech analysis.

4.1.1 Speech Coding and Storage

The purpose of speech coding is to reduce the information rate of the original speech signal, so that it can be stored and transmitted more efficiently. Within this context, the goal of speech analysis is to extract the most significant carriers of information, while discarding perceptually less relevant components of the signal. Hence, the governing constraints of information loss are determined by the human speech perception system (e.g., Atal and Schroeder 1979). Among the factors that result in information loss and contribute to information rate reduction are the assumptions and simplifications of the speech model employed, the artifacts of the analysis itself, and the noise introduced during storage or transmission (quantization, loss of data, etc.).

If the goal is to reduce the speech transmission rate by eliminating irrelevant information sources, it is of interest to know what are the dominant information carriers that need to be preserved. In a simple experiment, Isaac Newton (at the ripe old age of 24) noticed one such dominant source of linguistic information. He observed that while pouring liquid into a tall glass, it was possible to hear a series of sounds similar to the vowels [u], [o], [a], [e], and [i] (Ladefoged 1967, 1993). An interpretation of this remarkable observation is as follows. When the liquid stream hits the surface below, it generates an excitation signal, a direct analog of the glottal source. During the process of being filled the effective acoustic length of the glass is reduced, changing its resonant frequencies, as illustrated in Figure 2.3. If the excitation is assumed to be fixed, then any change in the resonant pattern of the glass (formants) will be analogous to the manner in which human articulatory movements change the resonances of the vocal tract during speech production (cf. Fant and Risberg 1962). The first and second formant frequencies of the primary vowels of American English are shown in Figure 2.4.

The first (mechanical) synthesizers of von Kempelen (1791) made evident that the speech signal could be decomposed into a harmonically rich excitation signal (he used a vibrating reed such as the one found in a bagpipe) and an envelope-shaping function as the main determinants of the

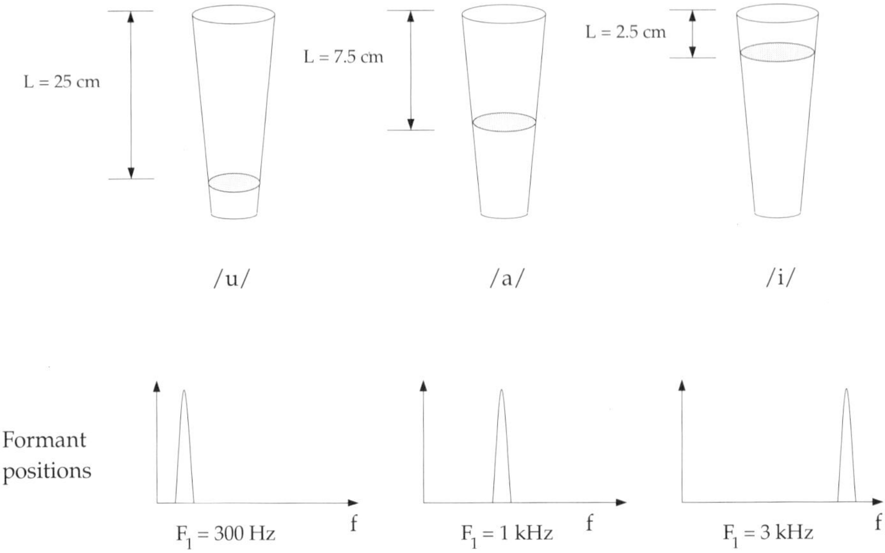

FIGURE 2.3. Newton's experiment. The resonances of the glass change as it is being filled. Below, the position of only the first resonance (F_1) is illustrated.

linguistic message. For shaping the spectral envelope, von Kempelen used a flexible mechanical resonator made out of leather. The shape of the resonator was modified by deforming it with one hand. He reported that his machine was able to produce a wide variety of sounds, sufficient to synthesize intelligible speech (Dudley and Tarnoczy 1950).

Further insight into the nature of speech and its frequency domain interpretation was provided by Helmholtz in the 19th century (Helmholtz 1863). He found that vowel-like sounds could be produced with a minimum number of tuning forks.

One hundred and fifty years after von Kempelen, the idea of shaping the spectral envelope of a harmonically rich excitation to produce a speech-like signal was used by Dudley to develop the first electronic synthesizer. His Voder used a piano-style keyboard that enabled a human operator to control the parameters of a set of resonant electric circuits capable of shaping the signal's spectral envelope. The excitation (source) was selected from a "buzz" or a "hiss" generator depending on whether the sounds were voiced or not.

The Voder principle was later used by Dudley (1939) for the efficient representation of speech. Instead of using human operators to control the resonant circuits, the parameters of the synthesizer were obtained directly from the speech signal. The fundamental frequency for the excitation source was obtained by a pitch extraction circuit. This same circuit contained a module whose function was to make decisions as to whether at any particu-

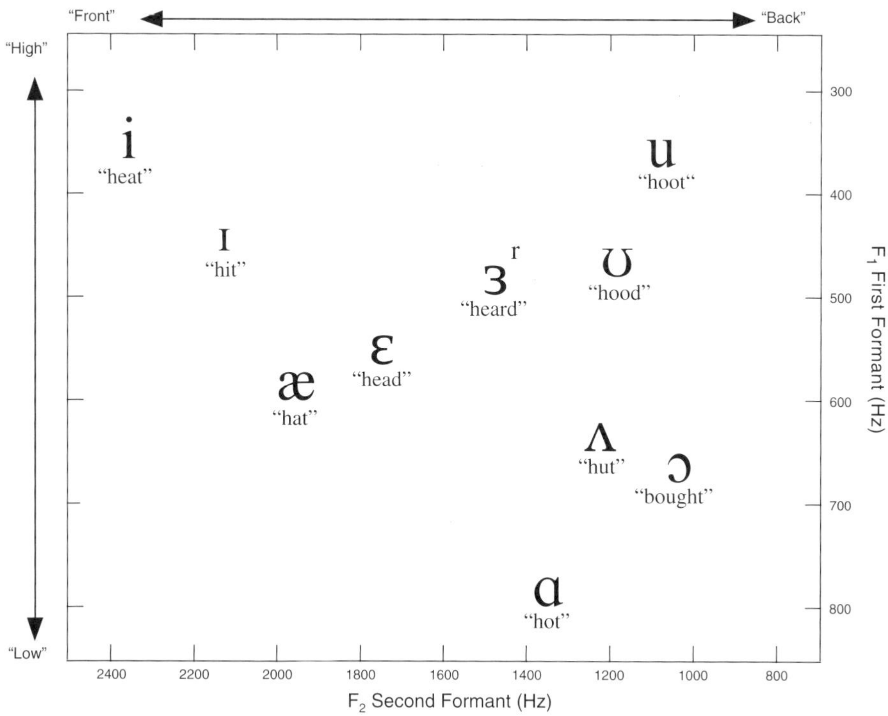

FIGURE 2.4. Average first and second formant frequencies of the primary vowels of American English as spoken by adult males (based on data from Hillenbrand et al., 1997). The standard pronunciation of each vowel is indicated by the word shown beneath each symbol. The relative tongue position ("Front," "Back," "High," "Low") associated with the pronunciation of each vowel is also shown.

lar time the speech was voiced or unvoiced. To shape the spectral envelope, the VOCODER (Voice Operated reCOrDER) used the outputs of a bank of bandpass filters, whose center frequencies were spaced uniformly at 300-Hz intervals between 250 and 2950 Hz (similar to the tuning forks used in Helmholtz's experiment). The outputs of the filters were rectified and low-pass filtered at 25 Hz to derive energy changes at "syllabic frequencies." Signals from the buzz and hiss generators were selected (depending on decisions made by the pitch-detector circuit) and modulated by the low-passed filtered waveforms in each channel to obtain resynthesized speech. By using a reduced set of parameters, i.e., pitch, voiced/unvoiced, and 10 spectral envelope energies, the VOCODER was able to efficiently represent an intelligible speech signal, reducing the data rate of the original speech.

After the VOCODER many variations of this basic idea occurred (see Flanagan 1972 for a detailed description of the channel VOCODER and its variations). The interest in the VOCODER stimulated the development of new signal analysis tools. While investigating computer implementations of the channel VOCODER (Gold and Rader 1969; Gold and Morgan 1999), demonstrated the feasibility of simulating discrete resonator circuits. Their contribution resulted in explosive development in the new area of digital signal processing. With the advent of the fast Fourier transform (FFT) (Cooley and Tukey 1965), further improvements in the efficiency of the VOCODER were obtained, increasing its commercial applications.

Another milestone in speech coding was the development of the linear prediction coder (LPC). These coders approximate the spectral envelope of speech with the spectrum of an all-pole model derived from linear predictive analysis (Atal and Schroeder 1968; Itakura and Saito 1970; see section 5.4.2) and can code efficiently the spectral envelope with just a few parameters.

Initially, the source signal for LPCs was obtained in the same fashion as the VOCODER (i.e., a buzz tone for voiced sounds and a hiss for the unvoiced ones). Major quality improvements were obtained by increasing the complexity of the excitation signal. Ideally, the increase in complexity should not increase the bit rate significantly, so engineers devised ingenious ways of providing low-order models of the excitation. Atal and Remde (1982) used an analysis-by-synthesis procedure to adjust the positions and amplitudes of a set of pulses to generate an optimal excitation for a given frame of speech. Based on a similar idea, Schroeder and Atal (1985), developed the code-excited linear prediction (CELP) coder, which further improved the quality of the synthetic speech at lower bit rates. In CELP the optimal excitation signal is derived from a precomputed set of signals stored in a code book. This coder and its variations are the most common types used in digital speech transmission today.

4.1.2 Display of Speech Signals

By observing the actions of the vocal organs during the production of different speech sounds, early speech researchers were able to derive models of speech production and construct speech synthesizers. A different perspective on the speech signal can be obtained through its visual display. Because such displays rely on the human visual cognitive system for further processing, the main concern is to preserve relevant information with as much detail as possible. Thus, an information rate reduction is not of primary concern.

The pioneer of developing display techniques was Scripture (1906), who studied gramophone recordings of speech. His first attempts for deriving meaningful features from the raw speech waveform were not very encouraging. However, he soon introduced features such as phone duration, signal energy, "melody" (i.e., fundamental frequency), as well as a form of short-

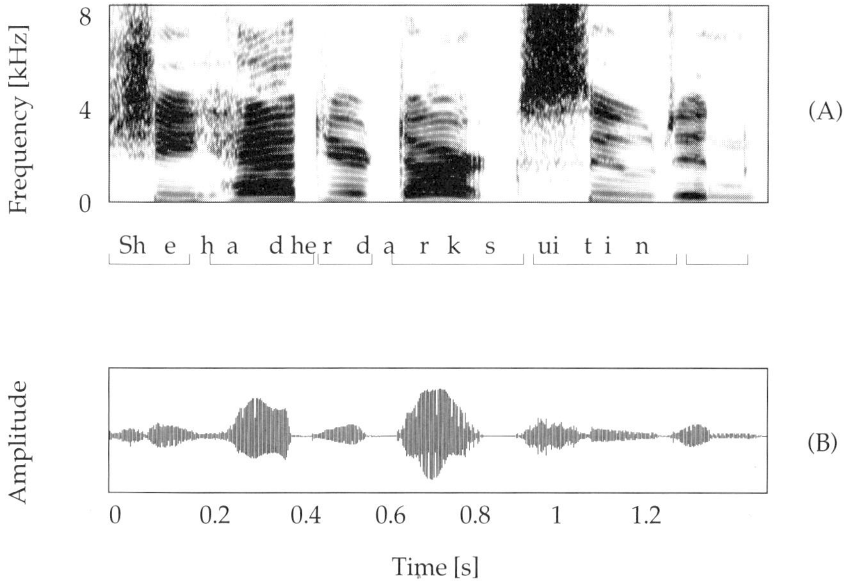

FIGURE 2.5. (A) Spectrogram and (B) corresponding time-domain waveform signal.

time Fourier analysis (see section 5.3.1) to derive the amplitude spectrum of the signal.

An important breakthrough came just after the Second World War, when the sound spectrograph was introduced as a new tool for audio signal analysis (Koenig et al. 1946; Potter et al. 1946). The sound spectrograph allowed for relatively fast spectral analysis of speech. Its spectral resolution was uniform at either 45 or 300 Hz over the frequency range of interest and was capable of displaying the lower four to five formants of speech.

Figure 2.5A shows a spectrogram of a female speaker uttering the sentence "She had her dark suit in . . ." The abscissa is time, the ordinate frequency, and the darkness level of the pattern is proportional to the intensity (logarithmic magnitude). The time-domain speech signal is shown below for reference.

Some people have learned to accurately decode ("read") such spectrograms (Cole et al. 1978). Although such capabilities are often cited as evidence for the sufficiency of the visual display representation for speech communication (or its applications), it is important to realize that all the generalizing abilities of the human visual language processing and cognition systems are used in the interpretation of the display. It is not a trivial task to simulate such human processes with signal processing algorithms.

5. Techniques for Speech Analysis

5.1 Speech Data Acquisition

The first step in speech data acquisition is the recording of the acoustic signal. A standard practice is to use a single microphone sensitive to the entire spectral range of speech (about 0 to 10kHz). Rapid advances in computational hardware makes it possible to conduct most (if not all) of the processing of speech in the digital domain. Thus, the second process in data acquisition is analog-to-digital conversion (ADC). During the sampling process a set of requirements known as the Nyquist criterion (e.g., Oppenheim and Schafer 1989) have to be met. To input the values of the signal samples in the computer, they need to be described by a finite number of bits, that is, with a finite precision. This process results in quantization noise, whose magnitude decreases as the number of bits increases (Jayant 1974).

In the rest of the chapter we only use discrete-time signals. The notation that we use for such signals is $s(n)$, with n representing discretely sampled time. This is done to distinguish it from an analog signal $s(t)$, where the variable t indicates continuous time.

5.2 Short-Time Analysis

The speech signal is nonstationary. Its nonstationarity originates from the fact that the vocal organs are continuously moving during speech production. However, there are physical limitations on the rate at which they can move. A segment of speech, if sufficiently short, can be considered equivalent to a stationary process. This short-time segment of speech can then be analyzed by signal processing techniques that assume stationarity.

An utterance typically needs to be subdivided into several short-time segments. One way of looking at this segmentation process is to think of each segment as a section of the utterance seen though a short-time window that isolates only a particular portion of speech. This perspective is illustrated in Figure 2.6.

Sliding the window across the signal results in a sequence of short-time segments, each having two time indices, one that describes its evolution within the window, and another that determines the position of the segment relative to the original time signal. In this fashion a two-dimensional representation of the original signal can be made.

The window function can have a fixed length and shape, or it can vary with time. The rate at which the window is moved across the signal depends on the window itself and the desired properties of the analysis.

Once the signal is divided into approximately stationary segments we can apply signal processing analysis techniques. We divide these techniques into three categories, depending of the specific goal of the analysis.

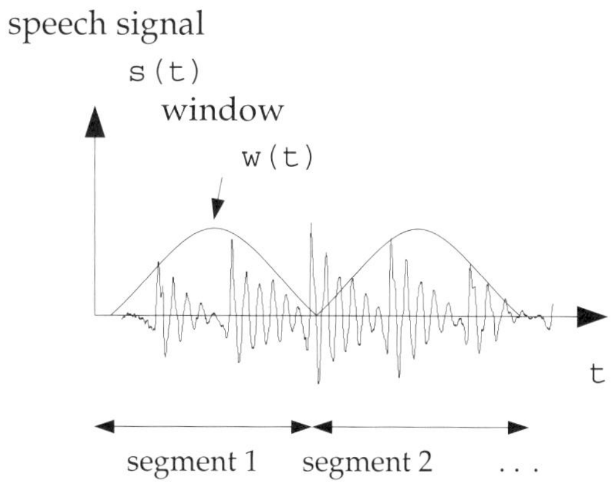

FIGURE 2.6. Short-time analysis of speech. The signal is segmented by the sliding short-time window.

5.3 Signal-Based Techniques

Signal-based analysis techniques describe the signal in terms of its fundamental components, paying no specific attention to how the signal was produced, or how it is processed by the human hearing system. In this sense the speech within a short segment is treated as if it were an arbitrary stationary signal. Several analysis techniques for stationary signals are available.

A basic signal analysis approach is Fourier analysis, which decomposes the signal into its sinusoidal constituents at various frequencies and phases. Applying Fourier analysis to the short duration speech segments yields a representation known as the short-time Fourier transform (STFT). While the STFT is not the only signal-based form of analysis, it has been extensively studied (e.g., Portnoff 1980) and is used in a wide variety of speech applications.

5.3.1 Short-Time Fourier Analysis

Once the signal is segmented, the next step in the STFT computation consists of performing a Fourier analysis of each segment. The Fourier analysis expands a signal into a series of harmonically related basis functions (sinusoids). A segment of length N of a (periodic) discrete-time signal $s(n)$ can be represented by N sinusoidal components:

$$s(n) = \frac{1}{N}\sum_{n=0}^{N-1} S(k)\cos\left(\frac{(2\pi kn)}{N}\right) + j\frac{1}{N}\sum_{n=0}^{N-1} S(k)\sin\left(\frac{2\pi kn}{N}\right) \qquad (1)$$

where the coefficients $S(k)$ are called the discrete Fourier coefficients and are obtained from the signal as

$$S(k) = \sum_{k=0}^{N-1} S(n)\cos\left(\frac{(2\pi kn)}{N}\right) - j\sum_{k=0}^{N-1} S(n)\sin\left(\frac{2\pi kn}{N}\right) \quad (2)$$

The magnitude of the Fourier coefficients determines the amplitude of the sinusoid at frequency $\omega_k = \frac{2\pi k}{N}$. The larger the magnitude, the stronger the signal component is (i.e., more energy) at a given frequency. The phase (angle) of the Fourier coefficients determines the amount of time each frequency component is shiftable relative to the others. Equations 1 and 2 are known as the discrete Fourier transform (DFT) pair.

Fourier analysis merely describes the signal in terms of its frequency components. However, its use in speech processing has sometimes been justified by the fact that some form of spectral analysis is being carried out by the mammalian peripheral auditory system (Helmholtz 1863; Moore 1989).

The STFT is a two-dimensional representation consisting of a sequence of Fourier transforms, each corresponding to a windowed segment of the original speech. The STFT is a particular instance of a more general class of representations called time-frequency transforms (Cohen 1995). In Figure 2.7 we illustrate the computation of the STFT. A plot of the logarithmic magnitude of the STFT results in a spectrogram (see Fig. 2.5).

The STFT can be inverted to recover the original time-domain signal $s(n)$. Inversion of the STFT can be accomplished in several ways, for example through overlap-and-add (OLA) or filter-bank summation (FBS) techniques, the two most commonly used methods (Portnoff 1980).

It is convenient to think of the short-time Fourier transform in terms of a filter bank, analogous in certain respects to the frequency analysis performed in the human auditory system. In the following section we interpret the frequency analysis capabilities of the STFT using an alternative representation based on a filter-bank structure.

5.3.2 Filter-Bank Interpretation of the STFT

One way to estimate the frequency content of a time varying signal is to pass it through a bank of bandpass filters, each with a different center frequency, covering the frequency range of interest. The STFT can be shown to be equivalent to a filter bank with certain properties related to the analysis window and Fourier basis functions. In this section we provide an informal but intuitive explanation of this equivalence. The reader who wishes to study these issues in depth is referred to Rabiner and Schafer (1978).

When the STFT is described in terms of a sliding window, we assume that the signal is static and that the windowing operation and Fourier analysis are applied as we "travel" across the length of the signal. The same operations involved in the computation of the STFT can be visualized from a dif-

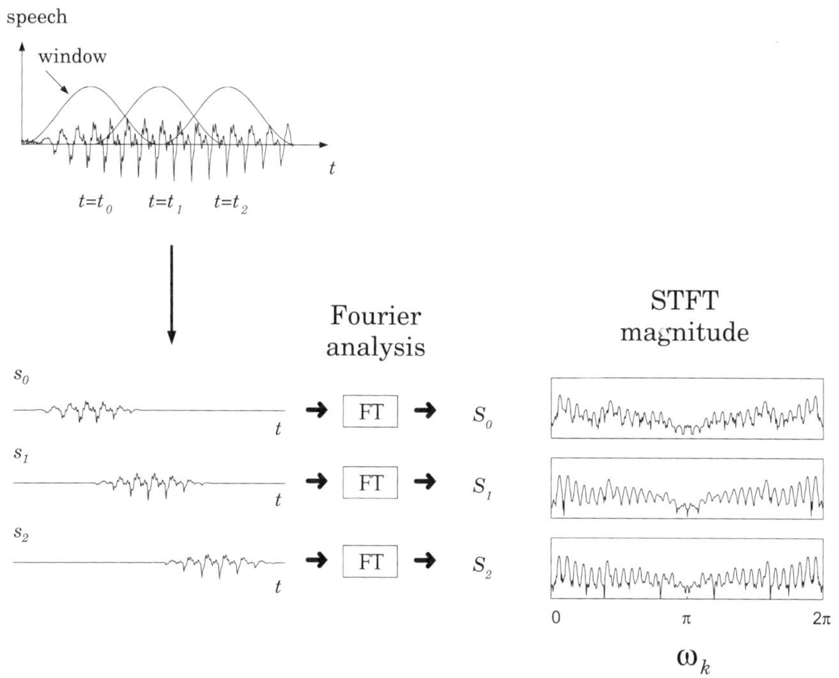

FIGURE 2.7. Short-time analysis of speech. The signal $s(n)$ is segmented by the sliding short-time window $w(n)$. Fourier analysis is applied to the resulting two-dimensional representation $s(n,m)$ to yield the short-time Fourier transform $S(n,\omega_k)$. Only the magnitude response in dB (and not the phase) of the STFT is shown. Note that the segments in this instance overlap with each other.

ferent perspective by choosing an alternative time reference. For example, instead of sliding the window across the signal we can fix the window and slide the signal across the window. With this new time reference the Fourier analysis appears to be static together with the window.

We can also reverse the order of these operations (as this is a linear system), by applying Fourier analysis to the window function to obtain a static system whose input signal is "traveling" in time. The fixed system (window/Fourier analysis) constitutes a bank of bandpass filters. This filter-bank is composed of bandpass filters having center frequencies equal to the frequencies of the basis functions of the Fourier analysis, i.e., $\omega_k = 2\frac{\pi}{N}k$ (Rabiner and Schafer 1978). The shapes of the bandpass filters are frequency-shifted copies of the transfer function of the analysis window function $w(n)$.

Thus, the STFT can be viewed from two different perspectives. We can view it either as a sequence of spectra corresponding to a series of short-time segments, or as a set of time signals that contain information about the original signal at each frequency band (i.e., filter bank outputs).

5.3.3 Time-Frequency Resolution Compromise

When we apply a finite-length rectangular window function to obtain a segment of speech, the Fourier series analysis (Equation 2) is a finite sum. For a window of length N, the Fourier analysis is a weighted summation of N basis functions.

According to the Nyquist criterion, the frequencies of the basis functions are multiples of the lowest component that can be resolved, i.e., $\Delta\omega = \dfrac{2\pi}{N}$. In other words, the frequency resolution of the analysis is inversely proportional to $\Delta\omega$, which depends on the length of the segment N.

Longer analysis windows yield spectra with finer frequency resolution. However, a longer analysis window averages speech over a longer time interval, and consequently the analysis cannot follow fast spectral changes within the signal (i.e., its time resolution is poor). Thus, increasing N to achieve better frequency resolution results in a decrease of time resolution. This trade-off is known as the time-bandwidth product, akin to the Heisenberg uncertainty principle originally formulated within quantum mechanics (see Cohen 1995 for a more detailed discussion), and it states that we cannot simultaneously make both time and frequency measures arbitrarily small. The more localized a signal is in frequency, the more spread out it is in time, and vice versa.

Quantitatively, the time-bandwidth product says that the product of duration and bandwidth of a signal is bounded by a constant, satisfying the inequality $\Delta\omega\Delta t \geq C$. The constant C depends on the definitions of effective duration Δt and effective bandwidth $\Delta\omega$ (Cohen 1995). These quantities vary for different window functions and play an important role in defining the properties of the STFT.

5.3.4 Effect of Windowing

As mentioned above, some properties of short-time analysis depend on the characteristics of the window function $w(n)$. The finite Fourier series analysis assumes that the signal is periodic, with the period equal to the length of the segment being analyzed. Any discontinuity resulting from the difference between the signal at the beginning and end of the segment will produce analysis artifacts (i.e., spectral leakage). To reduce the discontinuity, we apply a window function that attempts to match as many orders of derivatives at these points as possible (Harris 1978). This is easily achieved with the use of analysis window functions with tapered ends that bring the signal smoothly to zero at those points. With such a window function (e.g., Hamming window, Hanning window, Kaiser window), the difference between the segment and its periodic extensions at the boundaries is reduced and the discontinuity minimized.

For a given window length, the amount of admissible spectral leakage determines the particular choice of the window function. If the effective duration of the window is reduced (as is generally the case with tapered-end functions), the effective bandwidth increases and frequency resolution decreases.

An alternative interpretation of this process is that multiplication of the window function with the signal segment translates into a convolution of its Fourier transforms in the frequency domain. Thus, the convolution smears the spectral estimate of the signal.

From the Nyquist theorem it can be shown that the signal has to be sampled at a rate at least twice its bandwidth, i.e., $2B_w$ samples per second. For example, if we use a rectangular window function of length N, with the bandwidth approximately given by $B_w = sf/N$ (sf is the original sampling rate of the speech signal), then the signal has to be sampled with a period of $T = \frac{N}{2} sf$. It follows that the decimation factor for the STFT must be, at most, $M = \frac{N}{2}$ (i.e., 50% window overlap). The reduction of spectral leakage thus exacts a price in increasing the number of short-time segments required to be computed.

5.3.5 Relative Irrelevance of the Short-Time Phase

Fourier analysis provides not only the amplitude of a given frequency component of the signal, but also its phase (i.e., the amount of time a component is shifted relative to a given reference point). Within a short-time segment of speech the phase yields almost no useful information. It is therefore standard practice in applications that do not require resynthesis of the signal to disregard the short-time phase and to use only the short-time amplitude spectrum.

The irrelevance of the short-time phase is a consequence of our choice of an analysis window sufficiently short to assure stationarity of the speech segment. Had we attempted to perform the Fourier analysis of a much longer segment of speech (e.g., on the order of seconds), it would have been the phase spectrum that would have contained most of the relevant information (cf. Schroeder and Strube 1986).

5.3.6 Filter Bank and Wavelet Techniques

If the goal of speech analysis is to decompose the signal into its constituent frequency components, the more general way of achieving it is through a filter bank. In section 5.3.1 we described one of the most commonly used techniques of implementing a filter bank for speech analysis, the STFT. The obvious disadvantage of this analysis method is the inherent inflexibility of the design: all filters have the same shape, the center frequencies of the filters are equally spaced, and the properties of the window function limit the resolution of the analysis. However, since very efficient algorithms exist for computing the DFT, such as the fast Fourier transform, the FFT-based STFT is typically used for speech analysis.

Other filter bank techniques such as DFT-based filter banks, capitalize on the efficiency of the FFT (Crochiere and Rabiner 1983). While these filter banks suffer from some of the same restrictions as the STFT (e.g., equally spaced center frequencies), their design allows for improved spectral leakage rejection (sharper filter slopes and well-defined pass-bands) by allowing the effective length of the analysis filter to be larger than the analysis segment.

Alternative basis functions like cosines or sinusoids can also be used. The cosine modulated filter banks use the discrete cosine transform (DCT) and its FFT-based implementation for efficient realization (Vaidyanathan 1993).

There exist more general filter bank structures that possess perfect reconstruction properties and yet are not constrained to yield equally spaced center frequencies (and that provide for multiple resolution representations). One such structure can be implemented using wavelets. Wavelets have emerged as a new and powerful tool for nonstationary signal analysis (Vaidyanathan 1993). Many engineering applications of wavelets have benefited from this technique, ranging from video and audio coding to spread-spectrum communications (Akansu and Smith 1996).

One of the main properties of this technique is its ability to analyze a signal with different levels of resolution. Conceptually this is accomplished by using a sliding analysis window function that can dilate or contract, and that enables the details of the signal to be resolved depending on its temporal properties. Fast transients can be analyzed with short windows, while slowly varying phenomena can be observed with longer time windows.

From the time-bandwidth product (cf. the uncertainty principle, section 5.3.1), it can be demonstrated that this form of analysis is capable of providing good frequency resolution at the low end of the spectrum, but much poorer frequency resolution at the upper end of the spectrum. The use of this type of filter bank in speech analysis is motivated by the evidence that frequency analysis of the human auditory system behaves in a similar way (Moore 1989).

5.4 Production-Based Techniques

Speech is not an arbitrary signal, but rather is produced by a well-defined and constrained physical system (i.e., the human vocal apparatus). The process of speech generation is not simple, and deriving the state of the speech production system from the speech signal remains one of the main challenges of speech research. However, a crude model of the speech production process can provide certain useful constraints on the types of features derived from the speech signal. One of the most commonly used production models in speech analysis is the linear model described in section 3.5. Some of the speech analysis techniques that take advantage of this model are described in the following sections.

5.4.1 The Spectral Envelope

If we look at the short-time spectra of male and female speech with comparable linguistic messages, we can observe that the corresponding spectral envelopes reveal certain pattern similarities and differences (Fig. 2.8). The most obvious difference lies in the fine structure of the spectrum.

In the linear model of speech production, it is assumed that the filter properties (i.e., the spectral envelope) carry the bulk of the linguistic

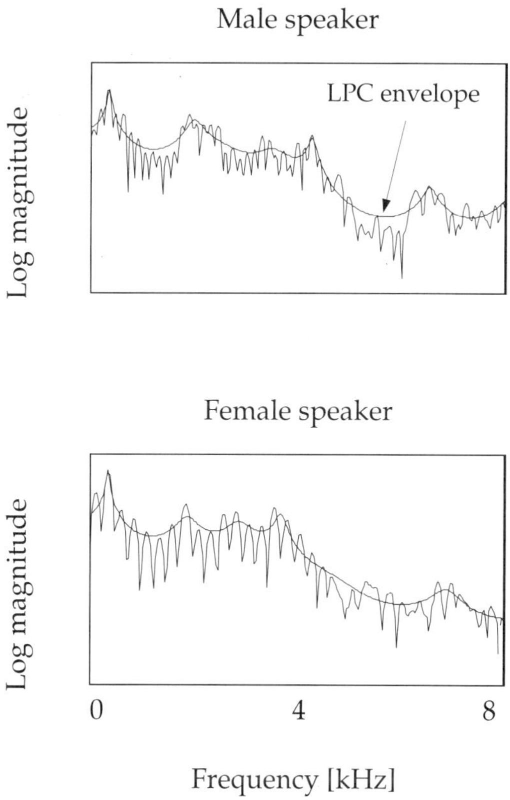

FIGURE 2.8. The short-time spectra of speech produced by a male and female speaker. The spectra correspond to a frame with a similar linguistic message.

message, while the main role of the source is to excite the filter so as to produce an audible acoustic signal. Thus, the task of many speech analysis techniques is to separate the spectral envelope (filter) from the fine structure (source).

The peaks of the spectral envelope correspond to the resonances of the vocal tract, (formants). The positions of the formants in the frequency scale (formant frequencies) are considered the primary carriers of linguistic information in the speech signal. However, formants are dependent on the inherent geometry of the vocal tract, which, in turn, is highly dependent on the speaker. Formant frequencies are typically higher for speakers with shorter vocal tracts (women and children). Also, gender-dependent formant scaling appears to be different for different phonetic segments (Fant 1965).

In Newton's early experiment (described in section 4.1.1), the glass resonances (formants) varied as the glass filled with beer. The distribution of formants along the frequency axis carries the linguistic information that enables one to hear the vowel sequences observed. Some later work supports this early notion that for decoding the linguistic message, the perception of speech effectively integrates several formant peaks (Fant and Risberg 1962; Chistovich 1985; Hermansky and Broad 1989).

5.4.2 LPC Analysis

Since its introduction to speech research in the early 1970s, linear prediction (LP) analysis has developed into one of the primary analysis techniques used in speech research. In its original formulation, LP analysis is a time-domain technique that attempts to predict "as well as possible" a speech sample through a linear combination of several previous signal samples:

$$\tilde{s}(n) = \sum_{k=1}^{p} -a_k s(n-k) \tag{3}$$

where $\tilde{s}(n)$ is the prediction. The number of previous signal samples used in the prediction determines the order of the LP model, denoted by p. The weights, a_k, of the linear combination are called predictive (or autoregressive) coefficients. To obtain these coefficients, the error between the speech segment and the estimate of the speech based on the prediction (Equation 3) is minimized in the least squares sense. This error can be expressed as

$$e(n) = s(n) - \tilde{s}(n) + \sum_{k=1}^{p} a_k s(n-k) \tag{4}$$

and the error minimization yields the least squares formulation

$$\min_{a_k} \sum_n e(n)^2 = \sum_n \left[s(n) + \sum_{k=1}^{p} a_k s(n-k) \right]^2 \tag{5}$$

The summation over n in Equation 5 pertains to the length of the data segment. The particular manner in which the data are segmented determines whether the covariance method, the autocorrelation method, or any of the lattice methods of LP analysis are used (Haykin 1991). Differences among methods are significant when the data window is very short. However, for a typical window length (about 20 to 25 ms, 160 to 200 data samples at an 8-kHz sampling rate), the differences among the LP methods are not substantial.

One way of interpreting Equation 4 is to look at the autoregressive coefficients, a_k, as the weights of a finite impulse response (FIR) filter. If we let $a_0 = 1$, then Equation 4 can be written as a filtering operation:

$$e(n) = \sum_{k=0}^{p} a_k s(n-k) \tag{6}$$

where the input to the filter is the speech signal $s(n)$ and the output is the error $e(n)$, also referred to as the residual signal. Figure 2.9A illustrates this operation. The formulation in Equation 5 attempts to generate an error signal with the smallest possible degree of correlation (i.e., flat spectrum) (Haykin 1991). Thus, the correlation with the speech signal is captured in the filter (via the autoregressive coefficients). For low-order models, the magnitude spectrum of the inverse filter (Fig. 2.9B) used to recover speech from the residual signal corresponds to the spectral envelope. Figure 2.8 shows the spectral envelopes for female and male speech obtained by a 14th order autocorrelation LP technique.

The solution to the autocorrelation LP method consists of solving a set of p linear equations. These equations involve the first $p + 1$ samples of the autocorrelation function of the signal segment. Since the autocorrelation function of the signal is directly related to the power spectrum through the Fourier transform, the autocorrelation LP model can also be directly derived in the frequency domain (e.g., Makhoul 1975 gives a more detailed description of this topic). The frequency domain formulation reveals some interesting properties of LP analysis. The average prediction error can be written in terms of the continuous Fourier transforms $S(\omega)$ and $\tilde{S}(\omega)$ of the signal $s(n)$ and the estimate $\tilde{s}(n)$, as

$$E = \frac{G^2}{2\pi} \int_{-\pi}^{\pi} \frac{|S(\omega)|^2}{|\tilde{S}(\omega)|^2} d\omega \tag{7}$$

where G is a constant gain factor. One consequence of LP modeling is that the spectrum of the LP model closely fits the peaks of the signal spectrum

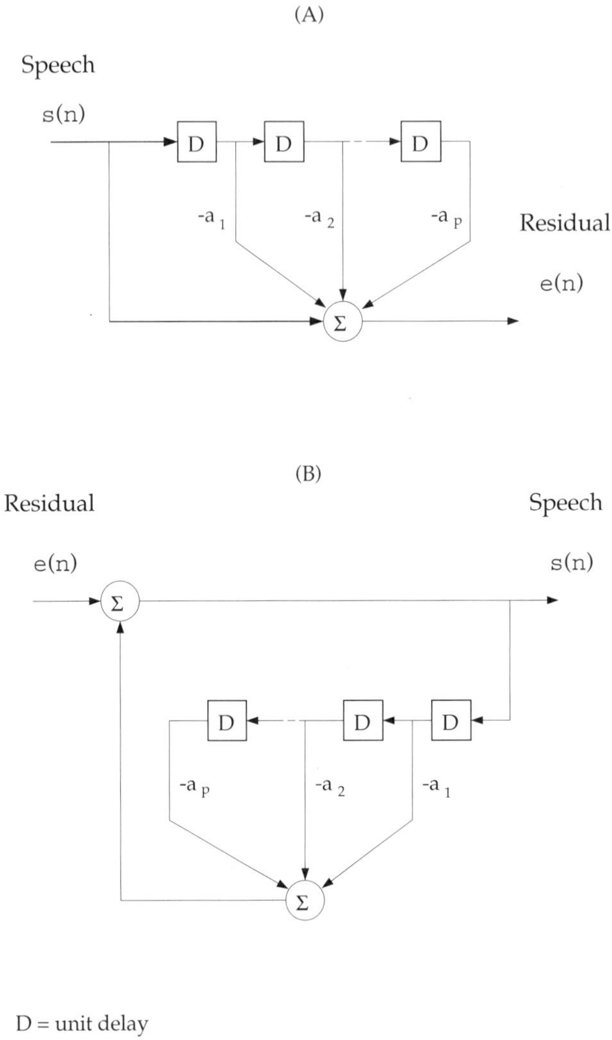

FIGURE 2.9. Linear prediction (LP) filter (A) and inverse filter (B).

at the expense of the fit at the spectral troughs, as observed in Equation 7. When the signal spectrum $S(\omega)$ exceeds the model spectrum, $\tilde{S}(\omega)$, the contribution to the error is greater than when the estimate exceeds the target spectrum, $S(\omega)$. Large differences contribute more to the error, and consequently the minimization of the error results in a better fit to the spectral peaks.

As the order of the LP model increases, more detail in the power spectrum of speech can be approximated (Fig. 2.10). The choice of the model

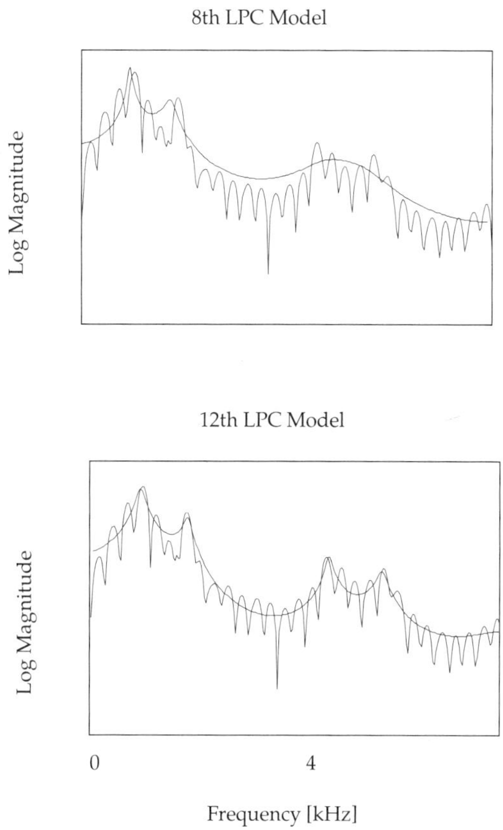

FIGURE 2.10. Spectrum of a short frame of speech. Superimposed are the spectra of the corresponding 8th- and 12th-order models.

order is an empirical issue. Typically, an 8th order model is used for analysis of telephone-quality speech sampled at 8 kHz. Thus, the spectral envelope can be efficiently represented by a small number of parameters (in this particular case by the autoregressive coefficients).

Besides the autoregressive coefficients, other parametric representations of the model can be used. Among these the most common are the following:

- Complex poles of the prediction polynomial describe the position and bandwidth of the resonance peaks of the model.
- The reflection coefficients of the model relate to the reflections of the acoustic wave inside a hypothetical acoustic tube whose frequency characteristic is equivalent to that of a given LP model.
- Area functions describe the shape of the hypothetical tube.

- Line spectral pairs relate to the positions and shapes of the peaks of the LP model.
- Cepstral coefficients of the LP model form a Fourier pair with the logarithmic spectrum of the model (they can be derived recursively from the prediction coefficients).

All of these parameters carry the same information and uniquely specify the LP model by $p + 1$ numbers. The analytic relationships among the different sets of LP parameters are described by Viswanathan and Makhoul (1975).

The LP analysis is neither optimal nor specific for speech signals, so it is to be expected that given the wide variety of sounds present in speech, some frames will not be well described by the model. For example, nasalized sounds are produced by a pole-zero system (the nasal cavity) and are poorly described by an all-pole model such as LP. Since the goal of the LP model is to approximate the spectral envelope, other problems may occur: the shapes of the spectral peaks (i.e., the bandwidths of complex roots of the LP model) are quite sensitive to the fine harmonic structure of high-pitched speech (e.g., woman or child) and to the presence of pole-zero pairs in nasalized sounds. The LP model is also vulnerable to noise present in the signal.

The LP modeling technique has been widely used in speech coding and synthesis (see section 4.1.1). The linear model of speech production (Fig. 2.2) allows for a significant reduction of bit rate by substituting the excitation (redundant part) with simple pulse trains or noise sequences (e.g., Atal and Hanauer 1971).

5.4.3 Cepstral Analysis

Another way of estimating the spectral envelope of speech is through cepstral analysis. The cepstrum of a signal is obtained in the following way. First, a Fourier analysis of the signal is performed. Then, the logarithm of this analysis is taken and an inverse Fourier transform is applied (Oppenheim and Schafer 1989).

Cepstral processing is a way of separating into additive terms components that have been convolved in the time domain. An example is the model of speech production illustrated in Figure 2.2, where the excitation signal (source) is convolved with the filter.

For a given frame of speech, it is assumed that the filter and source components are additive in the cepstral domain. The filter component is represented by the lower cepstral coefficients and the source by the higher components. Cepstral analysis then estimates the spectral envelope by truncating the cepstrum below a certain threshold. The threshold is set, based on assumptions about the duration of the filter's impulse response and the pitch (f_0) range of the speaker. Analogously, the fine structure can

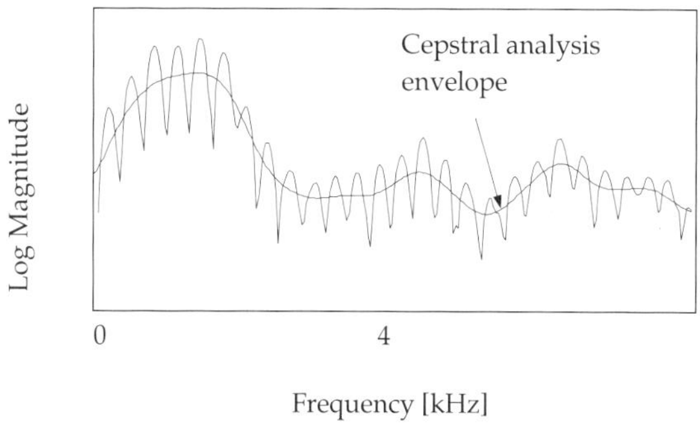

FIGURE 2.11. A frame of a speech segment (male speaker) and the spectral envelope estimated by cepstral analysis.

be separated by eliminating the coefficients below the threshold (Noll 1967).

Figure 2.11 shows a frame of speech and the estimate of the spectral envelope using cepstral analysis. We observe that the estimate is much smoother than the LP estimate, and that it does not necessarily fit all the peaks of the spectrum. Cepstral analysis has also been used to separate the source and filter components of the speech signal.

5.5 *Perception-Based Analysis Techniques*

Communication theory dictates that, in the presence of noise, most of the information should be transmitted through the least noisy locations (in frequency or time) in the transmission channel (e.g., Gallager 1968). It is likely that, in the same fashion, evolutionary processes provided the means by which the human speech production/perception apparatus with the means to optimally allocate its resources for speech communication through imperfect (albeit realistic) acoustic channels.

Perception-based analysis attempts to represent the speech signal from the perspective of the human speech processing apparatus. In section 4.1.2 we observed how visual displays could enhance the information necessary to understand some properties of speech and provide the human visual and language processing systems with sufficient information to decode the message itself. In a similar vein, it is possible to extract the information in speech relevant to the auditory system. For applications that require no human intervention to decode the message (such as automatic speech recognition), this second alternative may be advantageous. If speech evolved so that it would optimally use properties of the human auditory perception, then it makes sense that the analysis should attempt to emulate this perceptual process.

5.5.1 Analysis Techniques with a Nonlinear Frequency Scale

One potential problem (from the perceptual point of view) of the early sound spectrograph is the linear frequency scale employed, placing excessive emphasis on the upper end of the speech spectrum (from the auditory system's point of view). Several attempts to emulate this nonlinear frequency scaling property of human hearing for speech analysis have been proposed including the constant-Q filter bank (see section 5.3.6). The frequency resolution of such filter banks increases as a function of frequency (in linear frequency units) in such a fashion as to be constant on a logarithmic frequency scale.

Makhoul (1975) attempted to use nonlinear frequency resolution in LP analysis by introducing selective linear prediction. In this technique different parts of the speech spectra are approximated by LP models of variable order. Typically, the lower band of the speech spectrum is approximated by a higher order LP model, while the higher band is approximated by a low-order model, yielding reduced spectral detail at higher frequencics.

Itahashi and Yokoyama (1976), applied the Mel scale to LP analysis by first computing the spectrum of a relatively high LP model, warping it into Mel-scale coordinates, and then approximating this warped spectrum with that of a lower order LP model. Strube (1980) introduced Mel-like spectral warping into LP analysis by filtering the autocorrelation of the speech signal through a particular frequency-warping all-pass filter and using this all-pass filtered autocorrelation sequence to derive an LP model.

Bridle (personal communication, 1995), Mermelstein (1976), and Davis and Mermelstein (1980) have studied the use of the cosine transform on spectra with a nonlinear frequency scale. The cepstral analysis of Davis and Mermelstein uses the so-called Mel spectrum, derived by a weighted summation of the magnitude of the Fourier coefficients of speech. A triangular-shaped weighting function is used to approximate the hypothesized shapes of auditory filters.

Perceptual linear prediction (PLP) analysis (Hermansky 1990) simulates several well-known aspects of human hearing, and serves as a good example of the application of engineering approximations to perception-based analysis. PLP uses the Bark scale (Schroeder 1977) as the nonlinear frequency warping function. The critical-band integrated spectrum is obtained by a weighted summation of each frame of the squared magnitude of the STFT. The weighting function is derived from a trapezoid-shaped curve that approximates the asymmetric masking curve of Schroeder (1977). The critical-band integrated spectrum is then weighted by a fixed inverse equal-loudness function, simulating the equal-loudness characteristics at 40 dB SPL.

Frequency warping, critical band integration, and equal-loudness compensation are simultaneously implemented by applying a set of weighting

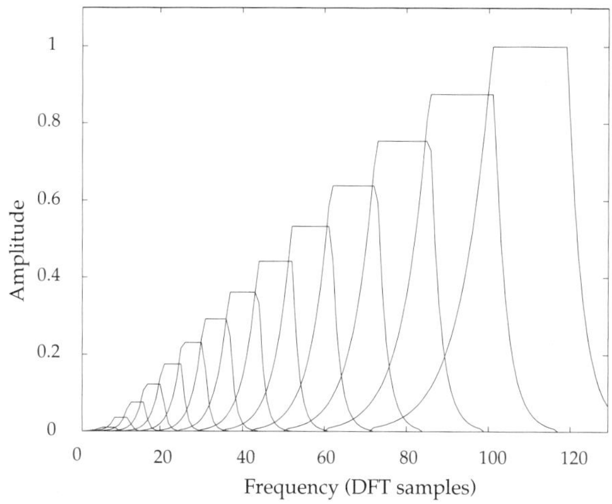

FIGURE 2.12. Perceptual linear prediction (PLP) weighting functions. The number of frequency points in this example correspond to a typical short-time analysis with a 256-point fast Fourier transform (FFT). Only the first 129 points of the even-symmetric magnitude spectrum are used.

functions to each frame of the squared magnitude of the STFT and adding the weighted values below each curve (Fig. 2.12).

To simulate the intensity-loudness power law of hearing (Stevens 1957), the equalized critical band spectrum is compressed by a cubic-root nonlinearity. The final stage in PLP approximates the compressed auditory-like spectrum by an LP model. Figure 2.13 gives an example of a voiced speech sound (the frequency scale of the plot is linear). Perceptual linear prediction fits the low end of the spectrum more accurately than the higher frequencies, where only a single peak represents the formants above 2 kHz.

Perceptual linear predictive and Mel cepstral analyses are currently the most widely used techniques for deriving features for automatic speech recognition (ASR) systems. Apart from minor differences in the frequency-warping function (e.g., Mel cepstrum uses the Mel scale) and auditory filter shapes, the main difference between PLP and Mel cepstral analysis is the method for smoothing the auditory-like spectrum. Mel cepstrum analysis truncates the cepstrum (see section 5.4.3), while PLP derives an all-pole LP model to approximate the dominant peaks of the auditory-like spectra.

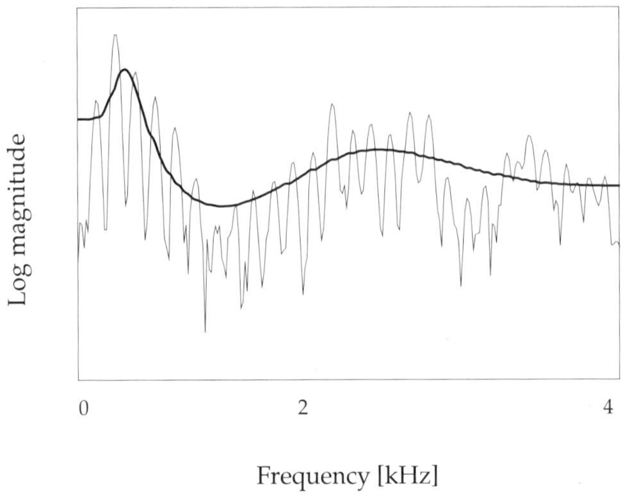

FIGURE 2.13. Spectrum of voiced speech and 7th-order PLP analysis (dark line).

5.5.2 Techniques Based on Temporal Properties of Hearing

The nonlinear frequency scale models discussed above consider only the static properties of human perception. There exist analysis techniques that also take into account temporal and dynamic properties of human auditory perception, such as temporal resolution, forward masking, temporal adaptation, and so on.

In speech recognition, Cohen (1989) used a feature-extraction module that simulates static, as well as dynamic perceptual properties. In addition to the nonlinear frequency scale and compressive nonlinearities, he used a short-term adaptation of the loudness-equalized filter bank outputs to simulate the onset and offset present in neural firing for different stimulus intensities.

Complex auditory representations based on physiological mechanisms underlying human perception have been suggested as possible feature extraction modules for ASR. Yang et al. (1992) have simulated the mechanical and neural processing in the early stages of the auditory system (Shamma 1985). Among other properties, they incorporate a long time constant integrator to simulate the limitation of auditory neurons to follow rapid temporal modulations. They claim that information integrity is maintained at several stages of the analysis and that resynthesized speech from the auditory representation is intelligible.

Perceptual phenomena pertaining to longer time intervals (150–250 ms), such as forward masking, have been simulated and used in ASR

(Hermansky and Morgan 1994; Hermansky and Pavel 1995; Kingsbury et al. 1997). Temporal masking has also been applied to increase the efficiency of music and speech coders (Johnston and Brandenburg 1992).

Kollmeier and Koch (1994) have devised a method for analyzing speech based on temporal information. They represented the temporal information in each frequency band by its Fourier components or modulation frequencies. This modulation spectrogram consists of a two-dimensional representation of modulation frequencies versus center frequency as a function of time.

The encoding of speech information in the slow modulations of the spectral envelope studied by Houtgast and Steeneken (1985) was used for speech analysis by Greenberg and Kingsbury (1997). They developed a speech visualization tool that represents speech in terms of the dominant modulation frequencies (around 2–8 Hz). Their modulation spectrogram uses a nonlinear frequency scale, and a much lower temporal resolution (higher modulation frequency resolution) than Kollmeier's. The ASR experiments confirm the utility of this perceptually based representation in automatic decoding of speech. The preservation (or enhancement) of the dominant modulation frequencies of the spectral envelope is advantageous in alleviating the effects of adverse environmental conditions in a variety speech applications (Avendano 1997; Greenberg and Kingsbury 1997; Kingsbury et al. 1997).

6. Summary

The basic concepts of speech production and analysis have been described. Speech is an acoustic signal produced by air pressure changes originating from the vocal production systems. The anatomical, physiological, and functional aspects of this process have been discussed from a quantitative perspective. With a description of various models of speech production, we have provided background information with which to understand the different components found in speech and the relevance of this knowledge for the design of analysis techniques.

The techniques for speech analysis can be divided into three major categories: signal-based, production-based, and perception-based. The choice of the appropriate speech analysis technique is dictated by the requirements of the particular application. Signal-based techniques permit the decomposition of speech into basic components, without regard to the signal's origin or destination. In production-based techniques emphasis is placed on models of speech production that describe speech in terms of the physical properties of the human vocal organs. Perception-based techniques analyze speech from the perspective of the human perceptual system.

List of Abbreviations

ADC analog-to-digital conversion
AM amplitude modulation
ASR automatic speech recognition
CELP code-excited linear prediction
DCT discrete cosine transform
DFT discrete Fourier transform
FBS filter-bank summation (waveform synthesis)
FFT fast Fourier transform
FIR finite impulse response (filter)
f_0 fundamental frequency
F_1 first format
F_2 second formant
LP linear prediction
LPC linear prediction coder
OLA overlap-and-add (waveform synthesis)
PLP perceptual linear prediction
STFT short-time Fourier transform

References

Akansu AN, Smith MJ (1996) Subband and Wavelet Transforms: Design and Applications. Boston: Kluwer Academic.

Atal BS, Hanauer SL (1971) Speech analysis and synthesis by linear prediction of the speech wave. J Acoust Soc Am 50:637–655.

Atal BS, Remde JR (1982) A new model of LPC excitation for producing natural sounding speech. Proc IEEE Int Conf Acoust Speech Signal Proc, pp. 614–618.

Atal BS, Schroeder MR (1979) Predictive coding of speech signals and subjective error criterion. IEEE Trans Acoust Speech Signal Proc 27:247–254.

Avendano C (1997) Temporal Processing of Speech in a Time-Feature Space. Ph.D. thesis, Oregon Graduate Institute of Science and Technology, Oregon.

Boubana S. Maeda S (1998) Multi-pulse LPC modeling of articulatory movements. Speech Comm 24:227–248.

Browman C, Goldstein L (1989) Articulatory gestures as phonological units. Phonology 6:201–251.

Browman C, Goldstein L (1992) Articulatory phonology: an overview. Phonetica 49:155–180.

Chistovich LA (1985) Central auditory processing of peripheral vowel spectra. J Acoust Soc Am 77:789–805.

Chistovich LA, Sheikin RL, Lublinskaja VV (1978) Centers of gravity and spectral peaks as the determinants of vowel quality. In: Lindblom B, Ohman S (eds) Frontiers of Speech Communication Research. London: Academic Press, pp. 143–157.

Cohen JR (1989) Application of an auditory model to speech recognition. J Acoust Soc Am 85:2623–2629.

Cohen L (1995) Time-Frequency Analysis. Englewoods Cliffs: Prentice Hall.

Cole RA, Zue V, Reddy R (1978) Speech as patterns on paper. In: Cole RA (ed) Perception and Production of Fluent Speech. Hillsdale, NJ: Lawrence Erlbaum.

Cooley JW, Tukey JW (1965) An algorithm for the machine computation of complex Fourier series. Math Comput 19:297–301.

Crochiere RE, Rabiner L (1983) Multirate Digital Signal Processing. Englewood Cliffs, NJ: Prentice Hall.

Davis SB, Mermelstein P (1980) Comparison of parametric representations for monosyllabic word recognition in continuously spoken sentences. IEEE Trans Acoust Speech Signal Proc 28:357–366.

Deng L, Sun D (1994) A statistical approach to automatic speech recognition using the atomic speech units constructed from overlapping articulatory features. J Acoust Soc Am 95:2702–2719.

Dudley H (1939) Remaking speech. J Acoust Soc Am 11:169–177.

Dudley H (1940) The carrier nature of speech. Bell System Tech J 19:495–513.

Dudley H, Tarnoczy TH (1950) The speaking machine of Wolfgang von Kempelen. J Acoust Soc Am 22:151–166.

Fant G (1960) Acoustic Theory of Speech Production. The Hague: Mouton.

Fant G (1965) Acoustic description and classification of phonetic units. Ericsson Technics 1. Reprinted in: Fant G (ed) Speech Sounds and Features. Cambridge: MIT Press.

Fant G, Risberg A (1962) Auditory matching of vowels with two formant synthetic sounds. Speech Transmission Laboratory Quarterly Progress Research Report (QPRS) 2–3. Stockholm: Royal Institute of Technology.

Flanagan J (1972) Speech Analysis, Synthesis and Perception. New York: Springer-Verlag.

Gallager RG (1968) Information Theory and Reliable Communication. New York: Wiley.

Gold B, Morgan N (1999) Speech and Audio Signal Processing: Processing and Perception of Speech and Music. New York: John Wiley & Sons.

Gold B, Rader CM (1969) Digital Processing of Signals. New York: McGraw-Hill.

Greenberg S (1996) Understanding speech understanding: towards a unified theory of speech perception. In: Ainsworth W, Greenberg S (eds) Proc ESCA Tutorial and Research Workshop on the Auditory Basis of Speech Recognition. United Kingdom: Keele University.

Greenberg S, Kingsbury B (1997) The modulation spectrogram: in pursuit of an invariant representation of speech. Proc IEEE Int Conf Acoust Speech Signal Proc, pp. 1647–1650.

Greenberg S, Hollenback J, Ellis D (1996) Insights into spoken language gleaned from phonetic transcription of the Switchboard corpus. Proc Fourth Int Conf on Spoken Lang (ICSLP): S24–27.

Harris FJ (1978) On the use of windows for harmonic analysis with discrete Fourier transform. IEEE Proc 66:51–83.

Haykin S (1991) Adaptive Filter Theory. Englewood Cliffs: Prentice Hall.

Helmholtz H (1863) On the Sensation of Tone. New York: Dover, 1954.
Hermansky H (1987) Why is the formant frequency DL curve asymmetric? J Acoust Soc Am 81:S18. (Full text in STL Research Reports 1, Santa Barbara, CA.)
Hermansky H (1990) Perceptual linear predictive (PLP) analysis of speech. J Acoust Soc Am 87:1738–1752.
Hermansky H, Broad D (1989) The effective second formant F2′ and the vocal front cavity. Proc IEEE Int Conf Acoust Speech Signal Proc, pp. 480–483.
Hermansky H, Morgan N (1994) RASTA processing of speech. IEEE Trans Speech Audio Proc 2:578–589.
Hermansky H, Pavel M (1995) Psychophysics of speech engineering. Proc Int Conf Phon Sci 3:42–49.
Hillenbrand J, Getty L, Clark MJ, Wheeler K (1995) Acoustic characteristics of American English vowels. J Acoust Soc Am 97:3099–3111.
Houtgast T, Steeneken HJM (1985) A review of the MTF concept in room acoustics and its use for estimating speech intelligibility. J Acoust Soc Am 77: 1069–1077.
Itahashi S, Yokoyama S (1976) Automatic formant extraction utilizing mel scale and equal loudness contour. Proc IEEE Int Conf Acoust Speech Signal Proc, pp. 310–313.
Itakura F, Saito S (1970) A statistical method for estimation of speech spectral density and formant frequencies. Electronics Commun Jpn 53-A:36–43.
Jayant NS (1974) Digital coding of speech waveforms: PCM, DPCM and DM quantizers. IEEE Proc 62:611–632.
Johnston JD, Brandenburg K (1992) Wideband coding: perceptual considerations for speech and music. In: Furui S, Sondhi MM (eds) Advances in Speech Signal Processing. New York: Dekker, pp. 109–140.
von Kempelen W (1791) Mechanismus der menschlichen Sprache nebst der Beschreibung seiner sprechenden Machine. Reprint of the German edition, with Introduction by Herbert E. Brekle and Wolfgang Wildgren (1970). Stuttgart: Frommann-Holzboog.
Keyser J, Stevens K (1994) Feature geometry and the vocal tract. Phonology 11: 207–236.
Kingsbury B, Morgan N, Greenberg S (1997) Improving ASR performance for reverberant speech. Proc ESCA Workshop on Robust Speech Recognition for Unknown Communication Channels, pp. 87–90.
Klatt D (1992) Review of selected models of speech perception. In: Marslen-Wilson W (ed) Lexical Representation and Processes. Cambridge: MIT Press, pp. 169–226.
Koenig W, Dunn HK, Lacey LY (1946) The sound spectrograph. J Acoust Soc Am 18:19–49.
Kollmeier B, Koch R (1994) Speech enhancement based on physiological and psychoacoustical models of modulation perception and binaural interaction. J Acoust Soc Am 95:1593–1602.
Ladefoged P (1967) Three Areas of Experimental Phonetics. Oxford: Oxford University Press.
Ladefoged P (1993) A Course in Phonetics. San Diego: Harcourt, Brace, Jovanovich.
Ladefoged P, Maddieson I (1990) Vowels of the world's languages. J Phonetics 18: 93–122.

Levelt W (1989) Speaking. Cambridge: MIT Press.
Makhoul J (1975) Spectral linear prediction properties and applications. IEEE Trans Acoust Speech Signal Proc 23:283–296.
McCarthy J (1988) Feature geometry and dependency: a review. Phonetica 43: 84–108.
Mermelstein P (1976) Distance measures for speech recognition, psychological and instrumental. In: Chen CH (ed) Pattern Recognition and Artificial Intelligence. New York: Academic Press, pp. 374–388.
Moore BCJ (1989) An Introduction to the Psychology of Hearing. London: Academic Press.
Noll AM (1967) Cepstrum pitch determination. J Acoust Soc Am 41:293–298.
Oppenheim AV, Schafer RW (1989) Discrete-Time Signal Processing. Englewood Cliffs, NJ: Prentice Hall.
Perkell JS (1969) Physiology of Speech Production: Results and Implications of a Quantitative Cineradiographic Study. Cambridge: M.I.T. Press.
Perkell JS (1980) Phonetic features and the physiology of speech production. In: Butterworth B (ed) Language Production. London: Academic Press.
Perkell JS, Matthies ML, Svirsky MA, Jordan MI (1995) Goal-based speech motor control: a theoretical framework and some preliminary data. J Phonetics 23:23–25.
Portnoff M (1980) Time-frequency representation of digital signals and systems based on short-time Fourier analysis. IEEE Trans Acoust Speech Signal Proc 28:55–69.
Potter RK, Kopp GA, Green HG (1946) Visible Speech. New York: Van Nostrand.
Rabiner LR, Schafer RW (1978) Digital Processing of Speech Signals. Englewood Cliffs, NJ: Prentice-Hall.
Saltzman E, Munhall K (1989) A dynamical approach to gestural patterning in speech production. Ecol Psychol 1:333–382.
Schroeder MR (1977) In: Bullock TH (ed) Recognition of Complex Acoustic Signals. Berlin: Abakon Verlag, p. 324.
Schroeder MR, Atal BS (1968) Predictive coding of speech signals. In: Kohashi Y (ed) Report of the 6th International Congress on Acoustics, Tokyo.
Schroeder M, Atal BS (1985) Code-excited linear prediction (CELP): high-quality speech at very low bit rates. Proc IEEE Int Conf Acoust Speech Signal Proc, pp. 937–940.
Schroeder MR, Strube HW (1986) Flat-spectrum speech. J Acoust Soc Am 79: 1580–1582.
Scripture C (1906) Researches in Experimental Phonetics. Washington, DC: Carnegie Institution of Washington.
Shamma SA (1985) Speech processing in the auditory system I: representation of speech sounds in the responses of the auditory nerve. J Acoust Soc Am 78: 1612–1621.
Smith CL, Browman CP, McGowan RS, Kay B (1993) Extracting dynamic parameters from speech movement data. J Acoust Soc Am 93:1580–1588.
Stevens SS (1957) On the psychophysical law. Psych Rev 64:153–181.
Strube HW (1980) Linear prediction on a warped frequency scale. J Acoust Soc Am 68:1071–1076.
Vaidyanathan, PP (1993) Multirate Systems and Filter Banks. Englewood Cliffs, NJ: Prentice-Hall.

Viswanathan R, Makhoul J (1975) Quantization properties of transmission parameters in linear predictive systems. IEEE Trans Acoust Speech Signal Proc 23: 587–596.
Yang X, Wang W, Shamma SA (1992) Auditory representations of acoustic signals. IEEE Trans Inform Theory 38:824–839.

3
Explaining the Structure of Feature and Phoneme Inventories: The Role of Auditory Distinctiveness

RANDY L. DIEHL and BJÖRN LINDBLOM

1. Introduction

Linguists and phoneticians have always recognized that the sounds of spoken languages—the vowels and consonants—are analyzable into component properties or features. Since the 1930s, these features have often been viewed as the basic building blocks of language, with sounds, or phonemes, having a derived status as feature bundles (Bloomfield 1933; Jakobson et al. 1963; Chomsky and Halle 1968). This chapter addresses the question: What is the explanatory basis of phoneme and feature inventories?

Section 2 presents a brief historical sketch of feature theory. Section 3 reviews some acoustic and articulatory correlates of important feature distinctions. Section 4 considers several sources of acoustic variation in the realization of feature distinctions. Section 5 examines a common tendency in traditional approaches to features, namely, to introduce features in an ad hoc way to describe phonetic or phonological data. Section 6 reviews two recent theoretical approaches to phonemes and features, quantal theory (QT) and the theory of adaptive dispersion (TAD). Contrary to most traditional approaches, both QT and TAD use facts and principles independent of the phonetic and phonological data to be explained in order to derive predictions about which phoneme and feature inventories should be preferred among the world's languages. QT and TAD are also atypical in emphasizing that preferred segments and features have *auditory* as well as articulatory bases. Section 7 looks at the issue of whether phonetic invariance is a design feature of languages. Section 8 presents a concluding summary.

An important theme of this chapter is that hypotheses about the auditory representation of speech sounds must play a central role in any explanatory account of the origin of feature and phoneme systems. Such hypotheses should be grounded in a variety of sources, including studies of identification and discrimination of speech sounds and analogous non-speech sounds by human listeners, studies of speech perception in nonhuman animals, electrophysiological investigations of auditory responses to speech, and computational modeling of speech processing in the auditory system.

2. Feature Theory: A Brief Historical Outline

As early as the investigations of Pānini (520–460 B.C.), it was understood that vowels and consonants are composed of more elementary properties that tend to recur across different speech sounds. This fundamental insight into the nature of spoken language served as the basis for A.M. Bell's (1867) "visible speech" alphabet, designed to help teach pronunciation to the deaf. The symbols of this alphabet consisted of combinations of markers each representing an independent vocal tract property such as narrow glottis (yielding vocal fold vibration, or voicing), soft palate depressed (yielding nasality), and lips as primary articulators. (This particular combination of properties corresponds to the consonant /m/.) Using the limited tools then available, Bell and other early phoneticians managed to perform a detailed and fairly accurate articulatory analysis of many sounds that confirmed their componential character. Thus, "features" in the sense of vocal tract properties or parameters assumed an important role in phonetic description of spoken languages. As techniques were developed to measure acoustic correlates of vocal tract properties, phonetic features could, in principle, be seen as either articulatory or acoustic, but articulatory properties remained the principal focus of phonetic description.

An important shift in the meaning of the term *feature* resulted from the adoption of the *phonemic principle* in early 20th century linguistics. This principle states that for any particular language only some phonetic differences are linguistically relevant in that they serve to distinguish meanings of words or morphemes. For example, the phonetic segments [b] and [p^h] correspond to different phonemes in English, symbolized as /b/ and /p/, because there are pairs of English words that differ only in the choice of those two segments at a particular position (e.g., "ban" versus "pan"). In contrast, the segments [p] and [p^h] are both phonetic variants of the same phoneme /p/ in English, since the presence or absence of aspiration (denoted by the superscript "h") associated with a [p] segment does not correspond to any minimal lexical pairs. With the adoption of the phonemic principle, the science of phonetics came to be seen as fundamentally distinct from the study of language. Phonetics was understood to be concerned with description of the articulatory, acoustic, and auditory properties of speech sounds (i.e., the *substance* of spoken language) irrespective of the possible role of those properties in specifying distinctions of meaning. Linguistics, particularly the subdiscipline that came to be known as "phonology," focused on sound patterns and the distinctive or phonemic use of speech sounds (i.e., the *form* of spoken language).

One of the founders of modern phonology, Trubetzkoy (1939) expressed the substance/form distinction as follows:

> The speech sounds that must be studied in phonetics possess a large number of acoustic and articulatory properties. All of these are important for the phonetician since it is possible to answer correctly the question of how a specific sound is produced only if all of these properties are taken into consideration. Yet most of these properties are quite unimportant for the phonologist. The latter needs to consider only *that aspect of sound which fulfills a specific function in the system of language.* [Italics are the author's]
>
> This orientation toward function is in stark contrast to the point of view taken in phonetics, according to which, as elaborated above, any reference to the meaning of the act of speech (i.e., any reference to signifier) must be carefully eliminated. This fact also prevents phonetics and phonology from being grouped together, even though both sciences appear to deal with similar matters. To repeat a fitting comparison by R. Jakobson, phonology is to phonetics what national economy is to market research, or financing to statistics (p. 11).

Trubetzkoy and Jakobson were leading members of the highly influential Prague Linguistic Circle, which developed the modern notion of features as phonological markers that contribute to phonemic distinctions. To demarcate features in this restricted sense from phonetic properties or parameters in general, Jakobson applied the adjective "distinctive." (Later theorists have sometimes preferred the adjective "phonological.") For the Prague phonologists, phonemic oppositions were seen as entirely reducible to contrasts between features, and so the phoneme itself became nothing more than a bundle of distinctive features. Thus, Jakobson (1932) wrote, "By this term [phoneme] we designate a set of those concurrent sound properties which are used in a given language to distinguish words of unlike meaning" (p. 231). Trubetzkoy (1939) similarly wrote, "One can say that the phoneme is the sum of the phonologically relevant properties of a sound" (p. 36).

Among Trubetzkoy's contributions to feature theory was a formal classification of various types of phonemic oppositions. He distinguished between *bilateral* oppositions, in which a single phonetic dimension yields only a contrasting pair of phonemes, and *multilateral* oppositions, in which more than two phonemes contrast along the same dimension. For example, in English the [voice] distinction is bilateral, whereas distinctions along the place-of-articulation dimension (the location along the vocal tract where the primary constriction occurs) correspond, according to Trubetzkoy, to a multilateral opposition among /b/, /d/, and /g/ or among /p/, /t/, and /k/. Also, Trubetzkoy made a three-way distinction among *privative* oppositions, based on the presence or absence of a feature (e.g., presence or absence of voicing), *equipollent* oppositions, in which the phonemes contain different features (e.g., front versus back vowels), and *gradual* oppositions, in which the phonemes contain different amounts of a feature (e.g., degree of openness of vowels).

Trubetzkoy's proposed set of features was defined largely in articulatory terms. For example, the primary vowel features were specified on the basis of place of articulation, or "localization" (e.g., front versus back), degree of jaw opening, or "aperture" (e.g., high versus low), and degree of lip rounding. However, some feature dimensions were also labeled in terms of auditory impressions. For example, degree of aperture corresponded to "sonority" or loudness, and localization corresponded to timbre, with back vowels such as /u/ being described as "dark" and front vowels such as /i/ being described as "clear." Significantly, Trubetzkoy noted that the combination of lip rounding and tongue retraction produced a maximally dark timbre, while the combination of unrounding and tongue fronting produced a maximally clear timbre. This implies that independent articulatory parameters do not necessarily map onto independent acoustic or auditory parameters.

Although Jakobson agreed with Trubetzkoy on several key issues (including the primacy of phonological criteria in defining features), he and later colleagues (Jakobson 1939; Jakobson et al. 1963; Jakobson and Halle 1971) introduced some important modifications of feature theory. First, Trubetzkoy's elaborate taxonomy of types of phonological oppositions was rejected in favor of a simplified system in which all feature contrasts are bilateral (later called "binary") and logically privative. One motivation for such a restriction is that the listener's task is thereby reduced to detecting the presence or absence of some property, a simple qualitative judgment, rather than having to make a quantitative judgment about relative location along some physical continuum, as in multilateral or gradual oppositions.

An obvious challenge for Jakobson was to show how an apparently multilateral opposition, such as consonant place of articulation, could be analyzed as a set of binary feature contrasts. His solution was to posit two contrasts: *grave* versus *acute*, and *compact* versus *diffuse*. Grave consonants are produced with a large, undivided oral cavity, resulting in a relative predominance of low or middle frequency energy in the acoustic spectrum. English examples are the labial sounds /b/ and /p/ (articulated at the lips) and the velar sounds /g/ and /k/ (articulated with the tongue body at the soft palate, or velum). Acute consonants are produced with a constriction that divides the oral cavity into smaller subcavities, yielding a relative predominance of high-frequency energy. Examples are the English alveolar sounds /d/ and /t/ (articulated with the tongue tip or blade at the alveolar ridge behind the upper incisors) and the German palatal fricative /x/, as in "ich," (articulated with middle part the tongue at the dome of the hard palate). Compact consonants are produced with the major constriction relatively retracted in the oral cavity (e.g., palatals and velars), producing a concentration of energy in the middle frequencies of the spectrum, whereas diffuse consonants are produced with a more anterior constriction (e.g., labials and alveolars) and lack such a concentration of middle frequency energy.

As the names of these features suggest, another point of difference with Trubetzkoy's system was Jakobson's emphasis on acoustic criteria in defining feature contrasts. Although articulatory correlates were also provided, Jakobson et al. (1963) saw the acoustic specification of features as theoretically primary "given the evident fact that we speak to be heard in order to be understood" (p. 13). Apart from the role of features in defining phonological distinctions, Jakobson recognized that they function more generally to pick out "natural classes" of speech sounds that are subject to the same phonological rules or that are related to each other through historical processes of sound change. (The distinctive function of features may be viewed as a special case of this larger classificatory function.) Jakobson argued that an acoustic specification of features allows certain natural classes to be characterized that are not easily defined in purely articulatory terms. For example, the feature *grave* corresponds to both labial and velar consonants, which do not appear to form an articulatorily natural class but which have acoustic commonalities (i.e., a predominance of energy in the low to middle frequencies of the speech spectrum). These acoustic commonalities help to account for sound changes such as the shift from the velar fricative /x/ (the final sound in German words such as "Bach") in Old English words such as "rough" and "cough" to the labial fricative /f/ in modern English (Ladefoged 1971). Another Jakobsonian feature that has an acoustically simple specification but that subsumes a number of articulatory dimensions is *flat*. This feature corresponds to a relative frequency lowering of spectral energy and is achieved by any or all of the following: lip rounding, retroflexion or bunching of the tongue, constriction in the velar region, and constriction of the pharynx.

An important aim of Jakobson was to define a small set of distinctive features that were used universally among the world's languages. The restriction to binary and privative features was one means to accomplish that end. Another means, as just noted, was to characterize features in acoustic rather than purely articulatory terms: "The supposed multiplicity of features proves to be largely illusory. If two or more allegedly different features never co-occur in a language, and if they, furthermore, yield a common property, distinguishing them from all other features, then they are to be interpreted as different implementations of one and the same feature" (Jakobson and Halle 1971, p. 39).

For example, Jakobson suggested that no language makes distinctive use of both lip rounding and pharyngeal constriction, and that fact, together with the common acoustic correlate of these two articulatory events (viz., formant frequency lowering), justifies application of the feature flat to both. A final means of restricting the size of the universal inventory of distinctive features was to define them in relative rather than absolute terms (see, for example, the acoustic definition of flat, above). This allowed the same features to apply to varying phonetic implementations across different phonological contexts and different languages. Accordingly, Jakobson was able to limit the size of the putative universal inventory of distinctive features to about 12 binary contrasts.

The next important developments in feature theory appeared in Chomsky and Halle's *The Sound Pattern of English* (SPE), published in 1968. Within the SPE framework, "features have a phonetic function and a classificatory function. In their phonetic function, they are [physical] scales that admit a fixed number of values, and they relate to independently controllable aspects of the speech event or independent elements of perceptual representation. In their classificatory function they admit only two coefficients ["+" or "–"] and they fall together with other categories that specify the idiosyncratic properties of lexical items" (p. 298).

Although this characterization of features has some similarities to Jakobson's (e.g., the binarity of phonological feature distinctions), there are several significant differences. First, by identifying features with physical scales that are at least in part independently controllable, Chomsky and Halle in effect expanded the range of potential features to be included in the universal inventory. The SPE explicitly proposed at least 22 feature scales, and given the known degrees of freedom of articulatory control, the above definition is actually compatible with a much higher number than that. Second, the relation between phonetic features (or features in their phonetic function) and phonological features (or features in their classificatory function) was assumed to be considerably more direct than in previous formulations. Both types of feature refer to the same physical variables; they differ only in whether those variables are treated as multivalued or binary.

A third point of difference with Jakobson is that Chomsky and Halle emphasized articulatory over acoustic specification of features. All but one of their feature labels are articulatory, and almost all of the accompanying descriptions refer to the way sounds are produced in the vocal tract. [The one exception, the feature *strident*, refers to sounds "marked acoustically by greater noisiness than their nonstrident counterparts" (p. 329).] Thus, for example, in place of Jakobson's acoustically based place of articulation features, grave and compact, Chomsky and Halle used the articulatorily based features *anterior* (produced with the major constriction in front of the palatoalveolar region of the mouth) and *coronal* (produced with the tongue blade raised above its neutral position). By this classification, /b/ is [+anterior] and [–coronal], /d/ is [+anterior] and [+coronal], and /g/ is [–anterior] and [–coronal]. In addition to these purely consonantal place features, Chomsky and Halle posited three tongue body positional features that apply to both consonants and vowels: *high* (produced by raising the tongue body from its neutral position), *low* (produced by lowering the tongue body from its neutral position), and *back* (produced by retracting the tongue body from its neutral position). For the tongue body, the neutral position corresponds roughly to its configuration for the English vowel /ɛ/, as in "bed."

Use of the same tongue body features for both consonants and vowels satisfied an important aim of the SPE framework, namely, that commonly attested phonological processes should be expressible in the grammar in a formally simple way. In various languages, consonants are produced with secondary articulations involving the tongue body, and often these are consistent with the positional requirements of the following vowel. Such a process of assimilation of consonant to the following vowel may be simply expressed by allowing the consonant to assume the same binary values of the features [high], [low], and [back] that characterize the vowel.

Several post-SPE developments are worth mentioning. Ladefoged (1971, 1972) challenged the empirical basis of several of the SPE features and proposed an alternative (albeit partially overlapping) feature system that was, in his view, better motivated phonetically. In Ladefoged's system, features, such as consonantal place of articulation, vowel height, and glottal constriction were multivalued, while most other features were binary. With a few exceptions (e.g., *gravity*, *sibilance*, and *sonorant*), the features were given articulatory definitions.

Venneman and Ladefoged (1973) elaborated this system by introducing a distinction between "prime" and "cover" features that departs significantly from the SPE framework. Recall that Chomsky and Halle claimed that phonetic and phonological features refer to the same independently controllable physical scales, but differ as to whether these scales are viewed as multivalued or binary. For Venneman and Ladefoged, a prime feature refers to "a single measurable property which sounds can have to a greater or lesser degree" (pp. 61–62), and thus corresponds to a phonetic feature in the SPE framework. A prime feature (e.g., nasality) can also be a phonological feature if it serves to form lexical distinctions and to define natural classes of sounds subject to the same phonological rules. This, too, is consistent with the SPE framework. However, at least some phonological features—the cover features—are not reducible to a single prime feature but instead represent a disjunction of prime features. An example is *consonantal*, which corresponds to any of a sizable number of different measurable properties or, in SPE terms, independently controllable physical scales. Later, Ladefoged (1980) concluded that all but a very few phonological features are actually cover features in the sense that they cannot be directly correlated with individual phonetic parameters.

Work in phonology during the 1980s led to an important modification of feature theory referred to as "feature geometry" (Clements 1985; McCarthy 1988). For Chomsky and Halle (1968), a segment is analyzed into a feature list without any internal structure. The problem with this type of formal representation is that there is no simple way of expressing regularities in which certain subsets of features are jointly and consistently affected by the same phonological processes. Consider, for example, the strong cross-linguistic tendency for nasal consonants to assimilate to the place of articulation value of the following consonant (e.g., in "lump," "lint," and "link"). In the SPE framework, this regularity is described as assimilation of the feature values for [anterior], [coronal], and [back]. However, if the feature lists corresponding to segments have no internal structure, then "this common process should be no more likely than an impossible one that assimilates any arbitrary set of three features, like [coronal], [nasal], and [sonorant]" (McCarthy 1988, p. 86). The solution to this problem offered in the theory of feature geometry is to posit for segments a hierarchical feature structure including a "place node" that dominates all place of articulation features such as [coronal], [anterior], and [back]. The naturalness of assimilation based on the above three features (and the unnaturalness of assimilation based on arbitrary feature sets) is then captured formally by specifying the place node as the locus of assimilation. In this way, all features dominated by the place node, and only those features, are subject to the assimilation process.

3. Some Articulatory and Acoustic Correlates of Feature Distinctions

This section reviews some of the principal articulatory and acoustic correlates of selected feature contrasts. The emphasis is on acoustic correlates that have been suggested to be effective perceptual cues to the contrasts.

3.1 The Feature [Sonorant]

Sonorant sounds include vowels, nasal stop consonants (e.g., /m/, /n/, and /ŋ/ as in "si*ng*," glides (e.g., /h/, /w/ and /j/, pronounced "y"), and liquids (e.g., /r/ and /l/), whereas nonsonorant, or obstruent, sounds include oral stop consonants (e.g., /b/, /p/, /d/, /t/, /g/, /k/), fricatives (e.g., /f/, /v/, /s/, /z/, /ʃ/ as in "*sh*ow," and /ʒ/ as in "bei*ge*"), and affricates (e.g., /ʧ/ as in "wat*ch*" and /ʤ/ as in "bu*dge*"). Articulatorily, [–sonorant] implies that the maximal vocal tract constriction is sufficiently narrow to yield an obstruction to the airflow (Stevens and Keyser 1989) and thus a significant pressure buildup behind the constriction (Halle 1992; Stevens 1998), while [+sonorant] implies the absence of this degree of constriction. Stevens and Keyser (1989) note that, acoustically, [+sonorant] is "characterized by continuity of the spectrum amplitude at low frequencies in the region of the first and second harmonics—a continuity of amplitude that extends into an adjacent vowel without substantial change" (p. 87). Figure 3.1 shows spectrograms of the disyllables /awa/, which displays the low-frequency continuity at the margin between vowel and consonant, and /afa/, which does not.

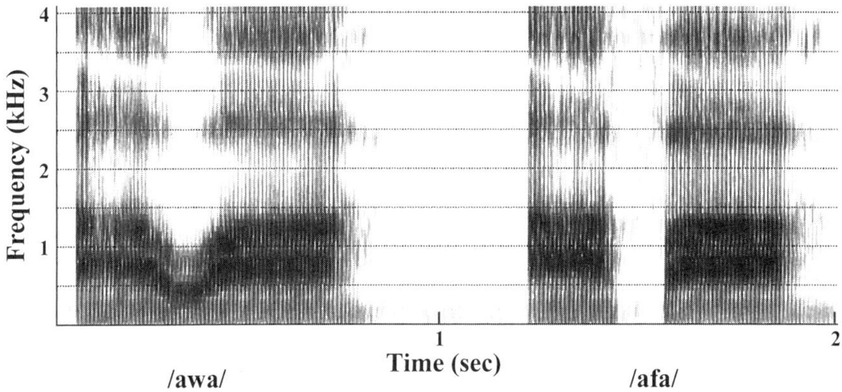

FIGURE 3.1. Spectrograms of /awa/ and /afa/.

3.2 The Feature [Continuant]

Sounds that are [–continuant] are produced with a complete blockage of the airflow in the oral cavity and include oral and nasal stop consonants and affricates; [+continuant] sounds are produced without a complete blockage of the oral airflow and include fricatives, glides, and most liquids. Stevens and Keyser (1989) suggest that the distinguishing acoustic property of [–continuant] sounds is "an abrupt onset of energy over a range of frequencies preceded by an interval of silence or of low amplitude" (p. 85). The onset of [+continuant] sounds is less abrupt because of energy present in the interval prior to the release of the consonant, and because the amplitude rise time at the release is longer.

Two examples of the [+/–continuant] distinction—/w/ versus /b/, and /ʃ/ versus /tʃ/—have been been the subject of various acoustic and perceptual investigations. Because the /w/-/b/ contrast also involves the feature distinction [+/–sonorant], we focus here mainly on the fricative/affricate contrast between /ʃ/ and /tʃ/. Acoustic measurements have generally shown that frication noise is longer and (consistent with the claim of Stevens and Keyser 1989) has a longer amplitude rise time in /ʃ/ than in /tʃ/ (Gerstman 1957; Rosen and Howell 1987). Moreover, in several perceptual studies (Gerstman 1957; Cutting and Rosner 1974; Howell and Rosen 1983), variation in rise time was reported to be an effective cue to the fricative/affricate distinction. However, in each of the latter studies, rise time was varied by removing successive portions of the initial frication segment such that rise time was directly related to frication duration. In a later study, Kluender and Walsh (1992) had listeners label several series of /ʃ/-/tʃ/ stimuli in which rise time and frication duration were varied independently. Whereas differences in frication duration were sufficient to signal the fricative/affricate distinction (with more fricative labeling

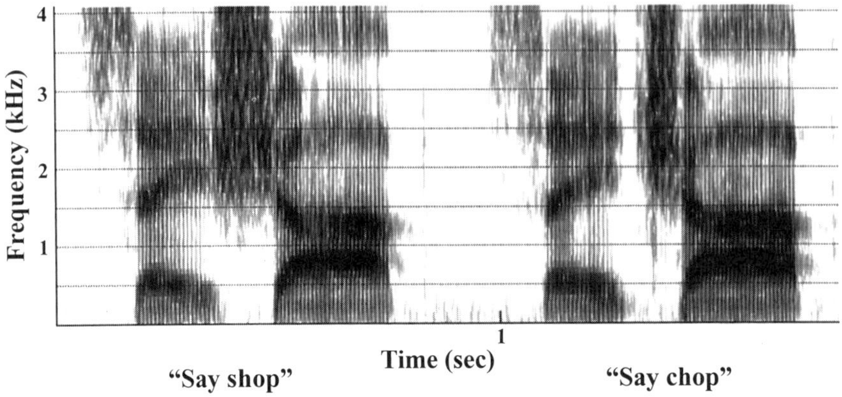

FIGURE 3.2. Spectrograms of "say shop" and "say chop."

responses occurring at longer frication durations), rise-time variation had only a small effect on labeling performance. [An analogous pattern of results was found by Walsh and Diehl (1991) for the distinction between /w/ and /b/. While formant transition duration was a robust cue for the contrast, rise time had a very small effect on identification.] Thus, contrary to the claim of Stevens and Keyser (1989), it appears unlikely that abruptness of onset, as determined by rise time, is the primary perceptual cue for the [+/–continuant] distinction. For the contrast between /ʃ/ and /ʧ/ the more important cues appear to be frication duration and, in noninitial position, the duration of silence prior to the onset of frication (Repp et al. 1978; Castleman and Diehl 1996a). Figure 3.2 displays spectrograms of "Say shop" and "Say chop." Note that frication duration is longer for the fricative than for the affricate and that there is an interval of silence associated with the affricate but not the fricative.

3.3 The Feature [Nasal]

Sounds that are [+nasal] are produced with the velum lowered, allowing airflow through the nasal passages. Sounds that are [–nasal] are produced with the velum raised, closing the velopharyngeal port, and allowing airflow only through the mouth. Many languages (e.g., French) have a phonemic contrast between nasalized and nonnasalized vowels, but here we focus on the [+/–nasal] distinction among consonants. Figure 3.3 shows spectrograms of the utterances "a mite," "a bite," and "a white." In each case, the labial articulation yields a rising, first-formant (F_1) frequency following the consonant release, although the rate of frequency change varies. The nasal stop /m/, like the oral stop /b/, shows a predominance of low-frequency energy and a marked discontinuity in spectral energy level (especially for the higher formants) before and after the release. By comparison, the glide /w/

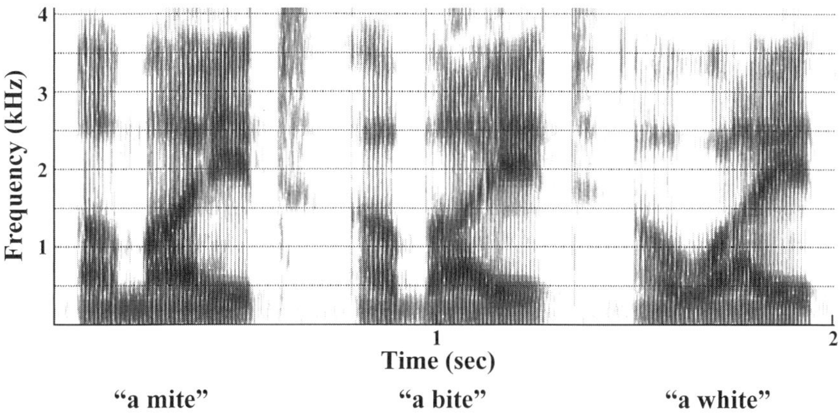

FIGURE 3.3. Spectrograms of "a mite," "a bite," and "a white."

shows greater energy in the second formant (F_2) during the constriction and greater spectral continuity before and after the release. The nasal stop differs from the oral stop in having greater amplitude during the constriction and in having more energy associated with F_2 and higher formants. Relatively little perceptual work has been reported on the [+/–nasal] distinction in consonants.

3.4 The Features for Place of Articulation

The features for consonant place of articulation have been the subject of a great many phonetic investigations. Articulatorily, these features refer to the location along the vocal tract where the primary constriction occurs. They involve both an active articulator (e.g., lower lip, tongue tip or blade, tongue dorsum, or tongue root) and a more nearly rigid anatomical structure (e.g., the upper lip, upper incisors, alveolar ridge, hard palate, velum, or pharyngeal wall), sometimes called the passive articulator, with which the active articulator comes into close proximity or contact. We focus here on the distinctions among English bilabial, alveolar, and velar oral stop consonants.

Figure 3.4 shows spectrograms of the syllables /ba/, /da/, and /ga/. At the moment of consonant release, there is a short burst of energy, the frequency characteristics of which depend on the place of articulation. The energy of this release burst is spectrally diffuse for both /b/ and /d/, with the bilabial consonant having more energy at lower frequencies and the alveolar consonant having more energy at higher frequencies. The spectrum of the /g/ release burst has a more compact energy distribution centered in the middle frequencies. Stevens and Blumstein (1978; Blumstein and Stevens 1979, 1980) suggested that these gross spectral shapes of the release burst are

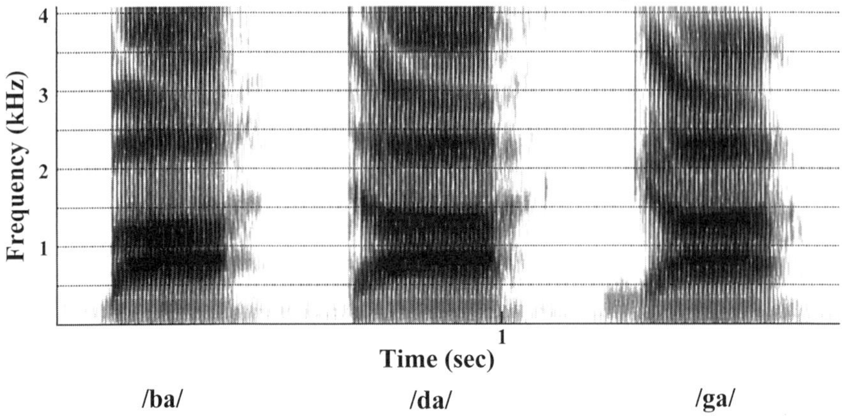

FIGURE 3.4. Spectrograms of /ba/, /da/, and /ga/.

invariant correlates of stop place and that they serve as the primary cues for place perception. More recently, Stevens and Keyser (1989) proposed a modified view according to which the gross spectral shape of the burst may be interpreted relative to the energy levels in nearby portions of the signal. Thus, for example, [+coronal] (e.g., /d/) is characterized as having "greater spectrum amplitude at high frequencies than at low frequencies, or at least an increase in spectrum amplitude at high frequencies relative to the high-frequency amplitude at immediately adjacent times" (p. 87).

After the consonant release, the formants of naturally produced stop consonants undergo quite rapid frequency transitions. In all three syllables displayed in Figure 3.4, the F_1 transition is rising. However, the directions of the F_2 and F_3 transitions clearly differ across the three place values: for /ba/ F_2 and F_3 are rising; for /da/ F_2 and F_3 are falling; and for /ga/ F_2 is falling, and F_3 is rising, from a frequency location near that of the release burst. Because formant transitions reflect the change of vocal tract shape from consonant to vowel (or vowel to consonant), it is not surprising that frequency extents and even directions of the transitions are not invariant properties of particular consonants. Nevertheless, F_2 and F_3 transitions are highly effective cues for perceived place of articulation (Liberman et al. 1954; Harris et al. 1958).

There have been several attempts to identify time-dependent or relational spectral properties that may serve as invariant cues to consonant place. For example, Kewley-Port (1983) described three such properties: tilt of the spectrum at burst onset (bilabials have a spectrum that falls or remains level at higher frequencies; alveolars have a rising spectrum); late onset of low-frequency energy (velars have a delayed F_1 onset relative to the higher formants; bilabials do not); mid-frequency peaks extending over time (velars have this property; bilabials and alveolars do not). Kewley-Port et al. (1983) reported that synthetic stimuli that preserved these dynamic properties were identified significantly better by listeners than stimuli that preserved only the static spectral properties proposed by Stevens and Blumstein (1978).

Sussman and his colleagues (Sussman 1991; Sussman et al. 1991, 1993) have proposed a different set of relational invariants for consonant place. From measurements of naturally produced tokens of /bVt/, /dVt/, and /gVt/ with 10 different vowels, they plotted F_2 onset frequency as a function of the F_2 value of the mid-vowel nucleus. For each of the three place categories, the plots were highly linear and showed relatively little scatter within or between talkers. Moreover, the regression functions, or "locus equations," for these plots intersected only in regions where there were few tokens represented. Thus, the slopes and y-intercepts of the locus equations define distinct regions in the F_2-onset × F_2-vowel space that are unique to each initial stop place category. Follow-up experiments with synthetic speech (Fruchter 1994) suggest that proximity to the relevant locus equation is a fairly good predictor of listeners' judgments of place.

3.5 *The Feature [Voice]*

The feature [voice] refers articulatorily to the presence or absence or, more generally, the relative timing of vocal fold vibration, or voicing. To produce voicing, the vocal folds must be positioned relatively close together and there must be sufficiently greater air pressure below the folds than above. In English, [voice] is a distinctive feature for oral stops, fricatives, and affricates, with the following [+/–voice] contrasting pairs: /b/ versus /p/, /d/ versus /t/, /g/ versus /k/, /f/ versus /v/, /θ/ (as in "*th*in") versus /ð/ (as in "*th*en"), /s/ versus /z/, /ʃ/ versus /ʒ/, and /ʧ/ versus /ʤ/.

In a cross-language study of word-initial stop consonants, Lisker and Abramson (1964) measured *voice onset time* (VOT), the interval between the consonant release and the onset of voicing. In all languages examined, the [+/–voice] distinction was acoustically well specified by differences in VOT. Moreover, across the entire data set, the VOT values were distributed into three distinct phonetic categories: (1) voicing onset significantly precedes the consonant release (conventionally represented as a negative VOT), producing a low-frequency "voice bar" during the consonant constriction interval; (2) voicing onset coincides with or lags shortly (under 30 ms) after the release; and (3) voicing onset lags significantly (over 40 ms) after the release. In some languages (e.g., Dutch, Spanish, and Tamil), [+voice] and [–voice] are realized as categories 1 and 2, respectively, whereas in others (e.g., Cantonese) they are realized as categories 2 and 3, respectively. Speakers of English use either category 1 or 2 to implement [+voice] and category 3 to implement [–voice], while Thai speakers have a three-way phonemic distinction among categories 1, 2, and 3. Figure 3.5

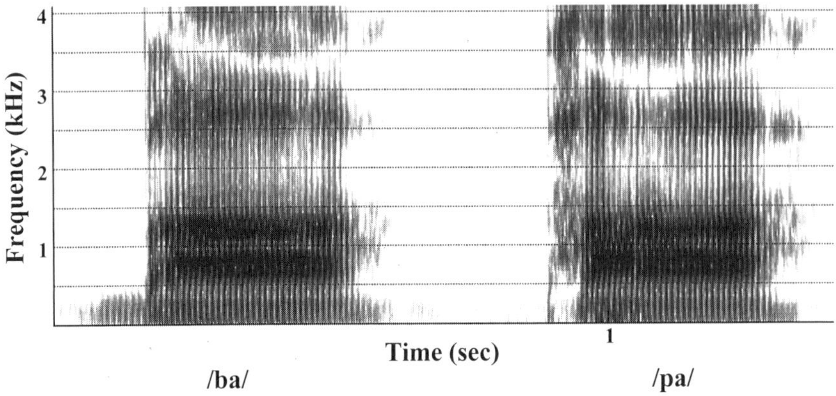

FIGURE 3.5. Spectrograms of /ba/ and /pa/.

shows spectrograms of the syllables /ba/ and /pa/, illustrating the differences in VOT for the English [+/–voice] contrast in word-initial position.

Although the mapping between the [+/–voice] distinction and phonetic categories is not invariant across languages, or even within a language across different utterance positions and stress levels, the [+voice] member of a minimal-pair contrast generally has a smaller VOT value (i.e., less positive or more negative) than the [–voice] member, all other things being equal (Kingston and Diehl 1994). In various studies (e.g., Lisker and Abramson 1970; Lisker 1975), VOT has been shown to be a highly effective perceptual cue for the [+/–voice] distinction.

There are at least four acoustic correlates of positive VOT values, where voicing onset follows the consonant release. First, there is no low-frequency energy (voice bar) during the consonant constriction interval, except perhaps as a brief carryover of voicing from a preceding [+voice] segment. Second, during the VOT interval the first formant is severely attenuated, delaying its effective onset to the start of voicing. Third, because of this delayed onset of F_1, and because the frequency of F_1 rises for stop consonants following the release, the onset frequency of F_1 tends to increase at longer values of VOT. Fourth, during the VOT interval the higher formants are excited aperiodically, first by the rapid lowering of oral air pressure at the moment of consonant release (producing the short release burst), next by frication noise near the point of constriction, and finally by rapid turbulent airflow through the open vocal folds (Fant 1973). (The term *aspiration* technically refers only to the third of these aperiodic sources, but in practice it is often used to denote the entire aperiodic interval from release to the onset of voicing, i.e., the VOT interval.) Perceptual studies have shown that each of these four acoustic correlates of VOT independently affects [+/–voice] judgments. Specifically, listeners make more [–voice] identification responses when voicing is absent or reduced during consonant constriction (Lisker 1986), at longer delays of F_1 onset (Liberman et al. 1958), at higher F_1 onset frequencies (Lisker 1975; Summerfield and Haggard 1977; Kluender 1991), and with longer or more intense intervals of aspiration (Repp 1979).

Several other acoustic correlates of the [voice] feature should be noted. First, across languages, differences in fundamental frequency (f_0) in the vicinity of the consonant are widely attested, with lower f_0 values for [+voice] than for [−voice] consonants (House and Fairbanks 1953; Lehiste and Peterson 1961; Kohler 1982; Petersen 1983; Silverman 1987). Correspondingly, in various perceptual studies, a lower f_0 near the consonant has been shown to increase [+voice] judgments of listeners (Fujimura 1971; Haggard et al. 1970, 1981; Diehl and Molis 1995). Second, in word-medial and -final poststress positions, [+voice] consonant constriction or closure intervals tend to be significantly shorter than those of their [−voice] counterparts (Lisker 1972; Pickett 1980), and closure duration is an effective perceptual cue for the [+/−voice] distinction (Lisker 1957; Parker et al. 1986). Third, in these same utterance positions, the preceding vowel tends to be longer for [+voice] than for [−voice] consonants (House and Fairbanks 1953; Peterson and Lehiste 1960; Chen 1970), and variation in vowel duration is sufficient to signal the [+/−voice] contrast (Denes 1955; Raphael 1972; Kluender et al. 1988). Fourth, F_1 tends to be lower in frequency during the preceding vowel when a syllable-final consonant is [+voice] rather than [−voice] (Summers 1987), and a lower vowel F_1 value correspondingly yields more syllable-final [+voice] judgments in perceptual experiments (Summers 1988). Figures 3.6 and 3.7 show spectrograms of the [+/−voice] distinction in the word pairs "rapid" versus "rabid" and "bus" versus "buzz."

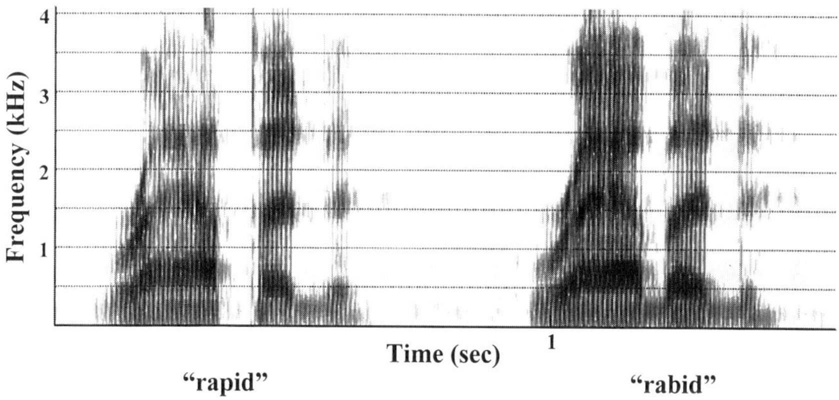

FIGURE 3.6. Spectrograms of "rapid" and "rabid."

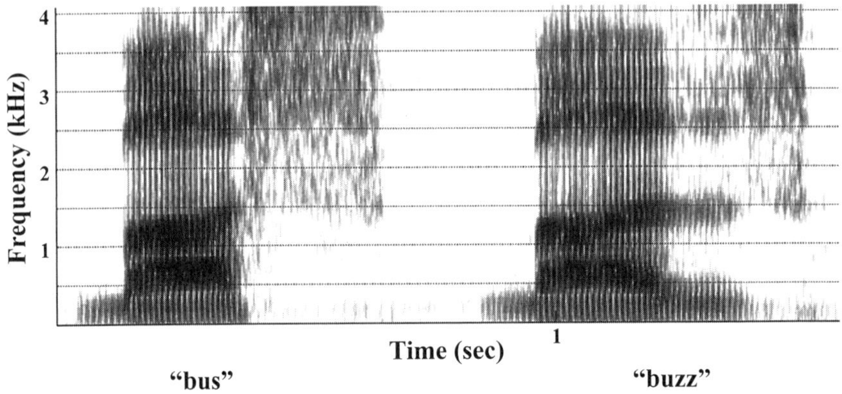

FIGURE 3.7. Spectrograms of "bus" and "buzz."

3.6 The Feature [Strident]

Chomsky and Halle (1968) described [+strident] sounds as "marked by greater noisiness" (p. 329) than their [–strident] counterparts. The greater noise intensity of [+strident] sounds is produced by a rapid airstream directed against an edge such as the lower incisors or the upper lip. The contrast between /s/ and /θ/ ("sin" vs "thin") is an example of the [+/–strident] distinction.

An important subclass of [+strident] sounds are the *sibilants*, which have the additional feature value of [+coronal]. They are characterized by frication noise of particularly high intensity and by a predominance of high spectral frequencies. Although English has an equal number of [+voice] and [–voice] sibilants (viz., /s/, /z/, /ʃ/, /ʒ/, /ʧ/, and /ʤ/), there is a strong cross-language tendency for sibilants to be [–voice] (Maddieson 1984). A likely reason for this is that the close positioning of the vocal folds required for voicing reduces the airflow through the glottis (the space between the folds), and this in turn reduces the intensity of the frication noise (Balise and Diehl 1994). Because high-intensity noisiness is the distinctive acoustic characteristic of [+strident] sounds, especially sibilants, the presence of the [+voice] feature reduces the contrast between these sounds and their [–strident] counterparts.

3.7 Vowel Features

Traditional phonetic descriptions of vowels have tended to focus on three articulatory dimensions: (1) vertical position of the tongue body relative to, say, the hard palate; (2) horizontal position of the tongue body relative to, say, the back wall of the pharynx; and (3) configuration of the lips as rounded or unrounded. These phonetic dimensions have typically been used to define phonological features such as [high], [low], [back], and [round]. (Recall that in the SPE framework, the tongue body features are applied to both vowels and consonants, in the latter case to describe secondary articulations.)

3.7.1 Vowel Height

Early work on the analysis and synthesis of vowel sounds showed that F_1 decreases with greater vowel height (Chiba and Kajiyama 1941; Potter et al. 1947; Potter and Steinberg 1950; Peterson and Barney 1952), and that variation in F_1 is an important cue to differences in perceived vowel height (Delattre et al. 1952; Miller 1953). It was also established that f_0 tends to vary directly with vowel height (Peterson and Barney 1952; House and Fairbanks 1953; Lehiste and Peterson 1961; Lehiste 1970) and that higher f_0 values in synthetic vowels produce upward shifts in perceived height (Potter and Steinberg 1950; Miller 1953).

Traunmüller (1981) presented synthetic vowels (in which F_1 and f_0 were varied independently) to speakers of a German dialect with five distinct height categories, and found that the distance between F_1 and f_0 in Bark units (Zwicker and Feldkeller 1967) was a nearly invariant correlate of perceived height, with smaller F_1-f_0 Bark distances yielding vowels of greater perceived height. Similar results were obtained for synthetic Swedish vowels (Traunmüller 1985). Consistent with these perceptual findings, Syrdal (1985) and Syrdal and Gopal (1986) analyzed two large data sets of American English vowels and reported that [–back] (/i/ as in "beet," /ɪ/ as in "bit," /ɛ/ as in "bet," and /æ/ as in "bat") and [+back] (/u/ as in "boot," /ʊ/ as in "book," /ɔ/ as in "bought," and /a/ as in "hot") vowel height series were both ordered monotonically with respect to mean F_1-f_0 Bark distance. They also noted that an F_1-f_0 Bark distance of about 3 Bark corresponds to the line of demarcation between [+high] vowels such as /ɪ/ and [–high] vowels such as /ɛ/.

3.7.2 Vowel Backness

The same early studies that showed F_1 to be a primary correlate of vowel height also showed F_2 to be an important correlate of vowel backness, with lower F_2 values corresponding to a more retracted tongue body position. Moreover, experiments with two-formant synthetic vowels showed that variation in F_2 alone was sufficient to cue distinctions in perceived backness (Delattre et al. 1952). F_3 also varies with vowel backness, being lower for the [+back] vowels /u/ and /ʊ/ than for the [–back] vowels /i/ and /ɪ/; however, the relative degree of variation is considerably smaller for F_3 than for F_2 (Peterson and Barney 1952). Syrdal (1985) and Syrdal and Gopal (1986) reported that for American English vowels, vowel backness is most clearly and invariantly related to the distance between F_3 and F_2 in Bark units, and that the line of demarcation between the [+back] and [–back] categories occurs at an F_3-F_2 distance of about 3 Bark.

3.7.3 The Feature [Round]

Lip rounding typically involves two components: a protrusion of the lips that lengthens the oral cavity, and a constriction of the lip aperture. Both of these components of the rounding gesture have the effect of lowering F_2 for back vowels (Stevens et al. 1986) and F_2 and F_3 for front vowels (Stevens 1989). Across languages, about 94% of front vowels are produced without lip rounding, and about the same percentage of back vowels are produced with rounding. As noted by Trubetzkoy (1939) and many others since, a likely reason for the strong covariation between tongue position and lip configuration is that the auditory distinctiveness of vowel categories is thereby enhanced. Specifically, a retracted tongue body and lip rounding yield maximal lowering of F_2 (what Trubetzkoy termed a maximally "dark" vowel timbre), while a fronted tongue and lip spreading produce maximal raising of F_2 (a maximally "clear" timbre), other parameters being equal.

3.7.4 Alternative Characterizations of Vowel Information

In the above discussion of vowel features, the important acoustic correlates were assumed to be frequencies of the lower formants (F_1, F_2, F_3) and f_0, or the relations among these frequencies. This assumption is widely held among speech researchers (e.g., Peterson and Barney 1952; Fant 1960; Chistovich et al. 1979; Syrdal and Gopal 1986; Miller 1989; Nearey 1989); however, it has not gone unchallenged. For example, Bladon (1982) criticized formant-based descriptions of speech on three counts: reduction, determinacy, and perceptual adequacy. According to the reduction objection, a purely formant-based description eliminates linguistically important information, such as formant-bandwidth cues for nasalization. The determinacy objection is based on the well-known difficulty of locating all and only the formant peaks by instrumental means. For example, two formant peaks that are very close together may not be physically distinguishable. According to the perceptual adequacy objection, listeners' judgments of perceptual distance among vowels are well predicted by distances among overall spectral shapes (Bladon and Lindblom 1981) but not necessarily by distances among formant frequencies (Longchamp 1981). In light of these difficulties, Bladon favored a whole-spectrum approach to the descriptions of speech sounds, including vowels.

Another argument against formant-based approaches, related to Bladon's determinacy objection, is due to Klatt (1982). He noted that when automatic formant trackers make errors, they are usually large ones based on omitting formants altogether or detecting spurious ones. In contrast, human errors in vowel identification almost always involve confusions between spectrally adjacent vowel categories. Such perceptual results are readily accommodated by a whole-spectrum approach to vowel description. For additional evidence favoring spectral shape properties over formant frequencies as appropriate descriptors for vowels, see Zahorian and Jagharghi (1993) and Ito et al. (2001).

However, whole-spectrum approaches are themselves open to serious criticisms. The most important of these is that although formant frequencies are not the only parameters that influence the phonetic quality of vowels, they appear to be the most important. For example, Klatt (1982) had listeners judge the psychophysical and phonetic distances among synthetic vowel tokens that differed in formant frequencies, formant bandwidths, phase relations of spectral components, spectral tilt, and several other parameters. Although most of these parameters had significant effects on judged psychophysical distance, only formant frequencies contributed significantly to judged phonetic distance.

Noting the problems associated with both the formant-based and whole-spectrum approaches, Hillenbrand and Houde (1995) proposed a compromise model that extracts properties of spectral shape rather than formants frequencies per se, but that weights the contributions of spectral peaks and shoulders (the edges of spectral plateaus) much more highly than other spectral properties. While quite promising, this approach remains to be thoroughly tested.

4. Sources of Acoustic Variation in the Realization of Feature Contrasts

In considering how feature contrasts are realized in terms of articulation, acoustics, or auditory excitation patterns, we immediately come up against one of the key characteristics of speech, namely, its variability. Pronunciations vary in innumerable ways and for a great many reasons.

Speaker identity is one source of variation. We are good at recognizing people by simply listening to them. Different speakers sound different, although they speak the same dialect and utter phonetic segments with identical featural specifications: the same syllables, words, and phrases. To the speech researcher, there is a fundamental challenge in the use of the word *same* here. For when analyzed physically, it turns out that "identical" speech samples from different speakers are far from identical. For one thing, speakers are built differently. Age and gender are correlated with anatomical and physiological variations, such as differences in the size of the vocal tract and the properties of the vocal folds. Was the person calling old or young, male or female? Usually, we are able to tell correctly from the voice alone, in other words, from cues in the acoustic signal impinging on our ears.

If the physical shape of speech is so strongly speaker-dependent, what are the acoustic events that account for the fact that we hear the "same words" whether they are produced by speaker A or speaker B? That is in fact a very important question in all kinds of contemporary research on speech. It should be noted that the problem of variability remains a major one even if we limit our focus to the speech of a single person. A few moments' reflection will convince us that there is an extremely large number of ways in which the syllables and phonemes of the "same" phonetic forms could be spoken.

For example, usually without being fully aware of it, we speak in a manner that depends on whom we are talking to and on the situation that we are in. Note the distinctive characteristics of various speaking styles such as talking to a baby or a dog, or to someone who is hard of hearing, or who has only a partial command of the language spoken or who speaks a different dialect. Moreover, we spontaneously raise vocal effort in response to noisy conditions. We articulate more carefully and more slowly in addressing a large audience under formal conditions than when chatting with an old acquaintance. In solving a problem, or looking for something, we tend to mumble and speak more to ourselves than to those present, often with drastic reduction of clarity and intelligibility as a result. The way we sound is affected by how we feel, our voices reflecting the state of our minds and bodies. Clearly, the speech of a given individual mirrors the intricate interplay of an extremely large number of communicative, social, cognitive, and physiological factors.

4.1 Phonetic Context: Coarticulation and Reduction

It is possible to narrow the topic still further by considering samples from a single person's speech produced in a specific speaking style, at a particular vocal effort and fixed tempo, and from a list of test items, not spontaneously generated by the speaker but chosen by the experimenter. With such a narrow focus the variability contributed by stylistic factors is kept at a minimum; such a speaking style is known as "laboratory speech," probably the type of spoken materials that has so far been studied the most by phoneticians. Even under restricted lab conditions, the articulatory and acoustic correlates of phonological units exhibit extensive variations arising from the fact that these entities are not clearly delimited one by one in a neat sequence, but are produced as a seamless stream of movements that overlap in time and whose physical shapes depend crucially on how the language in question builds its syllables and how it uses prosodic dimensions such as timing, stress, intonation, and tonal phenomena.

This temporal interaction between the realizations of successive units is present irrespective of the language spoken and whether we observe speech as an articulatory or acoustic phenomenon. It is known as *coarticulation* and is exemplified in the following comparisons. Phonologically, the first (and second) vowel of "blue wool" is classified as [+back], as is that of "choosy." However, phonetically, the /u/ in "choosy" is normally pronounced with a much more anterior variant owing to the influence of the surrounding sounds. Acoustically, such coarticulatory effects give rise to variations in the formant patterns, a posterior /u/ of "blue wool" showing an F_2 several hundred Hz lower than for the fronted /u/ of "choosy."

4.1.1 Effects in Stop Consonants

The situation is analogous for consonants. The /k/ of "key" comes out as fronted in the context of /i/, a [–back] vowel, whereas that of "coo" is more posterior in the environment of the [+back] vowel /u/. For a fronted /k/ as in "key," the noise burst would typically be found at about 3 kHz, near F_3 of the following /i/, whereas in /ku/ it would be located in the region around 1300 Hz, closer to F_2. These examples suggest the same mechanism for vowels and consonants. The movements associated with (the features of) a given phonetic segment are not completed before the articulatory activities for the next unit are initiated. As a result, there is overlap and contextual interaction, in other words "coarticulation."

Coarticulation is responsible for a large portion of intraspeaker phonetic variability. Although it is a much researched topic, so far no final account of its articulatory origins and perceptual function has been unanimously embraced by all investigators. We can illustrate its problematic nature by briefly reviewing a few classical studies. The investigation of Öhman (1966) is an early demonstration that the transitional patterns of consonants such as /b/, /d/, and /g/ exhibit strong and systematic coarticulation effects across symmetrical and asymmetrical vowel contexts. In parallel work based on cineradiographic observations, Öhman (1967) represented the vocal tract as an articulatory profile and proposed the following formula as an attempt to model coarticulation quantitatively:

$$s(x,t) = v(x) + k(t)[c(x) - v(x)]w_c(x) \tag{1}$$

Here x represents position along the vocal tract, and t is time. Equation 1 says that, at any given moment in time, the shape of the tongue, $s(x)$, is a linear combination of a vowel shape, $v(x)$, and a consonant shape, $c(x)$. As the interpolation term, $k(t)$, goes from 0 to 1, a movement is generated that begins with a pure vowel, $v(x)$, and then changes into a consonant configuration that will more and more retain aspects of the vowel contour as the value of a weighting function, $w_c(x)$ goes from 1 to 0. We can think of $s(x)$, $c(x)$, and $v(x)$ as tables that, for each position x, along the vocal tract, indicate the distance of the tongue contour from a fixed reference point. In the $w_c(x)$ table, each x value is associated with a coefficient ranging between 0 and 1 that describes the extent to which $c(x)$ resists coarticulation at the location specified by x. For example, at $k = 1$, we see from Equation 1 that $w_c(x) = 0$ reduces the expression to $v(x)$, but when $w_c(x) = 1$, it takes the value of $c(x)$. In VCV (i.e., vowel + consonant + vowel) sequences with $C = [d]$, $w_c(x)$ would be given the value of 1 (i.e., no coarticulation) at the place of articulation, but exhibit values in between 0 and 1 elsewhere along the tract.

With the aid of this model, Öhman succeeded in deriving observed context-dependent shape variations for each phoneme from a single, nonvarying description of the underlying vocal tract shape. That is, for a given [$V_1 dV_2$] sequence, each vowel had its unique $v(x)$ contour, and the consonant [d] was specified by a single context-independent $c(x)$ and its associated "coarticulation resistance" function $w_c(x)$.

Does the nature of coarticulation, as revealed by the preceding analysis, imply that speaking is structured primarily in terms of articulatory rather than acoustic/auditory goals? Does it suggest that "features" are best defined at an articulatory level rather than as perceptual attributes? The answer given in this chapter is an unequivocal no, but to some investigators, there is a definite possibility that phonetic invariance might be present at an articulatory level, but is absent in the acoustics owing to the coproduction of speech movements. Recent interest in articulatory recovery appears to have gained a lot of momentum from that idea. This is a paradigm (e.g., McGowan 1994) aimed at defining a general inverse mapping from spectral information to articulatory parameters, an operation that would seem to convert a highly context-dependent acoustic signal into another representation, which, by hypothesis, ought to exhibit less context-dependence and therefore improve the chances of correct recognition. Theoretical approaches such as the motor theory (Liberman and Mattingly 1985, 1989) and direct realism (Fowler 1986, 1994), as well as projects on speech recognition, low bit-rate coding, and text-to-speech systems (Schroeter and Sondhi 1992) have converged on this theme. (See also papers on "articulatory recovery" in J Acoust Soc Am, 1996;99:1680–1741, as well as Avendaño et al., Chapter 2; Morgan et al., Chapter 6).

The answer to be developed in this chapter is that "features" are neither exclusively articulatory nor exclusively perceptual. They are to be understood as products of both production and perception constraints. But before we reach that final conclusion, a more complete picture of coarticulation needs to be given. A crucial aspect has to do with its perceptual consequences. What task does coarticulation present to the listener? Is it generally true, as tacitly assumed in projects on articulatory recovery, that acoustic signals show more context-dependence than articulatory patterns?

For an answer, we return to the transitional patterns of /b/, /d/, and /g/. Öhman's 1966 study used V_1CV_2 sequences with all possible combinations of the segments /b d g/ and /y ø a o u/ as spoken by a single Swedish subject. It showed that the acoustic correlate of place of articulation is not a constant formant pattern, a fixed set of locus frequencies, as assumed by (the strong version of) the "locus theory" (Liberman et al. 1954), but that, for a given place, the F_2 and F_3 frequencies as observed at V_1C and CV_2 boundaries depend strongly on the identities of the V_1 and V_2 vowels. At the CV boundary, formant patterns depend not only on the identity of V_2 but also on V_1. Conversely at the VC boundary, they depend on both V_1 and V_2. Because of the strong vowel-dependence, the F_2 and F_3 ranges for /b/, /d/, and /g/ were found to overlap extensively, and, accordingly, it was not possible to describe each place with a single nonvarying formant pattern, a fact that would at first glance tend to support the view of Liberman and Mattingly (1985) "that there is simply no way to define a phonetic category in purely acoustic terms" (p. 12).

However, a detailed examination of the acoustic facts suggests that the conclusion of Liberman and Mattingly is not valid. It involves replotting Öhman's (1966) average values three-dimensionally (Lindblom 1996). This diagram has the onset of F_2 at the CV_2 boundary along the x-axis, the F_3 onset at the CV_2 boundary along the y-axis, and F_2 at the V_2 steady state along the z-axis. When the measurements from all the test words are included in the diagram and enclosed by smooth curves, three elongated cloud-like shapes emerge. Significantly, the three configurations do not overlap.

If we assume that a listener trying to identify VCV utterances has access to at least the above-mentioned three parameters, there ought to be sufficient information in the acoustic signal for the listener to be able to disambiguate the place of the consonants despite significant coarticulation effects. Needless to say, the three dimensions selected do not in any way constitute an exhaustive list of signal attributes that might convey place information. The spectral dynamics of the stop releases is one obvious omission. Presumably, adding such dimensions to the consonant space would be an effective means of further increasing the separation of the three categories. That assumption implies that the three-dimensional diagram underestimates the actual perceptual distinctiveness of stops in VCV contexts.

Coarticulation may thus eliminate absolute acoustically invariant correlates of the three stop categories, but, when represented in a multidimensional acoustic space, their phonetic correlates nevertheless remain distinct, meeting the condition of "sufficient contrast." Also, an articulatory account of coarticulation may at first seem conceptually simple and attractive, but the preceding analysis indicates that an equally simple and meaningful picture can be obtained at the acoustic level.

4.1.2 Effects on Vowel Formants

Above, we noted that the articulatory activity for one phonetic segment is never quite finished before movements for a following segment are begun. We saw how this general principle of overlap and anticipation gives rise to contextual variations in both the production and acoustics of consonants.

Vowels are subject to the same mode of motor organization. They too show the effect of their environment, particularly strongly when they have short durations and differ markedly as a function of context. In many Americans' slow and careful pronunciation of "New York," the /u/ of the first syllable would normally be said with the tongue in a back position and with rounded lips. In faster and more casual speech, however, the sequence is likely to come out as [nyjork] with a front variant of the [+back] /u/. The quality change can be explained as follows. The articulators' first task is the /n/, which is made with a lowered velum and the tongue tip in contact with the alveolar ridge. To accomplish the /n/ closure the tongue body synergistically cooperates with the tongue tip by moving forward. For /u/ it moves back and for /j/ it comes forward again. At slow speaking rates, the neural motor signals for /n/, /u/, and /j/ can be assumed to be sufficiently separated in time to allow the tongue body to approach rather closely the target configurations intended for the front-back-front movement sequence. However, when they arrive in close temporal succession, the overlap between the /n/, /u/, and /j/ gestures is increased. The tongue begins its front-back motion to /u/, but is interrupted by the command telling it to make the /j/ by once more assuming a front position. As a consequence the tongue undershoots its target, and, since during these events the lips remain rounded, the result is that the intended /u/ is realized as an [y].

The process just described is known as "vowel reduction." Its acoustic manifestations have been studied experimentally a great deal during the past decades and are often referred to as "formant undershoot," signifying failure of formants to reach underlying ideal "target" values. Vowel reduction can be seen as a consequence of general biomechanical properties that the speech mechanism shares with other motor systems. From such a vantage point, articulators are commonly analyzed as strongly damped mechanical oscillators (Laboissière et al. 1995; Saltzman 1995; Wilhelms-Tricarico and Perkell 1995). When activated by muscular forces, they do not respond instantaneously but behave as rather sluggish systems with virtual mass, damping, and elasticity, which determine the specific time constants of the individual articulatory structures (Boubana 1995). As a result, an articulatory movement from A to B typically unfolds gradually following a more or less S-shaped curve. Dynamic constraints of this type play an important role in shaping human speech both as an on-line phenomenon and at the level of phonological sound patterns. It is largely because of them, and their interaction with informational and communicative factors, that speech sounds exhibit such a great variety of articulatory and acoustic shapes.

The biomechanical perspective provides important clues as to how we should go about describing vowel reduction quantitatively. An early study (Lindblom 1963) examined the formant patterns of eight Swedish short vowels embedded in /b_b/, /d_d/, and /g_g/ frames and varied in duration by the use of carrier phrases with different stress patterns. For both F_1 and F_2, systematic undershoot effects were observed directed away from hypothetical target values toward the formant frequencies of the adjacent consonants. The magnitude of those displacements depended on two factors: the duration of the vowel and the extent of the CV formant transition (the "locus-target" distance). The largest shifts were thus associated with short durations and large "locus-target" distances. Similar undershoot effects were found in a comparison of stress and tempo with respect to their effect on vowel reduction. It was concluded that duration, whether stress- or tempo-controlled, seemed to be the primary determinant of vowel reduction.

However, subsequent biomechanical analyses (Lindblom 1983; Nelson 1983; Nelson et al. 1984) have suggested several refinements of the original duration- and context-dependent undershoot model. Although articulatory time constants indeed set significant limits on both extent and rate of movements, speakers do have a choice. They have the possibility of overcoming those limitations by varying how forcefully they articulate, which implies that a short vowel duration need not necessarily produce undershoot, if the articulatory movement toward the vowel is executed with sufficient force and, hence, with enough speed.

In conformity with that analysis, the primacy of duration as a determinant of formant undershoot has been challenged in a large number of studies, among others those of Kuehn and Moll (1976), Gay (1978), Nord (1975, 1986), Flege (1988), Engstrand (1988), Engstrand and Krull (1989), van Son and Pols (1990, 1992), and Fourakis (1991). Some have even gone so far as to suggest that vowel duration should not be given a causal role at all (van Bergem 1995).

Conceivably, the lack of reports in the literature of substantial duration-dependent formant displacement effects can be attributed to several factors. First, most of the test syllables investigated are likely to have had transitions covering primarily moderate "locus-target" distances. Second, to predict formant undershoot successfully, it is necessary to take movement/formant velocity into account as shown by Kuehn and Moll (1976), Flege (1988), and others, and as suggested by biomechanical considerations.

4.2 Variations of Speaking Style and Stress

Several attempts have been made to revise the undershoot model along these lines. Moon and Lindblom (1994) had five American English speakers produce words with one, two, and three syllables in which the initial stressed syllable was /wil/, /wɪl/, /wɛl/, or /weˡl/. The speakers were first asked to produce the test words in isolation at a comfortable rate and vocal effort ("citation-form speech"), and then to pronounce the same words "as clearly as possible" ("clear speech"). The differences in word length gave rise to a fairly wide range of vowel durations. Large undershoot effects were observed, especially at short durations, for an F_2 sampled at the vowel midpoint.

The original (Lindblom 1963) model was fitted to the measurements on the preliminary assumption that the degree of undershoot depends only on two factors: vowel duration and context. However, in clear speech undershoot effects were less marked, often despite short vowel durations. Speakers achieved this by increasing durations and by speeding up the F_2 transition from /w/ into the following vowel. In some instances they also chose to increase the F_2 target value. These findings were taken to suggest that speakers responded to the "clear speech" task by articulating more energetically, thereby generating faster formant transitions and thus compensating for undershoot. On the basis of these results, a model was proposed with three rather than two factors, namely, duration, context, and articulatory effort as reflected by formant velocity.

Two studies shed further light on that proposal. Brownlee (1996) investigated the role of stress in reduction phenomena. A set of /wil/, /wɪl/, and /wɛl/ test syllables were recorded from three speakers. Formant displacements were measured as a function of four degrees of stress. A substantial improvement in the undershoot predictions was reported when the original (Lindblom 1963) model was modified to include the velocity of the initial formant transition of the [wVl] syllables. Interestingly, there was a pattern of increasing velocity values for a given syllable as a function of increasing stress.

Lindblom et al. (1996) used three approximately 25-minute long recordings of informal spontaneous conversations from three male Swedish talkers. All occurrences of each vowel were analyzed regardless of consonantal context. Predictions of vowel formant patterns were evaluated taking a number of factors into account: (1) vowel duration, (2) onset of initial formant transition, (3) end point of final formant transition, (4) formant velocity at initial transition onset, and (5) final transition endpoint. Predictive performance improved as more factors were incorporated. The final model predicts the formant value of the vowel as equal to the formant target value (T) plus four correction terms associated with the effects of initial and final contexts and initial and final formant velocities. The original undershoot model uses only the first two of those factors. Adding the other terms improved predictions dramatically. Since only a single target value was used for each vowel phoneme (obtained from the citation forms), it can be concluded that the observed formant variations were caused primarily by the interaction of durational and contextual factors rather than by phonological allophone selections. It also lends very strong support to the view that vowel reduction can be modeled on the basis of biomechanical considerations.

The points we can make about vowels are similar to the ones we made in discussing consonants. Reduction processes eliminate absolute acoustic invariant correlates of individual vowel categories. Thus, one might perhaps be tempted to argue that invariance is articulatory, hidden under the phonetic surface and to be found only at the level of the talker's intended gestures. However, the evidence shows that speakers behave as if they realize the perceptual dangers of phonetic variations are becoming too extensive.

They adapt by speaking more clearly and by reorganizing their articulation according to listener-oriented criteria.

5. The Traditional Approach to Distinctive Features

The traditional approach to features may be classified as *axiomatic* in the sense that new features are proposed on the basis of how phonological contrasts pattern among the world's languages. This method runs as a common theme during the historical development of distinctive-feature theory from Trubetzkoy to present times (see the introductory historical outline in section 2). In other words, features are postulated rather than derived entities. The motivation for their existence in linguistic description is empirical, not theoretical.

The axiomatic approach can be contrasted with a *deductive* treatment of sound structure, which, so far, represents a less traveled road in linguistics. The deductive approach aims at providing theoretical motivations for features, seeking their origins in facts separate from the linguist's primary data. It derives features from behavioral constraints on language use as manifested in speaking, listening, and learning to speak. Accordingly, features are deduced and independently motivated entitities, rather than merely data-driven descriptors.

The distinction between axiomatic and deductive treatments of sound structure can be illuminated by making an analogy between "features" and "formants." Taking an axiomatic approach to formants, a phonetician would justify the use of this hypothetical notion in terms of spectral properties that can frequently be observed on acoustic records, whereas working deductively, he or she would derive the formant from physics and then apply it to the description of speech sounds.

Evidently, modern phonetics is in a position to benefit from the existence of a well-developed theory of acoustics, but cannot invoke the analogous theoretical support in phonology to the same extent. The reasons for this difference will not be discussed here. Suffice it to mention one important underlying factor: the form-substance distinction (see section 2), which assigns to phonology the task of extracting the minimum phonetic information needed to define the basic formal building blocks of sound structure. Phonetics, on the other hand, does its job by borrowing the entities of phonological analyses and by investigating how those units are actualized in phonetic behavior (production, perception, and development). By limiting observations to those phonetic attributes that are distinctive (in other words, to the properties that a language uses to support differences in meaning), phonologists have been able to solve, in a principled way, a number of problems associated with the details and variability of actual phonetic behavior. The solution implies a stripping away of phonetic substance thereby making it irrelevant to further analyses of phonological structure from that point on.[1] In developing this procedure, linguistics has obtained a powerful method for idealizing speech in a principled manner and for extracting a core of linguistically relevant information from phonetic substance. Descriptive problems are solved by substituting discreteness and invariance of postulated units for the continuous changes and the variability of observed speech patterns. Hence, the relationship between phonetics and phonology is not symmetrical. Phonological form takes precedence over phonetic substance. As a further consequence, the axiomatic approach becomes the prevailing method, whereas deductive frameworks are dismissed as being fundamentally at odds with the time-honored "inescapable" form-first, substance-later doctrine (cf. Chomsky 1964, p. 52).

A belief shared by most phoneticians and phonologists is that distinctive features are not totally arbitrary, empty logical categories, but are somehow linked to the production and perception of speech. Few phonologists would today seriously deny the possibility that perceptual, articulatory, and other behavioral constraints are among the factors that contribute to giving distinctive features the properties they exhibit in linguistic analyses. For instance, in Jakobson's vision, distinctive features represented the universal dimensions of phonetic perception available for phonological contrast. According to Chomsky and Halle (1968), in their phonetic function, distinctive features relate to independently controllable aspects of speech production. Accordingly, a role for phonetic substance is readily acknowledged with respect to sound structure and features. (For a more current discussion of the theoretical role of phonetics in phonology, see Myers 1997.) However, despite the in-principle recognition of the relevance of phonetics, the axiomatic strategy of "form first, substance later" remains the standard approach.

Few would deny the historical importance of the form-substance distinction (Saussure 1916). It made the descriptive linguistics of the 20th century possible. It is fundamental to an understanding of the traditional division of labor between phonetics and phonology. However, the logical priority of form (Chomsky 1964) is often questioned, at least implicitly, particularly by behaviorally and experimentally oriented researchers. As suggested above,

---

[1] In the opinion of contemporary linguists:

> The fundamental contribution which Saussure made to the development of linguistics [was] to focus the attention of the linguist on the system of regularities and relations which support the differences among signs, rather than on the details of individual sound and meaning in and of themselves. . . . For Saussure, the detailed information accumulated by phoneticians is of only limited utility for the linguist, since he is primarily interested in the ways in which sound images differ, and thus does not need to know everything the phonetician can tell him. By this move, then, linguists could be emancipated from their growing obsession with phonetic detail." [Anderson 1985, pp. 41–42]

the strengths of the approach accrue from abstracting away from actual language use, stripping away phonetic and other behavioral performance factors, and declaring them, for principled reasons, irrelevant to the study of phonological structure. A legitimate question is whether that step can really be taken with impunity.

Our next aim is to present some attempts to deal with featural structure deductively and to show that, although preliminary, the results exemplify an approach that not only is feasible and productive, but also shows promise of offering deeper explanatory accounts than those available so far within the axiomatic paradigm.

6. Two Deductive Approaches to Segments and Features: Quantal Theory and the Theory of Adaptive Dispersion

In contrast to what we have labeled axiomatic approaches that have traditionally dominated discussions of segments and features, there are two theories, developed during the last 30 years, that may properly be called deductive. In these theories, independently motivated principles are used to derive predictions about preferred segment and feature inventories among the world's languages. The two theories also differ from most traditional approaches in emphasizing that preferred segments and features have *auditory* as well as articulatory bases.

6.1 Quantal Theory (QT)

Stevens's quantal theory (1972, 1989, 1998) is grounded on the observation that nonlinearities exist in the relation between vocal-tract configurations and acoustic outputs, and also between speech signals and auditory responses.

6.1.1 Articulatory-to-Acoustic Transform

Along certain articulatory dimensions, such as length of the back cavity, there are regions where perturbations in that parameter cause small acoustic changes (e.g., in formant frequencies) and other regions where comparable articulatory perturbations produce large acoustic changes. Figure 3.8 presents this situation schematically. These alternating regions of acoustic stability and instability yield conditions for a kind of optimization of a language's phoneme or feature inventory. If a feature is positioned within an acoustically stable region, advantages accrue to both the talker and the listener. For the talker, phonetic realization of the feature requires only modest articulatory precision since a range of articulatory values will correspond to roughly the same acoustic output. For the listener, the output from an acoustically stable region is (approximately) invariant and the

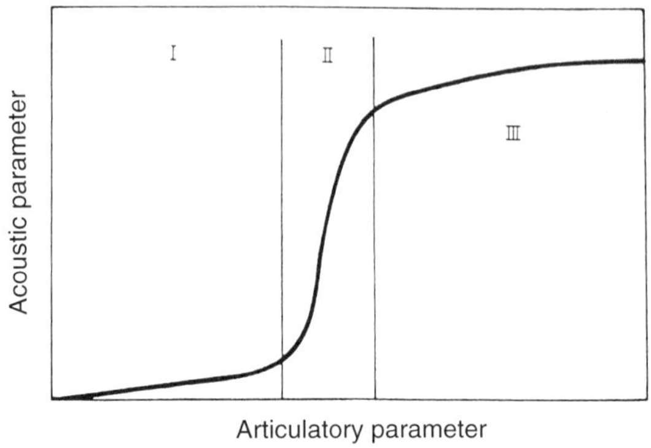

FIGURE 3.8. Schematic representation of a nonlinear relationship between variation of an articulatory parameter on the abscissa and the consequent variation of an acoustic parameter on the ordinate. (From Stevens 1989, with permission of Academic Press.)

perceptual task is therefore reduced to detecting the presence or absence of some invariant property. Another important advantage for the listener is that different feature values tend to be auditorily very distinctive because they are separated by regions of acoustic instability, that is, regions corresponding to a high rate of acoustic change. The convergence of both talker-oriented and listener-oriented selection criteria leads to cross-language preferences for certain "quantal" phonemes and features.

Consider, for example, the acoustic effects of varying back cavity length in the simplified two-tube vocal tract model illustrated in Figure 3.9 when the overall length of the configuration is held constant at 16 cm. The wide-diameter front tube (on the right) simulates an open oral cavity, while the narrow-diameter back tube simulates a constricted pharyngeal cavity. Such a configuration is produced by a low-back tongue body position and unrounded lips. Figure 3.10 shows the frequencies of the first four resonances or formants as the length of the back cavity (l_1) varies (cf. Avendaño et al., chapter 2). When the ratio of the cross-sectional areas of the two tubes, A_1/A_2, is very small, the tubes are decoupled acoustically so that the resonances of one cavity are independent of those of the other. This case is represented by the dashed lines in Figure 3.10. However, when the area ratio of the two tubes is somewhat larger, so that acoustic coupling is non-negligible, with the points of intersection between the front- and back-cavity resonances acoustically realized as formants spaced close together in frequency (see the solid frequency curves in Fig. 3.10). It may be seen that the regions of formant proximity are relatively stable, and intermediate regions are relatively unstable, with respect to small changes in back-cavity length. The region of greatest stability for the first two formants occurs near the point where the back and front cavities are equal in length (viz., 8 cm). Such a configuration corresponds closely to the vowel /a/, one of the three

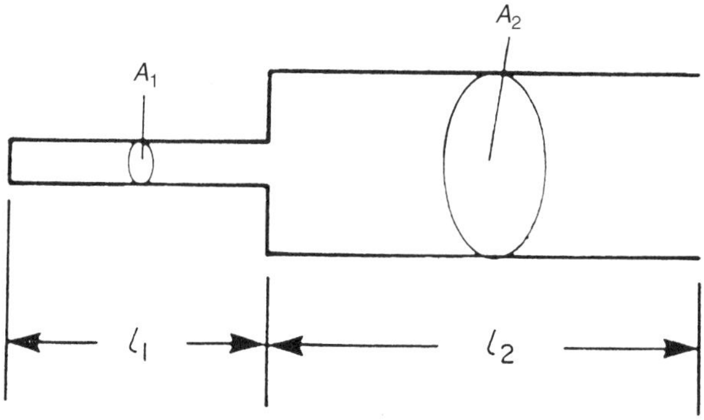

FIGURE 3.9. A two-tube model of the vocal tract. l_1 and l_2 correspond to the lengths, and A_1 and A_2 correspond to the cross-sectional areas, of the back and front cavities, respectively. (From Stevens 1989, with permission of Academic Press.)

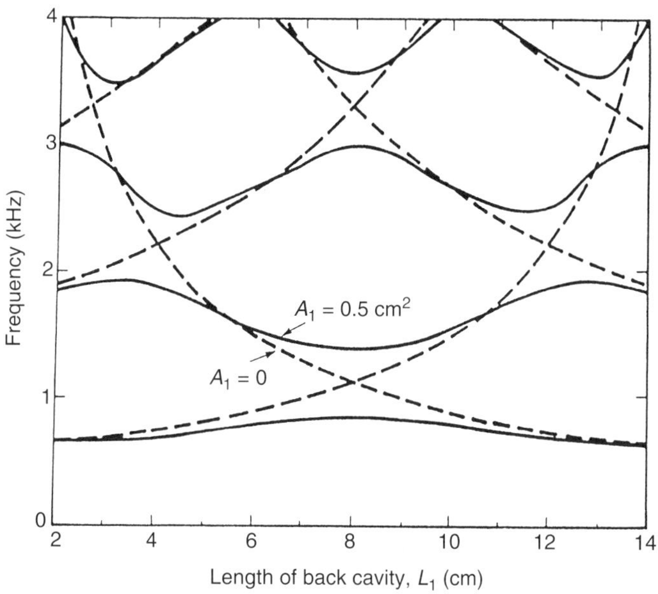

FIGURE 3.10. The first four resonant frequencies for the two-tube model shown in Figure 3.9, as the length l_1 of the back cavity is varied while holding overall length of the configuration constant at 16 cm. Frequencies are shown for two values of back cavity cross-sectional area: A_1 = 0 cm, 0.5 cm. (From Stevens 1989, with permission of Academic Press.)

most widely occurring vowels among the world's languages. The other two most common vowels, /i/ and /u/, similarly correspond to regions of formant stability (and proximity) that are bounded by regions of instability.

It must be emphasized that acoustic stability alone is not sufficient to confer quantal status upon a vowel. The listener-oriented selection criterion of distinctiveness must also be satisfied, which, in QT terms, requires that the vowel be bounded by regions of acoustic instability separating that vowel from adjacent vowels. Consider again the two-tube model of Figure 3.9 and the associated formant curves of Figure 3.10. The effect of enlarging A_1 while keeping A_2 constant is to increase the acoustic coupling between the two cavities, which in turn flattens out the peaks and troughs of the formant curves. In the limit, when A_1 is equal to A_2 (yielding a uniform tube corresponding to a schwa-like vowel, as in the first syllable of "about"), changes in the length of the "back" cavity obviously do not change the configuration at all, and the formant curves will be perfectly flat. Such a configuration is maximally stable, but it does not represent a quantal vowel because there are no bounding regions of instability that confer distinctiveness. An implication of this is that all vowels that are quantal with respect to variation in back cavity length must have fairly weak acoustic coupling between the front and back cavities. Such a condition is met only when some portion of the vocal tract is highly constricted relative to other portions.

Diehl (1989) noted two problems with QT as a basis for explaining preferred vowel categories. The first concerns the claim that quantal vowels are relatively stable. A strong version of this claim would be that there are vocal-tract configurations that are stable with respect to variation in all or at least most of the principal articulatory parameters. Stevens's claim is actually much weaker, defining stability with respect to variation in one particular articulatory dimension, namely, back-cavity length. It is reasonable to ask how stable the quantal configurations are with respect to other parameters. The answer, revealed in Stevens (1989, Figs. 3, 4, and 5), is that those regions that are most stable with respect to variation in back-cavity length turn out to be *least* stable with respect to variation in several other important parameters, including cross-sectional area of the back cavity in configurations such as that of Figure 3.9, and cross-sectional area and length of any constriction separating two larger cavities. This would appear to pose a significant problem for the claim that quantal vowels are relatively stable.

The second problem noted by Diehl (1989) is that QT is not altogether successful at predicting favored vowel inventories among the world's languages. As noted earlier, the three most common vowels are /i/, /a/, and /u/ (Crothers 1978; Maddieson 1984), often referred to as the "point vowels" since they occupy the most extreme points of the vowel space. These three vowels clearly satisfy the quantal criteria, and so their high frequency of occurrence is well predicted by QT. However, while the point vowels are paradigmatic quantal vowels, they are not the only quantal vowels. As indicated in Stevens (1989, Fig. 13), the high, front, unrounded vowel /i/ and the high, front, rounded vowel /y/ (as in the French word "lune") satisfy the quantal selection criteria equally well: each has relatively stable formant frequencies, with F_2 and F_3 in close proximity, and each is bounded by acoustically unstable regions, which enhances auditory distinctiveness. On the basis of QT alone, one would therefore expect /y/ to be about as common cross-linguistically as /i/. But, in fact, /y/ occurs at only about 8% of the frequency of /i/ (Maddieson 1984). Equally problematic for QT is the high frequency of /e/ (or /ɛ/), which, after the point vowels, is one of the most common vowels cross-linguistically (Crothers 1978; Maddieson 1984). For middle front vowels such as /e/, the front and back cavities have relatively similar cross-sectional areas, and they are not separated by a region of constriction (Fant 1960). Accordingly, there is a high degree of acoustic coupling between the front and back cavities. As was noted above, such vowels are relatively stable with respect to variation in back-cavity length, but they are not quantal because they are not bounded by regions of high acoustic instability. To summarize, while QT does a good job of predicting the preferred status of the point vowels, it fails to predict the rarity of /y/ and the high frequency of /e/.

6.1.2 Acoustic-to-Auditory Transform

Although the quantal notion originally applied only to the mapping between vocal tract shapes and acoustic outputs (Stevens 1972), it was later extended to the relation between acoustic signals and auditory responses (Stevens 1989). In the expanded version of QT, there are assumed to be nonlinearities in the auditory system, such that along certain acoustic dimensions, auditorily stable regions are separated from each other by auditorily unstable regions. Phonemes or feature values tend to be located in the stable regions, while an intervening unstable region corresponds to a kind of threshold between two qualitatively different speech percepts. Figure 3.8 schematically characterizes this situation if the abscissa is relabeled as an "acoustic parameter" and the ordinate as an "auditory response."

As an example of a quantal effect in auditory processing of speech, Stevens (1989) refers to the "center of gravity" effect reported by Chistovich and her colleagues (Chistovich and Lublinskaya 1979; Chistovich et al. 1979). In a typical experiment, listeners were asked to adjust the frequency of a single-formant comparison stimulus to match the perceptual quality of a two-formant standard. If the frequencies of the two formants in the standard were more than about 3 Bark apart, the listeners tended to adjust the comparison to equal either F_1 or F_2 of the standard. However, when the frequency distance between the two formants of the standard was less than 3 Bark, listeners set the comparison stimulus to a frequency value intermediate between F_1 and F_2, referred to as the "center of gravity." Chistovich et al. concluded that within 3 Bark, spectral peaks are averaged auditorily to produce a single auditory prominence, while beyond 3 Bark, spectral peaks remain auditorily distinct. Thus, a 3-Bark separation between formants or other spectral peaks appears to define a region of high auditory instability, that is, a quantal threshold.

As discussed earlier, Traunmüller (1981) reported evidence that the Bark distance between F_1 and F_0 (which typically corresponds to a spectral peak) is an invariant correlate of perceived vowel height, at least in a central Bavarian dialect of German. Moreover, Syrdal (1985) and Syrdal and Gopal (1986) found that the zone of demarcation between naturally produced [+high] and [–high] vowels in American English (i.e., between /ɪ/ and /ɛ/ in the front series, and between /ʊ/ and /o/ in the back series) occurs at an F_1-f_0 distance of 3 Bark. (Analogously, [+back] and [–back] vowels of American English were divided at a 3-Bark F_3-F_2 distance.) Consistent with QT, these results may be interpreted to suggest that speakers of American English tend to position their [+high] and [–high] vowel categories on either side of the 3 Bark F_1-f_0 distance so as to exploit the natural quantal threshold afforded by the center of gravity effect.

To provide a more direct perceptual test of this quantal interpretation, Hoemeke and Diehl (1994) had listeners identify three sets of synthetic front vowels varying orthogonally in F_1 and f_0, and ranging perceptually from /i/-/ɪ/, /ɪ/-/ɛ/, and /ɛ/-/æ/. For the /ɪ/-/ɛ/ set, corresponding to the [+high]/[–high] distinction, there was a relatively sharp labeling boundary located at an F_1-f_0 distance of 3 to 3.5 Bark. However, for the other two vowel sets, which occupied regions in which F_1-f_0 distance was always greater than or always less than 3 Bark, identification varied more gradually as a function of F_1-F_0 Bark distance. Hoemeke and Diehl interpreted their results as supporting the existence of a quantal boundary, related to the center of gravity effect, between the [+high] and [–high] vowel categories. However, in a follow-up study, Fahey et al. (1996) failed to find convincing evidence for a similar quantal boundary between [+high] and [–high] categories among back vowels. Instead, the results were consistent with the claim (Traunmüller 1984) that vowel category decisions (including height feature values) are determined by the Bark distances between *any* adjacent spectral peaks (e.g., F_3-F_2, F_2-F_1, and F_1-f_0), with greater perceptual weight given to smaller distances. Thus, evidence for the role of quantal boundaries in vowel perception is mixed (for a review, see Diehl 2000).

A more convincing case for quantal auditory effects may be made with respect to the perception of the [+/–voice] distinction in initial position of stressed syllables. As discussed earlier, VOT is a robust cue for the distinction across many languages. Lisker and Abramson (1970; Abramson and Lisker 1970) showed that perception of VOT is "categorical" in the sense that listeners exhibit (1) a sharp identification boundary between [+voice] and [–voice] categories, and (2) enhanced discriminability near the identification boundary. By itself, categorical perception of speech sounds by adult human listeners provides only weak evidence for the existence of quantal thresholds based on nonlinearities in auditory processing. Greater discriminability at phoneme boundaries might simply reflect listeners' experience in categorizing sounds of their own language. However, several lines of evidence suggest that the VOT boundary separating [+voice] from

[–voice] stops in English and certain other languages corresponds closely to a natural auditory boundary or quantal threshold.

One line of evidence comes from studies of labeling and discrimination of nonspeech analogs of VOT stimuli. Miller et al. (1976) and Pisoni (1977) created VOT analogs by varying the relative temporal onset of noise and buzz segments or of two tones. Labeling functions for these nonspeech stimuli showed sharp boundaries at relative onset values roughly comparable to the VOT category boundaries for speech. Moreover, discriminability of relative onset time was enhanced in the region of the labeling boundaries. Since the stimuli were not perceived as speech-like, performance does not appear to be attributable to the language experience of the listeners.

A second line of evidence for a natural auditory boundary along the VOT dimension comes from studies of speech perception in prelinguistic infants. In the earliest of these studies, Eimas et al. (1971) used a high-amplitude sucking procedure to test VOT perception in 1- and 4-month-old infants. Both age groups displayed elevated discriminability of VOT near the English [+/–voice] boundary relative to regions well within the [+voice] or [–voice] categories. Similar results were also obtained from infants being raised in a Spanish-speaking environment, despite the fact that the Spanish [+/–voice] boundary differs from that of English (Lasky et al. 1975).

Perhaps the most compelling evidence for a natural VOT boundary derives from studies using animal subjects. Kuhl and Miller (1978) trained chinchillas to respond differently to two end-point stimuli of a synthetic VOT series (/da/, 0 ms VOT; and /ta/, 80 ms VOT) and then tested them with stimuli at intermediate values. Identification corresponded almost exactly to that of English-speaking listeners. Further generalization tests with bilabial (/ba/-/pa/) and velar (/ga/-/ka/) VOT stimuli, as well as tests of VOT discriminability, also showed close agreement with the performance of English-speaking adults. Analogous perceptual results were also obtained with macaque monkeys (Kuhl and Padden 1982) and Japanese quail (Kluender 1991).

Figure 3.11 displays the chinchilla identification results from Kuhl and Miller (1978), along with best-fitting functions for adult English speakers. Notice that for both groups of subjects, the identification boundaries for the three place series occur at different locations along the VOT dimension. The most likely reason for this is that, for VOT values near the identification boundaries, the F_1 onset frequency is lowest for the velar series and highest for the bilabial series. Kluender (1991) showed that F_1 onset frequency is a critical parameter in determining the VOT category boundary.

A neural correlate of enhanced discriminability near the VOT category boundary was demonstrated by Sinex et al. (1991). They recorded auditory-nerve responses in chinchilla to stimuli from an alveolar VOT series. For stimuli that were well within either the [+voice] and [–voice] categories, there was considerable response variability across neurons of different characteristic frequencies. However, for a 40-ms VOT stimulus, located near

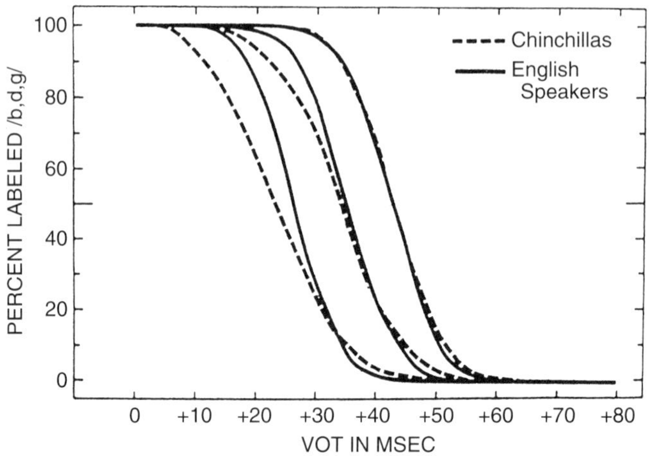

FIGURE 3.11. Human (English-speakers) and chinchilla identification functions for synthetic voice onset time (VOT) stimuli. Functions for both sets of subjects show sharp boundaries in approximately the same locations, with variation in the category boundary as a function of place of articulation. From left to right: labial, alveolar, and velar series. (From Kuhl and Miller 1978, with permission of the first author.)

the [+/–voice] boundary, the response to the onset of voicing was highly synchronized across the same set of neurons. The lower variability of auditory response near the category boundary is a likely basis for greater discriminability in that region. Figure 3.12 displays a comparison of neural population responses to pairs of stimuli for which the VOT difference was 10 ms. Notice the greater separation between the population responses to the pair (30-ms and 40-ms VOT) near the category boundary.

A natural quantal boundary in the 25 to 40-ms VOT region would enhance the distinctiveness of the [+/–voice] contrast for languages such as English and Cantonese, where the contrast is one of long-lag versus short-lag voicing onset (see section 3). However, in languages such as Dutch, Spanish, and Tamil, such a boundary falls well inside the [–voice] category and therefore would have no functional role. There is evidence from infant studies (Lasky et al. 1975; Aslin et al. 1979) and from studies of perception of non-speech VOT analogs (Pisoni 1977) that another natural boundary exists in the vicinity of –20-ms VOT. Although such a boundary location would be nonfunctional with respect to the [+/–voice] distinction in English and Cantonese, it would serve to enhance the contrast between long-lead ([+voice]) and short-lag ([–voice]) voicing onsets characteristic of Dutch, Spanish, and Tamil. Most of the world's languages make use of either of these two phonetic realizations of the [+/–voice] distinction (Maddieson 1984), and thus the quantal boundaries in the voicing lead and the voicing lag regions appear to have wide application.

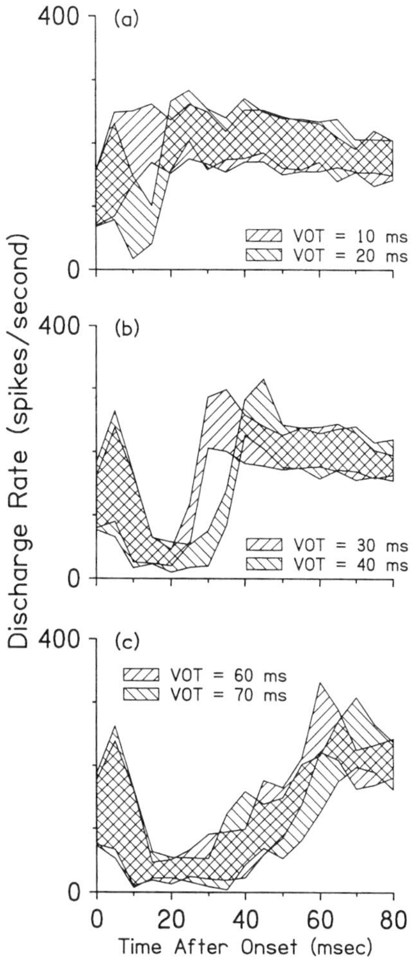

FIGURE 3.12. Auditory nerve responses in chinchilla to pairs of alveolar VOT stimuli in which the VOT difference was 10 ms. Each cross-hatched area encloses the mean ± 1 standard deviation of the average discharge rates of neurons. (From Sinex et al. 1991, with permission of the first author.)

6.2 The Theory of Adaptive Dispersion (TAD) and the Auditory Enhancement Hypothesis

Like the quantal theory, the theory of adaptive dispersion (Liljencrants and Lindblom 1972; Lindblom 1986; Lindblom and Diehl 2001) attempts to provide a deductive account of preferred phoneme and feature inventories among the world's languages. This theory is similar to QT in at least one additional respect: it assumes that preferred phoneme and feature inventories reflect both the listener-oriented selection criterion of auditory distinctiveness and the talker-oriented selection criterion of minimal articulatory effort. However, the specific content of these selection criteria differs between the two theories. Whereas in QT, distinctiveness is achieved

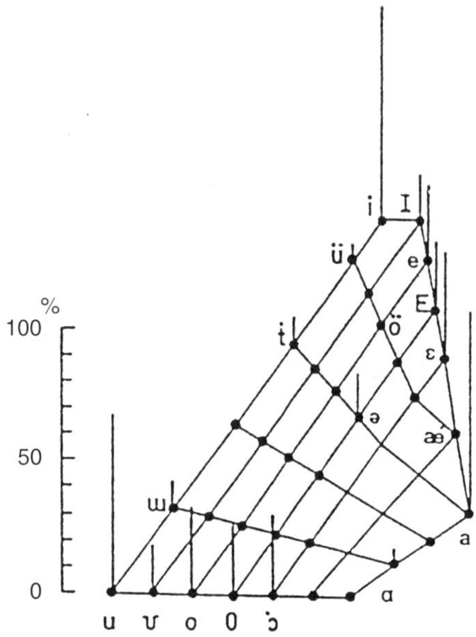

FIGURE 3.13. Frequency histogram of vowels from 209 languages. (Adapted from Crothers, 1978, Appendix III, with permission of Stanford University Press.)

through separation of articulatorily neighboring phonemes by regions of high acoustic instability, in TAD (at least in its earlier versions) it is achieved through maximal dispersion of phonemes in the available phonetic space. And whereas in QT the minimal effort criterion is satisfied by locating phonemes in acoustically stable phonetic regions so as to reduce the need for articulatory precision, in TAD this criterion is satisfied by selecting certain "basic" articulatory types (e.g., normal voicing, nonnasalization, and nonpharyngealization) as well as productions that require minimal deviation from an assumed neutral position.

6.2.1 Predicting Preferred Vowel Inventories

To date, TAD has been applied most successfully to the explanation of preferred vowel inventories. Figure 3.13 shows observations reported by Crothers (1978), who compared the vowel inventories of 209 languages. Occurrence of vowel categories (denoted by phonetic symbols) is expressed as a proportion of the total number of languages in the sample. There was a strong tendency for systems to converge on similar solutions. For example, out of 60 five-vowel systems, as many as 55 exhibit /ɪ e a o u/. The probability of such a pattern arising by chance is obviously negligible. Another clear tendency is the favoring of peripheral over central vowels; in particular, front unrounded and back rounded vowels dominate over central alternatives.

The larger UCLA Phonological Segment Inventory Database (UPSID) described by Maddieson (1984) yielded a very similar distributional pattern: relatively high frequencies occur for vowels on the peripheries (particularly along the front and back vowel height dimensions), and there is a sparse use of central vowels.

Following up on an idea discussed by linguists and phoneticians since the turn of the 20th century (Passy 1890; Jakobson 1941; Martinet 1955), Liljencrants and Lindblom (1972) proposed a *dispersion theory* of vowel systems. They investigated the hypothesis that languages tend to select systems with vowels that are perceptually "maximally distinct." To test this hypothesis they adopted quantitative definitions of three factors shown in Figure 3.14: first, the shape of the universal vowel space; second, a measure of perceptual contrast; and third, a criterion of optimal system. The vowel space was defined according to the two-dimensional configurations shown at the top of the figure. It was derived as a stylization of formant patterns (F_1, F_2, and F_3) generated by the articulatory model of Lindblom and Sundberg (1971). The degree of perceptual contrast between any two vowels was assumed to be modeled by the perceptual distance formula presented in the middle of the figure. It defines the perceptual distance, D_{ij}, between two vowels i and j as equal to the euclidean distance between their formant values. To simplify the computations, distances were restricted to two dimensions, M_1 (F_1 in Mel units) and M_2' (F_2, corrected to reflect the spectral contributions of F_3, in Mel units). Finally, the formula for system optimization, shown at the bottom of the figure, was derived by selecting vowels in such a way that the sum of the intervowel distances—inverted and squared—would be minimized for a given inventory.

Liljencrants and Lindblom predicted vowel formant patterns for three through 12 vowels per system. The results were plotted in the universal vowel space as a function of F_1 and F_2 (in kHz). These predicted vowel systems agreed with actual systems in that peripheral vowels are favored over central ones. Up through six vowels per system, the predicted systems were identical to the systems most favored cross-linguistically. However, for larger inventories the predicted systems were found to exhibit too many high vowels. A case in point is the nine-vowel system, which has five high vowels, rather than the four high vowels, plus a mid-central value that are typically observed for this inventory size.

Lindblom (1986) replicated these experiments using a better-motivated measure of perceptual distance and a slightly more realistic sampling of the vowel space. The distance measure was one originally proposed by Plomp (1970). Lindblom combined it with a computational model of critical-band analysis proposed by Schroeder et al. (1979). It had previously been shown experimentally that this combination could be used to make adequate predictions of perceptual distance judgments for vowel stimuli (Bladon and Lindblom, 1981). The new definition of distance was given by:

$$D_{ij} = \int_{0}^{24.5} \left(|E_i(z) - E_j(z)|^2 \, dx\right)^{1/2} \qquad (2)$$

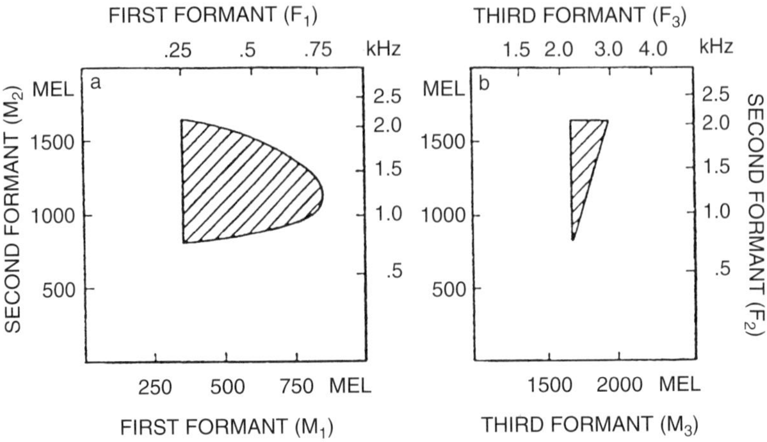

PERCEPTUAL DISTANCE

$$D_{ij} = \left((\Delta M_1)^2 + (\Delta M_2')^2 \right)^{1/2}$$

SYSTEM OPTIMIZATION

$$\sum_{i=2}^{k} \sum_{j=1}^{i-1} 1 / (D_{ij})^2 \text{ ------> } \text{MINIMIZED}$$

FIGURE 3.14. Assumptions of the dispersion theory of vowel inventory preferences (Liljencrants and Lindblom 1972). Top: A universal vowel space ($F_1 \times F_2$, $F_2 \times F_3$) defined by the range of possible vowel outputs of the articulatory model of Lindblom and Sundberg (1971). Middle: A measure of perceptual contrast or distance, D_{ij}, between two vowels, i and j, within the universal space. D_{ij} is assumed to be equal to the euclidean distance between their formant frequencies. M_1 is equal to F_1 in Mel units; M_2' is equal to F_2 in Mel units, corrected to reflect the spectral contribution of F_3. Bottom: A formula for system optimization derived by selecting vowels, such that the sum of the intervowel distances—inverted and squared—is minimized for a given vowel inventory.

The auditory model operated in the frequency domain, convolving an input spectrum with a critical-band filter whose shape was derived from masking data (Zwicker and Feldtkeller 1967). For the applications described here, the output "auditory spectra" [$E(z)$ in the formula] were calibrated in dB/Bark (E) versus Bark (z) units. Perceptual distances between two vowels i and j (D_{ij}) were estimated by integrating differences in critical-band excitation levels across a range of auditory frequencies spanning 0–24.5 Bark. The system optimization formula was the same as that used by Liljencrants and Lindblom (1972).

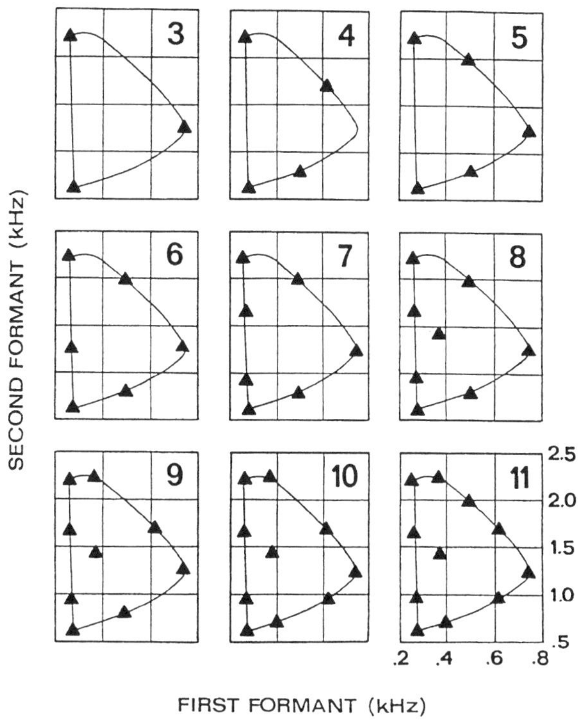

FIGURE 3.15. Results of vowel system predictions (for systems ranging in number from 3 to 11) plotted on an $F_1 \times F_2$ plane (from Lindblom 1986). The horseshoe-shaped vowel areas correspond to the possible outputs of the Lindblom and Sundberg (1971) articulatory model. Predicted inventories are based on the assumption that favored inventories are those that maximize auditory distances among the vowels.

Lindblom's (1986) predicted vowel systems are shown in Figure 3.15 for inventory sizes ranging from 3 to 11. As in the study by Liljencrants and Lindblom, peripheral vowels, especially along the F_1 dimension corresponding to vowel height, were favored over central qualities. And, again, for up to six vowels per system, the predicted sets were identical to the systems most common cross-linguistically. With respect to the problem of too many high vowels, there were certain improvements in this study. For instance, the system predicted for inventories of nine vowels is in agreement with typological data in that it shows four high vowels plus a mid-central vowel, whereas Liljencrants and Lindblom predicted five high vowels and no mid-central vowel. When the formant-based distances of Liljencrants and Lindblom are compared to the auditory-spectrum–based measures of the more recent study for identical spectra, it is clear that the spectrum-based distances leave less room for high vowels, and this accounts for the improved predictions.

In sum, both Liljencrants and Lindblom (1972) and Lindblom (1986) were reasonably successful in predicting the structure of favored vowel systems on the basis of a principle of auditory dispersion. The results encourage the belief that vowel systems are adaptations to mechanisms of auditory analysis that are common to the perception of speech and non-speech sounds alike.

Disner (1983, 1984) independently confirmed that the selection of vowel qualities is governed by a dispersion principle. She found that about 86% of the languages in the UPSID sample have vowel systems that are "built on a basic framework of evenly dispersed peripheral vowels" and that another 10% of the languages "approach this specification" (1984, p. 154). Partly owing to these latter cases, the more recent formulations of TAD (Lindblom 1986, 1990a; Lindblom and Engstrand 1989) have emphasized the dynamic trade-off between the listener-oriented and talker-oriented constraints in speech production and have recast the dispersion principle in terms of "sufficient" rather than "maximal" contrast. The weaker notion of sufficient contrast permits some variation in the phonetic implementation of the dispersion principle and explicitly predicts a tendency for languages with smaller vowel inventories to exploit a somewhat reduced range of the universal vowel space. (With fewer vowel categories, there is a decreased potential for auditory confusability, which permits the use of less peripheral vowels and hence some reduction in articulatory costs.) This prediction was confirmed by Flege (1989), who found that speakers of English (a language with at least 14 vowel categories) produce /i/ and /u/ with higher tongue positions, and /a/ with a lower tongue position, than speakers of Spanish (a language with only five categories).

6.2.2 Auditory Enhancement of Vowel Distinctions

Next, we consider how the strategy of auditory dispersion is implemented phonetically. One fairly simple approach is to use tongue body and jaw positions that approximate articulatory extremes, since articulatory distinctiveness tends to be correlated with acoustic and auditory distinctiveness. It is clear that the point vowels /i/, /a/, and /u/ do, in fact, represent articulatory extremes in this sense. However, auditory dispersion of the point vowels is only partly explained on the basis of the relatively extreme positions of the tongue body and jaw. A more complete account of the phonetic implementation of vowel dispersion is given by the *auditory enhancement hypothesis* (Diehl and Kluender 1989a,b; Diehl et al. 1990; Kingston and Diehl 1994). This hypothesis is an attempt to explain widely attested patterns of phonetic covariation that do not appear to derive from purely physical or physiological constraints on speech production. It states that phonetic features of vowels and consonants covary as they do largely because language communities tend to select features that have mutually enhancing auditory effects. For vowels, auditory enhancement is most characteristically achieved by combining articulatory properties that have similar acoustic consequences. Such reinforcement, when targeted on the most distinctive acoustic properties of vowels, results in an increased perceptual dispersion among the vowel sounds of a language. It should be noted that this hypothesis is closely related to the theory of redundant features independently developed by Stevens and his colleagues (Stevens et al. 1986; Stevens and Keyser 1989).

Consider, for example, how auditory enhancement works in the case of the [+high], [+back] vowel /u/, which occurs in most of the world's languages. The vowel /u/ is distinguished from [–high] vowels in having a low F_1, and from [–back] vowels in having a low F_2. Articulatory properties that contribute to a lowering of F_1 and F_2 thus enhance the distinctiveness of /u/. From the work of Fant (1960) and others, we know that for a tube-like configuration such as the vocal tract, there are theoretically several independent ways to lower a resonant frequency. These include (1) lengthening the tube at either end, (2) constricting the tube in any region where a maximum exists in the standing volume-velocity waveform corresponding to the resonance, and (3) dilating the tube in any region where a minimum exists in the same standing wave. It turns out that each of these theoretical options tends to be exploited when /u/ is produced in clear speech.

Vocal tract lengthening can be achieved by lip protrusion, which, as described earlier, is a typical correlate of [+high], [+back] vowels such as /u/. It can also be achieved by lowering the larynx, and this too has been observed during the production of /u/ (MacNeilage 1969; Riordan 1977). Each of these gestures lowers both F_1 and F_2. The pattern of vocal tract constrictions during the production of /u/ corresponds quite closely to the locations of volume-velocity maxima for F_1 and F_2, contributing further to their lowering. Lip constriction lowers both F_1 and F_2, since both of the underlying resonances have volume-velocity maxima at the lips. The tongue-body constriction occurs near the other volume-velocity maximum for the second resonance and thus effectively lowers F_2. The pattern of vocal tract dilations results in additional formant lowering, since these occur near a volume-velocity minimum at the midpalate corresponding to F_2 and near another minimum at the lower pharynx corresponding to both F_1 and F_2. The dilation of the lower pharynx is largely produced by tongue-root advancement, a gesture that is, anatomically speaking, at least partly independent of tongue height (Lindau 1979). In short, the shape of the vocal tract for /u/ produced in clear speech appears to be optimally tailored (within the physical limits that apply) to achieve a distinctive frequency lowering of the first two formants.

As discussed earlier, f_0 tends to vary directly with vowel height, such that /i/ and /u/ are associated with higher values of f_0 than /a/. Many phoneticians have viewed this covariation as an automatic consequence of anatomical coupling between the tongue and larynx via the hyoid bone (Ladefoged 1964; Honda 1983; Ohala and Eukel 1987). However, electromyographic studies of talkers from several language communities show that higher vowels are produced with greater activation of the cricothyroid muscle, the primary muscle implicated in the active control of f_0 (Honda and Fujimura 1991; Vilkman et al. 1989; Dyhr 1991).

If the anatomical coupling hypothesis cannot account for these findings, how may the covariation between vowel height and f_0 be explained? One hypothesis is suggested by the results of Traunmüller (1981) and Hoemeke and Diehl (1994), discussed earlier. Listeners were found to judge vowel height not on the basis of F_1 alone, but rather by the distance (in Bark units) between F_1 and f_0: the smaller this distance, the higher the perceived vowel. In view of the important cue value of F_1-f_0 distance, it is possible that talkers actively regulate both F_1 and F_0, narrowing the F_1-f_0 distance for higher vowels and expanding it for lower vowels, to enhance the auditory distinctiveness of height contrasts.

6.2.3 Auditory Enhancement of the [+/–Voice] Distinction

In section 3, we described various perceptually significant correlates of [+voice] consonants, including the presence of voicing during the consonant constriction, a low F_1 value near the constriction, a low f_0 in the same region, the absence of significant aspiration after the release, a short constriction interval, and a long preceding vowel. Kingston and Diehl (1994, 1995; see also Diehl et al. 1995) have argued that these correlates may be grouped into coherent subsets based on relations of mutual auditory enhancement. These subsets of correlates form "integrated perceptual properties" that are intermediate between individual phonetic correlates and full-fledged distinctive features.

One hypothesized integrated perceptual property includes consonant constriction duration and preceding vowel duration. Since these durations vary inversely in [+voice] and [–voice] word-medial consonants, both durations may be incorporated into a single measure that defines an overall durational cue for the [+/–voice] contrast, as proposed by both Kohler (1979) and Port and Dalby (1982). Port and Dalby suggested that the consonant/vowel duration ratio is the most relevant durational cue for the word-medial [+/–voice] contrast. A short consonant and a long preceding vowel enhance one another because they both contribute to a small C/V duration ratio typical of [+voice] consonants. Relative to either durational cue in isolation, a ratio of the two durations permits a considerably wider range of variation and hence greater potential distinctiveness. Evidence for treating the C/V duration ratio as an integrated perceptual property has been reviewed by Kingston and Diehl (1994).

Another possible integrated perceptual property associated with [+voice] consonants was identified by Stevens and Blumstein (1981). It consists of voicing during the consonant constriction, a low F_1 near the constriction, and a low f_0 in the same region. What these correlates have in common, according to Stevens and Blumstein, is that they all contribute to the presence of low-frequency periodic energy in or near the consonant constriction. We refer to this proposed integrated property as the "low-frequency property" and to Stevens and Blumstein's basic claim as the "low-frequency hypothesis."

Several predictions may be derived from the low-frequency hypothesis. One is that two stimuli in which separate subproperties of the low-frequency property are positively correlated (i.e., the subproperties are either both present or both absent) will be more distinguishable than two stimuli in which the subproperties are negatively correlated. This prediction was recently supported for stimulus arrays involving orthogonal variation in either f_0 and voicing duration or F_1 and voicing duration (Diehl et al. 1995; Kingston and Diehl 1995).

Another prediction of the low-frequency hypothesis is that the effects on [+/–voice] judgments of varying either f_0 and F_1 should pattern in similar ways for a given utterance position and stress pattern. Consider first the [+/–voice] distinction in utterance-initial prestressed position (e.g., "do" vs. "to"). As described earlier, variation in VOT is a primary correlate of the [+/–voice] contrast in this position, with longer, positive VOT values corresponding to the [–voice] category. Because F_1 is severely attenuated during the VOT interval and because F_1 rises after the consonant release, a longer VOT is associated with a higher F_1 onset frequency, all else being equal. The question of interest here is: What aspects of the F_1 trajectory help signal the [+/–voice] distinction in this position? The answer consistently found across several studies (Lisker 1975; Summerfield and Haggard 1977; Kluender 1991) is that only the F_1 value at voicing onset appears to influence utterance-initial prestressed [+/–voice] judgments.

Various production studies show that following voicing onset f_0 starts at a higher value for [–voice] than for [+voice] consonants and that this difference may last for some tens of milliseconds into the vowel. Interestingly, however, the perceptual influence of f_0, like that of F_1, appears to be limited to the moment of voicing onset (Massaro and Cohen 1976; Haggard et al. 1981). Thus, for utterance-initial, prestressed consonants, the effects of f_0 and F_1 on [+/–voice] judgments are similar in pattern.

Next, consider the [+/–voice] distinction in utterance-final poststressed consonants (e.g., "bid" vs. "bit"). Castleman and Diehl (1996b) found that the effects of varying f_0 trajectory on [+/–voice] judgments in this position patterned similarly to effects of varying F_1 trajectory. In both cases, lower frequency values during the vowel and in the region near the final consonant constriction yielded more [+voice] responses, and the effects of the frequency variation in the two regions were additive. The similar effects of F_1 and f_0 variation on final poststressed [+/–voice] judgments extend the parallel between the effects of F_1 and f_0 variation on initial prestressed [+/–voice] judgments. These findings are consistent with the claim that a low f_0 and a low F_1 both contribute to a single integrated low-frequency property.

Previously, we discussed one way in which the listener-oriented strategy of auditory dispersion (or sufficient contrast) is implemented phonetically, namely by combining articulatory events that have very similar, and hence mutually reinforcing, acoustic consequences. As just noted, another way of achieving dispersion is to regulate vocal tract activity to produce integrated perceptual properties such as the C/V duration ratio and the low-frequency property.

6.2.4 Phonological Assimilation: The Joint Role of Auditory and Articulatory Selection Criteria

In section 2, we referred to the theory of feature geometry (Clements 1985; McCarthy 1988), which modified the feature theory of SPE by positing a hierarchical feature structure for segments. The hierarchy includes, among other things, a "place node" that dominates the SPE place-of-articulation features such as [coronal], [anterior], and [back]. Recall that the theoretical motivation for a place node was to account for the frequent occurrence of phonological assimilation based on the above three features and the rare occurrence of assimilation based on arbitrary feature sets.

The account of place assimilation offered by feature geometry implicitly assumes that the process is conditioned by purely articulatory factors, since the phonetic content of the place node is a set of actual articulators (e.g., lips, tongue tip, tongue dorsum, tongue root). (This articulatory emphasis, as noted, was already present in the SPE approach to features.) Others, however, have suggested that perceptual factors may also have a conditioning role in place assimilation (Kohler 1990; Ohala 1990). Kohler pointed out that certain consonant classes (such as nasals and stops) are morely likely than other classes (such as fricatives) to undergo place assimilation to the following consonant. To account for this, he suggested that assimilation tends not to occur when the members of the consonant class are relatively distinctive perceptually, and their articulatory reduction would be particularly salient. This account presupposes that the stops and nasals that undergo place assimilation are less distinctive than fricatives, which tend not to assimilate. Hura et al. (1992) obtained evidence that supported Kohler's perceptual assumption. In the context of a following word-initial stop consonant, fricatives were more likely to be correctly identified than nasals or unreleased stops.

If advocates of feature geometry are correct in insisting that the naturalness of processes such as place assimilation should be captured by the posited feature representation of segments, and if Kohler (1990) is right in claiming that phonological assimilation is shaped by both articulatory and perceptual factors, it follows that features cannot, in general, be construed in purely articulatory terms. Rather they must be viewed as distinctive elements of spoken language that reflect both talker- and listener-oriented selection criteria.

7. Are There Invariant Physical Correlates of Features?

In section 4, we highlighted physical variability as one of the key characteristics of speech. We pointed out that some of it has to do with individual speaker differences, some of it is linked to style and situation, and a third type derives from the dynamics of speech motor control, which produces phenomena such as coarticulation and reduction. Linguistic analysis and psychological intuition tell us that phonemes, syllables, and other abstract building blocks of sound structure are discrete and invariant (context-independent). Phonetic measurements indicate that the realizations of these units are embedded in a continuous flow of sound and movement and are highly dependent on context. The overriding challenge is accordingly to make sense of this discrepancy between the physical and the psychological pictures of speech. Any comparison of phonetic and phonological evidence seems to lead, unavoidably, to an inconsistency: psychologically the same, but physically different. How do we best explain this paradox?

There are several routes that the quest for a resolution of the invariance issue can take. In the literature three main approaches can be discerned. An initial choice concerns the level at which invariants are expected to be found. Is invariance phonetic (articulatory and/or acoustic/auditory)? Or is it an emergent product of lexical access? (See the third alternative below.) Is it directly observable in the on-line phonetic behavior? If so, is it articulatory, acoustic, or auditory?

It appears fair to say that, in the past, much phonetic research has tacitly accepted the working hypothesis that invariants are indeed phonetic and do constitute measurable aspects of the experimental phonetic records. Within this tradition, researchers can be roughly grouped into those who favor "gesturalist" accounts and those who advocate acoustic/auditory invariance.

7.1 Gesturalist Theories

If it is true that speech sounds must inevitably bear the marks of coarticulation, one might reasonably ask whether there should not be a point in the speech production process where the coproduction of articulatory movements has not yet happened—conceivably, as Liberman and Mattingly (1985) have suggested, at the level of the talker's intention, a stage where phonetic units have not yet fused into patterns of complex interaction. If speech could be examined at such a level, perhaps that is where physical invariance of linguistic categories might be identified. Furthermore, if the units of speech are in fact phonetic gestures, which are invariant except for variations in timing and amplitude (Browman and Goldstein 1992), then the variability of speech could be explained largely as a consequence of gestural interactions.

The last paragraph strikes several prominent themes in gestural theories such as the motor theory (Liberman and Mattingly 1985, 1989), direct realism (Fowler 1986, 1994), articulatory phonology (Browman and Goldstein 1992), and the accounts of phonological development presented by Studdert-Kennedy (1987, 1989). Strange's dynamic specification (1989a,b) of vowels also belongs to this group.[2]

It should be pointed out that lumping all these frameworks together does not do justice to the differences that exist among them, particularly between direct realism and the motor theory. However, here we will concentrate on an aspect that they all share. That common denominator is as follows. Having embraced a gestural account of speech (whichever variant), one is immediately faced with the problem of specifying a perceptual mechanism capable of handling all the context-dependent variability. From the gesturalist perspective, underlying "phonetic gestures" are always drawn from the same small set of entities. Therefore, the task of the listener becomes that of perceiving the intended gestures on the basis of highly "encoded" and indirect acoustic information. What is the mechanism of "decoding"? What perceptual process do listeners access that makes the signal "transparent to phonemes"? That is the process that, as it were, does all the work. If such a mechanism could be specified, it would amount to "solving the invariance problem."

Gesturalists have so far had very little to say in response to those questions beyond proposing that it might either be a "biologically specialized module" (Liberman and Mattingly 1985, 1989), or a process of "direct per-

[2] "Phonetic perception is the perception of gesture. . . . The invariant source of the phonetic percept is somewhere in the processes by which the sounds of speech are produced" (Liberman and Mattingly 1985, p. 21). "The gestures have a virtue that the acoustic cues lack: instances of a particular gesture always have certain topological properties not shared by any other gesture" (Liberman and Mattlingly 1985, p. 22). "The gestures do have characteristic invariant properties, . . . though these must be seen, not as peripheral movements, but as the more remote structures that control the movements." "These structures correspond to the speaker's intentions" (Liberman and Mattingly 1985, p. 23). "The distal event considered locally is the articulating vocal tract" (Fowler 1986, p. 5). "An event theory of speech production must aim to characterize articulation of phonetic segments as overlapping sets of coordinated gestures, where each set of coordinated gestures conforms to a phonetic segment. By hypothesis, the organization of the vocal tract to produce a phonetic segment is invariant over variation in segmental and suprasegmental contexts" (Fowler 1986, p. 11). "It does not follow then from the mismatch between acoustic segment and phonetic segment, that there is a mismatch between the information in the acoustic signal and the phonetic segments in the talker's message. Possibly, in a manner as yet undiscovered by researchers but accessed by perceivers, the signal is transparent to phonetic segments" (Fowler 1986, p. 13). "Both the phonetically structured vocal-tract activity and the linguistic information . . . are directly perceived (by hypothesis) by the extraction of invariant information from the acoustic signal" (Fowler 1986, p. 24).

ception" (Fowler 1986, 1994). An algorithmic model for handling context-dependent signals has yet to be presented within this paradigm.

7.2 The Search for Acoustic Invariance

Several investigators have argued in favor of acoustic/auditory rather than gestural invariance. As already mentioned in section 3, the research by Stevens and Blumstein, Kewley-Port, Sussman and their associates points to the possibility of relational rather than absolute acoustic invariance. Other evidence, compatible with acoustic/auditory constancies in speech, includes data on "motor equivalence" (Maeda 1991; Perkell et al. 1993), articulatory covariation for the purpose of "auditory enhancement" (Diehl et al. 1990), "compensatory articulations" (e.g., Lindblom et al. 1979; Gay et al. 1981), demonstrations that the control of articulatory precision depends on constriction size (Gay et al. 1981; Beckman et al. 1995), the challenge of dynamic specification by "compound target theories" of vowel perception (Andruski and Nearey 1992; Nearey and Assmann 1986; Nearey 1989), and acoustically oriented control of vowel production (Johnson et al. 1994).

Evidence of this sort has led investigators to view speech production as a basically listener-oriented process (cf. Jakobson's formulation: "We speak to be heard in order to be understood" Jakobson et al. 1963, p 13). For instance, in Sussman's view, locus equations reflect an "orderly output constraint" that speakers honor in order to facilitate perceptual processing. That interpretation is reinforced by work on articulatory modeling (Stark et al. 1996) that shows that the space of possible locus patterns for alveolar consonants is not limited to the observed straight-line relationships between F_2-onset and F_2-vowel midpoint samples, but offers, at each F_2-vowel value, a sizable range of F_2-onsets arising from varying the degree of coarticulation between tongue body and tongue tip movements.

The same point can be made by considering the formant patterns of plain versus velarized consonants. For instance, "clear" and "dark" [l] sounds in English are both alveolars, but, in the context of the same following vowel, their F_2 onsets are very different because of the differences in underlying tongue body configuration. Such observations indicate that, in terms of production, there is nothing inevitable about the degree of coarticulation characterizing a given consonant production. Speakers, and languages, are free to vary that degree. Hence it follows that there is nothing inevitable about the locus line slopes and intercepts associated with a given place of articulation. The degrees of freedom of the vocal tract offer speakers many other patterns that they apparently choose not to exploit.

7.3 Reasons for Doubting that Invariance Is a Design Feature of Spoken Language: Implications of the Hyper-Hypo Dimension

In illustrating consonant and vowel variability in section 4, we acknowledged the gesturalist position in reviewing Öhman's numerical model, but we also contrasted that interpretation with an acoustic/auditory approach demonstrating the maintenance of sufficient contrast among place categories. Similarly, in discussing formant undershoot in vowels, we began by noting the possible correspondence between the inferred invariant formant targets and hypothetical underlying gestures. However, additional evidence on clear speech was taken to suggest listener-oriented adaptations of speech patterns, again serving the purpose of maintaining sufficient contrast.

Sufficient contrast and phonetic invariance carry quite different implications about the organization of speaker-listener interactions. In fact, sufficient contrast introduces a third way of looking at signal variability, which is aimed at making sense of variability rather than at making it disappear.

This viewpoint is exemplified by the so-called hyper and hypo (H & H) theory (Lindblom 1990b, 1996), developed from evidence supportive of the following claims: (1) The listener and the speaking situation make significant contributions to defining the speaker's task. (2) That task shows both long- and short-term variations. (3) The speaker adapts to it. (4) The function of that adaptation is not to facilitate the articulatory recovery of the underlying phonetic gestures, but to produce an auditory signal that can be rich or poor, but which, for successful recognition, must minimally possess sufficient discriminatory power. The H & H theory views speech production as an adaptive response to a variable task. On this view, the function of the speech signal is not to serve as a carrier of a core of constancies, which are embedded in signal variance, but that the listener nonetheless succeeds in "selecting."[3] The role it accords the signal is rather that of supplying "missing information." This perspective provides no reason for expecting invariance to be signal-based, or phonetic, at all. Rather, it assumes that *invariance is an emergent product of lexical access* and listener comprehension.

The H & H theory overlaps with accounts that view speech as organized around acoustic/auditory goals. Evidence for listener-oriented adaptations of speech movements provides support also for the H & H perspective. Reports indicate that "clear speech" exhibits systematic acoustic changes relative to casual patterns (Chen et al. 1983; Moon 1990, 1991; Moon and Lindblom 1994). And it possesses properties that make them perceptually more robust (Lively et al. 1993; Moon et al. 1995; Payton et al. 1994; Summers et al. 1988). Where H & H theory differs most from other descrip-

---

[3] "The perceived parsing must be in the signal; the special role of the perceptual system is not to create it, but only to select it" (Fowler 1986, p. 13).

tions of speech is with reference to the strong a priori claim it makes about the absence of signal invariance. Insofar as that claim is borne out by future tests of quantitatively defined H & H hypotheses, an explanation may be forthcoming as to why attempts to specify invariant physical correlates of features and other linguistic units have so far had very limited success.

8. Summary

Both phoneticians and phonologists traditionally have tended to introduce features in an ad hoc manner to describe the data in their respective domains of inquiry. With the important exception of Jakobson, theorists have emphasized articulatory over acoustic or auditory correlates in defining features. The set of features available for use in spoken languages are given, from a purely articulatory perspective, by the universal phonetic capabilities of human talkers. While most traditional theorists acknowledge that articulatorily defined features also have acoustic and auditory correlates, the latter usually have a descriptive rather than explanatory role in feature theory.

A problem for this traditional view of features is that, given the large number of degrees of freedom available articulatorily to talkers, it is unclear why a relatively small number of features and phonemes should be strongly favored cross-linguistically while many others are rarely attested. Two theories, QT and TAD, offer alternative solutions to this problem. Both differ from traditional approaches in attempting to derive preferred feature and phoneme inventories from independently motivated principles. In this sense, QT and TAD represent deductive rather than axiomatic approaches to phonetic and phonological explanation. They also differ from traditional approaches in emphasizing the needs of the talker *and the listener* as important constraints on the selection of feature and phoneme inventories.

The specific content of the posited talker- and listener-oriented selection criteria differ between QT and TAD. In QT, the talker-oriented criterion favors feature values and phonemes that are acoustically (auditorily) stable in the sense that small articulatory (acoustic) perturbations are relatively inconsequential. This stability reduces the demand on talkers for articulatory precision. The listener-oriented selection criteria in QT are that the feature values or phonemes have invariant acoustic (auditory) correlates and that they be separated from neighboring feature values or phonemes by regions of high acoustic (auditory) instability, yielding high distinctiveness. In TAD, the talker-oriented selection criterion also involves a "least effort" principle. In this case, however, effort is defined not in terms of articulatory precision but rather in terms of the "complexity" of articulation (e.g., whether only a single articulator is employed or are secondary articulators used as well) and the displacement and velocity requirements of the articulations). The listener-oriented selection criterion of TAD involves the notion of "sufficient contrast" and is implemented in the form of acoustic or auditory dispersion of feature and phoneme categories within the phonetic space available to talkers. According to the auditory enhancement hypothesis, such dispersion is achieved by combining articulatory events that have similar, or otherwise mutually reinforcing, acoustic and auditory consequences.

Although QT and TAD share the goal of explaining phonetic and phonological regularities (e.g., the structure of preferred phoneme and feature inventories) on the basis performance contraints on talkers and listeners, they differ crucially on the subject of phonetic invariance: QT assumes that there are invariant acoustic correlates of features and these play an important role in speech perception; TAD (with the associated H & H theory) makes no such assumption, stressing instead the perceptual requirement of sufficient contrast.

Further progress in explaining the structure of preferred phoneme and feature inventories will depend, among other things, on the development of better auditory models and distance metrics. A good deal has been learned in recent years about the response properties of several classes of neurons in the cochlear nucleus and other auditory regions (see Palmer and Shamma, Chapter 4), and it soon should be possible to extend currently available models to simulate processing of speech sounds by these classes of neurons. In general, auditory-distance metrics that are currently in use have been selected mainly on grounds of simplicity. Clearly, much additional research is needed to design distance metrics that are better motivated both empirically and theoretically.

Acknowledgments. This work was supported by research grants 5 R01 DC00427-10, -11, -12 from the National Institute on Deafness and Other Communication Disorders, National Institutes of Health, to the first author, and grant F 149/91 from the Council for Research in Humanities and the Social Sciences, Sweden, to the second author. We thank Steven Greenberg and Bill Ainsworth for very helpful comments on an earlier draft of the chapter.

List of Abbreviations

f_0	fundamental frequency
F_1	first format
F_2	second formant
F_3	third formant
QT	quantal theory
SPE	*The Sound Pattern of English*
TAD	theory of adaptive dispersion
UPSID	UCLA phonological segment inventory database
VOT	voice onset time

References

Abramson AS, Lisker L (1970) Discriminability along the voicing continuum: cross-language tests. Proceedings of the 6th International Congress of Phonetic Sciences, Prague, 1967. Prague: Academia, pp. 569–573.

Anderson SR (1985) Phonology in the Twentieth Century. Chicago: Chicago University Press.

Andruski J, Nearey T (1992) On the sufficiency of compound target specification of isolated vowels and vowels in /bVb/ syllables. J Acoust Soc Am 91:390–410.

Aslin RN, Pisoni DP, Hennessy BL, Perey AJ (1979) Identification and discrimination of a new linguistic contrast. In: Wolf JJ, Klatt DH (eds) Speech Communication: Papers Presented at the 97th Meeting of the Acoustical Society of America. New York: Acoustical Society of America, pp. 439–442.

Balise RR, Diehl RL (1994) Some distributional facts about fricatives and a perceptual explanation. Phonetica 51:99–110.

Beckman ME, Jung T-P, Lee S-H, et al. (1995) Variability in the production of quantal vowels revisited. J Acoust Soc Am 97:471–490.

Bell AM (1867) Visible Speech. London: Simpkin, Marshall.

Bergem van D (1995) Acoustic and lexical vowel reduction. Unpublished PhD dissertation, University of Amsterdam.

Bladon RAW (1982) Arguments against formants in the auditory representation of speech. In: Carlson R, Granstrom B (eds) The Representation of Speech in the Peripheral Auditory System. Amsterdam: Elsevier Biomedical Press, pp. 95–102.

Bladon RAW, Lindblom B (1981) Modeling the judgment of vowel quality differences. J Acoust Soc Am 69:1414–1422.

Bloomfield L (1933) Language. New York: Holt, Rinehart and Winston.

Blumstein SE, Stevens KN (1979) Acoustic invariance in speech production: evidence from measurements of the spectral characteristics of stop consonants. J Acoust Soc Am 72:43–50.

Blumstein SE, Stevens KN (1980) Perceptual invariance and onset spectra for stop consonants in different vowel environments. J Acoust Soc Am 67:648–662.

Boubana S (1995) Modeling of tongue movement using multi-pulse LPC coding. Unpublished Doctoral thesis, École Normale Supérieure de Télécommunications (ENST), Paris.

Browman C, Goldstein L (1992) Articulatory phonology: an overview. Phonetica 49:155–180.

Brownlee SA (1996) The role of sentence stress in vowel reduction and formant undershoot: a study of lab speech and spontaneous informal conversations. Unpublished Ph D dissertation, University of Texas at Austin.

Castleman WA, Diehl RL (1996a) Acoustic correlates of fricatives and affricates. J Acoust Soc Am 99:2546(abstract).

Castleman WA, Diehl RL (1996b) Effects of fundamental frequency on medial and final [voice] judgments. J Phonetics 24:383–398.

Chen FR, Zue VW, Picheny MA, Durlach NI, Braida LD (1983) Speaking clearly: acoustic characteristics and intelligibility of stop consonants. 1–8 in Working Papers II, Speech Communication Group, MIT.

Chen M (1970) Vowel length variation as a function of the voicing of the consonant environment. Phonetica 22:129–159.

Chiba T, Kajiyama M (1941) The Vowel: Its Nature and Structure. Tokyo: Tokyo-Kaiseikan. (Reprinted by the Phonetic Society of Japan, 1958).

Chistovich L, Lublinskaya VV (1979) The "center of gravity" effect in vowel spectra and critical distance between the formants: psychoacoustical study of the perception of vowel-like stimuli. Hear Res 1:185–195.
Chistovich LA, Sheikin RL, Lublinskaja VV (1979) "Centres of gravity" and spectral peaks as the determinants of vowel quality. In: Lindblom B, Ohman S (eds) Frontiers of Speech Communication Research. London: Academic Press, pp. 143–157.
Chomsky N (1964) Current trends in linguistic theory. In: Fodor JA, Katz JJ (eds) The Structure of Language, New York: Prentice-Hall, pp. 50–118.
Chomsky N, Halle M (1968) The Sound Pattern of English. New York: Harper & Row.
Clements GN (1985) The geometry of phonological features. Phonol Yearbook 2:223–250.
Crothers J (1978) Typology and universals of vowel systems. In: Greenberg JH, Ferguson CA, Moravcsik EA (eds) Universals of Human Language, vol. 2. Stanford, CA: Stanford University Press, pp. 99–152.
Cutting JE, Rosner BS (1974) Categories and boundaries in speech and music. Percept Psychophys 16:564–570.
Delattre PC, Liberman AM, Cooper FS, Gerstman LJ (1952) An experimental study of the acoustic determinants of vowel color: observations on one- and two-formant vowels synthesized from spectrographic patterns. Word 8:195–210.
Denes P (1955) Effect of duration on the perception of voicing. J Acoust Soc Am 27:761–764.
Diehl RL (1989) Remarks on Stevens' quantal theory of speech. J Phonetics 17:71–78.
Diehl RL (2000) Searching for an auditory description of vowel categories. Phonetica 57:267–274.
Diehl RL, Kluender KR (1989a) On the objects of speech perception. Ecolog Psychol 1:121–144.
Diehl RL, Kluender KR (1989b) Reply to commentators. Ecolog Psychol 1:195–225.
Diehl RL, Molis MR (1995) Effect of fundamental frequency on medial [voice] judgments. Phonetica 52:188–195.
Diehl RL, Castleman, WA, Kingston J (1995) On the internal perceptual structure of phonological features: The [voice] distinction. J Acoust Soc Am 97:3333–3334(abstract).
Diehl RL, Kluender KR, Walsh MA (1990) Some auditory bases of speech perception and production. In: Ainsworth WA (ed) Advances in Speech, Hearing and Language Processing. London: JAI Press, pp. 243–268.
Disner SF (1983) Vowel quality: the relation between universal and language-specific factors. Unpublished PhD dissertation, UCLA; also Working Papers in Phonetics 58.
Disner SF (1984) Insights on vowel spacing. In Maddieson I (ed) Patterns of Sound. Cambridge: Cambridge University Press, pp. 136–155.
Dyhr N (1991) The activity of the cricothyroid muscle and the intrinsic fundamental frequency of Danish vowels. J Acoust Soc Am 79:141–154
Eimas PD, Siqueland ER, Jusczyk P, Vigorito J (1971) Speech perception in infants. Science 171:303–306.
Engstrand O (1988) Articulatory correlates of stress and speaking rate in Swedish VCV utterances. J Acoust Soc Am 83:1863–1875.

Engstrand O, Krull D (1989) Determinants of spectral variation in spontaneous speech. In: Proc of Speech Research '89, Budapest, pp. 84–87.

Fahey RP, Diehl RL, Traunmüller H (1996) Perception of back vowels: effects of varying F_1-F_0 Bark distance. J Acoust Soc Am 99:2350–2357.

Fant G (1960) Acoustic Theory of Speech Production. The Hague: Mouton.

Fant G (1973) Speech Sounds and Features. Cambridge, MA: MIT Press.

Flege JE (1988) Effects of speaking rate on tongue position and velocity of movement in vowel production. J Acoust Soc Am 84:901–916.

Flege JE (1989) Differences in inventory size affect the location but not the precision of tongue positioning in vowel production. Lang Speech 32:123–147.

Fourakis M (1991) Tempo, stress, vowel reduction in American English. J Acoust Soc Am 90:1816–1827.

Fowler CA (1986) An event approach to the study of speech perception from a direct-realist perspective. J Phonetics 14:3–28.

Fowler CA (1994) Speech perception: direct realist theory. In: Asher RE (ed) Encyclopedia of Language and Linguistics. New York: Pergamon, pp. 4199–4203.

Fruchter DE (1994) Perceptual significance of locus equations. J Acoust Soc Am 95:2977(abstract).

Fujimura O (1971) Remarks on stop consonants: synthesis experiments and acoustic cues. In: Form and Substance: Phonetic and Linguistic Papers Presented to Eli Fischer-Jørgensen. Copenhagen: Akademisk Forlag, pp. 221–232.

Gay T (1978) Effect of speaking rate on vowel formant movements. J Acoust Soc Am 63:223–230.

Gay T, Lindblom B, Lubker J (1981) Production of bite-block vowels: acoustic equivalence by selective compensation. J Acoust Soc Am 69:802–810.

Gerstman LJ (1957) Perceptual dimensions for the friction portion of certain speech sounds. Unpublished PhD dissertation, New York University.

Haggard M, Ambler S, Callow M (1970) Pitch as a voicing cue. J Acoust Soc Am 47:613–617.

Haggard M, Summerfield Q, Roberts M (1981) Psychoacoustical and cultural determinants of phoneme boundaries: evidence from trading F_0 cues in the voiced-voiceless distinction. J Phonetics 9:49–62.

Halle M (1992) Phonological features. In: Bright W (ed) International Encyclopedia of Linguistics. New York: Oxford University Press, pp. 207–212.

Harris KS, Hoffman HS, Liberman AM, Delattre PC, Cooper FS (1958) Effect of third formant transitions on the perception of the voiced stop consonants. J Acoust Soc Am 30:122–126.

Hillenbrand J, Houde RA (1995) Vowel recognition: formants, spectral peaks, and spectral shape representations. J Acoust Soc Am 98:2949(abstract).

Hoemeke KA, Diehl RL (1994) Perception of vowel height: the role of F_1-F_0 distance. J Acoust Soc Am 96:661–674.

Honda K (1983) Relationship between pitch control and vowel articulation. In: Bless DM, Abbs JH (eds) Vocal Fold Physiology: Contemporary Research and Clinical Issues. San Diego, CA: College-Hill Press, pp. 286–297.

Honda K, Fujimura O (1991) Intrinsic vowel F_0 and phrase-final F_0 lowering: phonological versus biological explanations. In: Gauffin J, Hammarberg B (eds) Vocal Fold Physiology: Acoustic, Perceptual, and Physiological Aspects of Voice Mechanisms. San Dego, CA: Singular, pp. 149–157.

House AS, Fairbanks G (1953) The influence of consonant environment on the secondary acoustical characteristics of vowels. J Acoust Soc Am 25:105–135.

Howell P, Rosen S (1983) Production and perception of rise time in the voiceless affricate/fricative distinction. J Acoust Soc Am 73:976–984.

Hura SL, Lindblom B, Diehl RL (1992) On the role of perception in shaping phonological assimilation rules. Lang Speech 35:59–72.

Ito M, Tsuchida J, Yano M (2001) On the effectiveness of whole spectral shape for vowel perception. J Acoust Soc Am 110:1141–1149.

Jakobson R (1932) Phoneme and phonology. In the Second Supplementary Volume to the Czech Encyclopedia. Prague: Ottuv slovník naucny. (Reprinted in Jakobson R (1962) Selected Writings I. The Hague: Mouton, pp. 231–234.)

Jakobson R (1939) Zur Struktur des Phonems (based on two lectures at the University of Copenhagen). (Reprinted in Jakobson R (1962) Selected Writings I. The Hague: Mouton, pp. 280–311.)

Jakobson R (1941) Kindersprache, Aphasie und allgemeine Lautgesetze. Uppsala: Uppsala Universitets Arsskrift, pp. 1–83.

Jakobson R, Halle M (1971) Fundamentals of Language. The Hague: Mouton. (Originally published in 1956.)

Jakobson R, Fant G, Halle M (1963) Preliminaries to Speech Analysis. Cambridge, MA: MIT Press. (Originally published in 1951.)

Johnson K, Ladefoged P, Lindau M (1994) Individual differences in vowel production. J Acoust Soc Am 94:701–714.

Kewley-Port D (1983) Time-varying features as correlates of place of articulation in stop consonants. J Acoust Soc Am 73:322–335.

Kewley-Port D, Pisoni DB, Studdert-Kennedy M (1983) Perception of static and dynamic acoustic cues to place of articulation in initial stop consonants. J Acoust Soc Am 73:1779–1793.

Kingston J, Diehl RL (1994) Phonetic knowledge. Lang 70:419–454.

Kingston J, Diehl RL (1995) Intermediate properties in the perception of distinctive feature values. In: Connell B, Arvaniti A (eds) Phonology and Phonetic Evidence: Papers in Laboratory Phonology IV. Cambridge: Cambridge University Press, pp. 7–27.

Klatt DH (1982) Prediction of perceived phonetic distance from critical-band spectra: a first step. IEEE ICASSP, pp. 1278–1281.

Kluender KR (1991) Effects of first formant onset properties on voicing judgments result from processes not specific to humans. J Acoust Soc Am 90:83–96.

Kluender KR, Walsh MA (1992) Amplitude rise time and the perception of the voiceless affricate/fricative distinction. Percept Psychophys 51:328–333.

Kluender KR, Diehl RL, Wright BA (1988) Vowel-length differences before voiced and voiceless consonants: an auditory explanation. J Phonetics 16:153–169.

Kohler KJ (1979) Dimensions in the perception of fortis and lenis plosives. Phonetica 36:332–343.

Kohler KJ (1982) F_0 in the production of lenis and fortis plosives. Phonetica 39:199–218.

Kohler KJ (1990) Segmental reduction in connected speech: phonological facts and phonetic explanations. In: Hardcastle WJ, Marchal A (eds) Speech Production and Speech Modeling. Dordrecht: Kluwer, pp. 66–92.

Kuehn DP, Moll KL (1976) A cineradiographic study of VC and CV articulatory velocities. J Phonetics 4:303–320.

Kuhl PK, Miller JD (1978) Speech perception by the chinchilla: identification functions for synthetic VOT stimuli. J Acoust Soc Am 63:905–917.

Kuhl PK, Padden DM (1982) Enhanced discriminability at the phonetic boundaries for the voicing feature in Macaques. Percept Psychophys 32:542–550.

Laboissière R, Ostry D, Perrier P (1995) A model of human jaw and hyoid motion and its implications for speech production. In: Elenius K, Branderud P (eds) Proceedings ICPhS 95, Stockholm, vol 2, pp. 60–67.

Ladefoged P (1964) A Phonetic Study of West African Languages. Cambridge: Cambridge University Press.

Ladefoged P (1971) Preliminaries to Linguistic Phonetics. Chicago: University of Chicago Press.

Ladefoged P (1972) Phonetic prerequisites for a distinctive feature theory. In: Valdman A (ed) Papers in Linguistics and Phonetics to the Memory of Pierre Delattre. The Hague: Mouton, pp. 273–285.

Ladefoged P (1980) What are linguistic sounds made of? Lang 65:485–502.

Lasky RE, Syrdal-Lasky A, Klein RE (1975) VOT discrimination by four to six and a half month old infants from Spanish environments. J Exp Child Psychol 20:215–225.

Lehiste I (1970). Suprasegmentals. Cambridge, MA: MIT Press.

Lehiste I, Peterson GE (1961) Some basic considerations in the analysis of intonation. J Acoust Soc Am 33:419–425.

Liberman A, Mattingly I (1985) The motor theory of speech perception revised. Cognition 21:1–36.

Liberman A, Mattingly I (1989) A specialization for speech perception. Science 243:489–494.

Liberman AM, Delattre PC, Cooper FS (1958) Some cues for the distinction between voiced and voiceless stops in initial position. Lang Speech 1:153–167.

Liberman AM, Delattre PC, Cooper FS, Gerstman LJ (1954) The role of consonant-vowel transitions in the perception of the stop and nasal consonants. Psychol Monogr: Gen Applied 68:113.

Liljencrants J, Lindblom B (1972) Numerical simulation of vowel quality systems: the role of perceptual contrast. Lang 48:839–862.

Lindau M (1979) The feature expanded. J Phonetics 7:163–176.

Lindblom B (1963) Spectrographic study of vowel reduction. J Acoust Soc Am 35:1773–1781.

Lindblom B (1983) Economy of speech gestures. In: MacNeilage PF (ed) Speech Production. New York: Springer, pp. 217–245.

Lindblom B (1986) Phonetic universals in vowel systems. In: Ohala JJ, Jaeger JJ (eds) Experimental Phonology. Orlando, FL: Academic Press, pp. 13–44.

Lindblom B (1990a) On the notion of "possible speech sound." J Phonetics 18:135–152.

Lindblom B (1990b) Explaining phonetic variation: a sketch of the H&H theory. In: Hardcastle W, Marchal A (eds) Speech Production and Speech Modeling, Dordrecht: Kluwer, pp. 403–439.

Lindblom B (1996) Role of articulation in speech perception: clues from production. J Acoust Soc Am 99:1683–1692.

Lindblom B, Diehl RL (2001) Reconciling static and dynamic aspects of the speech process. J Acoust Soc Am 109:2380.

Lindblom B, Engstrand O (1989) In what sense is speech quantal? J Phonetics 17:107–121.

Lindblom B, Sundberg J (1971) Acoustical consequences of lip, tongue, jaw and larynx movement. J Acoust Soc Am 50:1166–1179.

Lindblom B, Lubker J F, Gay T (1979) Formant frequencies of some fixed-mandible vowels and a model of speech programming by predictive simulation. J Phonetics 7:147–161.

Lindblom B, Brownlee SA, Lindgren R (1996) Formant undershoot and speaking styles: an attempt to resolve some controversial issues. In: Simpson AP, Pätzold M (eds) Sound Patterns of Connected Speech: Description, Models and Explanation, Proceedings of the Symposium Held at Kiel University on 14–15 June 1996, Arbeitsberichte 31, Institut für Phonetik und digitale Sprachverarbeitung, Universität Kiel, pp. 119–129.

Lisker L (1957) Closure duration and the intervocalic voiced-voiceless distinctions in English. Lang 33:42–49.

Lisker L (1972) Stop duration and voicing in English. In: Valdman A (ed) Papers in Linguistics and Phonetics to the Memory of Pierre Delattre. The Hague: Mouton, pp. 339–343.

Lisker L (1975) Is it VOT or a first-formant transition detector? J Acoust Soc Am 57:1547–1551.

Lisker L (1986) "Voicing" in English: a catalogue of acoustic features signaling /b/ vsersus /p/ in trochees. Lang Speech 29:3–11.

Lisker L, Abramson A (1964) A cross-language study of voicing in initial stops: acoustical measurements. Word 20:384–422.

Lisker L, Abramson A (1970) The voicing dimension: some experiments in comparative phonetics. Proceedings 6th Intern Congr Phon Sci, Prague 1967. Prague: Academia, pp. 563–567.

Lively SE, Pisoni DB, Summers VW, Bernacki RH (1993) Effects of cognitive workload on speech production: acoustic analyses and perceptual consequences. J Acoust Soc Am 93:2962–2973.

Longchamp F (1981) Multidimensional vocalic perceptual space: How many dimensions? J Acoust Soc Am 69:S94(abstract).

MacNeilage PM (1969) A note on the relation between tongue elevation and glottal elevation in vowels. Monthly Internal Memorandum, University of California, Berkeley, January 1969, pp. 9–26.

Maddieson I (1984) Patterns of Sound. Cambridge: Cambridge University Press.

Maeda S (1991) On articulatory and acoustic variabilities. J Phonetics 19:321–331.

Martinet A (1955) Économie des Changements Phonétiques. Berne: Francke.

Massaro DW, Cohen MM (1976) The contribution of fundamental frequency and voice onset time to the /zi/-/si/ distinction. J Acoust Soc Am 60:704–717.

McCarthy JJ (1988) Feature geometry and dependency: a review. Phonetica 43:84–108.

McGowan RS (1994) Recovering articulatory movement from formant frequency trajectories using task dynamics and a genetic algorithm: preliminary model tests. Speech Comm 14:19–48.

Miller JD (1989) Auditory-perceptual interpretation of the vowel. J Acoust Soc Am 85:2088–2113.

Miller JD, Wier CC, Pastore RE, Kelly WJ, Dooling RJ (1976) Discrimination and labeling of noise-buzz sequences with varying noise-lead times: an example of categorical perception. J Acoust Soc Am 60:410–417.

Miller RL (1953) Auditory tests with synthetic vowels. J Acoust Soc Am 25:114–121.

Moon S-J (1990) Durational aspects of clear speech. Unpublished master's report, University of Texas at Austin.

Moon S-J (1991) An acoustic and perceptual study of undershoot in clear and citation-form speech. Unpublished PhD dissertation, University of Texas at Austin.

Moon S-J, Lindblom B (1994) Interaction between duration, context and speaking style in English stressed vowels. J Acoust Soc Am 96:40–55.

Moon S-J, Lindblom B, Lame J (1995) A perceptual study of reduced vowels in clear and casual speech. In: Elenius K, Branderud P (eds) Proceedings ICPhS 95 Stockholm, vol 2, pp. 670–677.

Myers S (1997) Expressing phonetic naturalness in phonology. In: Roca I (ed) Derivations and Constraints in Phonology. Oxford: Oxford University Press, pp. 125–152.

Nearey TM (1989) Static, dynamic, and relational properties in vowel perception. J Acoust Soc Am 85:2088–2113.

Nearey T, Assmann P (1986) Modeling the role of inherent spectral change in vowel identification. J Acoust Soc Am 80:1297–1308.

Nelson WL (1983) Physical principles for economies of skilled movments. Biol Cybern 46:135–147.

Nelson WL, Perkell J, Westbury J (1984) Mandible movements during increasingly rapid articulations of single syllables: preliminary observations. J Acoust Soc Am 75:945–951.

Nord L (1975) Vowel reduction—centralization or contextual assimilation? In: Fant G (ed) Proceedings of the Speech Communication Seminar, vol. 2, Stockholm: Almqvist &Wiksell, pp. 149–154.

Nord L (1986) Acoustic studies of vowel reduction in Swedish, 19–36 in STL-QPSR 4/1986, (Department of Speech Communication, RIT, Stockholm).

Ohala JJ (1990) The phonetics and phonology of assimilation. In: Kingston J, Beckman ME (eds) Papers in Laboratory Phonology I: Between the Grammar and Physics of Speech. Cambridge: Cambridge University Press, pp. 258–275.

Ohala JJ, Eukel BM (1987) Explaining the intrinsic pitch of vowels. In: Channon R, Shockey L (eds) In honor of Ilse Lehiste, Dordrecht: Foris, pp. 207–215.

Öhman S (1966) Coarticulation in VCV utterances: spectrographic measurements. J Acoust Soc Am 39:151–168.

Öhman S (1967) Numerical model of coarticulation. J Acoust Soc Am 41:310–320.

Parker EM, Diehl RL, Kluender KR (1986) Trading relations in speech and nonspeech. Percept Psychophys 34:314–322.

Passy P (1890) Études sur les Changements Phonétiques et Leurs Caractères Généraux. Paris: Librairie Firmin-Didot.

Payton KL, Uchanski RM, Braida LD (1994) Intelligibility of conversational and clear speech in noise and reverberation for listeners with normal and impaired hearing. J Acoust Soc Am 95:1581–1592.

Perkell JS, Matthies ML, Svirsky MA, Jordan MI (1993) Trading relations between tongue-body raising and lip rounding in production of the vowel /u/: a pilot "motor equivalence" study. J Acoust Soc Am 93:2948–2961.

Petersen, NR (1983) The effect of consonant type on fundamental frequency and larynx height in Danish. Annual Report of the Institute of Phonetics, University of Copenhagen 17:55–86.

Peterson GE, Barney HL (1952) Control methods used in the study of the vowels. J Acoust Soc Am 24:175–184.

Peterson GE, Lehiste I (1960) Duration of syllable nuclei in English. J Acoust Soc Am 32:693–703.

Pickett JM (1980) The Sounds of Speech Communication. Baltimore, MD: University Park.

Pisoni DB (1977) Identification and discrimination of the relative onset time of two component tones: implications for voicing perception in stops. J Acoust Soc Am 61:1352–1361.

Plomp R (1970) Timbre as multidimensional attribute of complex tones. In: Plomp R, Smoorenburg GF (eds) Frequency Analysis and Periodicity Detection in Hearing. Leiden: Sijthoff, pp. 397–414.

Port RF, Dalby J (1982) Consonant/vowel ratio as a cue for voicing in English. Percept Psychophys 32:141–152.

Potter RK, Steinberg JC (1950) Toward the specification of speech. J Acoust Soc Am 22:807–820.

Potter RK, Kopp G, Green H (1947) Visible Speech. New York: Van Nostrand Reinhold.

Raphael LF (1972) Preceding vowel duration as a cue to the perception of the voicing characteristic of word-final consonants in English. J Acoust Soc Am 51:1296–1303.

Repp BH (1979) Relative amplitude of aspiration noise as a voicing cue for syllable-initial stop consonants. Lang Speech 22:173–189.

Repp BH, Liberman AM, Eccardt T, Pesetsky D (1978) Perceptual integration of acoustic cues for stop, fricative, and affricate manner. J Exp Psychol 4:621–637.

Riordan CJ (1977) Control of vocal-tract length in speech. J Acoust Soc Am 62:998–1002.

Rosen S, Howell P (1987) Auditory, articulatory, and learning explanations of categorical perception in speech. In: Harnad S (ed) Categorical Perception. Cambridge: Cambridge University Press, pp. 113–160.

Saltzman E (1995) Intergestural timing in speech production: data and modeling. In: Elenius K, Branderud P (eds) Proceedings ICPhS 95 Stockholm, vol 2, pp. 84–91.

Saussure de F (1916) Cours de linguistic générale. Paris: Payot. (English translation by R. Harris: Course in General Linguistics. Lasalle, IL: Open Court, 1986.)

Schroeder MR, Atal BS, Hall JL (1979) Objective measure of certain speech signal degradations based on masking properties of human auditory perception. In: Lindblom B, Öhman S (eds) Frontiers of Speech Communication Research. London: Academic Press, pp. 217–229.

Schroeter J, Sondhi MM (1992) Speech coding based on physiological models of speech production. In: Furui S, Sondhi MM (eds) Advances in Speech Signal Processing, New York: M. Dekker, pp. 231–268.

Silverman K (1987) The structure and processing of fundamental frequency contours. Unpublished PhD dissertation, University of Cambridge.

Sinex DG, McDonald LP, Mott JB (1991) Neural correlates of nonmonotonic temporal acuity for voice onset time. J Acoust Soc Am 90:2441–2449.

Son van RJJH, Pols LCW (1990) Formant frequencies of Dutch vowels in a text, read at normal and fast rate. J Acoust Soc Am 88:1683–1693.

Son van RJJH, Pols LCW (1992) Formant movements of Dutch vowels in a text, read at normal and fast rate. J Acoust Soc Am 92:121–127.

Stark J, Lindblom B, Sundberg J (1996) APEX an articulatory synthesis model for experimental and computational studies of speech production. In: Fonetik 96, TMH-QPSR 2/1996, (KTH, Stockholm); pp. 45–48.

Stevens KN (1972) The quantal nature of speech: evidence from articulatory-acoustic data. In: David EE, Denes PB (eds) Human Communication: A Unified View. New York: McGraw-Hill, pp. 51–66.

Stevens KN (1989) On the quantal nature of speech. J Phonetics 17:3–45.

Stevens KN (1998) Acoustic Phonetics. Cambridge, MA: MIT Press.

Stevens KN, Blumstein SE (1978) Invariant cues for place of articulation in stop consonants. J Acoust Soc Am 64:1358–1368.

Stevens KN, Blumstein SE (1981) The search for invariant acoustic correlates of phonetic features. In: Eimas PD, Miller JL (eds) Perspectives on the Study of Speech. Hillsdale, NJ: Erlbaum, pp. 1–38.

Stevens KN, Keyser SJ (1989) Primary features and their enhancement in consonants. Lang 65:81–106.

Stevens KN, Keyser SJ, Kawasaki H (1986) Toward a phonetic and phonological theory of redundant features. In: Perkell JS, Klatt DH (eds) Invariance and Variability in Speech Processes. Hillsdale, NJ: Erlbaum, pp. 426–449.

Strange W (1989a) Evolving theories of vowel perception. J Acoust Soc Am 85:2081–2087.

Strange W (1989b) Dynamic specification of coarticulated vowels spoken in sentence context. J Acoust Soc Am 85:2135–2153.

Studdert-Kennedy M (1987) The phoneme as a perceptuo-motor structure. In: Allport A, MacKay D, Prinz W, Scheerer E (eds) Language, Perception and Production, New York: Academic Press.

Studdert-Kennedy M (1989) The early development of phonology. In: von Euler C, Forsberg H, Lagercrantz H (eds) Neurobiology of Early Infant Behavior. New York: Stockton.

Summerfield AQ, Haggard M (1977) On the dissociation of spectral and temporal cues to the voicing distinction in initial stop consonants. J Acoust Soc Am 62:435–448.

Summers WV (1987) Effects of stress and final consonant voicing on vowel production: articulatory and acoustic analyses. J Acoust Soc Am 82:847–863.

Summers WV (1988) F_1 structure provides information for final-consonant voicing. J Acoust Soc Am 84:485–492.

Summers WV, Pisoni DB, Bernacki RH, Pedlow RI, Stokes MA (1988) Effects of noise on speech production: acoustic and perceptual analyses. J Acoust Soc Am 84:917–928.

Sussman HM (1991) The representation of stop consonants in three-dimensional acoustic space. Phonetica 48:18–31.

Sussman HM, McCaffrey HA, Matthews SA (1991) An investigation of locus equations as a source of relational invariance for stop place categorization. J Acoust Soc Am 90:1309–1325.

Sussman HM, Hoemeke KA, Ahmed FS (1993) A cross-linguistic investigation of locus equations as a phonetic descriptor for place of articulation. J Acoust Soc Am 94:1256–1268.

Syrdal AK (1985) Aspects of a model of the auditory representation of American English vowels. Speech Comm 4:121–135.

Syrdal AK, Gopal HS (1986) A perceptual model of vowel recognition based on the auditory representation of American English vowels. J Acoust Soc Am 79:1086–1100.

Traunmüller H (1981) Perceptual dimension of openness in vowels. J Acoust Soc Am 69:1465–1475.

Traunmüller H (1984) Articulatory and perceptual factors controlling the age- and sex-conditioned variability in formant frequencies of vowels. Speech Comm 3:49–61.

Traunmüller H (1985) The role of the fundamental and the higher formants in the perception of speaker size, vocal effort, and vowel openness. Paper presented at the Franco-Swedish Seminar on Speech, SFA, Grenoble, France, April.

Trubetzkoy NS (1939) Grundzüge der Phonologie. Travaux du Cercle linguistique de Prague 7. (English translation by C. Baltaxe: Principles of Phonology. Berkeley: University of California Press, 1969.)

Vennemann T, Ladefoged P (1973) Phonetic features and phonological features. Lingua 32:61–74.

Vilkman E, Aaltonen O, Raimo L, Arajärvi P, Oksanen H (1989) Articulatory hyoid-laryngeal changes vs. cricothyroid activity in the control of intrinsic F_0 of vowels. J Phonetics 17:193–203.

Walsh MA, Diehl RL (1991) Formant transition duration and amplitude rise time as cues to the stop/glide distinction. Q J Exp Psychol 43A:603–620.

Wilhelms-Tricarico R, Perkell JS (1995) Towards a physiological model of speech production. In: Elenius K, Branderud P (eds) Proceedings Intern Congr Phon Sci 95 Stockholm, vol 2, pp. 68–75.

Zahorian S, Jagharghi A (1993) Spectral shape features versus formants as acoustic correlates for vowels. J Acoust Soc Am 94:1966–1982.

Zwicker E, Feldtkeller R (1967) Das Ohr als Nachrichtenempfänger. Stuttgart: Hirzel.

4
Physiological Representations of Speech

Alan Palmer and Shihab Shamma

1. Introduction

This chapter focuses on the physiological mechanisms underlying the processing of speech, particularly as it pertains to the signal's pitch and timbre, as well as its spectral shape and temporal dynamics (cf. Avendaño et al., Chapter 2). We will first describe the neural representation of speech in the peripheral and early stages of the auditory pathway, and then go on to present a more general perspective for central auditory representations. The utility of different coding strategies for various speech features will then be evaluated. Within this framework it is possible to provide a cohesive and comprehensive description of the representation of steady-state vowels in the early auditory stages (auditory nerve and cochlear nucleus) in terms of average-rate (spatial), temporal, and spatiotemporal representations. Similar treatments are also possible for dynamic spectral features such as voice onset time, formant transitions, sibilation, and pitch (cf. Avendano et al., Chapter 2; Diehl and Lindblom, Chapter 3, for discussion of these speech properties). These coding strategies will then be evaluated as a function of speech context and suboptimum listening conditions (cf. Assmann and Summerfield, Chapter 5), such as those associated with background noise and whispered speech. At more central stages of the auditory pathway, the physiological literature is less detailed and contains many gaps, leaving considerable room for speculation and conjecture.

1.1 The Anatomy and Connections of the Auditory Pathway

In this section we briefly review the anatomy of the auditory pathway, to provide an appropriate context for the physiological material that follows.

The sole afferent pathway from the cochlea to the central nervous system is the auditory nerve (AN). The cell bodies associated with the fibers of the AN are located in the spiral ganglion of the cochlea. They derive their afferent input from synapses under the inner and outer hair cells. Approximately

90% to 95% of the fibers in the mammalian AN innervate inner hair cells (Spoendlin 1972; Brown 1987). The spiral ganglion cells project centrally via the AN, innervating the principal cells of the cochlear nucleus complex (Ruggero et al. 1982; Brown et al. 1988; Brown and Ledwith 1990). Virtually all of our current knowledge concerning the activity of the AN derives from axons innervating solely the inner hair cells. The function of the afferents (type II fibers) innervating the outer hair cells is currently unknown.

The major connections of the auditory nervous system are illustrated in Figure 4.1. All fibers of the AN terminate and form synapses in the cochlear nucleus, which consists of three anatomically distinct divisions. On entry into the cochlear nucleus, the fibers of the AN bifurcate. One branch innervates the anteroventral cochlear nucleus (AVCN), while the other innervates both the posteroventral (PVCN) and dorsal (DCN) (Lorente de No 1933a,b) divisions of the same nucleus. The cochlear nucleus contains several principal cell types—spherical bushy cells, globular bushy cells, multipolar cells, octopus cells, giant cells, and fusiform cells (Osen 1969; Brawer et al. 1974)—that receive direct input from the AN, and project out of the cochlear nucleus in three separate fiber tracts: the ventral, intermediate, and dorsal acoustic striae. There are other cell types that have been identified as interneurons interconnecting cells in the dorsal, posteroventral, and ventral divisions. The cochlear nucleus is the first locus in the auditory pathway for transformation of AN firing patterns; its principal cells constitute separate, parallel processing pathways for encoding different properties of the auditory signal.

The relatively homogeneous responses characteristic of the AN are transformed in the cochlear nucleus by virtue of four physiological properties: (1) the pattern of afferent inputs, (2) the intrinsic biophysical properties of the cells, (3) the interconnections among cells within and between the cochlear nuclei, and (4) the descending inputs from inferior colliculus (IC), superior olive and cortex. The largest output pathway (the ventral acoustic stria), arising in the ventral cochlear nucleus from spherical cells, conveys sound-pressure-level information from one ear to the lateral superior olive (LSO) of the same side as well as timing information to the medial superior olive (MSO) of both sides (Held 1893; Lorente de No 1933a; Brawer and Morest 1975), where binaural cues for spatial sound location are processed. Axons from globular bushy cells also travel in the ventral acoustic stria to indirectly innervate the LSO to provide sound level information from the other ear. Octopus cells, which respond principally to the onset of sounds, project via the intermediate acoustic stria to the periolivary nuclei and to the ventral nucleus of the lateral lemniscus (VNLL); however, the function of this pathway is currently unknown. The dorsal acoustic stria carries axons from fusiform and giant cells of the dorsal cochlear nucleus directly to the central nucleus of the contralateral inferior colliculus, bypassing the superior olive. This pathway may be important for

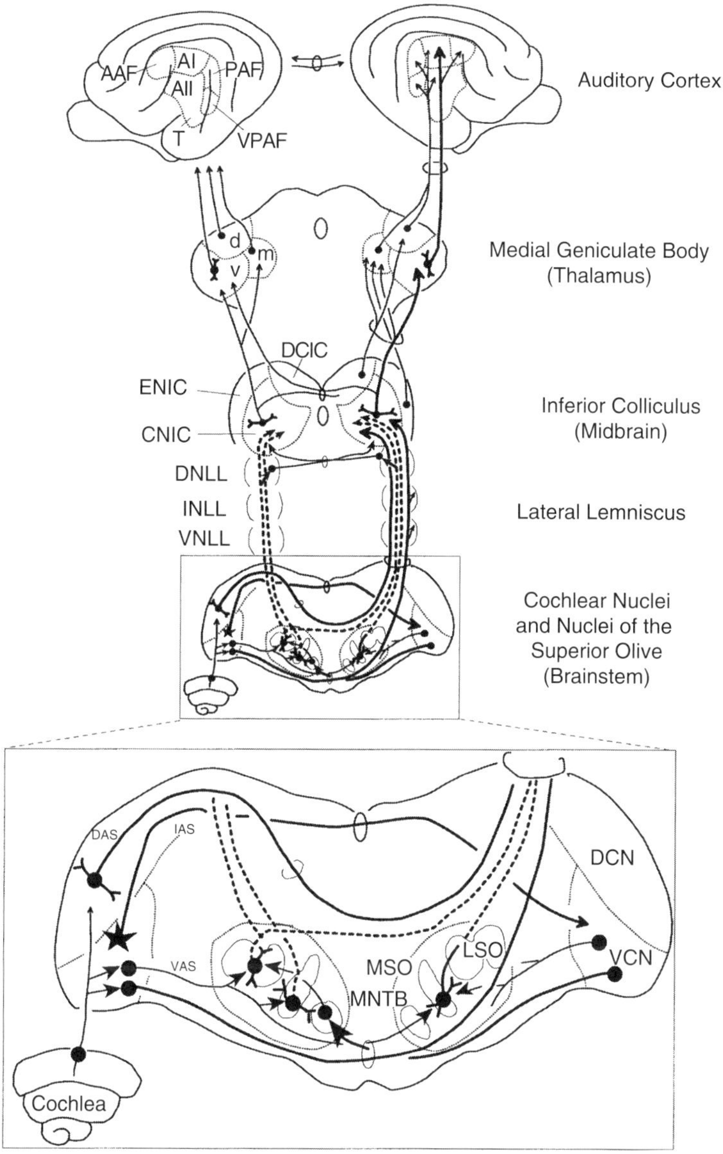

FIGURE 4.1. The ascending auditory pathway. AAF, anterior auditory field; PAF, posterior auditory field; AI, primary auditory cortex; AII, secondary auditory cortex; VPAF, ventroposterior auditory field; T, temporal; ENIC, external nucleus of the inferior colliculus; DCIC, dorsal cortex of the inferior colliculus; CNIC, central nucleus of the inferior colliculus; DNLL, dorsal nucleus of the lateral lemniscus; INLL, intermediate nucleus of the lateral lemniscus; VNLL, ventral nucleus of the lateral lemniscus; DAS, dorsal acoustic stria; IAS, intermediate acoustic stria; VAS, ventral acoustic stria; MSO, medial superior olive; MNTB, medial nucleus of the trapezoid body; LSO, lateral superior olive; DCN, dorsal cochlear nucleus; VCN, ventral cochlear nucleus. (Modified from Brodal 1981, with permission.)

processing spectral cues sculpted by the pinnae germane to localizing sound (Young et al. 1992). Some multipolar cells in the ventral cochlear nucleus also project directly to the IC via the ventral acoustic stria and lateral lemniscus. This pathway may convey a spatiotopic code associated with the spectra of complex sounds (Sachs and Blackburn 1991). An inhibitory commissural pathway, of unknown function, connects the cochlear nuclei of the opposing sides (Cant and Gaston 1982; Wenthold 1987; Shore et al. 1992).

The superior olivary complex is the site of the first major convergence of input from the two ears and is involved in the processing of cues for the localization of sounds in space. The cells of the MSO receive direct excitatory input from the large spherical bushy cells of both cochlear nuclei and project ipsilaterally to the central nucleus of the inferior colliculus (CNIC) (Stotler 1953; Harrison and Irving 1965; Warr 1966, 1982; Adams 1979; Brunso-Bechtold et al. 1981; Henkel and Spangler 1983). The LSO receives direct excitatory input from the spherical bushy cells on the same side and indirect inhibitory input from the globular bushy cells on the other side (Stotler 1953; Warr 1966, 1972, 1982; Cant and Casseday 1986). The LSO projects bilaterally to the CNIC (Stotler 1953; Adams 1979; Brunso-Bechtold et al. 1981; Glendenning and Masterton 1983; Shneiderman and Henkel 1987).

The output of the superior olive joins the fibers ascending from the cochlear nucleus to the inferior colliculus to form the lateral lemniscus tract. The nuclei of the lateral lemniscus are composed of neurons among the fibers of the tract, which are innervated both directly and via collateral branches from the ascending axons. The lemniscal outputs innervate the CNIC and continue to the medial geniculate body (Adams 1979; Brunso-Bechtold et al. 1981; Kudo 1981).

The inferior colliculus consists of several cytoarchitecturally distinct regions, the most important of which, for the present purposes, is the central nucleus (CNIC) (Morest and Oliver 1984; Oliver and Shneiderman 1991). Almost all ascending (or descending) pathways synapse in the inferior colliculus. The IC thus represents a site of extreme convergence of information that has been processed in parallel in various brain stem nuclei. The CNIC has a complicated structure of laminae, formed of disk-shaped cells, interconnected by stellate cells. All of the afferent pathways project topographically onto segregated parts of this laminar structure, which must therefore determine the way in which information from the lower levels is combined (Roth et al. 1978; Aitkin and Schuck 1985; Maffi and Aitkin 1987; Oliver and Shneiderman 1991). Both of the principal cell types in CNIC project to the ventral division of the medial geniculate body, which in turn projects to the primary and secondary auditory cortex (AI and AII).

The auditory cortex has been divided into a number of adjacent auditory fields using cytoarchitectural and electrophysiological criteria (see Fig. 4.1).

More detailed accounts of the neuroanatomy and neural processing may be found in several review volumes (Berlin 1984; Irvine 1986; Edelman et al. 1988; Pickles 1988; Altschuler et al. 1991; Popper and Fay 1992; Webster et al. 1992; Moore 1995; Eggermont 2001).

1.2 Overview of the Analysis of Sound by the Auditory Periphery

Acoustic signals pass through the outer ear on their journey through the auditory pathway. At these most peripheral stations the magnitude of the incoming signal is altered at certain frequencies as a consequence of the resonance structure of both the pinna (Batteau 1967) and external meatus (Shaw 1974; Rosowski 1995). Frequencies between 2.5 and 5 kHz are amplified by as much as 20 dB as a result of such resonances (Shaw 1974), accounting for the peak sensitivity of human audibility. The magnitude of energy below 500 Hz is significantly attenuated as a consequence of impedance characteristics of the middle ear, accounting for the reduction in sensitivity in this lowest segment of the speech spectrum (cf. Rosowski 1995; but cf. Ruggero and Temchin 2002 for a somewhat different perspective). The function of the outer and middle ears can be approximated by a simple bandpass filter (cf. Hewitt et al. 1992; Nedzelnitsky 1980) simulating much of the frequency-dependent behavior of the auditory periphery.

The cochlea serves as a spectral analyzer of limited precision, separating complex signals (i.e., containing many frequencies) into their constituent components. The sharply tuned mechanical vibrations of the cochlear partition produce sharply tuned receptor potentials in both the inner and outer hair cells (Russell and Sellick 1978). The activity of the inner hair cells is transmitted to the brain via the depolarizing effect on the AN, while the outer hair cells appear to exert most (if not all) of their influence by modifying the manner in which the underside of the tectorial membrane articulates with cilia of the inner hair cells (thus providing some measure of amplification and possibly a sharpening in spectral tuning as well) (Pickles 1988).

In AN fibers, sinusoidal stimulation produces an increase in the discharge rate above the resting or spontaneous level (Kiang et al. 1965; Ruggero 1992). Each fiber responds to only a limited range of frequencies. Its tuning is determined in large part by the position of the fiber along the cochlear partition. Fibers innervating the basal end are most sensitive to high frequencies (in the human, above 10 kHz), while fibers located in the apex are most sensitive to frequencies below 1 kHz. In between, fibers exhibit a graded sensitivity.

A common means with which to quantify the spectral sensitivity of a fiber is by varying both frequency and sound pressure level (SPL) of a sinusoidal signal and measuring the resulting changes in discharge activity relative to its background ("spontaneous") level. If the measurement is in terms of a fixed quantity (typically 20%) above this spontaneous level (often referred to as an "iso-rate" curve), the result is referred to as a frequency threshold (or "tuning") curve (FTC). The frequency at the intensity minimum of the FTC is termed the best or characteristic frequency (CF) and is an indication of the position along the cochlear partition of the hair cell that it innervates (see Liberman and Kiang 1978; Liberman 1982; Greenwood 1990).

Alternatively, if the frequency selectivity of the fiber is measured by keeping the SPL of the variable-frequency signal constant and measuring the absolute magnitude of the fiber's firing rate, the resulting function is referred to as an "iso-input" curve or "response area" (cf. Brugge et al. 1969; Ruggero 1992; Greenberg 1994). Typically, the fiber's discharge is measured in response to a broad range of input levels, ranging between 10 and 80 dB above the unit's rate threshold.

The FTCs of the fibers along the length of the cochlea can be thought of as an overlapping series of bandpass filters encompassing the hearing range of the animal. The most sensitive AN fibers exhibit minimum thresholds matching the behavioral audiogram (Kiang 1968; Liberman 1978). The frequency tuning observed in FTCs of AN fibers is roughly commensurate with behavioral measures of frequency selectivity (Evans et al. 1992).

The topographic organization of frequency tuning along the length of the cochlea gives rise to a tonotopic organization of responses to single tones in every major nucleus along the auditory pathway from cochlea to cortex (Merzenich et al. 1977). In the central nervous system large areas of tissue may be most sensitive to the same frequency, thus forming isofrequency laminae in the brain stem, midbrain and thalamus, and isofrequency bands in the cortex. It is this topographic organization that underlies the classic "place" representation of the spectra of complex sounds. In this representation, the relative spectral amplitudes are reflected in the strength of the activation (i.e., the discharge rates) of the different frequency channels along the tonotopic axis.

Alternatively, the spectral content of a signal may be encoded via the timing of neuronal discharges (rather than by the identity of the location along the tonotopic axis containing the most prominent response in terms of average discharge rate). Impulses are initiated in AN fibers when the hair cell is depolarized, which only occurs when their stereocilia are bent toward the longest stereocilium. Bending in this excitatory direction is caused by viscous forces when the basilar membrane moves toward the scala vestibuli. Thus, in response to low-frequency sounds the impulses in AN fibers do not occur randomly in time, but rather at particular times or phases with respect to the waveform. This phenomenon has been termed phase locking (Rose et al. 1967), and has been demonstrated to occur in all vertebrate auditory systems (see Palmer and Russell 1986, for a review). In the cat, the precision of phase locking begins to decline at about 800 Hz and is altogether absent for signals higher than 5 kHz (Kiang et al. 1965; Rose et al. 1967; Johnson 1980). Phase locking can be detected as an temporal entrainment of spontaneous activity up to 20 dB below the threshold for discharge rate, and persists with no indication of clipping at levels above the saturation of the fiber discharge rate (Rose et al. 1967; Johnson 1980; Evans 1980; Palmer and Russell 1986).

Phase locking in AN fibers gives rise to the classic temporal theory of frequency representation (Wundt 1880; Rutherford 1886). Wever (1949) suggested that the signal's waveform is encoded in terms of the timing pattern of an ensemble of AN fibers (the so-called volley principle) for frequencies below 5 kHz (with time serving a principal role below 400 Hz and combining with "place" for frequencies between 400 and 5000 Hz).

Most phase-locked information must be transformed to another representation at some level of the auditory pathway. There is an appreciable decline in neural timing information above the level of the cochlear nucleus and medial superior olive, with the upper limit of phase-locking being about 100 Hz at the pathway's apex in the auditory cortex (Schreiner and Urbas 1988; Phillips et al. 1991).

Already at the level of the cochlear nucleus there is a wide variability in the ability of different cell populations to phase lock. Thus, a certain proportion of multipolar cells (which respond most prominently to tone onsets) and spherical bushy cells (whose firing patterns are similar in certain respects to AN fibers) phase lock in a manner not too dissimilar from that of AN fibers (Lavine 1971; Bourk 1976; Blackburn and Sachs 1989; Winter and Palmer 1990a; Rhode and Greenberg 1994b). Other multipolar cells (which receive multiple synaptic contacts and manifest a "chopping" discharge pattern) have a lower cut-off frequency for phase locking than do AN fibers; the decline in synchrony starts at a few hundred hertz and falls off to essentially nothing at about 2 kHz (in cat—van Gisbergen et al. 1975; Bourk 1976; Young et al. 1988; in guinea pig—Winter and Palmer 1990a; Rhode and Greenberg 1994b). While few studies have quantified phase locking in the DCN, it appears to only occur to very low frequencies (Lavine 1971; Goldberg and Brownell 1973; Rhode and Greenberg 1994b). In the inferior colliculus only 18% of the cells studied by Kuwada et al. (1984) exhibited an ability to phase lock, and it was seldom observed in response to frequencies above 600 Hz. Phase locking has not been reported to occur in the primary auditory cortex to stimulating frequencies above about 100 Hz (Phillips et al. 1991).

2. Representations of Speech in the Early Auditory System

Spectrally complex signals, such as speech, can be quantitatively described in terms of a linear summation of its frequency constituents. This "linear" perspective has been used to good effect, particularly for describing the response of AN fibers to such signals (e.g., Ruggero 1992). However, it is becoming increasingly apparent that the behavior of neurons at even this peripheral level is not always predictable from knowledge of the response to sinusoidal signals. Phenomena such as two-tone suppression (Sachs and Kiang 1968; Arthur et al. 1971) and the nonlinearities involved in the representation of more than a single frequency component in the phase-locked discharge (Rose et al. 1971) point to the interdependence of fiber responses to multiple stimulus components. Extrapolations based on tonal responses, therefore, are not always adequate for understanding how multicomponent stimuli are represented in the AN (cf. Ruggero 1992).

Nevertheless, it is useful to initially describe the basic response patterns of auditory neurons to such "simple" signals as sinusoids (Kiang et al. 1965; Evans 1972) in terms of saturation (Kiang et al. 1965; Sachs and Abbas 1974; Evans and Palmer 1979), adaptation (Kiang et al. 1965; Smith 1979), and phase locking (Rose et al. 1971; Johnson 1980; Palmer and Russell 1986), and then extend these insights to more complex signals such as speech, noise, and complex tones.

Over the years there has been a gradual change in the type of acoustic signals used to characterize the response properties of auditory neurons. Early on, sinusoidal signals and clicks were used almost exclusively. In recent years spectrally complex signals, containing many frequency components have become more common. Speech sounds are among the most spectrally complex signals used by virtue of their continuously (and often rapidly) changing spectral characteristics.

In the following sections we describe various properties of neural responses to steady-state speech stimuli. In reality, very few elements of the speech signal are in fact steady state; however, for the purpose of analytical tractability, we will assume that under limiting conditions many spectral elements of speech are in fact steady state in nature.

Virtually all nuclei of the central auditory pathway are tonotopically organized (i.e., exhibit a spatial gradient of neural activity correlated with the signal frequency). Speech features (such as those described by Diehl and Lindblom Chapter 3) distinguished principally by spectral properties should therefore exhibit different representations across the tonotopic axes of the various auditory nuclei, whether in terms of average or synchronized discharge rate.

2.1 Encoding the Shape of the Acoustic Spectrum in the Early Auditory System

How does the auditory system encode and utilize response patterns of the auditory nerve to generate stable and robust representations of the acoustic spectrum? This question is difficult to answer because of the multiplicity of cues available, particularly those of a spectral and temporal nature. This duality arises because the basilar membrane segregates responses tonotopically, functioning as a distributed, parallel bank of bandpass filters. The filter outputs encode not only the magnitude of the response, but also its waveform by phase-locked patterns of activity. The frequency of a signal or a component in a complex is available both from the tonotopic place that is driven most actively (the "place code"), as well as from the periodicity pattern associated with phase-locked responses of auditory neurons (the "temporal code").

Various schemes that utilize one or both of these response properties have been proposed to explain the encoding of the acoustic spectrum. These are described below, together with findings from a variety of studies that have focused on speech signals as the experimental stimuli. Whatever encoding strategy is employed to signal the important elements of speech in AN fibers, the neurons of the cochlear nucleus must either faithfully transmit the information or perform some kind of transformation. The earliest studies (Moore and Cashin 1974) at the cochlear nucleus level suggested that the temporal responses were sharpened (subjectively assessed from time histograms of the responses to vowels), and that the effect depended on stimulus context and the relative energy within excitatory and inhibitory parts of the unit response area (Rupert et al. 1977). Subsequent studies (Moore and Cashin 1976; Caspary et al. 1977) found that units that phase locked to pure tones also phase locked to the fundamental frequency as well as the lower two formants of vowel sounds. Other units that responded similarly to tones at their best frequency did not necessarily respond similarly to speech (Rupert et al. 1977). Many of these early results appear to be consistent with later studies, but because they did not employ rigorous classification or histological controls, they are often difficult to compare directly with more recent studies that have measured the speech-driven responses of fully classified unit response types (Palmer et al. 1986; Kim et al. 1986; Kim and Leonard 1988; Blackburn and Sachs 1990; Winter and Palmer 1990b; Palmer and Winter 1992, Mandava et al. 1995; Palmer et al. 1996b; Recio and Rhode 2000).

2.1.1 Place Representations

In normal speech, vowels are often relatively stable and periodic over a limited interval of time. For this reason, it is possible to use synthetic stimuli that are completely stable as a coarse approximation. The vocalic spectrum consists of harmonics of the fundamental frequency (in the range of 80 to 300 Hz). Certain harmonics have greater amplitude than others, producing peaks (formants) that correspond to the resonant frequencies of the vocal tract (see Avendaño et al., Chapter 2). It is the frequency of the first and second formants that largely determines vowel identity (Peterson and Barney 1952; see Avendaño et al., Chapter 2; Diehl and Lindblom, Chapter 3). A major issue is whether the pattern of gross neural activity evoked at "places" within the tonotopically organized populations of neurons is sufficient for vowel identification, or whether the fine timing (phase locking) of the discharges must also be required (see detailed reviews by Sachs 1985; Sachs et al. 1988; Sachs and Blackburn 1991).

The most successful attempt to explore the adequacy of a mean rate-place code recorded the discharge rate of large numbers of AN fibers from a single ear of the cat in response to synthetic voiced vowels at different SPLs (Sachs and Young 1979). At moderate SPLs a clear place representation was evident; the vowels evoked more discharges in fibers with CFs near the formants than in those with CFs remote from the formants, as shown in Figure 4.2. In Figure 4.2A each symbol represents the mean discharge rate of a single AN fiber evoked by the vowel /ɛ/. The vast majority of these fibers are low-threshold, high-spontaneous-rate fibers (crosses). The continuous line in the figure is a moving window average of the discharge rates of the high-spontaneous-rate fibers. At lower stimulus levels (up to 48 dB SPL), the frequency positions of the first two or three formants (shown by arrows) are clearly signaled by regions of increased discharge. However, at higher presentation levels (58 dB SPL and above), the fibers with CFs between the formants increase their discharge, while fibers at the formant frequencies reach saturation, causing the formant-related peaks to lose definition and eventually to become obscured. An additional factor in this loss of definition is rate suppression of fibers, with CFs above the first formant, by energy at the first formant (Sachs and Young 1979). The progression of these level dependent changes in the population response is highlighted in Figure 4.2B, where only the moving window averages are shown superimposed. The loss of definition of the formant peaks in the place representation provided by the mean discharge rates of the most sensitive fibers was similar for all vowels presented (/ɪ/, /a/, /ɛ/).

Human vowel identification is unchanged at the highest sound levels. Data, such as those in Figure 4.2, suggest that the distribution of mean discharge rate across the tonotopic axis is inadequate, by itself, as an internal representation of the vowel at all sound levels. However, this is far too simplistic a conclusion, for several reasons. First, these plots of mean discharge rate include only fibers with high rates of spontaneous discharge; such fibers have low discharge-rate thresholds and narrow dynamic ranges of response. If a similar plot is made for fibers with low rates of spontaneous discharge (fewer in number, but having higher thresholds and wider dynamic ranges), formant-related peaks are still discernible at the highest sound levels used (see Young and Sachs 1979; Silkes and Geisler 1991). This is important because explanations of the responses to vowels found at the cochlear nucleus seem to require a contribution from AN fibers with low rates of spontaneous discharge and high thresholds (see below). Second, the mean rates shown in Figure 4.2 are for steady-state vowels. The dynamic range of AN fibers is wider for the onset component of discharge (Smith 1979), providing some extension to the range over which the mean rates signal the formant frequencies (Sachs et al. 1982). Third, the data have been collected in anesthetized animals; data suggest that the action of various feedback pathways (e.g., the middle-ear muscles and the efferent fibers innervating the cochlea, whose function is compromised under anesthesia) may

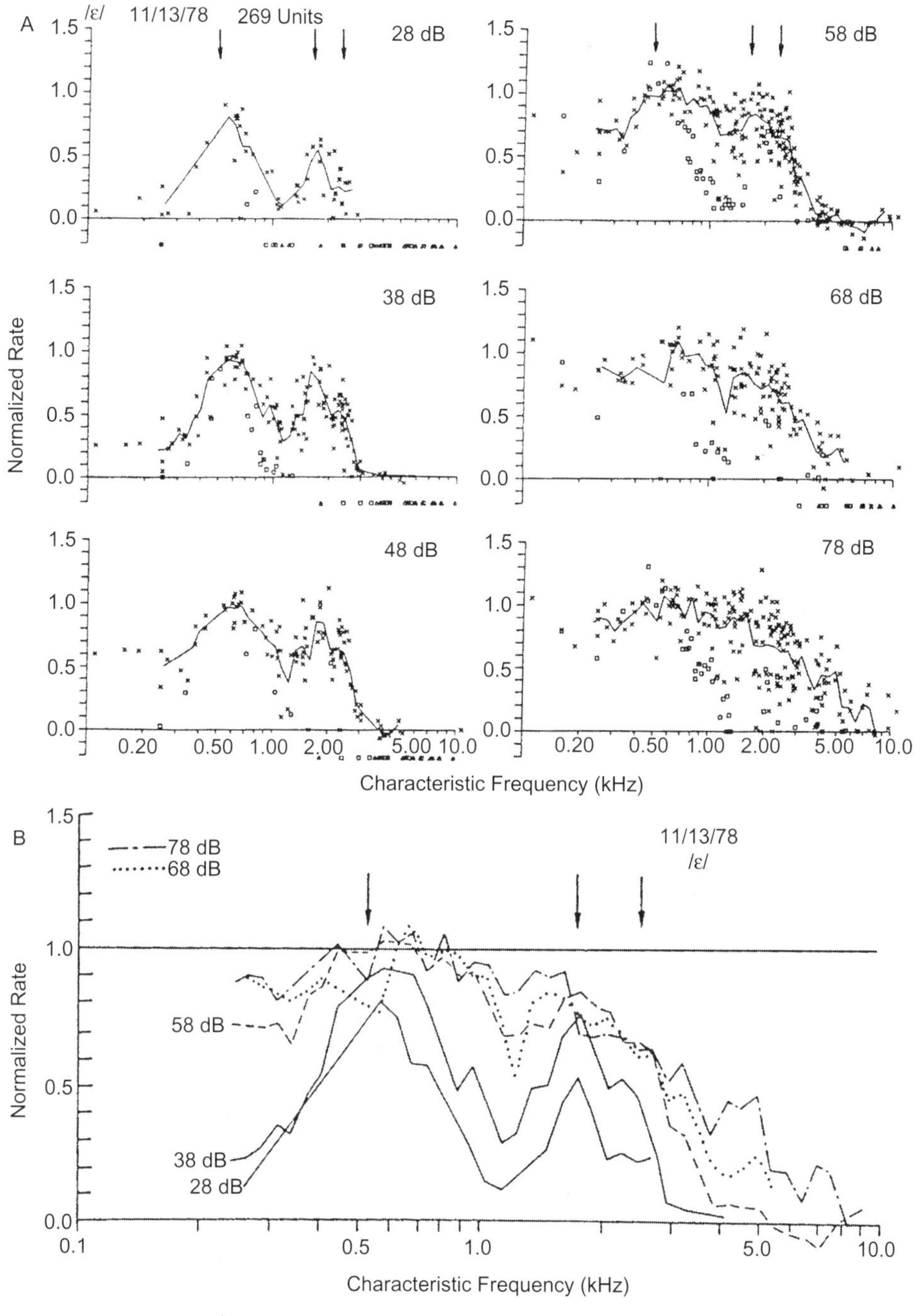

FIGURE 4.2. A: Plots of normalized rate vs. the fiber characteristic frequency for 269 fibers in response to the vowel /ε/. Units with spontaneous rates of less than 10/s are plotted as squares; others as crosses. The lines are the triangularly weighted moving window average taken across fibers with spontaneous rates greater than 10/s. The normalization of the discharge rate was achieved by subtracting the spontaneous rate and dividing by the driven rate (the saturation rate minus the spontaneous rate). B: Average curves from A. (From Sachs and Young 1979, with permission.)

preserve the ability of fibers to signal changes in signal level by changes in their discharge rate at high sound levels (Winslow and Sachs 1988; May and Sachs 1992). Fourth, a study of the distribution of mean discharge rates in response to a series of /ε/ vowels, differing only in the frequency of the second formant, concluded that, at least for SPLs of 70 dB, there was sufficient information to allow discrimination of different second formant frequencies among the vowels (Conley and Keilson 1995). Further, taking differences on a fiber-by-fiber basis between the discharge rates to different vowels provided very precise information that would provide for vowel identification performance better than that shown psychophysically (although obviously some form of precise internal memory would be required for this to operate). Finally, even at SPLs at which the formant structure is no longer evident (in terms of discharge rate), the distribution of mean rate across the population varies for different vowels, and hence discrimination could be made on the basis of the gross tonotopic profile (Winslow 1985).

Continuous background noise appears to adapt all AN fibers and thus potentially eliminates (or reduces) the contribution made by the wider dynamic range component of discharge near stimulus onset. Furthermore, at levels of background noise insufficient to prevent detection of second formant alterations, the low-spontaneous rate fibers seem no longer capable of sustaining an adequate mean-rate representation of the formant structure (Sachs et al. 1983). This set of results would seem to argue against the reliance on any simple place coding scheme under all listening conditions.

However, studies at the level of the cochlear nucleus have revealed a most intriguing finding. The distribution of mean discharge rates across populations of chopper units exhibits peaks at the positions of the formants even at sound levels at which such peaks are no longer discernible in the responses of AN fibers with low thresholds and high rates of spontaneous discharge (Fig. 4.3) (Blackburn and Sachs 1990; Recio and Rhode 2000). This spectral representation observed in chopper units also contains temporal information that "tags" the spectrum with f_0-relevant information useful for segregation of the source (Keilson et al. 1997).

At low SPLs the mean-rate profiles of choppers resembled the near-threshold profiles of high-spontaneous-rate AN fibers, and at high SPLs they resembled the profiles of low-spontaneous-rate fibers (compare equivalent levels in Figs. 4.2 and 4.3). This led Sachs and his colleagues (Sachs and Blackburn 1991) to suggest that the choppers respond differentially to low- and high-spontaneous-rate AN fibers as a function of sound level, and to propose possible mechanisms for this "selective listening" hypothesis (Winslow et al. 1987, Wang and Sachs 1995). It would be of considerable interest to know the effect of noise backgrounds on the representation of vowel spectra across the chopper population.

The place encoding of various other stationary speech spectra has also been investigated. Delgutte and Kiang (1984b) found that the distribution

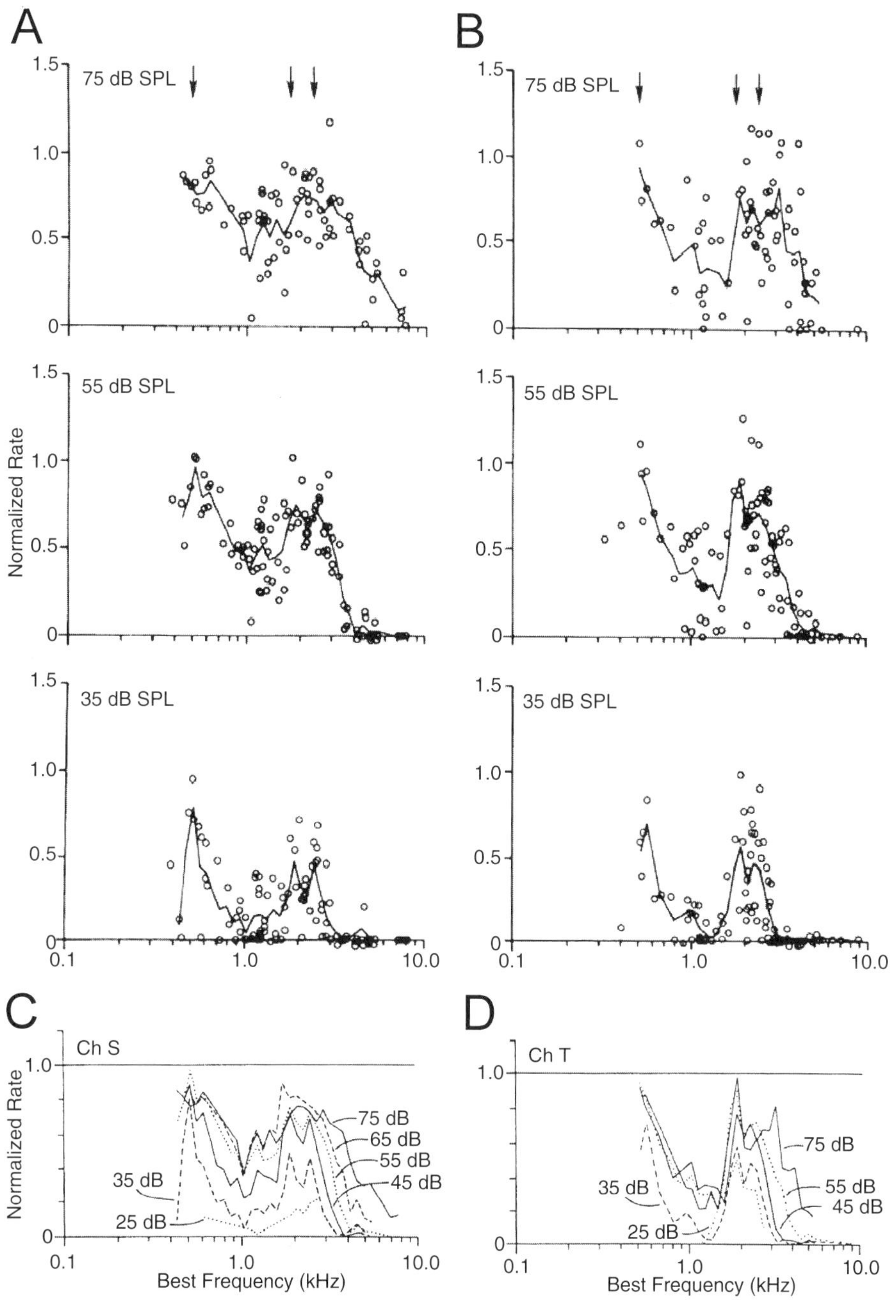

FIGURE 4.3. A: Plots of normalized rate vs. best frequency for sustained chopper (Ch S) units in response to the vowel /ɛ/. The lines show the moving window average based on a triangular weighting function 0.25 octaves wide. B: Plots as in A for transient chopper (Ch T) units. C: Average curves from A. D: Average curves from B. (From Blackburn and Sachs 1990, with permission.)

of mean discharge rate across the tonotopically ordered array of the most sensitive AN fibers was distinctive for each of four fricatives. The frequency range in which the mean discharge rates were highest corresponded to the spectral regions of maximal stimulus energy, a distinguishing characteristic of fricatives. One reason why this scheme is successful is that the SPL of fricatives in running speech is low compared to that of vowels. Because much of the energy in most fricatives is the portion of the spectrum above the limit of phase locking, processing schemes based on distribution of temporal patterns (see below) were less successful for fricatives (only /x/ and /s/ had formants within the range of phase locking; Delgutte and Kiang 1984b).

2.1.2 Temporal Representations

The distribution of mean discharge rates takes no account of the temporal pattern of the activity of each nerve fiber. Since the spectra of voiced speech sounds are largely restricted to the frequency range below 5 kHz, individual nerve fibers are phase locked to components of voiced speech sounds that fall within their response area. Fibers with CFs near a formant phase lock to that formant frequency (Hashimoto et al. 1975; Young and Sachs 1979; Sinex and Geisler 1983; Carney and Geisler 1986; Delgutte and Kiang 1984a; Palmer et al. 1986; Palmer 1990). This locking to the formants excludes phase-locked responses to other weaker components, as a result of synchrony suppression (Javel 1981), and the formant frequencies dominate the fiber activity. Fibers with CFs remote from formant frequencies at low (below the first formant) and middle (between the first and second formants) frequencies are dominated either by the harmonic closest to the fiber CF or by modulation at the voice pitch (see below), indicating a beating of the harmonics of the voice pitch that fall within their response area. At CFs above the second formant, the discharge is dominated either by the second formant or again by the modulations at the voice pitch caused by interactions of several harmonics. This is illustrated diagrammatically in Figure 4.4, in which the frequency dominating the fiber discharge, determined as the largest component in the Fourier transform of peristimulus time histograms, is plotted against fiber CF.

The frequencies of the first two formants and the fundamental are clearly evident in Figure 4.4 and could easily be extracted by taking the frequencies that dominate the most fibers. This purely temporal form of representation does change with SPL as the first formant frequency spreads across the fiber array, but the effect of this is generally to reduce or eliminate the phase locking to the weaker CF components.

2.1.3 Mixed Place-Temporal Representations

Purely temporal algorithms do not require in any essential way cochlear filter analysis or the tonotopic order of the auditory pathway. Furthermore,

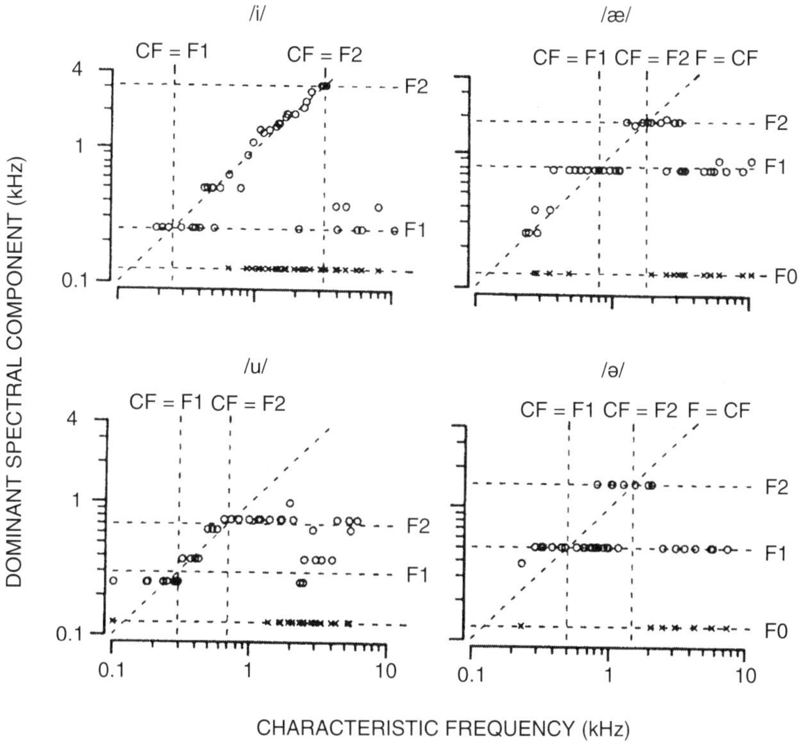

FIGURE 4.4. Plots of the frequency of the component of four vowels that evoked the largest phase-locked responses from auditory-nerve fibers. The frequency below (crosses) and above (open circles) 0.2 kHz evoking the largest response are plotted separately. Dashed vertical lines mark the positions of the formant frequencies with respect to fiber CFs, while the horizontal dashed lines show dominant components at the formant or fundamental frequencies. The diagonal indicates activity dominated by frequencies at the fiber CF. (From Delgutte and Kiang 1984a, with permission.)

they cannot be extended to the higher CF regions (>2–3 kHz) where phase locking deteriorates significantly. If phase locking carries important information, it must be reencoded into a more robust representation, probably at the level of the cochlear nucleus. For these reasons, and because of the lack of anatomical substrates in the AVCN that could accomplish the implied neuronal computations, alternative hybrid, place-temporal representations have been proposed that potentially shed new light on the issue of place versus time codes by combining features from both schemes.

One example of these algorithms has been used extensively by Sachs and colleagues, as well as by others, in the analysis of responses to vowels (Young and Sachs 1979; Delgutte 1984; Sachs 1985; Palmer 1990). The analyses generally utilize a knowledge of the periodicity of speech sounds (Fig. 4.5B), but can also be performed using only interspike intervals (Sachs and Young 1980; Voigt et al. 1982; Palmer 1990). The first stage involves construction of histograms of the fiber responses to vowel stimuli, which are then Fourier transformed to provide measures of the phase locking to individual harmonic components (Fig. 4.5B). These analyses revealed that phase locking to individual harmonics occurred at the appropriate "place" (i.e., in fibers whose CFs are equal or close to the harmonic frequency). As the SPL increases, phase locking to the strong harmonics near the formant frequency spreads from their place to dominate the temporal response patterns of fibers associated with other CFs and suppresses the temporal responses to the weaker harmonics in those fibers. By forming an average of the phase locking to each harmonic in turn, in fibers at the appropriate place for that harmonic, Young and Sachs (1979) were able to compare the amount of temporal response to various harmonics of the signal. The average localized synchronized rate (ALSR) functions so derived for the vowels /ε/, /a/, and /ɪ/ are shown in Fig. 4.5C for a series of sound levels. (The computation of the ALSR involves a "matched filtering" process, but the details of this sort of filter are largely irrelevant to the outcome, see Delgutte 1984; Palmer 1990).

The ALSR functions in Fig. 4.5C bear a striking resemblance to the spectra of these vowels (Fig. 4.5A) over a wide range of SPLs. Thus, it is evident that this form of representation (which combines phase locking, cochlear place, and discharge rate) is robust and shows little loss of the well-defined peaks at formant-related frequencies, even at high stimulus levels. This form of internal representation is unaffected by background noise (Sachs et al. 1983; Delgutte and Kiang 1984d; Geisler and Gamble 1989; Silkes and Geisler 1991), is capable of being computed for unvoiced vowels (Voigt et al. 1982), preserves the details of the spectra of two simultaneously presented vowels with different voice pitch (See Fig. 4.7C,D, below) (Palmer 1990), and can represent formant transitions (Fig. 4.7A,B) (Miller and Sachs 1984). It is salutatory to remember, however, that there is, as yet, no evidence to suggest that there are mechanisms in the central nervous system that are able to make direct use of (or transform) the information about the vowel spectrum contained in the variation in phase locking across the population of AN fibers (see below).

Only the spherical cells in the AVCN (which phase lock as well as do AN fibers) faithfully transmit the temporal activity and show population temporal responses (quantified by ALSR functions) similar to those in the AN (Blackburn and Sachs 1990; Winter and Palmer 1990b; Recio and Rhode 2000). This population of cells is thought to be involved in the localization of low-frequency sounds via binaural comparisons at the level of the superior olive. The ALSR functions computed for the chopper units (which retain a simple "place" representation; see Fig. 4.3) do not retain the information about formant-related peaks (Blackburn and Sachs 1990). Units in

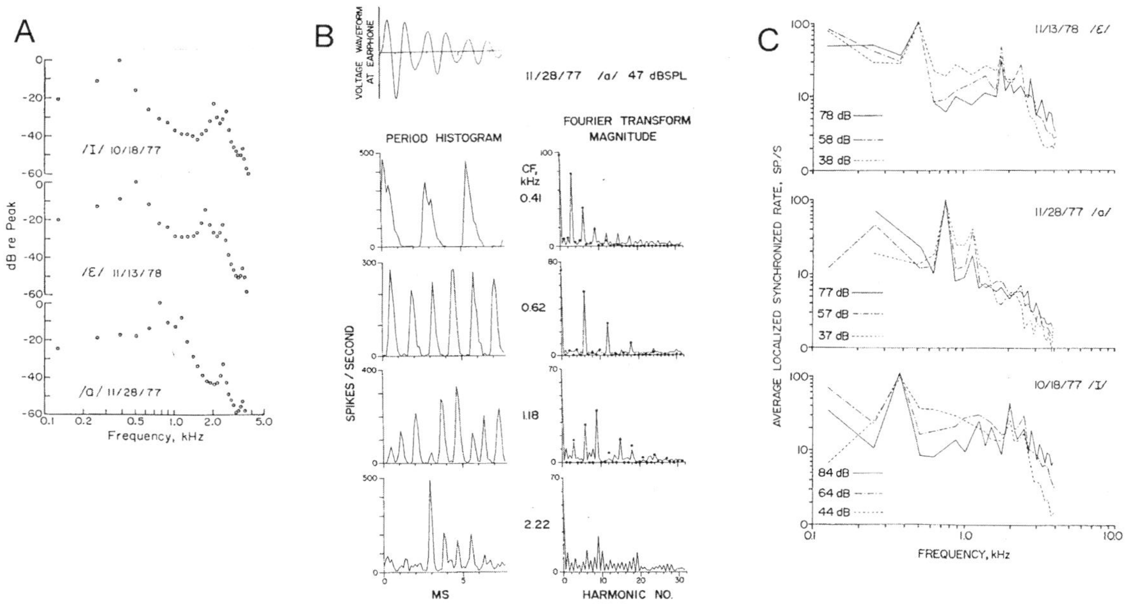

FIGURE 4.5. A: Spectra of the vowels /ɪ/, /ε/, and /a/ with fundamental frequency 128 Hz. B: Top the periodic waveform of the synthetic token of the vowel /a/. First column: Period histograms of the responses of single auditory-nerve fibers to the vowel /a/. Second column: Fourier spectra of the period histograms. C: Average localized synchronized rate functions of the responses of populations of auditory-nerve fibers to the three vowels. Each point is the number of discharges synchronized to each harmonic of the vowel averaged across nerve fibers with CFs within + 0.5 octaves of the harmonic frequency. (From Young and Sachs 1979, with permission.)

the dorsal cochlear nucleus do not phase lock well to pure tones, and thus no temporal representation of the spectrum of speech sounds is expected to be observed in this locus (Palmer et al. 1996b).

Finally, we consider a class of algorithms that make use of correlations or discontinuities in the phase-locked firing patterns of nearby (local) AN fibers to derive estimates of the acoustic spectrum. The LIN algorithm (Shamma 1985b) is modeled after the function of the *l*ateral *i*nhibitory *n*etworks, which are well known in the vision literature. In the retina, this sort of network enhances the representation of edges and peaks and other regions in the image that are characterized by fast transitions in light intensity (Hartline 1974). In audition, the same network can extract the spectral profile of the stimulus by detecting edges in the patterns of activity across the AN fiber array (Shamma 1985a,b, 1989).

The function of the LIN can be clarified if we examine the detailed spatiotemporal structure of the responses of the AN. Such a natural view of the response patterns on the AN (and in fact in any other neural tissue) has been lacking primarily because of technical difficulties in obtaining recordings from large populations of nerve cells. Figure 4.6 illustrates this view of the response of the ordered array of AN fibers to a two-tone stimulus (600 and 2000 Hz).

In Figure 4.6A, the basilar-membrane traveling wave associated with each signal frequency synchronizes the responses of a different group of fibers along the tonotopic axis. The responses reflect two fundamental properties of the traveling wave: (1) the abrupt decay of the wave's amplitude, and (2) the rapid accumulation of phase lag near the point of resonance (Shamma 1985a). These features are, in turn, manifested in the spatiotemporal response patterns as edges or sharp discontinuities between the response regions phase locked to different frequencies (Fig. 4.6B). Since the saliency and location of these edges along the tonotopic axis are dependent on the amplitude and frequency of each stimulating tone, a spectral estimate of the underlying complex stimulus can be readily derived by detecting these spatial edges. This is done using algorithms performing a derivative-like operation with respect to the tonotopic axis, effectively locally subtracting out the response waveforms. Thus, if the responses are identical, they are canceled out by the LIN, otherwise they are enhanced (Shamma 1985b). This is the essence of the operation performed by lateral inhibitory networks of the retina (Hartline 1974). Although discussed here as a derivative operation with respect to the tonotopic axis, the LIN can be similarly described using other operations, such as multiplicative correlation between responses of neighboring fibers (Deng et al. 1988).

Lateral inhibition in varying strengths is found in the responses of most cell types in all divisions of the cochlear nucleus (Evans and Nelson 1973; Young 1984; Rhode and Greenberg 1994a). If phase-locked responses (Fig. 4.5) are used to convey spectral information, then it is at the cochlear nucleus that time-to-place transformations must occur. Transient choppers

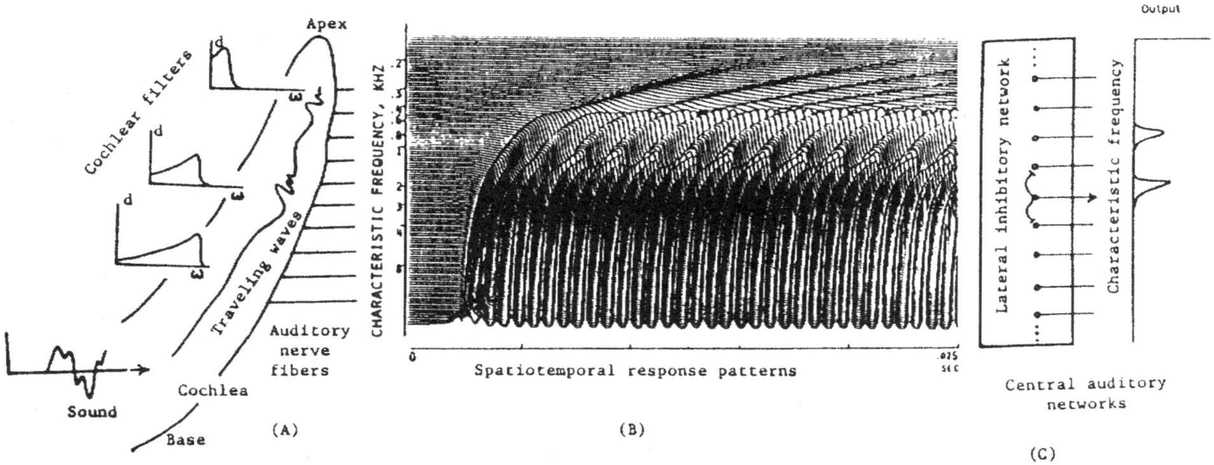

FIGURE 4.6. A schematic of early auditory processing. A: A two-tone stimulus (600 and 2000 Hz) is analyzed by a model of the cochlea (Shamma et al. 1986). Each tone evokes a traveling wave along the basilar membrane that peaks at a specific location reflecting the frequency of the tone. The responses at each location are transduced by a model of inner hair cell function, and the output is interpreted as the instantaneous probability of firing of the auditory nerve fiber that innervates that location. B: The responses thus computed are organized spatially according to their point of origin. This order is also tonotopic due to the frequency analysis of the cochlea, with apical fibers being most sensitive to low frequencies, and basal fibers to high frequencies. The characteristic frequency (CF) of each fiber is indicated on the spatial axis of the responses. The resulting total spatiotemporal pattern of responses reflects the complex nature of the stimulus, with each tone dominating and entraining the activity of a different group of fibers along the tonotopic axis. C: The lateral inhibitory networks of the central auditory system detect the discontinuities in the spatiotemporal pattern and generate an estimate of the spectrum of the stimulus.

exhibit strong sideband inhibition and, as described above (see Fig. 4.3), in response to vowels the pattern of their average rate responses along the tonotopic axis displays clear and stable representations of the acoustic spectral profile at all stimulus levels. Selective listening to the low- and high-spontaneous-rate AN fibers is one plausible mechanism for the construction of this place representation. However, these cells do receive a variety of inhibitory inputs (Cant 1981; Tolbert and Morest 1982; Smith and Rhode 1989) and therefore could be candidates for the operation of inhibition-mediated processes such as the LIN described above (Winslow et al. 1987; Wang and Sachs 1994, 1995).

2.2 Encoding Spectral Dynamics in the Early Auditory System

In speech, envelope amplitude fluctuations or formant transitions occur at relatively slow rates (<20 Hz), corresponding to the rate of different speech segments such as phonemes, syllables, and words. Few studies have been concerned with studying the sensitivity to such slow modulations, either in amplitude or frequency, of the components of the spectrum. Faster modulations at rates much higher than 50 Hz give rise to the percept of pitch (see section 2.2.3).

Our goal in this section is to understand the way in which the auditory system encodes these slow spectral modulations and the transient (nonperiodic) changes in the shape of the spectral profile. Such equivalent information is embodied in responses to clicks, click trains, frequency modulation (FM) sweeps, and synthetic speech-like spectra exhibiting nonperiodic cues such as voice onset cue, noise bursts, and formant transitions.

2.2.1 Sensitivity to Frequency Sweeps

The responses of AN fibers to FM tones are predictable from their response areas, rate-level functions, and adaptation to stationary pure tones (Britt and Starr 1975; Sinex and Geisler 1981). The fibers do not respond in a manner consistent with long-term spectral characteristics, but rather to any instantaneous frequency energy falling within the response area. The direction of frequency change has little effect on AN fibers other than a shift in the frequency evoking the maximum firing rate, in a manner consistent with adaptation by earlier components of the sweep. In response to low frequencies, the fibers phase lock to individual cycles associated with the instantaneous frequency at any point in time (Sinex and Geisler 1981).

In many cases, the responses of cochlear nucleus neurons to FM tones are also consistent with their responses to stationary pure tones (Watanabe and Ohgushi 1968; Britt and Starr 1975; Evans 1975; Rhode and Smith 1986a,b). However, some cochlear nucleus neurons respond differently, depending on the sweep direction (Erulkar et al. 1968; Britt and Starr 1975; Evans 1975; Rhode and Smith 1986a,b), in a manner consistent with the asymmetrical disposition of inhibitory regions of their response area.

Cochlear nucleus responses to frequency sweeps are maximal for frequency sweeps changing at a rate of 10 to 30 Hz/s (Møller 1977).

2.2.1 Representation of Speech and Speech-Like Stimuli

Synthetic and natural speech stimuli, or portions of them such as syllabic (e.g., /da/ and /ta/) or phonetic segments, have been used extensively to investigate the encoding of speech in both the peripheral and the central auditory system. Other types of stimuli mimic important acoustic features of speech, such as voice onset time (VOT) and formant transitions. Although the use of such relatively complex stimuli has been motivated by various theoretical considerations, it is fair to say that they all have an intuitive rather than an analytical flavor. Consequently, it is often difficult, particularly in more central nuclei, to relate these results to those obtained from simpler stimuli, such as pure tones and noise, or to other complex spectra such as moving ripples.

2.2.1.1 Consonant Formant Transitions

Phase locking across populations of AN fibers generally well represents the major spectral components of nasal and voiced stop consonants (Delgutte 1980; Miller and Sachs 1983; Sinex and Geisler 1983; Delgutte and Kiang 1984c; Carney and Geisler 1986; Deng and Geisler 1987). In addition, during formant transitions the mean discharge rates are also able to signal the positions of the formants (Miller and Sachs 1983; Delgutte and Kiang 1984c), even at sound levels where the mean rate distributions to vowels do not have formant-related peaks. Since the transitions are brief and occur near the start of the syllable, it is presumably the wider dynamic range of the onset discharge that underlies this increased mean rate signaling capacity (Miller and Sachs 1983).

In the cochlear nucleus, responses to formant transitions have been studied using both speech-like stimuli and tone-frequency sweeps (see previous section). The presence of strong inhibitory sidebands in many neuron types (Rhode and Greenberg 1994a) and the asymmetry of responses to frequency sweeps could provide some basis for differential responses to consonants in which the formant transitions sweep across the response areas of the units. The only detailed study of the responses of identified dorsal cochlear nucleus neurons to consonant-vowel (CV) syllables failed to detect any specific sensitivity to the particular formant transitions used (in such CV syllables as /ba/, /da/, and /ga/) over and above a linear summation of the excitation and inhibition evoked by each of the formants separately (Stabler 1991; Palmer et al. 1996b).

2.2.1.2 Voice Onset Time

Stop consonants in word- or syllable-initial positions are distinguished by interrelated acoustic cues, which include a delay, after the release, in the onset of voicing and of first formant energy (cf. Diehl and Lindblom,

Chapter 3). These and other related changes are referred to as voice onset time (VOT) and have been used to demonstrate categorical perception of stop consonants that differ only with respect to VOT. The categorical boundary lies between 30 and 40 ms for both humans (Abrahamson and Lisker 1970) and chinchillas (Kuhl and Miller 1978). However, the basis of this categorical perception is not evident at the level of the AN (Carney and Geisler 1986; Sinex and McDonald 1988). When a continuum of syllables along the /ga/-/ka/ or /da/-/ta/ dimension is presented, there is little discharge rate difference found for VOTs less than 20 ms. Above this value, low-CF fibers (near the first formant frequency) showed accurate signaling of the VOT, while high-CF fibers (near the second and third formant frequencies) did not. These discharge rate changes were closely related to changes in the spectral amplitudes that were associated with the onset of voicing. Sinex and McDonald (1988) proposed a simple model for the detection of VOT simply on the basis of a running comparison of the current discharge rate with that immediately preceding. There are also changes in the synchronized activity of AN fibers correlated with the VOT. At the onset of voicing, fibers with low-CFs produce activity synchronized to the first formant, while the previously ongoing activity of high-CF fibers, which during the VOT interval are synchronized to stimulus components associated with the second and third formants near CF, may be captured and dominated by components associated with the first formant. In the mid- and high-CF region, the synchronized responses provide a more accurate signaling of VOTs longer than 50 ms than do mean discharge rates. However, although more information is certainly available in the synchronized activity, the best mean discharge rate measures appear to provide the best estimates of VOT (Sinex and McDonald 1989). Neither the mean rate nor the synchronized rate changes appeared to provide a discontinuous representation consistent with the abrupt qualitative change in stimulus that both humans and chinchillas perceive as the VOT is varied.

In a later study Sinex et al. (1991) studied the discharge characteristics of low-CF AN fibers in more detail, specifically trying to find a basis for the nonmonotonic temporal acuity for VOT (subjects can discriminate small VOT differences near 30 to 40 ms, but discrimination is significantly less acute for shorter or longer VOTs). They found that the peak discharge rate and latency of populations of low-CF AN fibers in response to syllables with different VOTs were most variable for the shortest and longest VOTs. For VOTs near 30 to 40 ms, the peak responses were largest and the latencies nearly constant. Thus, variation in magnitude and latency varied nonmonotonically with VOT in a manner consistent with psychophysical acuity for these syllables. The variabilities in the fiber discharges were a result of the changes in the energy passing through the fiber's filter. It was concluded that correlated or synchronous activity was available to the auditory system over a wider bandwidth for syllables with VOTs of 30 to 40 ms than for other VOTs; thus, the pattern of response latencies in the auditory periphery could be an important factor limiting psychophysical performance. Pont (1990; Pont and Damper 1991) has used a model of the auditory periphery up to the dorsal acoustic stria as a means of investigating the categorical representation of English stop consonants (which differed only in VOT) in the cochlear nucleus.

2.2.1.3 Context Effects

In neural systems the response to a stimulus is often affected by the history of prior stimulation. For example, suppression of response to a following stimulus is commonly observed in the AN (see, for example, Harris and Dallos 1979). Both suppression and facilitation can be observed at the cochlear nucleus (Palombi et al. 1994; Mandava et al. 1995; Shore 1995). Clearly in the case of a continuous stimulation stream, typical of running speech, the context may radically alter the representation of speech sounds.

The effect of context on the response of AN fibers was investigated by Delgutte and Kiang (1984c) for the consonant-vowel syllable /da/. The context consisted of preceding the /da/ by another sound so that the entire complex sounded like either /da/, /ada/, /na/, /sa/, /sa/, /sta/, or /sta/. Both temporal and mean rate measures of response depended on the context. The largest context-dependent changes occurred in those frequency regions in which the context stimulus had significant energy. The average rate profile was radically altered by the context, but the major components of the synchronized response were little affected.

The cues available for discrimination of stop consonants are also context dependent, and their neural encoding has been separately measured for word-medial and for syllable-final consonants (Sinex 1993; Sinex and Narayan 1994). An additional spectral cue for discriminating voiced and unvoiced consonants when not at the initial position is the "voice bar," a low-amplitude component of voicing that is present during the interval of articulatory closure (cf. Diehl and Lindblom, Chapter 3). Responses to the voice bar were more prominent and occurred over a wider range of CF for /d/ than for /t/. The presence of the response to the voice bar led to a greater difference in the mean discharge rates to /d/ and /t/ at high sound levels (in contrast to a reduction in mean rate differences to vowels at high stimulus levels, as described above). Two similar interval measures were computed from the responses to medial and final consonants: the closure interval (from the onset of closure to articulatory release) for syllable-final consonants, and the consonant duration (from closure onset to the onset of the following vowel) for word-medial consonants. For the stimuli /d/ (voiced) and /t/ (unvoiced), large differences (relative to the neural variability observed) were found in these encoded intervals between the two consonants. However, the differences in these measures between the same consonants spoken by different talkers or with different vowel contexts were small compared to the neural variability. As with the VOT studies, it was suggested that the neural variability is large compared to within-category acoustic differences and thus could contribute to the formation of perceptual categories (Sinex and McDonald 1989; Sinex 1993; Sinex and Narayan 1994).

Strings of speech sounds have also been employed in studies at the level of the cochlear nucleus (Caspary et al. 1977; Rupert et al. 1977), where it was found that the temporal patterning and mean discharge rates were often profoundly affected by where a vowel appeared in a sequence. More recently, these experiments have been repeated with more careful attention to unit classification and stimulus control using only pairs of vowels (Mandava et al. 1995). This study reported considerable diversity of responses to single and paired vowels even among units with similar characteristics. Both mean discharge-rate changes and alterations in the temporal patterning of the discharge were found as a result of the position of the vowel in a vowel pair. For primary-like and chopper units, both enhancement and depression of the responses to the following vowel were found, while onset units showed only depression. It seems likely that much of the depression of the responses can be attributed to factors such as the adaptation, which is known to produce such effects in the AN. Facilitation of responses has been suggested to be a result of adaptation of mutual inhibitory inputs (see, for example, Viemeister and Bacon 1982) and would depend on the spectral composition of both vowels as well as the response area of the neuron.

2.3 Encoding of Pitch in the Early Auditory System

Complex sounds, including voiced speech, consisting of many harmonics, are heard with a strong pitch at the fundamental frequency, even if energy is physically lacking at that frequency. This phenomenon has been variously called the "missing" fundamental, virtual pitch, or residue pitch (see Moore 1997 for review), and typically refers to periodicities in the range of 70 to 1000 Hz. A large number of psychoacoustical experiments have been carried out to elucidate the nature of this percept, and its relationship to the physical parameters of the stimulus. Basically, all theories of pitch perception fall into one of two camps. The first, *the spectral pitch hypothesis*, states that the pitch is extracted explicitly from the harmonic spectral pattern. This can be accomplished in a variety of ways, for instance, by finding the best match between the input pattern and a series of harmonic templates assumed to be stored in the brain (Goldstein 1973). The second view, *the temporal pitch hypothesis*, asserts that pitch is extracted from the periodicities in the time waveform of responses in the auditory pathway, which can be estimated, for example, by computing their autocorrelation functions (Moore 1997). In these latter models, some form of organized delay lines or internal clocks are assumed to exist in order to perform the computations.

The encoding of pitch in the early auditory pathway supports either of the two basic representations outlined above. We review in this section the physiology of pitch in the early stages of auditory processing, highlighting the pros and cons of each perspective.

2.3.1 The Spectral Pitch Hypothesis

The primary requirement of this hypothesis is that the input spectral profile be extracted, and that its important harmonics be at least partially resolved (Plomp 1976). Note that this hypothesis is ambivalent about how the spectral profile is derived (i.e., whether it has a place, temporal, or mixed representation); rather, it simply requires that the profile be available centrally. In principle, the cochlear filters are narrow enough to resolve (at least partially) up to the 5th or 6th harmonic of a complex in terms of a simple place representation of the spectrum (Moore 1997); significantly better resolution can be expected if the profile is derived from temporal or mixed place-temporal algorithms (up to 30 harmonics, see below).

Pitch can be extracted from the harmonics of the spectral profile in various ways (Goldstein 1973; Wightman 1973; Terhardt 1979). A common denominator in all of these models is a pattern-matching step in which the resolved portion of spectrum is compared to "internal" stored templates of various harmonic series in the central auditory system. This can be done explicitly on the spectral profile (Goldstein 1973), or on some transformation of the spectral representation (e.g., its autocorrelation function, as performed by Wightman 1973). The perceived pitch is computed to be the fundamental of the best matching template. Such processing schemes predict well the perceived pitches and their relative saliency (Houtsma 1979).

The major shortcoming of the spectral pitch hypothesis is the lack of any biological evidence in support of harmonic-series templates (Javel and Mott 1988; Schwartz and Tomlinson 1990). Highly resolved spectral profiles are not found in pure-place representations (Miller and Sachs 1983). They can, however, be found at the level of the auditory nerve in the ALSR mixed place-temporal representation (Fig. 4.7), for single and paired vowels, providing ample information for tracking the pitch during formant transitions and for segregating two vowels (Miller and Sachs 1983; Palmer 1990).

Although such high-resolution spectra could exist in the output of spherical bushy cells of the cochlear nucleus, which act as relays of AN activity, they have yet to be observed in other types of cochlear nucleus cells (Blackburn and Sachs 1990), or at higher levels, including the cortex (Schwartz and Tomlinson 1990). This result is perhaps not surprising, given the reduction in phase locking and increased convergence observed in higher auditory nuclei.

2.3.2 The Temporal Pitch Hypothesis

Since the periodicities in the response waveforms of each channel are directly related to the overall repetitive structure of the stimulus, it is possible to estimate pitches perceived directly from the interval separating the peaks of the responses or from their autocorrelation functions. An intuitive

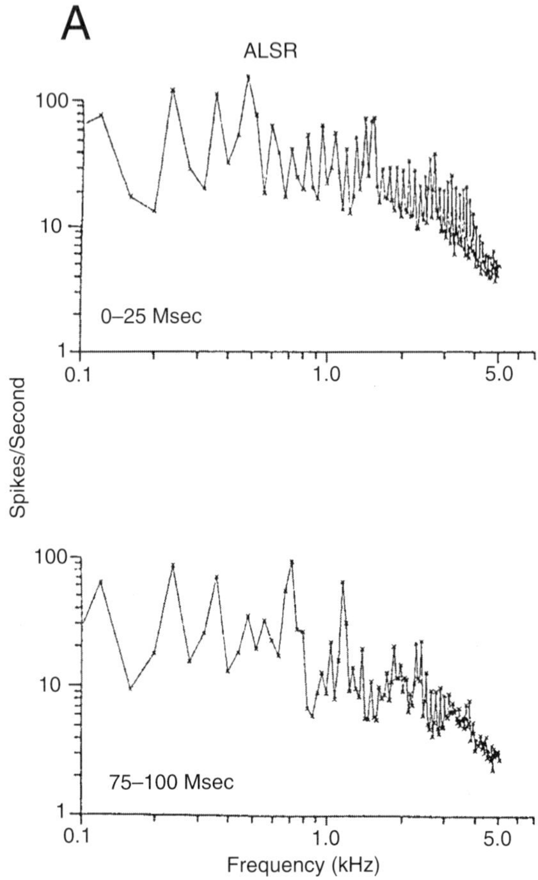

FIGURE 4.7. Average localized rate functions (as in Fig. 4.5C) for the responses to (A) the first and last 25 ms of the syllable /da/ (from Miller and Sachs 1984, with permission) and (B) pairs of simultaneously presented vowels; /i/ + /a/ and /ɔ/ + /i/ (from Palmer, 1990, with permission).

and computationally accurate scheme to achieve this operation is the correlogram method illustrated in Figure 4.8 (Slaney and Lyon 1990).

The correlogram sums the autocorrelation functions computed from all fibers (or neurons) along the tonotopic axis to generate a final profile from which the pitch value can be derived. Figure 4.8 shows three examples of steady-state sounds analyzed into the correlogram representation (with the time dimension not shown, since the sounds are steady and there is no good way to show it). In each case a clear vertical structure occurs in each channel that is associated with the pitch period. This structure also manifests itself as a prominent peak in the summary correlogram.

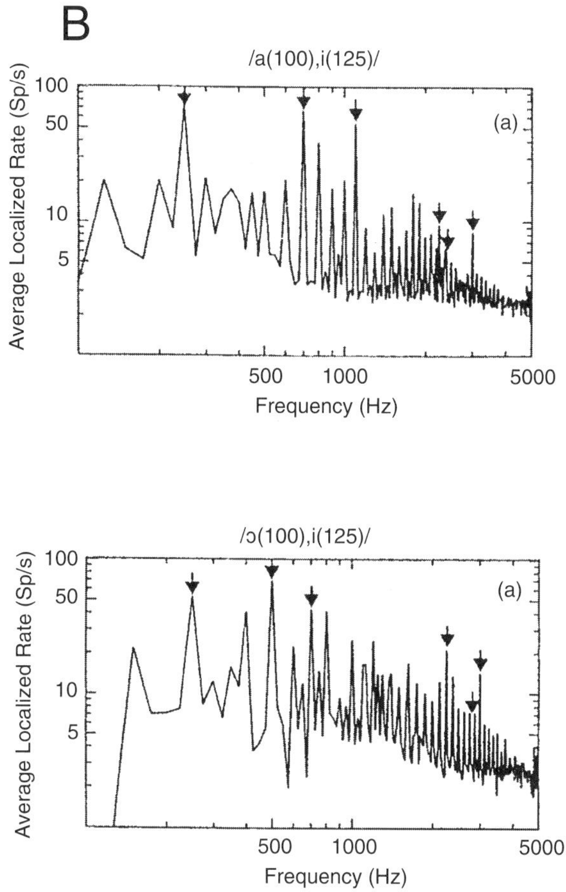

FIGURE 4.7. *Continued*

This same basic correlogram approach has been used as a method for the display of physiological responses to harmonic series and speech (Delgutte and Cariani 1992; Palmer 1992; Palmer and Winter 1993; Cariani and Delgutte 1996). Using spikes recorded from AN fibers over a wide range of CFs, interval-spike histograms are computed for each fiber and summed into a single autocorrelation-like profile. Stacking these profiles together across time produces a two-dimensional plot analogous to a spectrogram, but with a pitch-period axis instead of a frequency axis, as shown in Figure 4.9. Plots from single neurons across a range of CFs show a clear representation of pitch, as does the sum across CFs. The predominant interval in the AN input provides an estimate of pitch that is robust and comprehensive, explaining a very wide range of pitch phenomena: the missing fundamental, pitch invariance with respect to level, pitch equivalence of

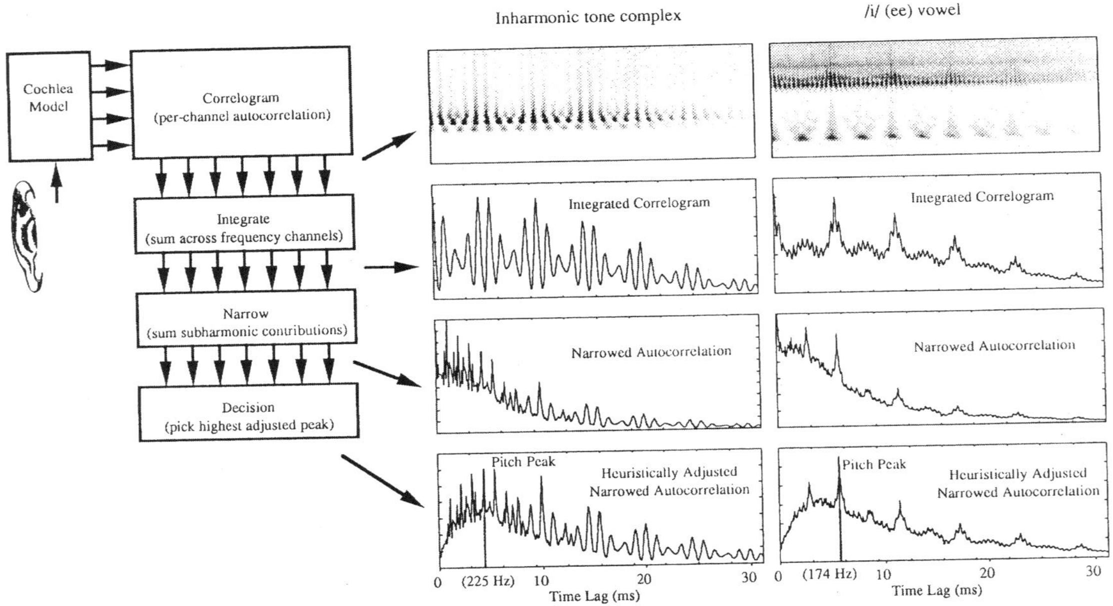

FIGURE 4.8. Schematic of the Slaney-Lyon pitch detector. It is based on the correlogram of the auditory nerve responses. (From Lyon and Shamma 1996, with permission.)

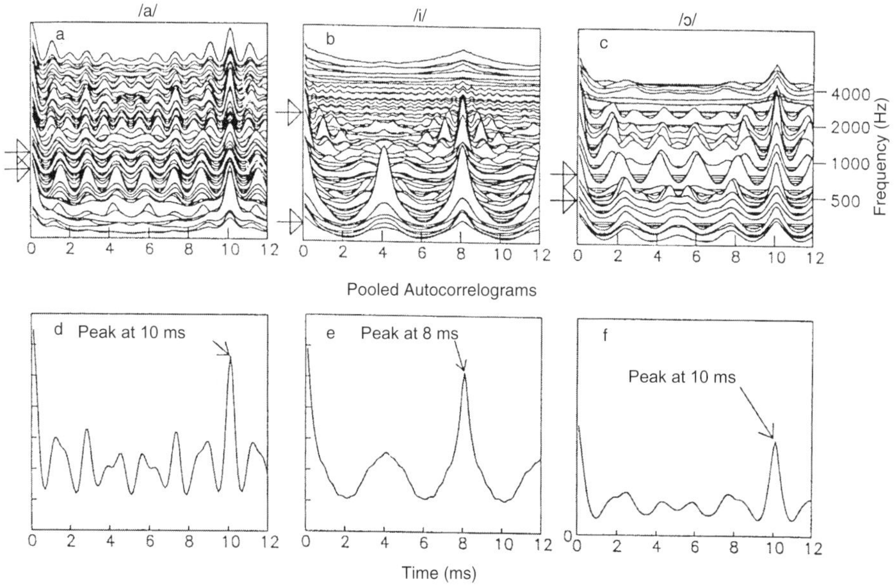

FIGURE 4.9. The upper plots show autocorrelograms of the responses of auditory nerve fibers to three three-formant vowels. Each line in the plot is the autocorrelation function of a single fiber plotted at its CF. The frequencies of the first two formants are indicated by arrows against the left axis. The lower plots are summations across frequency of the autocorrelation functions of each individual fiber. Peaks at the delay corresponding to the period of the voice pitch are indicated with arrows. (From Palmer 1992, with permission.)

spectrally diverse stimuli, the pitch of unresolved harmonics, the pitch of amplitude modulation (AM) noise, pitch salience, the pitch shift of inharmonic AM tones, pitch ambiguity, phase insensitivity of pitch, and the dominance region for pitch. It fails to account for the rate pitches of alternating click trains (cf. Flanagan and Guttman 1960), and it underestimates the salience of low-frequency tones (Cariani and Delgutte 1996).

The summation of intervals method is the basis for several temporal pitch models (van Noorden 1982; Moore 1997), with the main difference between them being the way interval measurements are computed. For instance, one method (Langner 1992) uses oscillators instead of delay lines as a basis for interval measurement. The physiological reality of this interval measurement (in any form), however, remains in doubt, because there is no evidence for the required organized delay lines, oscillators, or time constants to carry out the measurements across the wide range of pitches perceived. Such substrates would have to exist early in the monaural auditory pathway since the phase locking that they depend on deteriorates significantly in central auditory nuclei (cf. the discussion above as well as Langner 1992).

An analogous substrate for binaural cross-correlation is known in the MSO (Goldberg and Brown 1969; Yin and Chan 1990), but operates with maximum delays of only a few milliseconds, which are too low for monaural pitch and timbre processing below 1 kHz.

2.3.3 Correlates of Pitch in the Responses to Periodic Spectral Modulations

The most general approach to the investigation of temporal correlates of pitch has been to use steady-state periodic AM or FM signals, click trains, or moving spectral ripples to measure the modulation transfer functions (MTFs). These MTFs are then used as indicators of the speed of the response and, assuming system linearity, as predictors of the responses to any arbitrary dynamic spectra. At levels above the AN, neurons have band-pass MTFs, and some authors have taken this to indicate a substrate for channels signaling specific pitches.

2.3.3.1 Sensitivity to Amplitude Modulations

Auditory neurons throughout the auditory pathway respond to amplitude modulations by modulation of their discharge, which is dependent on signal modulation depth as well as modulation rate. For a fixed modulation depth, a useful summary of the neural sensitivity is given by the MTF, which plots the modulation of the neural discharge as a function of the modulation rate, often expressed as a decibel gain (of the output modulation depth relative to the signal modulation depth).

The MTFs of AN fibers are low-pass functions for all modulation depths (Møller 1976; Javel 1980; Palmer 1982; Frisina et al. 1990a,b; Kim et al. 1990; Joris and Yin, 1992). The modulation of the discharge is maximal at about 10 dB above rate threshold and declines as the fiber is driven into saturation (Evans and Palmer 1979; Smith and Brachman 1980; Frisina et al. 1990a,b; Kim et al. 1990; Joris and Yin 1992). For AN fibers the mean discharge rate does not vary with the modulation rate; the discharge modulation represents the fine temporal patterning of the discharges. The cut-off frequency of the MTF increases with fiber CF, probably reflecting the attenuation of the signal sidebands by cochlear filtering (Palmer 1982). However, increases in fiber bandwidth beyond 4 kHz are not accompanied by increases in MTF cut-off frequency, thus implying some additional limitation on response modulation in these units such as the phase-locking capability of the fibers (Joris and Yin 1992; Cooper et al. 1993; Rhode and Greenberg 1994b; Greenwood and Joris 1996). The best modulation frequency (BMF) (i.e., the modulation rate associated with the greatest discharge modulation) ranges between 400 and 1500 Hz for AN fibers (Palmer 1982; Kim et al. 1990; Joris and Yin 1992).

Wang and Sachs (1993, 1994) used single-formant synthetic speech stimuli with carriers matched to the best frequencies of AN fibers and ventral cochlear nucleus units. This stimulus is therefore midway between the sinusoidal AM used by the studies above and the use of full speech stimuli in that the envelope was modulated periodically but not sinusoidally. Their findings in the AN are like those of sinusoidal AM and may be summarized as follows. As the level of the stimuli increases, modulation of the fiber discharge by single formant stimuli increases, then peaks and ultimately decreases as the fiber is driven into discharge saturation. This occurred at the same levels above threshold for fibers with high- and low-spontaneous discharge rates. However, since low-spontaneous-rate fibers have higher thresholds, they were able to signal the envelope modulation at higher SPLs than high-spontaneous-rate fibers.

It is a general finding that most cochlear nucleus cell types synchronize their responses better to the modulation envelope than do AN fibers of comparable CF, threshold, and spontaneous rate (Frisina et al. 1990a,b; Wang and Sachs 1993, 1994; Rhode 1994; Rhode and Greenberg 1994b). This synchronization, however, is not accompanied by a consistent variation in the mean discharge rate with modulation frequency (Rhode 1994).

The MTFs of cochlear nucleus neurons are more variable than those of AN fibers. The most pronounced difference is that they are often bandpass functions showing large amounts of gain (10 to 20 dB) near the peak in the amount of response modulation (Møller 1972, 1974, 1977; Frisina et al. 1990a,b; Rhode and Greenberg 1994b). Some authors have suggested that the bandpass shape is a consequence of constructive interference between intrinsic oscillations that occur in choppers and certain dorsal cochlear nucleus units and the envelope modulation (Hewitt et al. 1992). In the ventral cochlear nucleus the degree of enhancement of the discharge modulation varies for different neuronal response types, although the exact hierarchy is debatable (see Young 1984 for details of cochlear nucleus response types). Frisina et al. (1990a) suggested that the ability to encode amplitude modulation (measured by the amount of gain in the MTF) is best in onset units followed by choppers, primary-like-with-a-notch units, and finally primary-like units and AN fibers (which show very little modulation gain at the peak of the MTF). Rhode and Greenberg (1994b) studied both dorsal and ventral cochlear nucleus and found synchronization in primary-like units equal to that of the AN. Synchronization in choppers, on-L, and pause/build units were found to be superior or comparable to that of low-spontaneous-rate AN fibers, while on-C and primary-like-with-a-notch units exhibited synchronization superior to other unit types (at least in terms of the magnitude of synchrony observed). In the study of Frisina et al. (1990a) in the ventral cochlear nucleus (VCN), the BMFs varied over different ranges for the various units types. The BMFs of onset units varied from 180 to 240 Hz, those associated with primary-like-with-a-notch units varied from 120 to 380 Hz, chopper BMFs varied from 80 to 520 Hz, and primary-like BMFs varied from 80 to 700 Hz. Kim et al. (1990) studied the responses to AM tones in the DCN and PVCN of unanesthetized, decerebrate cats. Their results are consistent with those of Møller (1972, 1974, 1977) and Frisina et al. (1990a,b) in that they found both low-pass and bandpass MTFs, with BMFs ranging from 50 to 500 Hz. The MTFs also changed from low-pass at low SPLs to bandpass at high SPLs for pauser/buildup units in DCN. Rhode and Greenberg (1994b) investigated the high-frequency limits of synchronization to the envelope by estimating the MTF cut-off frequency and found a rank ordering as follows: high-CF AN fibers >1.5 kHz, onset and primary-like units >1 kHz, and choppers and pauser/buildup units >600 Hz.

Using single-formant stimuli Wang and Sachs (1994) demonstrated a significant enhancement in the ability of all ventral cochlear nucleus units, except primary-likes, to signal envelope modulations relative to that observed in the AN, as is clearly evident in the raw histograms shown in Figure 4.10. Not only was the modulation depth increased, but the units were able to signal the modulations at higher SPLs. They suggested the following hierarchy (from best to worst) for the ability to signal the envelope at high sound levels: onsets > on-C > primary-like-with-a-notch, choppers > primary-likes. A very similar hierarchy was also found by Rhode (1994) using 200% AM stimuli.

Rhode (1995) employed quasi-frequency modulation (QFM) and 200% AM stimuli in recording from the cochlear nucleus in order to test the time-coding hypothesis of pitch. He found that units in the cochlear nucleus are relatively insensitive to the carrier frequency, which means that AM responses to a single frequency will be widespread. Furthermore, for a variety of response types, the dependencies on the stimulus of many psychophysical pitch effects could be replicated by taking the intervals between different peaks in the interspike interval histograms. Pitch representation in the timing of the discharges gave the same estimates for the QFM and AM signals, indicating the temporal coding of pitch was phase insensitive.

The enhancement of modulation in the discharge of cochlear nucleus units at high sound levels can be explained in a number of ways, such as cell-membrane properties, intrinsic inhibition, and convergence of low-spontaneous-rate fibers or off-CF inputs at high sound levels (Rhode and Greenberg 1994b; Wang and Sachs 1994). These conjectures were quantitatively examined by detailed computer simulations published in a subsequent article (Wang and Sachs 1995).

In summary, there are several properties of interest with respect to pitch encoding in the cochlear nucleus. First, all unit types in the cochlear nucleus respond to the modulation that would be created at the output of AN filters by speech-like stimuli. This modulation will be spread across a wide tonotopic range even for a single carrier frequency. Second, there are clearly mechanisms at the level of the cochlear nucleus that enhance the representation of the modulation and that operate to varying degrees in different cell types.

2.3.3.2 Sensitivity to Frequency Modulation

Modulation of CF tone carriers by small amounts allows construction of an MTF for FM signals. Given the similarity of the spectra of such sinusoidal

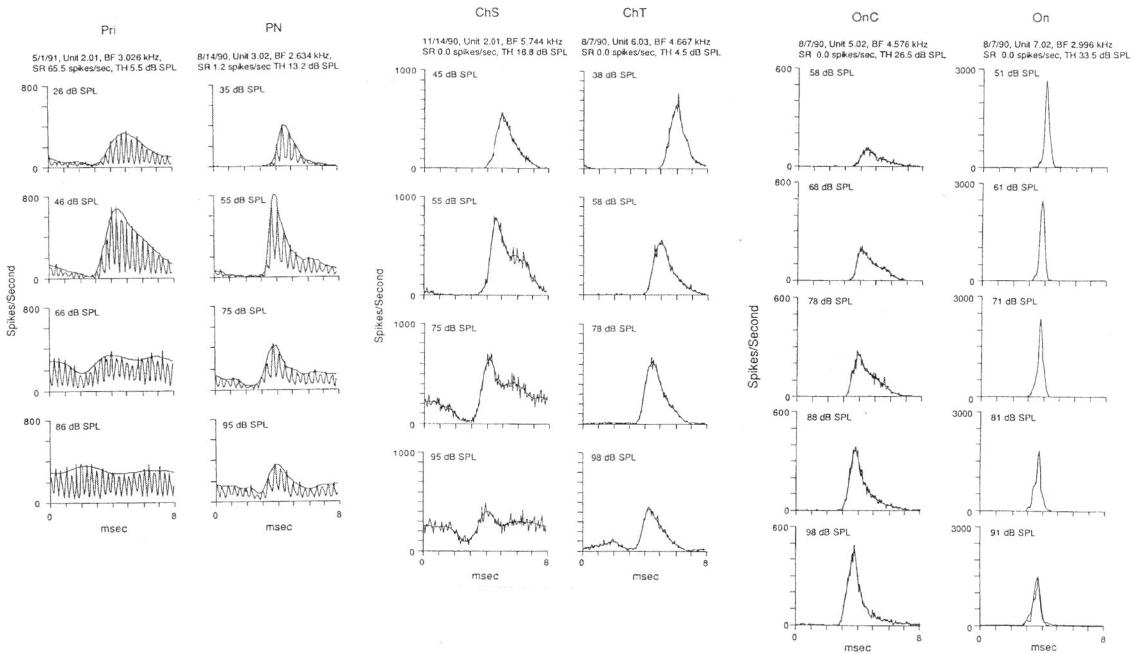

FIGURE 4.10. Period histograms of cochlear nucleus unit types in response to a single-formant stimulus as a function of sound level. Details of the units are given above each column, which shows the responses of a single unit of type primary-like (Pri), primary-like-with-a-notch (PN), sustained chopper (ChS), transient chopper (ChT), onset chopper (OnC) and onset (On) units. (From Wang and Sachs 1994 with permission.)

AM and FM stimuli, it is not surprising that the MTFs in many cases appear qualitatively and quantitatively similar to those produced by amplitude modulation of a CF carrier [as described above, i.e., having a BMF in the range of 50 to 300 Hz (Møller 1972)].

2.3.2.3 Responses to the Pitch of Speech and Speech-Like Sounds

The responses of most cell types in the cochlear nucleus to more complex sounds such as harmonic series and full synthetic speech sounds are modulated at the fundamental frequency of the complex. In common with the simpler AM studies (Frisina et al. 1990a,b; Kim et al. 1990; Rhode and Greenberg 1994b; Rhode 1994, 1995) and the single formant studies (Wang and Sachs 1994), it was found that onset units locked to the fundamental better than did other units types (Kim et al. 1986; Kim and Leonard 1988; Palmer and Winter 1992, 1993). All evidence points to the fact that, in onset units and possibly in some other cochlear nucleus cell types, the enhanced locking to AM and to the fundamental frequency of harmonic complexes is achieved by a coincidence detection mechanism following very wide convergence across the frequency (Kim et al. 1986; Rhode and Smith 1986a; Kim and Leonard 1988; Winter and Palmer 1995; Jiang et al. 1996; Palmer et al. 1996a; Palmer and Winter 1996).

3. Representations of Speech in the Central Auditory System

The central auditory system refers to the auditory midbrain (the IC), the thalamus [medial geniculate body (MGB)], and the auditory cortex (with its primary auditory cortex, AI, and its surrounding auditory areas), and is illustrated in Figure 4.1. Much less is known about the encoding of speech spectra and of other broadband sounds in these areas relative to what is known about processing in the early stages of the auditory pathway. This state of affairs, however, is rapidly changing as an increasing number of investigators turn their attention to these more central structures, and as new recording technologies and methodologies become available. In this section we first discuss the various representations of the acoustic spectrum that have been proposed for the central pathway, and then address the encoding of dynamic, broadband spectra, as well as speech and pitch.

3.1 Encoding of Spectral Shape in the Central Auditory System

The spectral pattern extracted early in the auditory pathway (i.e., the cochlea and cochlear nucleus) is relayed to the auditory cortex through several stages of processing associated with the superior olivary complex, nuclei of the lateral lemniscus, the inferior colliculus, and the medial geniculate body (Fig. 4.1). The core of this pathway, passing through the CNIC and the ventral division of the MGB, and ending in AI (Fig. 4.1), remains strictly tonotopically organized, indicating the importance of this structural axis as an organizational feature. However, unlike its essentially one-dimensional spread along the length of the cochlea, the tonotopic axis takes on an ordered two-dimensional structure in AI, forming arrays of neurons with similar CFs (known as isofrequency planes) across the cortical surface (Merzenich et al. 1975). Similarly, organized areas (or auditory fields) surround AI (Fig. 4.1), possibly reflecting the functional segregation of different auditory tasks into different auditory fields (Imig and Reale 1981).

The creation of an isofrequency axis suggests that additional features of the auditory spectral pattern are perhaps explicitly analyzed and mapped out in the central auditory pathway. Such an analysis occurs in the visual and other sensory systems and has been a powerful inspiration in the search for auditory analogs. For example, an image induces retinal response patterns that roughly preserve the form of the image or the outlines of its edges. This representation, however, becomes much more elaborate in the primary visual cortex, where edges with different orientations, asymmetry, and widths are extracted, and where motion and color are subsequently represented preferentially in different cortical areas. Does this kind of analysis of the spectral pattern occur in AI and other central auditory loci?

In general, there are two ways in which the spectral profile can be encoded in the central auditory system. The first is *absolute*, that is, to encode the spectral profile in terms of the absolute intensity of sound at each frequency, in effect combining both the shape information and the overall sound level. The second is *relative*, in which the spectral profile shape is encoded separately from the overall level of the stimulus.

We review below four general ideas that have been invoked to account for the physiological responses to spectral profiles of speech and other stimuli in the central auditory structures: (1) the simple place representation; (2) the best-intensity or threshold model; (3) the multiscale representation; and (4) the categorical representation. The first two are usually thought of as encoding the absolute spectrum; the others are relative. While many other representations have been proposed, they mostly resemble one of these four representational types.

3.1.1 The Simple Place Representation

Studies of central auditory physiology have emphasized the use of pure tones to measure unit response areas, with the intention of extrapolating from such data to the representation of complex broadband spectra. However, tonal responses in the midbrain, thalamus, and cortex are often complex and nonlinear, and thus not readily interpretable within the context of speech and complex environmental stimuli. For instance, single units may have response areas with multiple excitatory and inhibitory fields, and various asymmetries and bandwidths about their BFs (Shamma and Symmes 1985; Schreiner and Mendelson 1990; Sutter and Schreiner 1991; Clarey et al. 1992; Shamma et al. 1995a). Furthermore, their rate-level functions are commonly nonmonotonic, with different thresholds, saturation levels, and dynamic ranges (Ehret and Merzenich 1988a,b; Clarey et al. 1992). When monotonic, rate-level functions usually have limited dynamic ranges, making differential representation of the peaks and valleys in the spectral profile difficult.

Therefore, these response areas and rate-level functions preclude the existence of a simple place representation of the spectral profile. For instance, Heil et al. (1994) have demonstrated that a single tone evokes an alternating excitatory/inhibitory pattern of activity in AI at low SPLs. When tone intensity is moderately increased, the overall firing rate increases without change in topographic distribution of the pattern.

This is an instance of a place code in the sense used in this section, although not based on simple direct correspondence between the shape of the spectrum and the response distribution along the tonotopic axis. In fact, Phillips et al. (1994) go further, by raising doubts about the significance of the isofrequency planes as functional organizing principles in AI, citing the extensive cross-frequency spread and complex topographic distribution of responses to simple tones at different sound levels.

3.1.2 The Best-Intensity Model

This hypothesis is motivated primarily by the strongly nonmonotonic rate-level functions observed in many cortical and other central auditory cells (Pfingst and O'Connor 1981; Phillips and Irvine 1981). In a sense, one can view such a cell's response as being selective for (or encoding) a particular tone intensity. Consequently, a population of such cells, tuned to different frequencies (along the tonotopic axis) and intensities (along the isofrequency plane), can provide an explicit representation of the spectral profile by its spatial pattern of activity (Fig. 4.11). This scheme is not a transformation of the spectral features represented (which is the amplitude of the spectrum at a single frequency); rather, it is simply a change in the means of the representation (i.e., from simple spike rate to best intensity in the rate-level function of the neuron).

The most compelling example of such a representation is that of the doppler-shifted-constant-frequency area of AI in the mustache bat, where the best intensity of the hypertrophied (and behaviorally significant) region is 62 to 63 kHz and is mapped out in regular concentric circles (Suga and Manabe 1982). However, an extension of this hypothesis to multicomponent stimuli (i.e., as depicted in Fig. 4.11) has not been demonstrated in any species. In fact, several findings cast doubt on any simple form of this hypothesis (and on other similar hypotheses postulating maps of other rate-level function features such as threshold). These negative findings are (1)

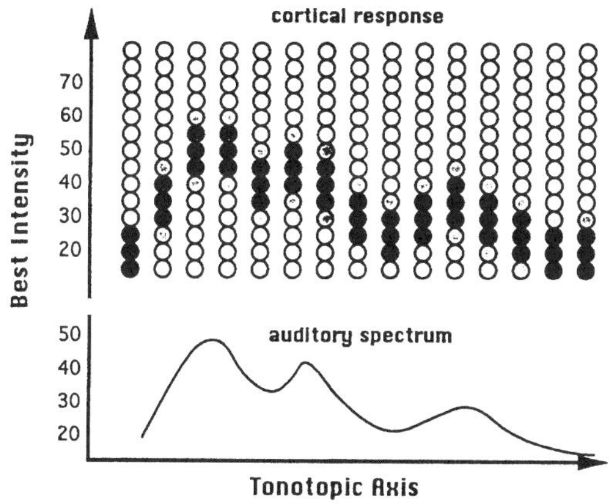

FIGURE 4.11. Top: A schematic representation of the encoding of a broadband spectrum according to the best-intensity model. The dots represent units tuned to different frequencies and intensities as indicated by the ordinate. Only those units at any frequency with best intensities that match those in the spectrum (bottom) are strongly activated (black dots).

the lack of spatially organized maps of best intensity (Heil et al. 1994), (2) the volatility of the best intensity of a neuron with stimulus type (Ehret and Merzenich 1988a), and (3) the complexity of the response distributions in AI as a function of pure-tone intensity (Phillips et al. 1994). Nevertheless, one may argue that a more complex version of this hypothesis might be valid. For instance, it has been demonstrated that high-intensity tones evoke *different patterns* of activation in the cortex, while maintaining a *constant* overall firing rate (Heil et al. 1994). It is not obvious, however, how such a scheme could be generalized to broadband spectra characteristic of speech signals.

3.1.3 The Multiscale Representation

This hypothesis is based on physiological measurements of response areas in cat and ferret AI (Shamma et al. 1993, 1995a; Schreiner and Calhoun 1995), coupled with psychoacoustical studies in human subjects (Shamma et al. 1995b). The data suggest a substantial transformation in the central representation of a spectral profile. Specifically, it has been found that, besides the tonotopic axis, responses are topographically organized in AI along two additional axes reflecting systematic changes in bandwidth and asymmetry of the response areas of units in this region (Fig. 4.12A) (Schreiner and Mendelson 1990; Versnel et al. 1995). Having a range of response areas with different widths implies that the spectral profile is represented repeatedly at different degrees of resolution (or different scales). Thus, fine details of the profile are encoded by units with narrower response

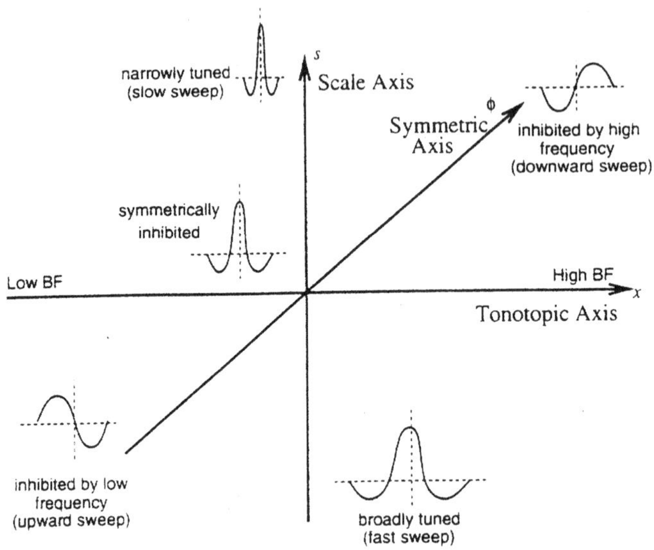
narrowly tuned
(slow sweep)
s
Scale Axis
ϕ
Symmetric
Axis
inhibited by high
frequency
(downward sweep)
symmetrically
inhibited
Low BF
High BF
x
Tonotopic Axis
inhibited by low
frequency
(upward sweep)
broadly tuned
(fast sweep)

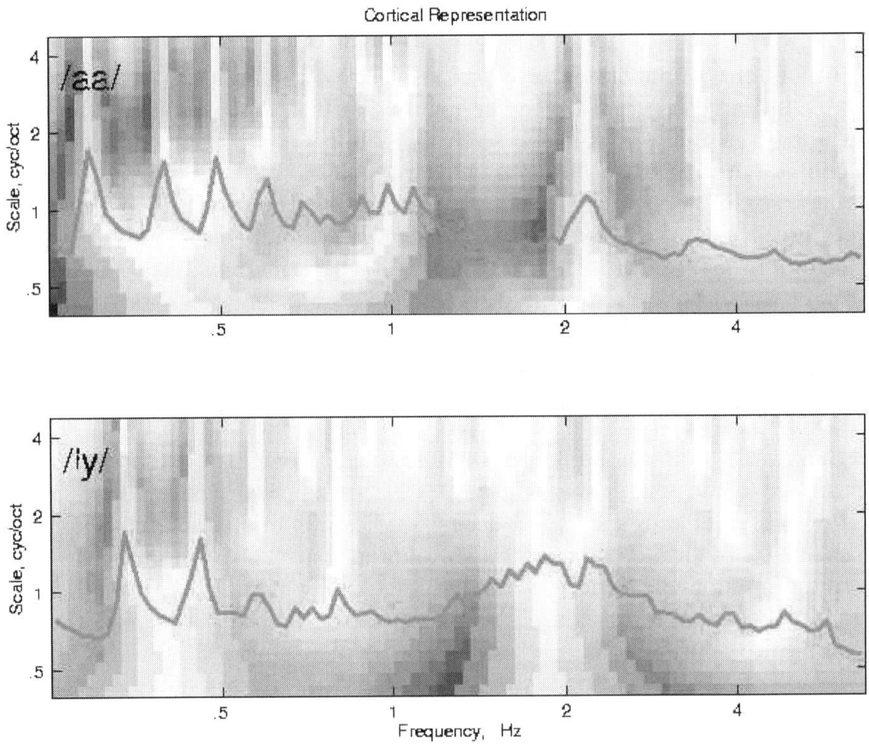
Cortical Representation
/aa/
/iy/
Scale, cyc/oct
4
2
1
.5
.5
1
2
4
Frequency, Hz

areas, whereas coarser outlines of the profile are encoded by broadly tuned response areas.

Response areas with different asymmetries respond differentially, and respond best to input profiles that match their asymmetry. For instance, an odd-symmetric response area would respond best if the input profile had the same local odd symmetry, and worst if it had the opposite odd symmetry. Therefore, a range of response areas of different symmetries (the symmetry axis in Fig. 4.12A) is capable of encoding the shape of a local region in the profile.

Figure 4.12B illustrates the responses of a model of an array of such cortical units to a broadband spectrum such as the vowel /a/. The output at each point represents the response of a unit whose CF is indicated along the abscissa (tonotopic axis), its bandwidth along the ordinate (scale axis), and its symmetry by the color. Note that the spectrum is represented repeatedly at different scales. The formant peaks of the spectrum are relatively broad in bandwidth and thus appear in the low-scale regions, generally <2 cycles/octave (indicated by the activity of the symmetric yellow units). In contrast, the fine structure of the spectral harmonics is only visible in high-scale regions (usually >1.5–2 cycles/octave; upper half of the plots). More detailed descriptions and analyses of such model representations can be found in Wang and Shamma (1995).

The multiscale model has a long history in the visual sciences, where it was demonstrated physiologically in the visual cortex using linear systems analysis methods and sinusoidal visual gratings (Fig. 4.13A) to measure the receptive fields of type VI units (De Valois and De Valois 1990). In the auditory system, the rippled spectrum (peaks and valleys with a sinusoidal spectral profile, Fig. 4.13B) provides a one-dimensional analog of the grating and has been used to measure the ripple transfer functions and response areas in AI, as illustrated in Figure 4.13E–M. Besides measuring the different response areas and their topographic distributions, these studies have also revealed that cortical responses are rather linear in character, satisfying the *superposition principle* (i.e., the response to a complex spectrum composed of several ripples is the same as the sum of the responses to the individual ripples). This finding has been used to predict the response of AI

FIGURE 4.12. A: The three organizational axes of the auditory cortical response areas: a tonotopic axis, a bandwidth axis, and an asymmetry axis. B: The cortical representations of spectral profiles of naturally spoken vowel /a/ and /iy/ and the corresponding cortical representations. In each panel, the spectral profiles of the vowels are superimposed upon the cortical representation. The abscissa indicates the CF in kHz (the tonotopic axis). The ordinate indicates the bandwidth or scale of the unit. The symmetry index is represented by shades in the following manner: White or light shades are symmetric response areas (corresponding to either peaks or valleys); dark shades are asymmetric with inhibition from either low or from high frequencies (corresponding to the skirts of the peaks).

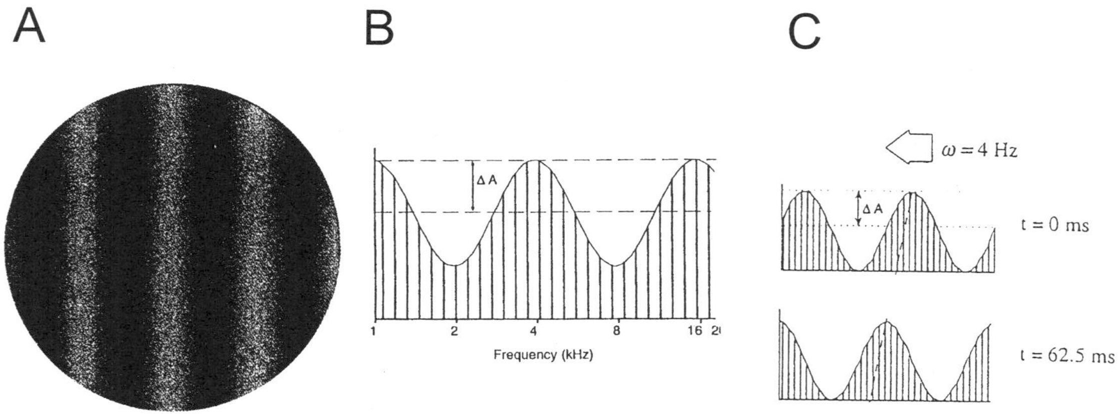

FIGURE 4.13. The sinusoidal profiles in vision and hearing. A: The two-dimensional grating used in vision experiments. B: The auditory equivalent of the grating. The ripple profile consists of 101 tones equally spaced along the logarithmic frequency axis spanning less than 5 octaves (e.g., 1–20 kHz or 0.5–10 kHz). Four independent parameters characterize the ripple spectrum: (1) the overall level of the stimulus, (2) the amplitude of the ripple (ΔA), (3) the ripple frequency (Ω) in units of cycles/octave, and (4) the phase of the ripple. C: Dynamic ripples travel to the left at a constant velocity defined as the number of ripple cycles traversing the lower edge of the spectrum per second (ω). The ripple is shown at the onset ($t = 0$) and 62.5 ms later.

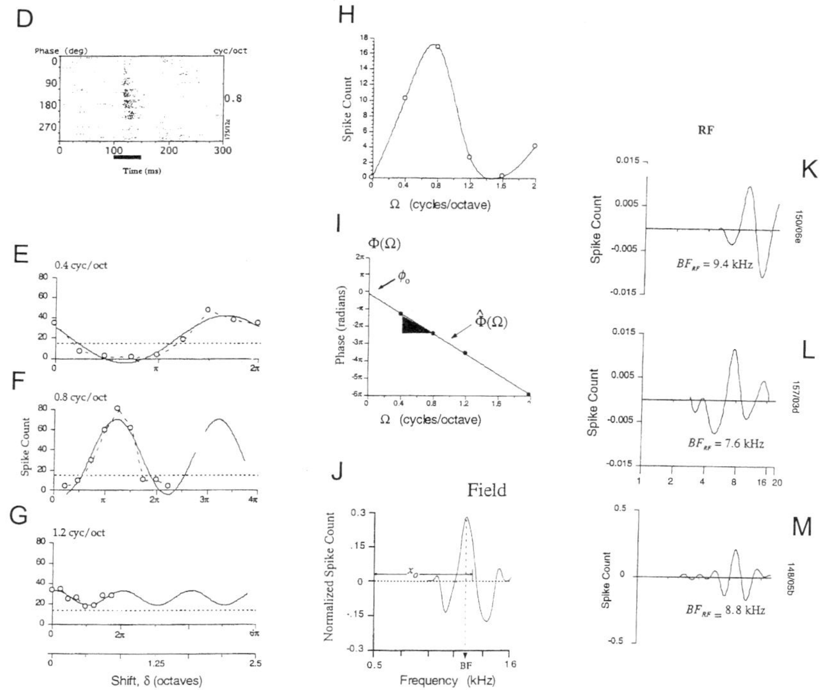

FIGURE 4.13. Analysis of responses to stationary ripples. Panel D shows raster responses of an AI unit to a ripple spectrum ($\Omega = 0.8$ cycle/octave) at various ripple phases (shifted from 0° to 315° in steps of 45°). The stimulus burst is indicated by the bar below the figure, and was repeated 20 times for each ripple phase. Spike counts as a function of the ripple are computed over a 60-ms window starting 10 ms after the onset of the stimulus. Panels E–G show measured (circles) and fitted (solid line) responses to single ripple profiles at various ripple frequencies. The dotted baseline is the spike count obtained for the flat-spectrum stimulus. Panels H–I show the ripple transfer function $T(\Omega)$. H represents the weighted amplitude of the fitted responses as a function of ripple frequency Ω. I represents the phases of the fitted sinusoids as a function of ripple frequency. The characteristic phase, F_0, is the intercept of the linear fit to the data. Panel J shows the response field (RF) of the unit computed as the inverse Fourier transform of the ripple transfer function $T(\Omega)$. Panels K–M show examples of RFs with different widths and asymmetries measured in AI.

units to natural vowel spectra (Shamma and Versnel 1995; Shamma et al. 1995b; Kowalski et al. 1996a,b; Versnel and Shamma 1998; Depireux et al. 2001).

Finally, responses in the anterior auditory field (AAF; see Fig. 4.1) resemble closely those observed in AI, apart from the preponderance of the much broader response areas. Ripple responses in the IC are quite different from those in the cortex. Specifically, while responses are linear in character (in the sense of superposition), ripple transfer functions are mostly low pass in shape, exhibiting little ripple selectivity. Therefore, it seems that ripple selectivity emerges in the MGB or the cortex. Ripple responses have not yet been examined in other auditory structures.

3.1.4 The Categorical Representation

The basic hypothesis underlying the categorical representation is that single units or restricted populations of neurons are selective to specific spectral profiles (e.g., corresponding to different steady-state vowels), especially within the species-specific vocalization repertoire (Winter and Funkenstein 1973; Glass and Wollberg 1979, 1983). An example of such highly selective sensitivity to a complex pattern in another sensory system is that of face-feature recognition in the inferotemporal lobe (Poggio et al. 1994). More generally, the notion of the so-called grandmother cell may include both the spectral shape and its dynamics, and hence imply selectivity to a whole call, call segment, or syllable (as discussed in the next section). With few exceptions (such as in birds, cf. Margoliash 1986), numerous studies in the central auditory system over the last few decades have failed to find evidence for this and similar hypotheses (Wang et al. 1995). Instead, the results suggest that the encoding of complex sounds involves relatively large populations of units with overlapping stimulus domains (Wang et al. 1995).

3.2 Encoding of Spectral Dynamics in the Central Auditory System

Responses of AN fibers tended to reflect the dynamics of the stimulus spectrum in a relatively simple and nonselective manner. In the cochlear nucleus, more complex response properties emerge such as the bandpass MTFs and FM directional selectivity. This trend, for increasing specificity to the parameters of spectral shape and dynamics, continues with ascent toward the more central parts of the auditory system, as we shall elaborate in the following sections.

3.2.1 Sensitivity to Frequency Sweeps

The degree and variety of asymmetries in the response to upward and downward frequency transitions increases from the IC (Nelson et al. 1966; Covey and Cassiday 1991) to the cortex (Whitfield and Evans 1965; Phillips et al. 1985). The effects of manipulating two specific parameters of the FM sweep—its direction and rate—have been well studied. In several species, and at almost all central auditory stages, cells can be found that are selectively sensitive to FM direction and rate. Most studies have confirmed a qualitative theory in which directional selectivity arises from an asymmetric pattern of inhibition in the response area of the cell, whereas rate sensitivity is correlated with the bandwidth of the response area (Heil et al. 1992; Kowalski et al. 1995). Furthermore, there is accumulating evidence that these two parameters are topographically mapped in an orderly fashion in AI (Schreiner and Mendelson 1990; Shamma et al. 1993).

Frequency modulation responses, therefore, may be modeled as a temporal sequential activation of the excitatory and inhibitory portions of the response area (Suga 1965; Wang and Shamma 1995). If an FM sweep first traverses the excitatory response area, discharges will be evoked that cannot be influenced by the inhibition activated later by the ongoing sweep. Conversely, if an FM sweep first traverses the inhibitory area, the inhibition may still be effective at the time the tone sweeps through the excitatory area. If the response is the result of a temporal summation of the instantaneous inputs, then it follows that it will be smaller in this latter direction of modulation. This theory also explains why the response area bandwidth is correlated with the FM rate preference (units with broad response areas respond best to very fast sweeps), and why FM directional selectivity decreases with FM rate.

Nevertheless, while many FM responses in cortical neurons are largely predictable from responses to stationary tones, some units show responses to FM tones even though they do not respond to stationary tones. Thus, some cortical units respond to frequency sweeps that are entirely outside the unit's response area (as determined with pure tones). For many cells, only one direction of frequency sweep is effective irrespective of the relationship of the sweep to the cells' CF (Whitfield and Evans 1965). For others, responses are dependent on whether the sweep is narrow or wide, or on the degree of overlap with the response area (Phillips et al. 1985).

3.2.2 Representation of Speech and Species-Specific Stimuli

3.2.2.1 The Categorical Representation

Most complex sounds are dynamic in nature, requiring both temporal and spectral features to characterize them fully. Central auditory units have been shown, in some cases, to be highly selective to the complex spectrotemporal features of the stimulus (e.g., in birds; see Margoliash 1986). Units can also be classified in different regions depending on their stimulus selectivity, response pattern complexity, and topographic organization (Watanabe and Sakai 1973, 1975, 1978; Steinschneider et al. 1982, 1990; Newman 1988; Wang et al. 1995).

Mammalian cortical units, however, largely behave as general spectral and temporal filters rather than as specialized detectors for particular categories of sounds or vocal repertoire. For instance, detailed studies of the responses of monkey cortical cells (e.g., Wang et al. 1995) to conspecific vocalizations have suggested that, rather than responding to the spectra of the sounds, cells follow the time structure of individual stimulus components in a very context-dependent manner. The apparent specificity of some cells for particular vocalizations may result from overlap of the spectra of transient parts of the stimulus with the neuron's response area (see Phillips et al. 1991, for a detailed review).

A few experiments have been performed in the midbrain and thalamus to study the selective encoding of complex stimuli, such as speech and species-specific vocalizations (Symmes et al. 1980; Maffi and Aitkin 1987; Tanaka and Taniguchi 1987, 1991; Aitkin et al. 1994). The general finding is that in most divisions of the IC and MGB, responses are vigorous but nonselective to the calls. For instance, it is rare to find units in the IC that are selective to only one call, although they may exhibit varying preferences to a single or several elements of a particular call. Units in different regions of the IC and MGB also differ in their overall responses to natural calls (Aitkin et al. 1994), being more responsive to pure tones and to noise in the CNIC, and to vocal stimuli in other subdivisions of the IC (i.e., the external nucleus and dorsal IC). It has also been shown that the temporal patterns of responses are more complex and faithfully correlated to those of the stimulus in the ventral division of the MGB than in other divisions or in the auditory cortex (Creutzfeldt et al. 1980; Clarey et al. 1992).

The one significant mammalian exception, where high stimulus specificity is well established and understood, is in the encoding of echolocation signals in various bat species (Suga 1988). Echolocation, however, is a rather specialized task involving stereotypical spectral and temporal stimulus cues that may not reflect the situation for more general communication signals.

3.2.2.2 Voice Onset Time

The VOT cue has been shown to have a physiological correlate at the level of the primary auditory cortex in the form of a reliable "double-on" response, reflecting the onset of the noise burst followed by the onset of the periodic portion of the stimulus. This response can be detected in evoked potential records, in measurements of current source density, as well as in multi- and single-unit responses (Steinschneider et al. 1994, 1995; Eggermont 1995). The duration of the VOT is normally perceived categorically and evoked potentials in AI have been reported to behave similarly (Steinschneider et al. 1994). However, these findings are contradicted by AI single and multiunit records that encode the VOT in a monotonic continuum (Eggermont 1995). Consequently, it seems that processes responsible for the categorical perception of speech sounds may reside in brain structures beyond the primary auditory cortex.

3.2.3 The Multiscale Represention of Dynamic Spectra

The multiscale representation of the spectral profile outlined earlier can be extended to dynamic spectra if they are thought of as being composed of a weighted sum of moving ripples with different ripple frequencies, ripple phases, and ripple velocities. Thus, assuming linearity, cortical responses to such stimuli can be weighted and summed in order to predict the neural responses to any arbitrary spectrum (Kowalski et al. 1996a).

Cortical units in AI and AAF exhibit responses that are selective for moving ripples spanning a broad range of ripple parameters (Kowalski et al. 1996b). Using moving ripple stimuli, two different transfer functions can be measured: (1) *a temporal transfer function* by keeping the ripple density constant and varying the velocity at which the ripples are moved (Fig. 4.14), and (2) a *ripple transfer function* by keeping the velocity constant and varying the ripple density (Fig. 4.15). These transfer functions can be inverse Fourier transformed to obtain the corresponding response fields (RFs) and the temporal impulse responses (IRs) as shown in Figures 4.14E and 4.15E.

Both the RFs and IRs derived from transfer function measurements such as those in Figures 4.14 and 4.15 have been found to exhibit a wide variety of shapes (widths, asymetries, and polarities) that suggest that a multiscale analysis is taking place not only along the frequency axis but also in time. Thus, for any given RF, there are units with various IR shapes, each encoding the local dynamics of the spectrum at a different time scales (i.e., there are units exclusively sensitive to slow modulations in the spectrum, and others tuned only to moderate or fast spectral changes). This temporal decomposition is analogous to (and complements) the multiscale representation of the shape of the spectrum produced by the RFs. Such an analysis may underlie many important perceptual invariances, such as the ability to recognize speech and melodies despite large changes in rate of delivery (Julesz and Hirsh 1972), or to perceive continuous music and speech through gaps, noise, and other short-duration interruptions in the sound stream. Furthermore, the segregation into different time scales such as fast and slow corresponds to the intuitive classification of many natural sounds and music into transient and sustained, or into stops and continuents in speech.

3.3 Encoding of Pitch in the Central Auditory System

Regardless of how pitch is encoded in the early auditory pathway, one implicit or explicit assumption is that pitch values should be finally representable as a spatial map centrally. Thus, in temporal and mixed place-temporal schemes, phase-locked information on the AN is used before it

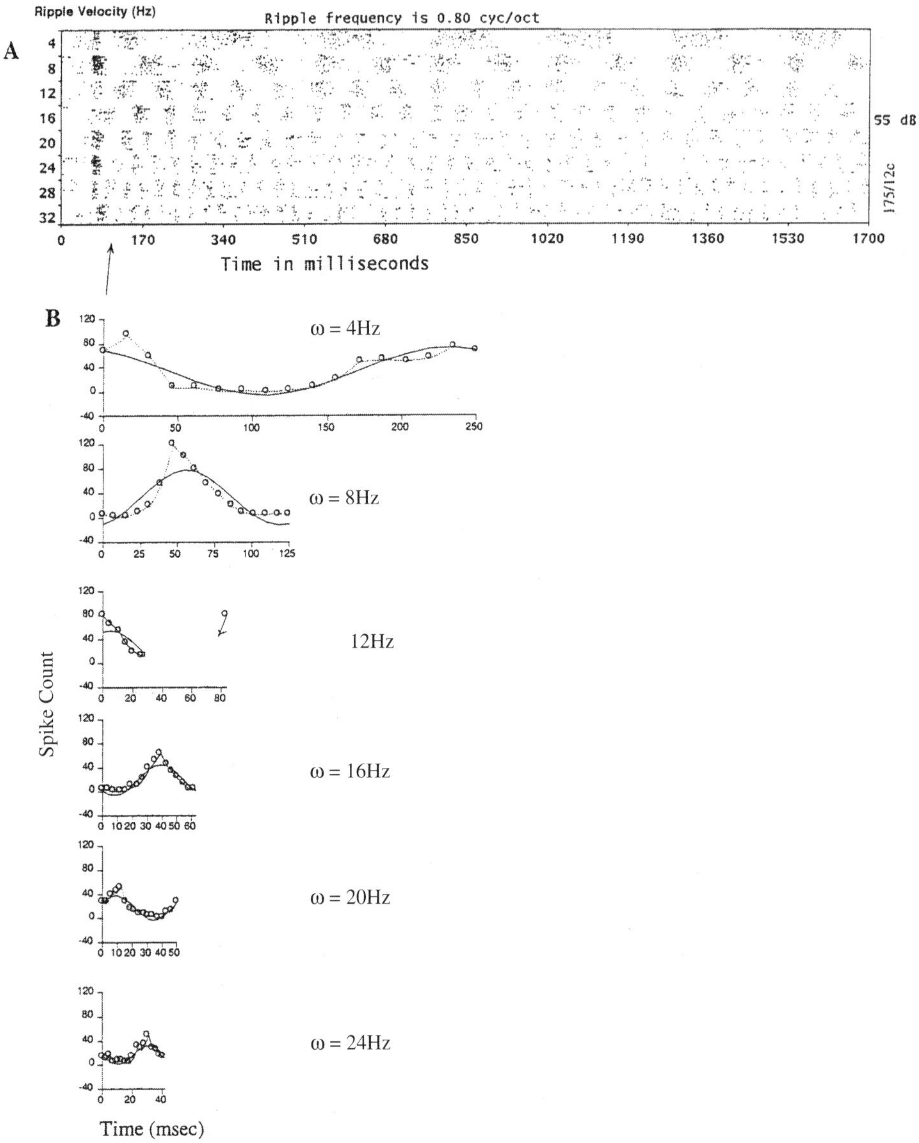

FIGURE 4.14. Measuring the dynamic response fields of auditory units in AI using ripples moving at different velocities. A: Raster responses to a ripple ($\Omega = 0.8$ cycle/octave) moving at different velocities, ω. The stimulus is turned on at 50 ms. Period histograms are constructed from responses starting at $t = 120$ ms (indicated by the arrow). B: 16-bin period histograms constructed at each ω. The best fit to the spike counts (circles) in each histogram is indicated by the solid lines.

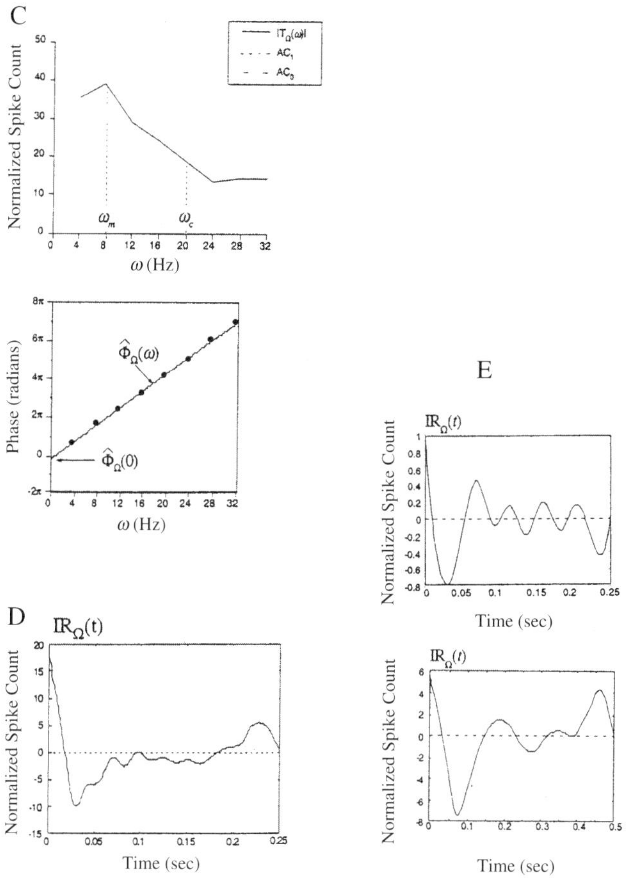

FIGURE 4.14. C: The amplitude (dashed line in top plot) and phase (bottom data points) of the best fit curves plotted as a function of ω. Also shown in the top plot is the normalized transfer function magnitude ($|T_\Omega(\omega)|$) and the average spike count as functions of *w*. A straight line fit of the phase data points is also shown in the lower plot. D: The inverse Fourier transform of the ripple transfer function $T_\Omega(\omega)$ giving the impulse response of the cell IR_Ω. E: Two further examples of impulse responses from different cells.

deteriorates in its journey through multiple synapses to higher centers of the brain. In fact, many studies have confirmed that synchrony to the repetitive features of a stimulus, be it the waveform of a tone or its amplitude modulations, becomes progressively poorer toward the cortex. For instance, while maximum synchronized rates in the cochlear nucleus cells can be as high as in the auditory nerve (4 kHz), they rarely exceed 800 to 1000 Hz in the IC (Langner 1992), and are under 100 Hz in the anterior auditory cortical field (Schreiner and Urbas 1988). Therefore, it seems inescapable that pitch be represented by a spatial (place) map in higher auditory centers if

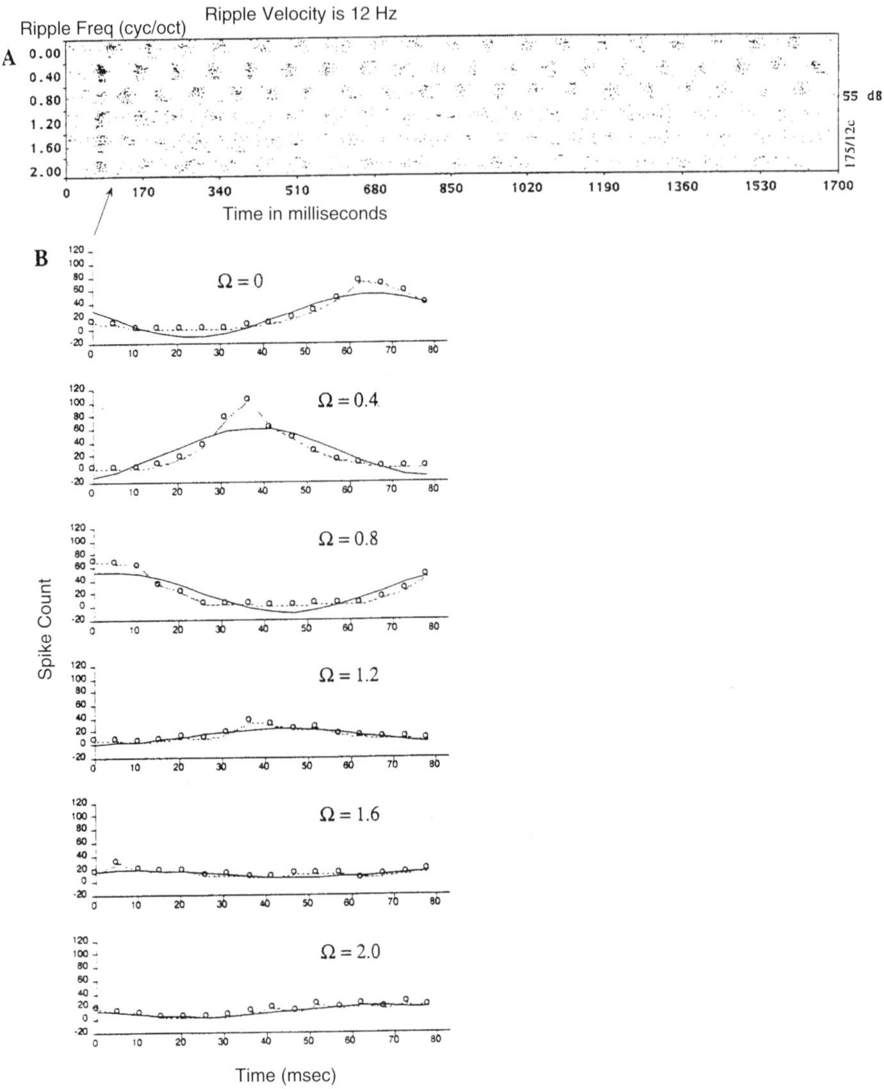

FIGURE 4.15. Measuring the dynamic response fields of auditory units in AI using different ripple frequencies moving at at the same velocity. A: Raster responses to a moving ripple ($\omega = 12$ Hz) with different ripple frequencies $\Omega = 0$–2 cycle/octave. The stimulus is turned on at 50 ms. Period histograms are constructed from responses starting at $t = 120$ ms (indicated by the arrow). B: 16-bin period histograms constructed at each Ω. The best fit to the spike counts (circles) in each histogram is indicated by the solid lines.

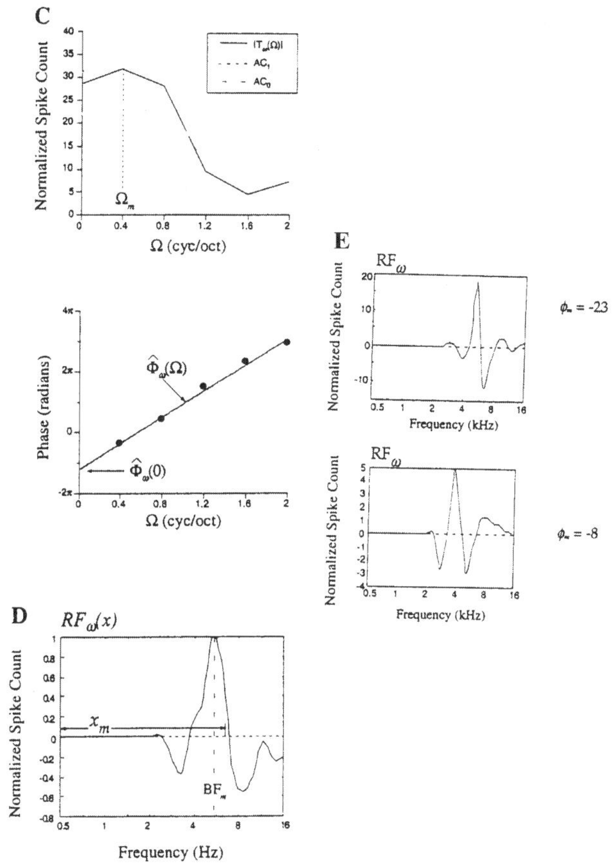

FIGURE 4.15. C: The amplitude (dashed line in top plot) and phase (bottom data points) of the best fit curves plotted as a function of W. Also shown in the top plot is the normalized transfer function magnitude ($|T_w(W)|$) and the average spike count as functions of W. A straight line fit of the phase data points is also shown in the lower plot. D: The inverse Fourier transform of the ripple transfer function $T_w(W)$ giving the response field of the cell Rf_w. E: Two further examples of response fields from different cells showing different widths and asymmetries.

they are involved in the formation of this percept. Here we review the sensitivity to modulated stimuli in the central auditory system and examine the evidence for the existence of such maps.

3.3.1 Sensitivity to AM Spectral Modulations

The MTFs of units in the IC are low-pass in shape at low SPLs, becoming bandpass at high SPLs (Rees and Møller 1983, 1987; Langner and Schreiner 1988; Rees and Palmer 1989). The BMFs in the IC are generally lower than those in the cochlear nucleus. In both rat and guinea pig, IC BMFs are less than 200 Hz. In cat, the vast majority of neurons (74%) had BMFs below 100 Hz. However, about 8% of the units had BMFs of 300 to 1000 Hz (Langner and Schreiner 1988). The most striking difference at the level of the IC compared to lower levels is that for some neurons the MTFs are similar whether determined using synchronized activity or the mean discharge rate (Langner and Schreiner 1988; Rees and Palmer 1989; but also see Müller-Preuss et al. 1994; Krishna and Semple 2000), thus suggesting that a significant recoding of the modulation information has occurred at this level.

While at lower anatomical levels there is no evidence for topographic organization of modulation sensitivity, in the IC of the cat there is evidence of topographic ordering producing "contour maps" of modulation sensitivity within each isofrequency lamina (Schreiner and Langner 1988a,b). Such detailed topographical distributions of BMFs have only been found in the cat IC, and while their presence looks somewhat unlikely in the IC of rodents and squirrel monkeys (Müller-Preuss et al. 1994; Krishna and Semple 2000), there is some evidence that implies the presence of such an organization in the gerbil and chinchilla (Albert 1994; Heil et al. 1995). The presence of modulation maps remains highly controversial, for it is unclear why such maps are to be found in certain mammalian species and not in others (certain proposals have been made, including the variability in sampling resolution through lamina, and the nature of the physiological recording methodology used). In our view it would be surprising if the manner of modulation representation in IC were not similar in all higher animals.

In many studies of the auditory cortex, the majority of neurons recorded are unable to signal envelope modulation at rates more than about 20 Hz (Whitfield and Evans 1965; Ribaupierre et al. 1972; Creuzfeldt et al. 1980; Gaese and Ostwald 1995). Eighty-eight percent of the population of cortical neurons studied by Schreiner and Urbas (1986, 1988) showed bandpass MTFs, with BMFs ranging between 3 and 100 Hz. The remaining 12% had low-pass MTFs, with a cut-off frequency of only a few hertz. These authors failed to find any topographic organization with respect to the BMF. They did, however, demonstrate different distributions of BMFs within the various divisions of the auditory cortex. While neurons in certain cortical fields (AI, AAF) had BMFs of 2 to 100 Hz, the majority of neurons in other cortical fields [secondary auditory cortex (AII), posterior auditory field (PAF), ventroposterior auditory field (VPAF)] had BMFs of 10 Hz or less.

However, evidence is accumulating, particularly from neural recordings obtained from awake monkeys, that amplitude modulation may be represented in more than one way at the auditory cortex. Low rates of AM, below 100 Hz, are represented by locking of the discharges to the modulated envelope (Bieser and Müller-Preuss 1996; Schulze and Langner 1997, 1999; Steinschneider et al. 1998; Lu and Wang 2000). Higher rates of AM are represented by a mean rate code (Bieser and Müller-Preuss 1996; Lu and Wang 2000). The pitch of harmonic complexes with higher fundamental frequencies is also available from the appropriate activation pattern across the tonotopic axis (i.e., a spectral representation; Steinschneider et al. 1998). Most striking of all is the result of Schulze and Langner in gerbil cortex using AM signals in which the spectral components were completely outside the cortical cell response area, demonstrating a periodotopic representation in the gerbil cortex. A plausible explanation for this organization is a response by the cells to distortion products, although the authors present arguments against this and in favor of broad spectral integration.

3.3.2 Do Spatial Pitch Maps Exist?

Despite its fundamental role in auditory perception, only a few reports exist of physiological evidence of a spatial pitch map, and none has been independently and unequivocally confirmed. For example, nuclear magnetic resonance (NMR) scans of human primary auditory cortex (e.g., Pantev et al. 1989) purport to show that low-CF cells in AI can be activated equally by a tone at the BF of these cells, or by higher-order harmonics of this tone. As such, it is inferred that the tonotopic axis of the AI (at least among lower CFs) essentially represents the frequency of the "missing" fundamental, in addition to the frequency of a pure tone. Another study in humans using magnetoencephalography (MEG) has also reported a "periodotopic" organization in auditory cortex (Langner et al. 1997). Attempts at confirming these results, using higher resolution single- and multiunit recordings in animals, have generally failed (Schwartz and Tomlinson 1990). For such reasons, these and similar evoked-potential results should be viewed either as experimental artifacts or as evidence that pitch coding in humans is of a different nature than in nonhuman primates and other mammals. As of yet, the only detailed evidence for pitch maps are those described above in the IC of the cat and auditory cortex of gerbil using AM tones, and these results have not yet been fully duplicated in other mammals (or by other research groups).

Of course, it is indeed possible that pitch maps don't exist beyond the level of the IC. However, this possibility is somewhat counterintuitive, given the results of ablation studies showing that bilateral cortical lesions in the auditory cortex severely impair the perception of pitch associated with complex sounds (Whitfield 1980), without affecting the fine frequency and intensity discrimination of pure tones (Neff et al. 1975). The difficulty so far in demonstrating spatial maps of pitch in the cortex may also be due to the fact that the maps sought are not as straightforwardly organized as researchers have supposed. For instance, it is conceivable that a spatial map of pitch can be derived from the cortical representation of the spectral profile discussed in the preceding sections. In this case, no simple explicit mapping of the BMFs would be found. Rather, pitch could be represented in terms of more complicated spatially distributed patterns of activity in the cortex (Wang and Shamma 1995).

4. Summary

Our present understanding of speech encoding in the auditory system can be summarized by the following sketches for each of the three basic features of the speech signal: spectral shape, spectral dynamics, and pitch.

Spectral shape: Speech signals evoke complex spatiotemporal patterns of activity in the AN. Spectral shape is well represented in both the distribution of AN fiber responses (in terms of discharge rate) along the tonotopic axis, as well as their phase-locked temporal structure. However, representations of spectrum in terms of the temporal fine structure seems unlikely at the level of the cochlear nucleus output (to various brain stem nuclei), with the exception of the pathway to the superior olivary binaural circuits. The spectrum is well represented by the average rate response profile along the tonotopic axis in at least one of the output pathways of the cochlear nucleus. At more central levels, the spectrum is further analyzed into specific shape features representing different levels of abstraction. These range from the intensity of various spectral components, to the bandwidth and asymmetry of spectral peaks, and perhaps to complex spectrotemporal combinations such as segments and syllables of natural vocalizations as in the birds (Margoliash 1986).

Spectral dynamics: The ability of the auditory system to follow the temporal structure of the stimulus on a cycle-by-cycle basis decreases progressively at more central nuclei. In the auditory nerve the responses are phase locked to frequencies of individual spectral components (up to 4–5 kHz) and to modulations reflecting the interaction between these components (up to several hundred Hz). In the midbrain, responses mostly track the modulation envelope up to about 400 to 600 Hz, but rarely follow the frequencies of the underlying individual components. At the level of the auditory cortex only relatively slow modulations (on the order of tens of Hertz) of the overall spectral shape are present in the temporal structure of the responses (but selectivity is exhibited to varying rates, depths of modulation, and directions of frequency sweeps). At all levels of the auditory pathway these temporal modulations are analyzed into narrower ranges that are encoded in different channels. For example, AN fibers respond to modulations over a range determined by the tuning of the unit and its phase-locking capabilities. In the midbrain, many units are selectively responsive to different narrow ranges of temporal modulations, as reflected by the broad range of BMFs to AM stimuli. Finally, in the cortex, units tend to be selectively responsive to different overall spectral modulations as revealed by their tuned responses to AM tones, click trains, and moving rippled spectra.

Pitch: The physiological encoding of pitch remains controversial. In the early stages of the auditory pathway (AN and cochlear nucleus) the fine-tune structure of the signal (necessary for mechanisms involving spectral template matching) is encoded in temporal firing patterns, but this form of temporal activity does not extend beyond this level. Purely temporal correlates of pitch (i.e., modulation of the firing) are preserved only up to the IC or possibly the MGB, but not beyond. While place codes for pitch may exist in the IC or even in the cortex, data in support of this are still equivocal or unconfirmed.

Overall, the evidence does not support any one simple scheme for the representation of any of the major features of complex sounds such as speech. There is no unequivocal support for simple place, time, or place/time codes beyond the auditory periphery. There is also little indication, other than in the bat, that reconvergence at high levels generates specific sensitivity to features of communication sounds. Nevertheless, even at the auditory cortex spatial frequency topography is maintained, and within this structure the sensitivities are graded with respect to several metrics, such as bandwidth and response asymmetry. Currently available data thus suggest a rather complicated form of distributed representation not easily mapped to individual characteristics of the speech signal. One important caveat to this is our relative lack of knowledge about the responses of secondary cortical areas to communication signals and analogous sounds. In the bat it is in these, possibly higher level, areas that most of the specificity to ethologically important features occurs (cf., Rauschecker et al. 1995).

List of Abbreviations

AAF	anterior auditory field
AI	primary auditory cortex
AII	secondary auditory cortex
ALSR	average localized synchronized rate
AM	amplitude modulation
AN	auditory nerve
AVCN	anteroventral cochlear nucleus
BMF	best modulation frequency
CF	characteristic frequency
CNIC	central nucleus of the inferior colliculus
CV	consonant-vowel
DAS	dorsal acoustic stria
DCIC	dorsal cortex of the inferior colliculus
DCN	dorsal cochlear nucleus
DNLL	dorsal nucleus of the lateral lemniscus
ENIC	external nucleus of the inferior colliculus
FM	frequency modulation
FTC	frequency threshold curve
IAS	intermediate acoustic stria
IC	inferior colliculus
INLL	intermediate nucleus of the lateral lemniscus

IR	impulse response
LIN	lateral inhibitory network
LSO	lateral superior olive
MEG	magnetoencephalography
MGB	medial geniculate body
MNTB	medial nucleus of the trapezoid body
MSO	medial superior olive
MTF	modulation transfer function
NMR	nuclear magnetic resonance
On-C	onset chopper
PAF	posterior auditory field
PVCN	posteroventral cochlear nucleus
QFM	quasi-frequency modulation
RF	response field
SPL	sound pressure level
VAS	ventral acoustic stria
VCN	ventral cochlear nucleus
VNLL	ventral nucleus of the lateral lemniscus
VOT	voice onset time
VPAF	ventroposterior auditory field

References

Abrahamson AS, Lisker L (1970) Discriminability along the voicing continuum: cross-language tests. Proc Sixth Int Cong Phon Sci, pp. 569–573.

Adams JC (1979) Ascending projections to the inferior colliculus. J Comp Neurol 183:519–538.

Aitkin LM, Schuck D (1985) Low frequency neurons in the lateral central nucleus of the cat inferior colliculus receive their input predominantly from the medial superior olive. Hear Res 17:87–93.

Aitkin LM, Tran L, Syka J (1994) The responses of neurons in subdivisions of the inferior colliculus of cats to tonal noise and vocal stimuli. Exp Brain Res 98:53–64.

Albert M (1994) Verarbeitung komplexer akustischer signale in colliculus inferior des chinchillas: functionelle eigenschaften und topographische repräsentation. Dissertation, Technical University Darmstadt.

Altschuler RA, Bobbin RP, Clopton BM, Hoffman DW (eds) (1991) Neurobiology of Hearing: The Central Auditory System. New York: Raven Press.

Arthur RM, Pfeiffer RR, Suga N (1971) Properties of "two-tone inhibition" in primary auditory neurons. J Physiol (Lond) 212:593–609.

Batteau DW (1967) The role of the pinna in human localization. Proc R Soc Series B 168:158–180.

Berlin C (ed) (1984) Hearing Science. San Diego: College-Hill Press.

Bieser A, Müller-Preuss P (1996) Auditory responsive cortex in the squirrel monkey: neural responses to amplitude-modulated sounds. Exp Brain Res 108:273–284.

Blackburn CC, Sachs MB (1989) Classification of unit types in the anteroventral cochlear nucleus: PST histograms and regularity analysis. J Neurophysiol 62: 1303–1329.

Blackburn CC, Sachs MB (1990) The representation of the steady-state vowel sound /e/ in the discharge patterns of cat anteroventral cochlear nucleus neurons. J Neurophysiol 63:1191–1212.

Bourk TR (1976) Electrical responses of neural units in the anteroventral cochlear nucleus of the cat. Ph.D. thesis, Massachusetts Institute of Technology, Cambridge, MA.

Brawer JR, Morest DK (1975) Relations between auditory nerve endings and cell types in the cats anteroventral cochlear nucleus seen with the Golgi method and Nomarski optics. J Comp Neurol 160:491–506.

Brawer J, Morest DK, Kane EC (1974) The neuronal architecture of the cat. J Comp Neurol 155:251–300.

Britt R, Starr A (1975) Synaptic events and discharge patterns of cochlear nucleus cells. II. Frequency-modulated tones. J Neurophysiol 39:179–194.

Brodal A (1981) Neurological Anatomy in Relation to Clinical Medicine. Oxford: Oxford University Press.

Brown MC (1987) Morphology of labelled afferent fibers in the guinea pig cochlea. J Comp Neurol 260:591–604.

Brown MC, Ledwith JV (1990) Projections of thin (type II) and thick (type I) auditory-nerve fibers into the cochlear nucleus of the mouse. Hear Res 49:105–118.

Brown M, Liberman MC, Benson TE, Ryugo DK (1988) Brainstem branches from olivocochlear axons in cats and rodents. J Comp Neurol 278:591–603.

Brugge JF, Anderson DJ, Hind JE, Rose JE (1969) Time structure of discharges in single auditory-nerve fibers of squirrel monkey in response to complex periodic sounds. J Neurophysiol 32:386–401.

Brunso-Bechtold JK, Thompson GC, Masterton RB (1981) HRP study of the organization of auditory afferents ascending to central nucleus of inferior colliculus in cat. J Comp Neurol 197:705–722.

Cant NB (1981) The fine structure of two types of stellate cells in the anteroventral cochlear nucleus of the cat. Neuroscience 6:2643–2655.

Cant NB, Casseday JH (1986) Projections from the anteroventral cochlear nucleus to the lateral and medial superior olivary nuclei. J Comp Neurol 247:457–476.

Cant NB, Gaston KC (1982) Pathways connecting the right and left cochlear nuclei. J Comp Neurol 212:313–326.

Cariani PA, Delgutte B (1996) Neural correlates of the pitch of complex tones 2. Pitch shift, pitch ambiguity, phase invariance, pitch circularity, rate pitch and the dominance region for pitch. J Neurophysiol 76:1717–1734.

Carney LH, Geisler CD (1986) A temporal analysis of auditory-nerve fiber responses to spoken stop consonant-vowel syllables. J Acoust Soc Am 79:1896–1914.

Caspary DM, Rupert AL, Moushegian G (1977) Neuronal coding of vowel sounds in the cochlear nuclei. Exp Neurol 54:414–431.

Clarey J, Barone P, Imig T (1992) Physiology of thalamus and cortex. In: Popper AN, Fay RR (eds) The Mammalian Auditory Pathway: Neurophysiology. New York: Springer-Verlag, pp. 232–334.

Conley RA, Keilson SE (1995) Rate representation and discriminability of second formant frequencies for /e/-like steady-state vowels in cat auditory nerve. J Acoust Soc Am 98:3223–3234.

Cooper NP, Robertson D, Yates GK (1993) Cochlear nerve fiber responses to amplitude-modulated stimuli: variations with spontaneous rate and other response characteristics. J Neurophysiol 70:370–386.

Covey E, Casseday JH (1991) The monaural nuclei of the lateral lemniscus in an echolating bat: parallel pathways for analyzing temporal features of sound. Neuroscience 11:3456–3470.

Creutzfeldt O, Hellweg F, Schreiner C (1980) Thalamo-cortical transformation of responses to complex auditory stimuli. Exp Brain Res 39:87–104.

De Valois R, De Valois K (1990) Spatial Vision. Oxford: Oxford University Press.

Delgutte B (1980) Representation of speech-like sounds in the discharge patterns of auditory nerve fibers. J Acoust Soc Am 68:843–857.

Delgutte B (1984) Speech coding in the auditory nerve: II. Processing schemes for vowel-like sounds. J Acoust Soc Am 75:879–886.

Delgutte B, Cariani P (1992) Coding of the pitch of harmonic and inharmonic complex tones in the interspike intervals of auditory nerve fibers. In: Schouten MEH (ed) The Auditory Processing of Speech. Berlin: Mouton-De-Gruyer, pp. 37–45.

Delgutte B, Kiang NYS (1984a) Speech coding in the auditory nerve: I. Vowel-like sounds. J Acoust Soc Am 75:866–878.

Delgutte B, Kiang NYS (1984b) Speech coding in the auditory nerve: III. Voiceless fricative consonants. J Acoust Soc Am 75:887–896.

Delgutte B, Kiang NYS (1984c) Speech coding in the auditory nerve: IV. Sounds with consonant-like dynamic characteristics. J Acoust Soc Am 75:897–907.

Delgutte B, Kiang NYS (1984d) Speech coding in the auditory nerve: V. Vowels in background noise. J Acoust Soc Am 75:908–918.

Deng L, Geisler CD (1987) Responses of auditory-nerve fibers to nasal consonant-vowel syllables. J Acoust Soc Am 82:1977–1988.

Deng L, Geisler CD, Greenberg S (1988) A composite model of the auditory periphery for the processing of speech. J Phonetics 16:93–108.

Depireux DA, Simon JZ, Klein DJ, Shamma SA (2001) Spectro-temporal response field characterization with dynamic ripples in ferret primary auditory cortex. J Neurophysiol 85:1220–1234.

Edelman GM, Gall WE, Cowan WM (eds) (1988) Auditory Function. New York: John Wiley.

Eggermont JJ (1995) Representation of a voice onset time continuum in primary auditory cortex of the cat. J Acoust Soc Am 98:911–920.

Eggermont JJ (2001) Between sound and perception: reviewing the search for a neural code. Hear Res 157:1–42.

Ehret G, Merzenich MM (1988a) Complex sound analysis (frequency resolution filtering and spectral integration) by single units of the IC of the cat. Brain Res Rev 13:139–164.

Ehret G, Merzenich M (1988b) Neuronal discharge rate is unsuitable for coding sound intensity at the inferior colliculus level. Hearing Res 35:1–8.

Erulkar SD, Butler RA, Gerstein GL (1968) Excitation and inhibition in the cochlear nucleus. II. Frequency modulated tones. J Neurophysiol 31:537–548.

Evans EF (1972) The frequency response and other properties of single fibres in the guinea pig cochlear nerve. J Physiol 226:263–287.

Evans EF (1975) Cochlear nerve and cochlear nucleus. In: Keidel WD, Neff WD (eds) Handbook of Sensory Physiology, vol. 5/2. Berlin: Springer-Verlag, pp. 1–108.

Evans EF (1980) "Phase-locking" of cochlear fibres and the problem of dynamic range. In: Brink, G van den, Bilsen FA (eds) Psychophysical, Physiological and Behavioural Studies in Hearing. Delft: Delft University Press, pp. 300–311.

Evans EF, Nelson PG (1973) The responses of single neurones in the cochlear nucleus of the cat as a function of their location and anaesthetic state. Exp Brain Res 17:402–427.

Evans EF, Palmer AR (1979) Dynamic range of cochlear nerve fibres to amplitude modulated tones. J Physiol (Lond) 298:33–34P.

Evans EF, Palmer AR (1980) Relationship between the dynamic range of cochlear nerve fibres and their spontaneous activity. Exp Brain Res 40:115–118.

Evans EF, Pratt SR, Spenner H, Cooper NP (1992) Comparison of physiological and behavioural properties: auditory frequency selectivity. In: Cazals Y, Demany L, Horner K (eds) Auditory Physiology and Perception. Oxford: Pergamon Press.

Flanagan JL, Guttman N (1960) Pitch of periodic pulses without fundamental component. J Acoust Soc Am 32:1319–1328.

Frisina RD, Smith RL, Chamberlain SC (1990a) Encoding of amplitude modulation in the gerbil cochlear nucleus: I. A hierarchy of enhancement. Hear Res 44:99–122.

Frisina RD, Smith RL, Chamberlain SC (1990b) Encoding of amplitude modulation in the gerbil cochlear nucleus: II. Possible neural mechanisms. Hear Res 44:123–142.

Gaese B, Ostwald J (1995) Temporal coding of amplitude and frequency modulations in rat auditory cortex. Eur J Neurosci 7:438–450.

Geisler CD, Gamble T (1989) Responses of "high-spontaneous" auditory-nerve fibers to consonant-vowel syllables in noise. J Acoust Soc Am 85:1639–1652.

Glass I, Wollberg Z (1979) Lability in the responses of cells in the auditory cortex of squirrel monkeys to species-specific vocalizations. Exp Brain Res 34:489–498.

Glass I, Wollberg Z (1983) Responses of cells in the auditory cortex of awake squirrel monkeys to normal and reversed species-species vocalization. Hear Res 9:27–33.

Glendenning KK, Masterton RB (1983) Acoustic chiasm: efferent projections of the lateral superior olive. J Neurosci 3:1521–1537.

Goldberg JM, Brown PB (1969) Response of binaural neurons of dog superior olivary complex to dichotic tonal stimuli: some physiological mechanisms of sound localization. J Neurophysiol 32:613–636.

Goldberg JM, Brownell WE (1973) Discharge characteristics of neurons in the anteroventral and dorsal cochlear nuclei of cat. Brain Res 64:35–54.

Goldstein JL (1973) An optimum processor theory for the central formation of pitch complex tones. J Acoust Soc Am 54:1496–1516.

Greenberg SR (1994) Speech processing: auditory models. In: Asher RE (ed) The Encyclopedia of Language and Linguistics. Oxford: Pergamon, pp. 4206–4227.

Greenwood DD (1990) A cochlear frequency-position function for several species—29 years later. J Acoust Soc Am 87:2592–2605.

Greenwood DD, Joris PX (1996) Mechanical and "temporal" filtering as codeterminants of the response by cat primary fibers to amplitude-modulated signals. J Acoust Soc Am 99:1029–1039.

Harris DM, Dallos P (1979) Forward masking of auditory nerve fiber responses. J Neurophysiol 42:1083–1107.

Harrison JM, Irving R (1965) The anterior ventral cochlear nucleus. J Comp Neurol 126:51–64.

Hartline HK (1974) Studies on Excitation and Inhibition in the Retina. New York: Rockefeller University Press.
Hashimoto T, Katayama Y, Murata K, Taniguchi I (1975) Pitch-synchronous response of cat cochlear nerve fibers to speech sounds. Jpn J Physiol 25:633–644.
Heil P, Rajan R, Irvine D (1992) Sensitivity of neurons in primary auditory cortex to tones and frequency-modulated stimuli. II. Organization of responses along the isofrequency dimension. Hear Res 63:135–156.
Heil P, Rajan R, Irvine D (1994) Topographic representation of tone intensity along the isofrequency axis of cat primary auditory cortex. Hear Res 76:188–202.
Heil P, Schulze H, Langner G (1995) Ontogenetic development of periodicity coding in the inferior colliculus of the mongolian gerbil. Audiol Neurosci 1:363–383.
Held H (1893) Die centrale Gehorleitung. Arch Anat Physiol Anat Abt 17:201–248.
Henkel CK, Spangler KM (1983) Organization of the efferent projections of the medial superior olivary nucleus in the cat as revealed by HRP and autoradiographic tracing methods. J Comp Neurol 221:416–428.
Hewitt MJ, Meddis R, Shackleton TM (1992) A computer model of the cochlear nucleus stellate cell: responses to amplitude-modulated and pure tone stimuli. J Acoust Soc Am 91:2096–2109.
Houtsma AJM (1979) Musical pitch of two-tone complexes and predictions of modern pitch theories. J Acoust Soc Am 66:87–99.
Imig TJ, Reale RA (1981) Patterns of cortico-cortical connections related to tonotopic maps in cat auditory cortex. J Comp Neurol 203:1–14.
Irvine DRF (1986) The Auditory Brainstem. Berlin: Springer-Verlag.
Javel E (1980) Coding of AM tones in the chinchilla auditory nerve: implication for the pitch of complex tones. J Acoust Soc Am 68:133–146.
Javel E (1981) Suppression of auditory nerve responses I: temporal analysis intensity effects and suppression contours. J Acoust Soc Am 69:1735–1745.
Javel E, Mott JB (1988) Physiological and psychophysical correlates of temporal processes in hearing. Hear Res 34:275–294.
Jiang D, Palmer AR, Winter IM (1996) The frequency extent of two-tone facilitation in onset units in the ventral cochlear nucleus. J Neurophysiol 75:380–395.
Johnson DH (1980) The relationship between spike rate and synchrony in responses of auditory nerve fibers to single tones. J Acoust Soc Am 68:1115–1122.
Joris PX, Yin TCT (1992) Responses to amplitude-modulated tones in the auditory nerve of the cat. J Acoust Soc Am 91:215–232.
Julesz B, Hirsh IJ (1972) Visual and auditory perception—an essay of comparison In: David EE Jr, Denes PB (eds) Human Communication: A Unified View. New York: McGraw-Hill, pp. 283–340.
Keilson EE, Richards VM, Wyman BT, Young ED (1997) The representation of concurrent vowels in the cat anesthetized ventral cochlear nucleus: evidence for a periodicity-tagged spectral representation. J Acoust Soc Am 102:1056–1071.
Kiang NYS (1968) A survey of recent developments in the study of auditory physiology. Ann Otol Rhinol Larnyngol 77:577–589.
Kiang NYS, Watanabe T, Thomas EC, Clark LF (1965) Discharge patterns of fibers in the cat's auditory nerve. Cambridge, MA: MIT Press.
Kim DO, Leonard G (1988) Pitch-period following response of cat cochlear nucleus neurons to speech sounds. In: Duifhuis H, Wit HP, Horst JW (eds) Basic Issues in Hearing. London: Academic Press, pp. 252–260.

Kim DO, Rhode WS, Greenberg SR (1986) Responses of cochlear nucleus neurons to speech signals: neural encoding of pitch intensity and other parameters In: Moore BCJ, Patterson RD (eds) Auditory Frequency Selectivity. New York: Plenum, pp. 281–288.

Kim DO, Sirianni JG, Chang SO (1990) Responses of DCN-PVCN neurons and auditory nerve fibers in unanesthetized decerebrate cats to AM and pure tones: analysis with autocorrelation/power-spectrum. Hear Res 45:95–113.

Kowalski N, Depireux D, Shamma S (1995) Comparison of responses in the anterior and primary auditory fields of the ferret cortex. J Neurophysiol 73:1513–1523.

Kowalski N, Depireux D, Shamma S (1996a) Analysis of dynamic spectra in ferret primary auditory cortex 1. Characteristics of single-unit responses to moving ripple spectra. J Neurophysiol 76:3503–3523.

Kowalski N, Depireux DA, Shamma SA (1996b) Analysis of dynamic spectra in ferret primary auditory cortex 2. Prediction of unit responses to arbitrary dynamic spectra. J Neurophysiol 76:3524–3534.

Krishna BS, Semple MN (2000) Auditory temporal processing: responses to sinusoidally amplitude-modulated tones in the inferior colliculus. J Neurophysiol 84:255–273.

Kudo M (1981) Projections of the nuclei of the lateral lemniscus in the cat: an autoradiographic study. Brain Res 221:57–69.

Kuhl PK, Miller JD (1978) Speech perception by the chinchilla: identification functions for synthetic VOT stimuli. J Acoust Soc Am 63:905–917.

Kuwada S, Yin TCT, Syka J, Buunen TJF, Wickesberg RE (1984) Binaural interaction in low frequency neurons in inferior colliculus of the cat IV. Comparison of monaural and binaural response properties. J Neurophysiol 51:1306–1325.

Langner G (1992) Periodicity coding in the auditory system. Hear Res 60:115–142.

Langner G, Schreiner CE (1988) Periodicity coding in the inferior colliculus of the cat. I. Neuronal mechanisms. J Neurophysiol 60:1815–1822.

Langner G, Sams M, Heil P, Schulze H (1997) Frequency and periodicity are represented in orthogonal maps in the human auditory cortex: evidence from magnetoencephalography. J Comp Physiol (A) 181:665–676.

Lavine RA (1971) Phase-locking in response of single neurons in cochlear nuclear complex of the cat to low-frequency tonal stimuli. J Neurophysiol 34:467–483.

Liberman MC (1978) Auditory nerve responses from cats raised in a low noise chamber. J Acoust Soc Am 63:442–455.

Liberman MC (1982) The cochlear frequency map for the cat: labeling auditory-nerve fibers of known characteristic frequency. J Acoust Soc Am 72:1441–1449.

Liberman MC, Kiang NYS (1978) Acoustic trauma in cats—cochlear pathology and auditory-nerve activity. Acta Otolaryngol Suppl 358:1–63.

Lorente de No R (1933a) Anatomy of the eighth nerve: the central projections of the nerve endings of the internal ear. Laryngoscope 43:1–38.

Lorente de No R (1933b) Anatomy of the eighth nerve. III. General plan of structure of the primary cochlear nuclei. Laryngoscope 43:327–350.

Lu T, Wang XQ (2000) Temporal discharge patterns evoked by rapid sequences of wide- and narrowband clicks in the primary auditory cortex of cat. J Neurophysiol 84:236–246.

Lyon R, Shamma SA (1996) Auditory representations of timbre and pitch. In: Hawkins H, Popper AN, Fay RR (eds) Auditory Computation. New York: Springer-Verlag.

Maffi CL, Aitkin LM (1987) Diffential neural projections to regions of the inferior colliculus of the cat responsive to high-frequency sounds. J Neurohysiol 26:1–17.
Mandava P, Rupert AL, Moushegian G (1995) Vowel and vowel sequence processing by cochlear nucleus neurons. Hear Res 87:114–131.
Margoliash D (1986) Preference for autogenous song by auditory neurons in a song system nucleus of the white-crowned sparrow. J Neurosci 6:1643–1661.
May BJ, Sachs MB (1992) Dynamic-range of neural rate responses in the ventral cochlear nucleus of awake cats, J Neurophysiol 68:1589–1602.
Merzenich M, Knight P, Roth G (1975) Representation of cochlea within primary auditory cortex in the cat. J Neurophysiol 38:231–249.
Merzenich MM, Roth GL, Andersen RA, Knight PL, Colwell SA (1977) Some basic features of organisation of the central auditory nervous system In: Evans EF, Wilson JP (eds) Psychophysics and Physiology of Hearing. London: Academic Press, pp. 485–497.
Miller MI, Sachs MB (1983) Representation of stop consonants in the discharge patterns of auditory-nerve fibers. J Acoust Soc Am 74:502–517.
Miller MI, Sachs MB (1984) Representation of voice pitch in discharge patterns of auditory-nerve fibers. Hear Res 14:257–279.
Møller AR (1972) Coding of amplitude and frequency modulated sounds in the cochlear nucleus of the rat. Acta Physiol Scand 86:223–238.
Møller AR (1974) Coding of amplitude and frequency modulated sounds in the cochlear nucleus. Acoustica 31:292–299.
Møller AR (1976) Dynamic properties of primary auditory fibers compared with cells in the cochlear nucleus. Acta Physiol Scand 98:157–167.
Møller AR (1977) Coding of time-varying sounds in the cochlear nucleus. Audiology 17:446–468.
Moore BCJ (ed) (1995) Hearing. London: Academic Press.
Moore BCJ (1997) An Introduction to the Psychology of Hearing, 4th ed. London: Academic Press.
Moore TJ, Cashin JL (1974) Response patterns of cochlear nucleus neurons to excerpts from sustained vowels. J Acoust Soc Am 56:1565–1576.
Moore TJ, Cashin JL (1976) Response of cochlear-nucleus neurons to synthetic speech. J Acoust Soc Am 59:1443–1449.
Morest DK, Oliver DL (1984) The neuronal architecture of the inferior colliculus of the cat: defining the functional anatomy of the auditory midbrain. J Comp Neurol 222:209–236.
Müller-Preuss P, Flachskamm C, Bieser A (1994) Neural encoding of amplitude modulation within the auditory midbrain of squirrel monkeys. Hear Res 80:197–208.
Nedzelnitsky V (1980) Sound pressures in the basal turn of the cochlea. J Acoust Soc Am 698:1676–1689.
Neff WD, Diamond IT, Casseday JH (1975) Behavioural studies of auditory discrimination: central nervous system. In: Keidel WD, Neff WD (eds) Handbook of Sensory Physiology, vol. 5/2. Berlin: Springer-Verlag, pp. 307–400.
Nelson PG, Erulkar AD, Bryan JS (1966) Responses of units of the inferior colliculus to time-varying acoustic stimuli. J Neurophysiol 29:834–860.
Newman J (1988) Primate hearing mechanisms. In: Steklis H, Erwin J (eds) Comparative Primate Biology. New York: Wiley, pp. 469–499.

Oliver DL, Shneiderman A (1991) The anatomy of the inferior colliculus—a cellular basis for integration of monaural and binaural information. In: Altschuler RA, Bobbin RP, Clopton BM, Hoffman DW (eds) Neurobiology of Hearing: The Central Auditory System. New York: Raven Press, pp. 195–222.

Osen KK (1969) Cytoarchitecture of the cochlear nuclei in the cat. Comp Neurol 136:453–483.

Palmer AR (1982) Encoding of rapid amplitude fluctuations by cochler-nerve fibres in the guinea-pig. Arch Otorhinolaryngol 236:197–202.

Palmer AR (1990) The representation of the spectra and fundamental frequencies of steady-state single and double vowel sounds in the temporal discharge patterns of guinea-pig cochlear nerve fibers. J Acoust Soc Am 88:1412–1426.

Palmer AR (1992) Segregation of the responses to paired vowels in the auditory nerve of the guinea pig using autocorrelation In: Schouten MEG (ed) The Auditory Processing of Speech. Berlin: Mouton de Gruyter, pp. 115–124.

Palmer AR, Evans EF (1979) On the peripheral coding of the level of individual frequency components of complex sounds at high levels. In: Creutzfeldt O, Scheich H, Schreiner C (eds) Hearing Mechanisms and Speech. Berlin: Springer-Verlag, pp. 19–26.

Palmer AR, Russell IJ (1986) Phase-locking in the cochlear nerve of the guinea-pig and its relation to the receptor potential of inner hair cells. Hear Res 24:1–15.

Palmer AR, Winter IM (1992) Cochlear nerve and cochlear nucleus responses to the fundamental frequency of voiced speech sounds and harmonic complex tones In: Cazals Y, Demany L, Horner K (eds) Auditory Physiology and Perception. Oxford: Pergamon Press, pp. 231–240.

Palmer AR, Winter IM (1993) Coding of the fundamental frequency of voiced speech sounds and harmonic complex tones in the ventral cochlear nucleus. In: Merchan JM, Godfrey DA, Mugnaini E (eds) The Mammalian Cochlear Nuclei: Organization and Function. New York: Plenum, pp. 373–384.

Palmer AR, Winter IM (1996) The temporal window of two-tone facilitation in onset units of the ventral cochlear nucleus. Audiol Neuro-otol 1:12–30.

Palmer AR, Winter IM, Darwin CJ (1986) The representation of steady-state vowel sounds in the temporal discharge patterns of the guinea-pig cochlear nerve and primarylike cochlear nucleus neurones. J Acoust Soc Am 79:100–113.

Palmer AR, Jiang D, Marshall DH (1996a) Responses of ventral cochlear nucleus onset and chopper units as a function of signal bandwidth. J Neurophysiol 75:780–794.

Palmer AR, Winter IM, Stabler SE (1996b) Responses to simple and complex sounds in the cochlear nucleus of the guinea pig. In: Ainsworth WA, Hackney C, Evans EF (eds) Cochlear Nucleus: Structure and Function in Relation to Modelling. London: JAI Press.

Palombi PS, Backoff PM, Caspary D (1994) Paired tone facilitation in dorsal cochlear nucleus neurons: a short-term potentiation model testable in vivo. Hear Res 75:175–183.

Pantev C, Hoke M, Lutkenhoner B Lehnertz K (1989) Tonotopic organization of the auditory cortex: pitch versus frequency representation. Science 246:486–488.

Peterson GE, Barney HL (1952) Control methods used in the study of vowels. J Acoust Soc Am 24:175–184.

Pfingst BE, O'Connor TA (1981) Characteristics of neurons in auditory cortex of monkeys performing a simple auditory task. J Neurophysiol 45:16–34.

Phillips DP, Irvine DRF (1981) Responses of single neurons in a physiologically defined area of cat cerebral cortex: sensitivity to interaural intensity differences. Hear Res 4:299–307.

Phillips DP, Mendelson JR, Cynader JR, Douglas RM (1985) Responses of single neurons in the cat auditory cortex to time-varying stimuli: frequency-modulated tone of narrow excursion. Exp Brain Res 58:443–454.

Phillips DP, Reale RA, Brugge JF (1991) Stimulus processing in the auditory cortex. In: Altschuler RA, Bobbin RP, Clopton BM, Hoffman DW (eds) Neurobiology of Hearing: The Central Auditory System. New York: Raven Press, pp. 335–366.

Phillips DP, Semple MN, Calford MB, Kitzes LM (1994) Level-dependent representation of stimulus frequency in cat primary auditory cortex. Exp Brain Res 102:210–226.

Pickles JO (1988) An Introduction to the Physiology of Hearing, 2nd ed. London: Academic Press.

Plomp R (1976) Aspects of Tone Sensation. London: Academic Press.

Pont MJ (1990) The role of the dorsal cochlear nucleus in the perception of voicing contrasts in initial English stop consonants: a computational modelling study. PhD dissertation, Department of Electronics and Computer Science, University of Southampton, UK.

Pont MJ, Damper RI (1991) A computational model of afferent neural activity from the cochlea to the dorsal acoustic stria. J Acoust Soc Am 89:1213–1228.

Poggio A, Logothetis N, Pauls J, Bulthoff H (1994) View-dependent object recognition in monkeys. Curr Biol 4:401–414.

Popper AN, Fay RR (eds) (1992) The Mammalian Auditory Pathway: Neurophysiology. New York: Springer-Verlag.

Rauschecker JP, Tian B, Hauser M (1995) Processing of complex sounds in the macaque nonprimary auditory cortex. Science 268:111–114.

Recio A, Rhode WS (2000) Representation of vowel stimuli in the ventral cochlear nucleus of the chinchilla. Hear Res 146:167–184.

Rees A, Møller AR (1983) Responses of neurons in the inferior colliculus of the rat to AM and FM tones. Hear Res 10:301–330.

Rees A, Møller AR (1987) Stimulus properties influencing the responses of inferior colliculus neurons to amplitude-modulated sounds. Hear Res 27:129–143.

Rees A, Palmer AR (1989) Neuronal responses to amplitude-modulated and pure-tone stimuli in the guinea pig inferior colliculus and their modification by broadband noise. J Acoust Soc Am 85:1978–1994.

Rhode WS (1994) Temporal coding of 200% amplitude modulated signals in the ventral cochlear nucleus of cat. Hear Res 77:43–68.

Rhode WS (1995) Interspike intervals as a correlate of periodicity pitch. J Acoust Soc Am 97:2414–2429.

Rhode WS, Greenberg S (1994a) Lateral suppression and inhibition in the cochlear nucleus of the cat. J Neurophysiol 71:493–514.

Rhode WS, Greenberg S (1994b) Encoding of amplitude modulation in the cochlear nucleus of the cat. J Neurophysiol 71:1797–1825.

Rhode WS, Smith PH (1986a) Encoding timing and intensity in the ventral cochlear nucleus of the cat. J Neurophysiol 56:261–286.

Rhode WS, Smith PH (1986b) Physiological studies of neurons in the dorsal cochlear nucleus of the cat. J Neurophysiol 56:287–306.

Ribaupierre F de, Goldstein MH, Yeni-Komishan G (1972) Cortical coding of repetitive acoustic pulses. Brain Res 48:205–225.

Rose JE, Brugge JF, Anderson DJ, Hind JE (1967) Phase-locked response to low-frequency tones in single auditory nerve fibers of the squirrel monkey. J Neurophysiol 30:769–793.

Rose JE, Hind JE, Anderson DJ, Brugge JF (1971) Some effects of stimulus intensity on responses of auditory nerve fibers in the squirrel monkey. J Neurophysiol 34:685–699.

Rosowski JJ (1995) Models of external- and middle-ear function. In: Hawkins HL, McMullen TA, Popper AN, Fay RR (eds) Auditory Computation. New York: Springer-Verlag, pp. 15–61.

Roth GL, Aitkin LM, Andersen RA, Merzenich MM (1978) Some features of the spatial organization of the central nucleus of the inferior colliculus of the cat. J Comp Neurol 182:661–680.

Ruggero MA (1992) Physiology and coding of sound in the auditory nerve. In: Popper AN, Fay RR (eds) The Mammalian Auditory System. New York: Springer-Verlag, pp. 34–93.

Ruggero MA, Temchin AN (2002) The roles of the external middle and inner ears in determining the bandwidth of hearing. Proc Natl Acad Sci USA 99: 13206–13210.

Ruggero MA, Santi PA, Rich NC (1982) Type II cochlear ganglion cells in the chinchilla. Hear Res 8:339–356.

Rupert AL, Caspary DM, Moushegian G (1977) Response characteristics of cochlear nucleus neurons to vowel sounds. Ann Otol 86:37–48.

Russell IJ, Sellick PM (1978) Intracellular studies of hair cells in the mammalian cochlea. J Physiol 284:261–290.

Rutherford W (1886) A new theory of hearing. J Anat Physiol 21 166–168.

Sachs MB (1985) Speech encoding in the auditory nerve. In: Berlin CI (ed) Hearing Science. London: Taylor and Francis, pp. 263–308.

Sachs MB, Abbas PJ (1974) Rate versus level functions for auditory-nerve fibers in cats: tone-burst stimuli. J Acoust Soc Am 56:1835–1847.

Sachs MB, Blackburn CC (1991) Processing of complex sounds in the cochlear nucleus. In: Altschuler RA, Bobbin RP, Clopton BM, Hoffman DW (eds) Neurobiology of Hearing: The Central Auditory System. New York: Raven Press, pp. 79–98.

Sachs MB, Kiang NYS (1968) Two-tone inhibition in auditory nerve fibers. J Acoust Soc Am 43:1120–1128.

Sachs MB, Young ED (1979) Encoding of steady-state vowels in the auditory nerve: representation in terms of discharge rate. J Acoust Soc Am 66:470–479.

Sachs MB, Young ED (1980) Effects of nonlinearities on speech encoding in the auditory nerve. J Acoust Soc Am 68:858–875.

Sachs MB, Young ED, Miller M (1982) Encoding of speech features in the auditory nerve. In: Carlson R, Grandstrom B (eds) Representation of Speech in the Peripheral Auditory System. Amsterdam: Elsevier.

Sachs MB, Voigt HF, Young ED (1983) Auditory nerve representation of vowels in background noise. J Neurophysiol 50:27–45.

Sachs MB, Winslow RL, Blackburn CC (1988) Representation of speech in the auditory periphery In: Edelman GM, Gall WE, Cowan WM (eds) Auditory Function. New York: John Wiley, pp. 747–774.

Schreiner C, Calhoun B (1995) Spectral envelope coding in cat primary auditory cortex. Auditory Neurosci 1:39–61.

Schreiner CE, Langner G (1988a) Coding of temporal patterns in the central auditory nervous system. In: Edelman GM, Gall WE, Cowan WM (eds) Auditory Function. New York: John Wiley, pp. 337–361.

Schreiner CE, Langner G (1988b) Periodicity coding in the inferior colliculus of the cat. II. Topographical organization. J Neurophysiol 60:1823–1840.

Schreiner CE, Mendelson JR (1990) Functional topography of cat primary auditory cortex: distribution of integrated excitation. J Neurophysiol 64:1442–1459.

Schreiner CE, Urbas JV (1986) Representation of amplitude modulation in the auditory cortex of the cat I. Anterior auditory field. Hear Res 21:277–241.

Schreiner CE, Urbas JV (1988) Representation of amplitude modulation in the auditory cortex of the cat II. Comparison between cortical fields. Hear Res 32:59–64.

Schulze H, Langner G (1997) Periodicity coding in the primary auditory cortex of the Mongolian gerbil (*Meriones unguiculatus*): two different coding strategies for pitch and rhythm? J Comp Physiol (A) 181:651–663.

Schulze H, Langner G (1999) Auditory cortical responses to amplitude modulations with spectra above frequency receptive fields: evidence for wide spectral integration. J Comp Physiol (A) 185:493–508.

Schwartz D, Tomlinson R (1990) Spectral response patterns of auditory cortex neurons to harmonic complex tones in alert monkey (*Macaca mulatta*). J Neurophysiol 64:282–299.

Shamma SA (1985a) Speech processing in the auditory system I: the representation of speech sounds in the responses of the auditory nerve. J Acoust Soc Am 78:1612–1621.

Shamma SA (1985b) Speech processing in the auditory system II: lateral inhibition and central processing of speech evoked activity in the auditory nerve. J Acoust Soc Am 78:1622–1632.

Shamma SA (1988) The acoustic features of speech sounds in a model of auditory processing: vowels and voiceless fricatives. J Phonetics 16:77–92.

Shamma SA (1989) Spatial and temporal processing in central auditory networks. In: Koch C, Segev I (eds) Methods in Neuronal Modelling. Cambridge, MA: MIT Press.

Shamma SA, Symmes D (1985) Patterns of inhibition in auditory cortical cells in the awake squirrel monkey. Hear Res 19:1–13.

Shamma SA, Versnel H (1995) Ripple analysis in ferret primary auditory cortex. II. Prediction of single unit responses to arbitrary spectra. Auditory Neurosci 1:255–270.

Shamma S, Chadwick R, Wilbur J, Rinzel J (1986) A biophysical model of cochlear processing: intensity dependence of pure tone responses. J Acoust Soc Am 80:133–144.

Shamma SA, Fleshman J, Wiser P, Versnel H (1993) Organization of response areas in ferret primary auditory cortex. J Neurophysiol 69:367–383.

Shamma SA, Vranic S, Wiser P (1992) Spectral gradient columns in primary auditory cortex: physiological and psychoacoustical correlates. In: Cazals Y, Demany

L, Horner K (eds) Auditory Physiology and Perception. Oxford: Pergamon Press, pp. 397–406.

Shamma SA, Versnel H, Kowalski N (1995a) Ripple analysis in ferret primary auditory cortex I. Response characteristics of single units to sinusoidally rippled spectra. Auditory Neurosci 1:233–254.

Shamma S, Vranic S, Versnel H (1995b) Representation of spectral profiles in the auditory system: theory, physiology and psychoacoustics. In: Manley G, Klump G, Köppl C, Fastl H, Oeckinhaus H (eds) Physiology and Psychoacoustics. Singapore: World Scientific, pp. 534–544.

Shaw EAG (1974) The external ear. In: Keidel WD, Neff WD (eds) Handbook of Sensory Physiology, vol. 5/2. Berlin: Springer-Verlag, pp. 445–490.

Shore SE (1995) Recovery of forward-masked responses in ventral cochlear nucleus neurons. Hear Res 82:31–34.

Shore SE, Godfrey DA, Helfert RH, Altschuler RA, Bledsoe SC (1992) Connections between the cochlear nuclei in the guinea pig. Hear Res 62:16–26.

Shneiderman A, Henkel CA (1987) Banding of lateral superiory olivary nucleus afferents in the inferior colliculus: a possible substrate for sensory integration. J Comp Neurol 266:519–534.

Silkes SM, Geisler CD (1991) Responses of lower-spontaneous-rate auditory-nerve fibers to speech syllables presented in noise 1. General-characteristics. J Acoust Soc Am 90:3122–3139.

Sinex DG (1993) Auditory nerve fiber representation of cues to voicing in syllable-final stop consonants. J Acoust Soc Am 94:1351–1362.

Sinex DG, Geisler CD (1981) Auditory-nerve fiber responses to frequency-modulated tones. Hear Res 4:127–148.

Sinex DG, Geisler CD (1983) Responses of auditory-nerve fibers to consonant-vowel syllables. J Acoust Soc Am 73:602–615.

Sinex DG, McDonald LP (1988) Average discharge rate representation of voice onset time in the chinchilla auditory nerve. J Acoust Soc Am 83:1817–1827.

Sinex DG, McDonald LP (1989) Synchronized discharge rate representation of voice-onset time in the chinchilla auditory nerve. J Acoust Soc Am 85:1995–2004.

Sinex DG, Narayan SS (1994) Auditory-nerve fiber representation of temporal cues to voicing in word-medial stop consonants. J Acoust Soc Am 95:897–903.

Sinex DG, McDonald LP, Mott JB (1991) Neural correlates of nonmonotonic temporal acuity for voice onset time. J Acoust Soc Am 90:2441–2449.

Slaney M, Lyon RF (1990) A perceptual pitch detector. Proceedings, International Conference on Acoustics Speech and Signal Processing, Albuquerque, NM.

Smith PH, Rhode WS (1989) Structural and functional properties distinguish two types of multipolar cells in the ventral cochlear nucleus. J Comp Neurol 282:595–616.

Smith RL (1979) Adaptation saturation and physiological masking in single auditory-nerve fibers. J Acoust Soc Am 65:166–179.

Smith RL, Brachman ML (1980) Response modulation of auditory-nerve fibers by AM stimuli: effects of average intensity. Hear Res 2:123–133.

Spoendlin H (1972) Innervation densities of the cochlea. Acta Otolaryngol 73:235–248.

Stabler SE (1991) The neural representation of simple and complex sounds in the dorsal cochlear nucleus of the guinea pig. MRC Institute of Hearing Research, University of Nottingham.

Steinschneider M, Arezzo J, Vaughan HG (1982) Speech evoked activity in the auditory radiations and cortex of the awake monkey. Brain Res 252:353–365.

Steinschneider M, Arezzo JC, Vaughan HG (1990) Tonotopic features of speech-evoked activity in primate auditory cortex. Brain Res 519:158–168.

Steinschneider M, Schroeder CE, Arezzo JC, Vaughan HG (1994) Speech-evoked activity in primary auditory cortex—effects of voice onset time. Electroencephalogr Clin Neurophysiol 92:30–43.

Steinschneider M, Reser D, Schroeder CE, Arezzo JC (1995) Tonotopic organization of responses reflecting stop consonant place of articulation in primary auditory cortex (Al) of the monkey. Brain Res 674:147–152.

Steinschneider M, Reser DH, Fishman YI, Schroeder CE, Arezzo JC (1998) Click train encoding in primary auditory cortex of the awake monkey: evidence for two mechanisms subserving pitch perception. J Acoust Soc Am 104:2935–2955.

Stotler WA (1953) An experimental study of the cells and connections of the superior olivary complex of the cat. J Comp Neurol 98:401–432.

Suga N (1965) Analysis of frequency modulated tones by auditory neurons of echo-locating bats. J Physiol 200:26–53.

Suga N (1988) Auditory neuroethology and speech processing; complex sound processing by combination-sensitive neurons. In: Edelman GM, Gall WE, Cowan WM (eds) Auditory Function. New York: John Wiley, pp. 679–720.

Suga N, Manabe T (1982) Neural basis of amplitude spectrum representation in the auditory cortex of the mustached bat. J Neurophysiol 47:225–255.

Sutter M, Schreiner C (1991) Physiology and topography of neurons with multi-peaked tuning curves in cat primary auditory cortex. J Neurophysiol 65: 1207–1226.

Symmes D, Alexander G, Newman J (1980) Neural processing of vocalizations and artificial stimuli in the medial geniculate body of squirrel monkey. Hear Res 3:133–146.

Tanaka H, Taniguchi I (1987) Response properties of neurons in the medial geniculate-body of unanesthetized guinea-pigs to the species-specific vocalized sound. Proc Jpn Acad (Series B) 63:348–351.

Tanaka H, Taniguchi I (1991) Responses of medial geniculate neurons to species-specific vocalized sounds in the guinea-pig. Jpn J Physiol 41:817–829.

Terhardt E (1979) Calculating virtual pitch. Hear Res 1:155–182.

Tolbert LP, Morest DK (1982) The neuronal architecture of the anteroventral cochlear nucleus of the cat in the region of the cochlear nerve root: Golgi and Nissl methods. Neuroscience 7:3013–3030.

Van Gisbergen JAM, Grashuis JL, Johannesma PIM, Vendrif AJH (1975) Spectral and temporal characteristics of activation and suppression of units in the cochlear nuclei of the anesthetized cat. Exp Brain Res 23:367–386.

Van Noorden L (1982) Two channel pitch perception. In: Clynes M (ed) Music Mind and Brain. New York: Plenum.

Versnel H, Shamma SA (1998) Spectral-ripple representation of steady-state vowels in primary auditory cortex. J Acoust Soc Am 103:2502–2514.

Versnel H, Kowalski N, Shamma S (1995) Ripple analysis in ferret primary auditory cortex III. Topographic distribution of ripple response parameters. Audiol Neurosci 1:271–285.

Viemeister NF, Bacon SP (1982) Forward masking by enhanced components in harmonic complexes. J Acoust Soc Am 71:1502–1507.

Voigt HF, Sachs MB, Young ED (1982) Representation of whispered vowels in discharge patterns of auditory nerve fibers. Hear Res 8:49–58.

Wang K, Shamma SA (1995) Spectral shape analysis in the primary auditory cortex. IEEE Trans Speech Aud 3:382–395.

Wang XQ, Sachs MB (1993) Neural encoding of single-formant stimuli in the cat. I. Responses of auditory nerve fibers. J Neurophysiol 70:1054–1075.

Wang XQ, Sachs MB (1994) Neural encoding of single-formant stimuli in the cat. II. Responses of anteroventral cochlear nucleus units. J Neurophysiol 71:59–78.

Wang XQ, Sachs MB (1995) Transformation of temporal discharge patterns in a ventral cochlear nucleus stellate cell model—implications for physiological mechanisms. J Neurophysiol 73:1600–1616.

Wang XQ, Merzenich M, Beitel R, Schreiner C (1995) Representation of a species-specific vocalization in the primary auditory cortex of the common marmoset: temporal and spectral characteristics. J Neurophysiol 74:2685–2706.

Warr WB (1966) Fiber degeneration following lesions in the anterior ventral cochlear nucleus of the cat. Exp Neurol 14:453–474.

Warr WB (1972) Fiber degeneration following lesions in the multipolar and globular cell areas in the ventral cochlear nucleus of the cat. Brain Res 40:247–270.

Warr WB (1982) Parallel ascending pathways from the cochlear nucleus: neuroanatomical evidence of functional specialization. Contrib Sens Physiol 7:1–38.

Watanabe T, Ohgushi K (1968) FM sensitive auditory neuron. Proc Jpn Acad 44:968–973.

Watanabe T, Sakai H (1973) Responses of the collicular auditory neurons to human speech. I. Responses to monosyllable /ta/. Proc Jpn Acad 49:291–296.

Watanabe T, Sakai H (1975) Responses of the collicular auditory neurons to connected speech. J Acoust Soc Jpn 31:11–17.

Watanabe T, Sakai H (1978) Responses of the cat's collicular auditory neuron to human speech. J Acoust Soc Am 64:333–337.

Webster D, Popper AN, Fay RR (eds) (1992) The Mammalian Auditory Pathway: Neuroanatomy. New York: Springer-Verlag.

Wenthold RJ, Huie D, Altschuler RA, Reeks KA (1987) Glycine immunoreactivity localized in the cochlear nucleus and superior olivary complex. Neuroscience 22:897–912.

Wever EG (1949) Theory of Hearing. New York: John Wiley.

Whitfield I (1980) Auditory cortex and the pitch of complex tones. J Acoust Soc Am 67:644–467.

Whitfield IC, Evans EF (1965) Responses of auditory cortical neurons to stimuli of changing frequency. J Neurophysiol 28:656–672.

Wightman FL (1973) The pattern transformation model of pitch. J Acoust Soc Am: 54:407–408.

Winslow RL (1985) A quantitative analysis of rate coding in the auditory nerve. Ph.D. thesis, Department of Biomedical Engineering, Johns Hopkins University, Baltimore, MD.

Winslow RL, Sachs MB (1988) Single tone intensity discrimination based on auditory-nerve rate responses in background of quiet noise and stimulation of the olivocochlear bundle. Hear Res 35:165–190.

Winslow RL, Barta PE, Sachs MB (1987) Rate coding in the auditory nerve. In: Yost WA, Watson CS (eds) Auditory Processing of Complex Sounds. Hillsdale, NJ: Lawrence Erbaum, pp. 212–224.

Winter P, Funkenstein H (1973) The effects of species-specific vocalizations on the discharges of auditory cortical cells in the awake squirrel monkeys. Exp Brain Res 18:489–504.

Winter IM, Palmer AR (1990a) Responses of single units in the anteroventral cochlear nucleus of the guinea pig. Hear Res 44:161–178.

Winter IM, Palmer AR (1990b) Temporal responses of primary-like anteroventral cochlear nucleus units to the steady state vowel /i/. J Acoust Soc Am 88:1437–1441.

Winter IM, Palmer AR (1995) Level dependence of cochlear nucleus onset unit responses and facilitation by second tones or broadband noise. J Neurophysiol 73:141–159.

Wundt W (1880) Grundzu ge der physiologischen Psychologie 2nd ed. Leipzig.

Yin TCT, Chan JCK (1990) Interaural time sensitivity in medial superior olive of cat. J Neurophysiol 58:562–583.

Young ED (1984) Response characteristics of neurons of the cochlear nuclei. In: Berlin C (ed) Hearing Science. San Diego: College-Hill Press, pp. 423–446.

Young ED, Sachs MB (1979) Representation of steady-state vowels in the temporal aspects of the discharge patterns of populations of auditory-nerve fibers. J Acoust Soc Am 66:1381–1403.

Young ED, Robert JM, Shofner WP (1988) Regularity and latency of units in ventral cochlea nucleus: implications for unit classification and generation of response properties. J Neurophysiol 60:1–29.

Young ED, Spirou GA, Rice JJ, Voigt HF (1992) Neural organization and responses to complex stimuli in the dorsal cochlear nucleus. Philos Trans R Soc Lond B 336:407–413.

5
The Perception of Speech Under Adverse Conditions

Peter Assmann and Quentin Summerfield

1. Introduction

Speech is the primary vehicle of human social interaction. In everyday life, speech communication occurs under an enormous range of different environmental conditions. The demands placed on the process of speech communication are great, but nonetheless it is generally successful. Powerful selection pressures have operated to maximize its effectiveness.

The adaptability of speech is illustrated most clearly in its resistance to distortion. In transit from speaker to listener, speech signals are often altered by background noise and other interfering signals, such as reverberation, as well as by imperfections of the frequency or temporal response of the communication channel. Adaptations for robust speech transmission include adjustments in articulation to offset the deleterious effects of noise and interference (Lombard 1911; Lane and Tranel 1971); efficient acoustic-phonetic coupling, which allows evidence of linguistic units to be conveyed in parallel (Hockett 1955; Liberman et al. 1967; Greenberg 1996; see Diehl and Lindblom, Chapter 3); and specializations of auditory perception and selective attention (Darwin and Carlyon 1995).

Speech is a highly efficient and robust medium for conveying information under adverse conditions because it combines strategic forms of redundancy to minimize the loss of information. Coker and Umeda (1974, p. 349) define redundancy as "any characteristic of the language that forces spoken messages to have, on average, more basic elements per message, or more cues per basic element, than the barest minimum [necessary for conveying the linguistic message]." This definition does not address the function of redundancy in speech communication, however. Coker and Umeda note that "redundancy can be used effectively; or it can be squandered on uneven repetition of certain data, leaving other crucial items very vulnerable to noise. . . . But more likely, if a redundancy is a property of a language and has to be learned, then it has a purpose." Coker and Umeda conclude that the purpose of redundancy in speech communication is to provide a basis for error correction and resistance to noise.

We shall review evidence suggesting that redundancy contributes to the perception of speech under adverse acoustic conditions in several different ways:

1. by limiting perceptual confusion due to errors in speech production;
2. by helping to bridge gaps in the signal created by interfering noise, reverberation, and distortions of the communication channel; and
3. by compensating for momentary lapses in attention and misperceptions on the part of the listener.

Redundancy is present at several levels in speech communication—acoustic, phonetic, and linguistic. At the acoustic level it is exemplified by the high degree of covariation in the pattern of amplitude modulation across frequency and over time. At the phonetic level it is illustrated by the many-to-one mapping of acoustic cues onto phonetic contrasts and by the presence of cue-trading relationships (Klatt 1989). At the level of phonology and syntax it is illustrated by the combinatorial rules that organize sound sequences into words, and words into sentences. Redundancy is also provided by semantic and pragmatic context.

This chapter discusses the ways in which acoustic, phonetic, and lexical redundancy contribute to the perception of speech under adverse conditions. By "adverse conditions" we refer to any perturbation of the communication process resulting from either an error in production by the speaker, channel distortion or masking in transmission, or a distortion in the auditory system of the listener. Section 2 considers the design features of speech that make it well suited for transmission in the presence of noise and distortion. The primary aim of this section is to identify perceptually salient properties of speech that underlie its robustness. Section 3 reviews the literature on the intelligibility of speech under adverse listening conditions. These include background noise of various types (periodic/random, broadband/narrowband, continuous/fluctuating, speech/nonspeech), reverberation, changes in the frequency response of the communication channel, distortions resulting from pathology of the peripheral auditory system, and combinations of the above. Section 4 considers strategies used by listeners to maintain, preserve, or enhance the intelligibility of speech under adverse acoustic conditions.

2. Design Features of Speech that Contribute to Robustness

We begin with a consideration of the acoustic properties of speech that make it well suited for transmission in adverse environments.

2.1 The Spectrum

The traditional starting point for studying speech perception under adverse conditions is the long-term average speech spectrum (LTASS) (Dunn and White 1940; French and Steinberg 1947; Licklider and Miller 1951; Fletcher 1953; Kryter 1985). A primary objective of these studies has been to characterize the effects of noise, filtering, and channel distortion on the LTASS in order to predict their impact on intelligibility. The short-term amplitude spectrum (computed over a time window of 10 to 30 ms) reveals the acoustic cues for individual vowels and consonants combined with the effects of distortion. The long-term spectrum tends to average out segmental variations. Hence, a comparison of the LTASS obtained under adverse conditions with the LTASS obtained in quiet can provide a clearer picture of the effects of distortion.

Figure 5.1 (upper panel) shows the LTASS obtained from a large sample of native speakers of 12 different languages reading a short passage from a story (Byrne et al. 1994). The spectra were obtained by computing the root mean square (rms) level in a set of one-third-octave-band filters over 125-ms segments of a 64-second recorded passage spoken in a "normal" speaking style. There are three important features of the LTASS. First, there is a 25-dB range of variation in average level across frequency, with the bulk of energy below 1 kHz, corresponding to the frequency region encompassing the first formant. Second, there is a gradual decline in spectrum level for frequencies above 0.5 kHz. Third, there is a clear distinction between males and females in the low-frequency region of the spectrum. This difference is attributable to the lower average fundamental frequency (f_0) of male voices. As a result, the first harmonic of a male voice contributes appreciable energy between 100 and 150 Hz, while the first harmonic of a female voice makes a contribution between 200 and 300 Hz.

The lower panel of Figure 5.1 shows the LTASS obtained using a similar analysis method from a sample of 15 American English vowels and diphthongs. After averaging, the overall spectrum level was adjusted to match that of the upper panel at 250 Hz in order to facilitate comparisons between panels. Compared to continuous speech, the LTASS of vowels shows a more pronounced local maximum in the region of f_0 (close to 100 Hz for males and 200 Hz for females). However, in other respects the pattern is similar, suggesting that the LTASS is dominated by the vocalic portions of the speech signal. Vowels and other voiced sounds occupy about half of the time waveform of connected speech, but dominate the LTASS because such segments contain greater power than the adjacent aperiodic segments.

The dashed line in each panel illustrates the variation in absolute sensitivity as a function of frequency for young adult listeners with normal hearing (Moore and Glasberg 1987). Comparison of the absolute threshold function with the LTASS shows that the decline in energy toward lower frequencies is matched by a corresponding decline in sensitivity. However, the

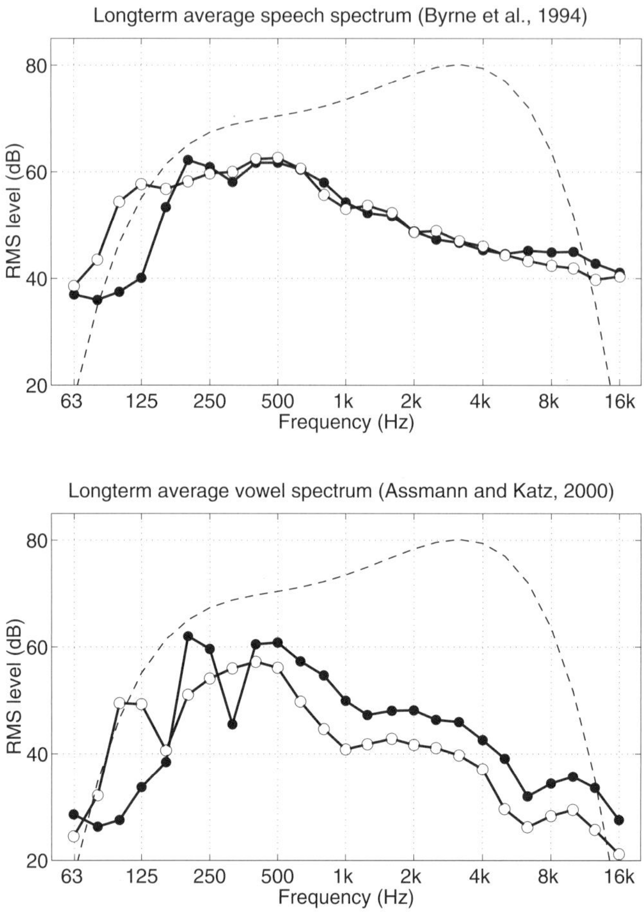

FIGURE 5.1. The upper panel shows the long-term average speech spectrum (LTASS) for a 64-second segment of recorded speech from 10 adult males and 10 adult females for 12 different languages (Byrne et al. 1994). The vertical scale is expressed in dB SPL (linear weighting). The lower panel shows the LTASS for 15 vowels and diphthongs of American English (Assmann and Katz 2000). Filled circles in each panel show the LTASS for adult males; unfilled circles show the LTASS for adult females. To facilitate comparisons, these functions were shifted along the vertical scale to match those obtained with continuous speech in the upper panel. The dashed line in each panel indicates the shape of the absolute threshold function for listeners with normal hearing (Moore and Glasberg 1987). The absolute threshold function is expressed on an arbitrary dB scale, with larger values indicating greater sensitivity.

speech spectrum has a shallower roll-off in the region above 4 kHz than the absolute sensitivity function and the majority of energy in the speech spectrum encompasses frequencies substantially lower than the peak in pure-tone sensitivity. This low-frequency emphasis may be advantageous for the transmission of speech under adverse conditions for several reasons:

1. The lowest three formants of speech, F_1 to F_3, generally lie below 3 kHz. The frequencies of the higher formants do not vary as much, and contribute much less to intelligibility (Fant 1960).
2. Phase locking in the auditory nerve and brain stem preserves the temporal structure of the speech signal in the frequency range up to about 1500 Hz (Palmer 1995). Greenberg (1995) has suggested that the low-frequency emphasis in speech may be linked to the greater reliability of information coding at low frequencies via phase locking.
3. To separate speech from background sounds, listeners rely on cues, such as a common periodicity and a common pattern of interaural timing (Summerfield and Culling 1995), that are preserved in the patterns of neural discharge only at low frequencies (Cariani and Delgutte 1996a,b; Joris and Yin 1995).
4. Auditory frequency selectivity is sharpest (on a linear frequency scale) at low frequencies and declines with increasing frequency (Patterson and Moore 1986).

The decline in auditory frequency selectivity with increasing frequency has several implications for speech intelligibility. First, auditory filters have larger bandwidths at higher frequencies, which means that high-frequency filters pass a wider range of frequencies than their low-frequency counterparts. Second, the low-frequency slope of auditory filters becomes shallower with increasing level. As a consequence, low-frequency maskers are more effective than high-frequency maskers, leading to an "upward spread of masking" (Wegel and Lane 1924; Trees and Turner 1986; Dubno and Ahlstrom 1995). In their studies of filtered speech, French and Steinberg (1947) observed that the lower speech frequencies were the last to be masked as the signal-to-noise ratio (SNR) was decreased.

Figure 5.2 illustrates the effects of auditory filtering on a segment of the vowel [I] extracted from the word "hid" spoken by an adult female talker. The upper left panel shows the conventional Fourier spectrum of the vowel in quiet, while the upper right panel shows the spectrum of the same vowel embedded in pink noise at an SNR of +6 dB. The lower panels show the "auditory spectra" or "excitation patterns" of the same two sounds. An excitation pattern is an estimate of the distribution of auditory excitation across frequency in the peripheral auditory system generated by a specific signal. The excitation patterns shown here were obtained by plotting the rms output of a set of gammatone filters[1] as a function of filter center frequency.

---

[1] The gammatone is a bandpass filter with an impulse response composed of two terms, one derived from the gamma function, and the other from a cosine function

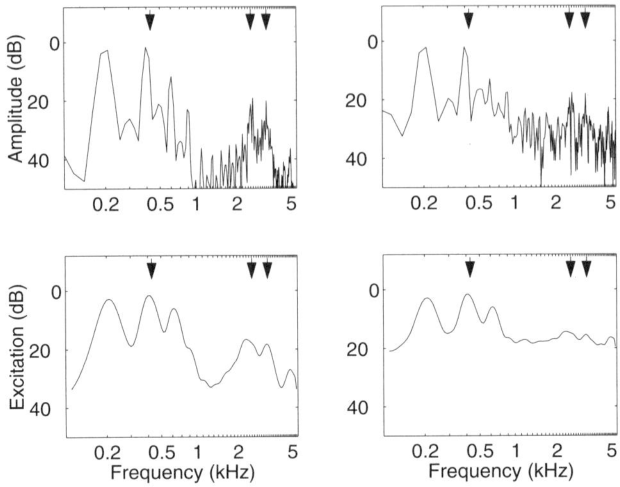

FIGURE 5.2. The upper left panel shows the Fourier amplitude spectrum of a 102.4-ms segment of the vowel [I] spoken by an adult female speaker of American English. The upper right panel shows the same segment embedded in pink noise at a signal-to-noise ratio (SNR) of +6 dB. Below each amplitude spectrum is its auditory excitation pattern (Moore and Glasberg 1983, 1987) simulated using a gammatone filter analysis (Patterson et al. 1992). Fourier spectra and excitation patterns are displayed on a log frequency scale. Arrows show the frequencies of the three lowest formants (F_1–F_3) of the vowel.

The three lowest harmonics are "resolved" as distinct peaks in the excitation pattern, while the upper harmonics are not individually resolved. In this example, the first formant (F_1) lies close to the second harmonic but does not coincide with it. In general, F_1 in voiced segments is not represented by a distinct peak in the excitation pattern and hence its frequency must be inferred, in all likelihood from the relative levels of prominent harmonics in this appropriate region (Klatt 1982; Darwin 1984; Assmann and Nearey 1986). The upper formants (F_2–F_4) give rise to distinct peaks in the excitation pattern when the vowel is presented in quiet. The addition of noise leads to a greater spread of excitation at high frequencies, and the spectral contrast (peak-to-valley ratio) of the upper formants is reduced.

The simulation in Figure 5.2 is based on data from listeners with normal hearing whose audiometric thresholds fall within normal limits and who

or "tone" (Patterson et al. 1992). The bandwidths of these filters increase with increasing center frequency, in accordance with estimates of psychophysical measures of auditory frequency selectivity (Moore and Glasberg 1983, 1987). Gammatone filters have been used to model aspects of auditory frequency selectivity as measured psychophysically (Moore and Glasberg 1983, 1987; Patterson et al. 1992) and physiologically (Carney and Yin 1988), and can be used to simulate the effects of auditory filtering on speech signals.

possess normal frequency selectivity. Sensorineural hearing impairments lead to elevated thresholds and are often associated with a reduction of auditory frequency selectivity. Psychoacoustic measurements of auditory filtering in hearing-impaired listeners often show reduced frequency selectivity compared to normal listeners (Glasberg and Moore 1986), and consequently these listeners may have difficulty resolving spectral features that could facilitate making phonetic distinctions among similar sounds. The reduction in spectral contrast can be simulated by broadening the bandwidths of the filters used to generate excitation patterns, such as those shown in Figure 5.2 (Moore 1995). Support for the idea that impaired frequency selectivity can result in poorer preservation of vocalic formant structure and lower identification accuracy comes from studies of vowel masking patterns (Van Tasell et al. 1987a; Turner and Henn 1989). In these studies, forward masking patterns were obtained by measuring the threshold of a brief sinusoidal probe at different frequencies in the presence of a vocalic masker to obtain an estimate of the "internal representation" of the vowel. Hearing-impaired listeners generally exhibit less accurate representations of the signal's formant peaks in their masking patterns than do normal-hearing listeners.

Many studies have shown that the intelligibility of masked, filtered, or distorted speech depends primarily on the proportion of the speech spectrum available to the listener. This principle forms the basis for the articulation index (AI), a model developed by Fletcher and his colleagues at Bell Laboratories in the 1920s to predict the effects of noise, filtering, and communication channel distortion on speech intelligibility (Fletcher 1953). Several variants of the AI have been proposed over the years (French and Steinberg 1947; Kryter 1962; ANSI S3.5 1969, 1997; Müsch and Buus 2001a,b).

The AI is an index between 0 and 1 that describes the effectiveness of a speech communication channel. An "articulation-to-intelligibility" transfer function can be applied to convert this index to predicted intelligibility in terms of percent correct. The AI model divides the speech spectrum into a set of up to 20 discrete frequency bands, taking into account the absolute threshold, the masked threshold imposed by the noise or distortion, and the long-term average spectrum of the speech. The AI has two key assumptions:

1. The contribution of any individual channel is independent of the contribution of other bands.
2. The contribution of a channel depends on the SNR within that band.

The predicted intelligibility depends on the proportion of time the speech signal exceeds the threshold of audibility (or the masked threshold, in conditions where noise is present) in each band. The AI is expressed by the following equation (Pavlovic 1984):

$$AI = P\int_0^\infty I(f)W(f)\mathrm{d}f \tag{1}$$

The term $I(f)$ is the importance function, which reflects the significance of different frequency bands to intelligibility. $W(f)$ is the audibility or weighting function, which describes the proportion of information associated with $I(f)$ available to the listener in the testing environment. The term P is the proficiency factor and depends on the clarity of the speaker's articulation and the experience of the listener (including such factors as the familiarity of the speaker's voice and dialect). Computation of the AI typically begins by dividing the speech spectrum into a set of n discrete frequency bands (Pavlovic 1987):

$$AI = P\sum_{i=1}^{n} I_i W_i \tag{2}$$

The AI computational procedure developed by French and Steinberg (1947) uses 20 frequency bands between 0.15 and 8 kHz, with the width of each band adjusted to make the bands equal in importance. These adjustments were made on the basis of intelligibility tests with low-pass and high-pass filtered speech, which revealed a maximum contribution from the frequency region around 2.5 kHz. Later methods have employed one-third octave bands (e.g., ANSI 1969) or critical bands (e.g., Pavlovic 1987) with nonuniform weights.[2]

The audibility term, W_i, estimates the proportion of the speech spectrum exceeding the masked threshold in the ith frequency band. The ANSI S3.5 model assumes that speech intelligibility is determined over a dynamic range of 30 dB, with the upper limit determined by the "speech peaks" (the sound pressure level exceeded 1% of the time by the speech energy integrated over 125-ms intervals—on average, about 12 dB above the mean level). The lower limit (representing the speech "valleys") is assumed to lie 18 dB below the mean level. The AI assumes a value of 1.0 under conditions of maximum intelligibility (i.e., when the 30-dB speech range exceeds the absolute threshold, as well as the masked threshold, if noise is present in every frequency band). If any part of the speech range lies below the threshold across frequency channels, or is masked by noise, the AI is reduced by the percentage of the area covered. The AI assumes a value of 0 when the speech is completely masked, or is below threshold, and hence

[2] Several studies have found that the shape of the importance function varies as a function of speaker, gender and type of speech material (e.g., nonsense CVCs versus continuous speech), and the procedure used (French and Steinberg 1947; Beranek 1947; Kryter 1962; Studebaker et al. 1987). Recent work (Studebaker and Sherbecoe 2002) suggests that the 30-dB dynamic range assumed in standard implementations may be insufficient, and that the relative importance assigned to different intensities within the speech dynamic range varies as a function of frequency.

unintelligible. As a final step, the value of the AI can be used to predict intelligibility with the aid of an empirically derived articulation-to-intelligibility transfer function (Pavlovic and Studebaker 1984). The shape of the transfer function differs for different speech materials and testing conditions (Kryter 1962; Studebaker et al. 1987).

The AI generates accurate predictions of average speech intelligibility over a wide range of conditions, including high- and low-pass filtering (French and Steinberg 1947; Fletcher and Galt 1950), different types of broadband noise (Egan and Wiener 1946; Miller 1947), bandpass-filtered noise maskers (Miller et al. 1951), and various distortions of the communication channel (Beranek 1947). It has also been used to model binaural masking level differences for speech (Levitt and Rabiner 1967) and loss of speech intelligibility resulting from sensorineural hearing impairments (Fletcher 1952; Humes et al. 1986; Pavlovic et al. 1986; Ludvigsen 1987; Rankovic 1995, 1998). The success of the AI model is consistent with the idea that speech intelligibility under adverse conditions is strongly affected by the audibility of the speech spectrum.[3] However, the AI was designed to accommodate linear distortions and additive noises with continuous spectra. It is less effective for predicting the effects of nonlinear or time-varying distortions, transmission channels with sharp peaks and valleys, masking noises with line spectra, and time-domain distortions, such as those created by echoes and reverberation. Some of these difficulties are overcome by a reformulation of AI theory—the speech transmission index—described below.

2.2 *Formant Peaks*

The vocal tract resonances (or "formants") provide both phonetic information (signaling the identity of the intended vowel or consonant) and source information (signaling the identity of the speaker). The frequencies of the lowest three formants, as well as their pattern of change over time, provide cues that help listeners ascertain the phonetic identities of vowels and consonants. Vocalic contrasts, in particular, are determined primarily by differences in the formant pattern (e.g., Peterson and Barney 1952; Nearey 1989; Hillenbrand et al. 1995; Hillenbrand and Nearey 1999; Assmann and Katz, 2000; see Diehl and Lindblom, Chapter 3).

---

[3] The AI generates a single number that can be used to predict the overall or average intelligibility of specified speech materials for a given communication channel. It does not predict the identification of individual segments, syllables, or words, nor does it predict the pattern of listeners' errors. Calculations are typically based on speech spectra accumulated over successive 125-ms time windows. A shorter time window and a short-time running spectral analysis (Kates 1987) would be required to predict the identification of individual vowels and consonants (and the confusion errors made by listeners) in tasks of phonetic perception.

The formant representation provides a compact description of the speech spectrum. Given an initial set of assumptions about the glottal source and a specification of the damping within the supralaryngeal vocal tract (in order to determine the formant bandwidths), the spectrum envelope can be predicted from a knowledge of the formant frequencies (Fant 1960). A change in formant frequency leads to correlated changes throughout the spectrum, yet listeners attend primarily to the spectral peaks in order to distinguish among different vocalic qualities (Carlson et al. 1979; Darwin 1984; Assmann and Nearey 1986; Sommers and Kewley-Port 1996).

One reason why spectral peaks are important is that spectral detail in the region of the formant peaks is more likely to be preserved in background noise. The strategy of attending primarily to spectral peaks is robust not only to the addition of noise, but also to changes in the frequency response of a communication channel and to some deterioration of the frequency resolving power of the listener (Klatt 1982; Assmann and Summerfield 1989; Roberts and Moore 1990, 1991a; Darwin 1984, 1992; Hukin and Darwin 1995). In comparison, a whole-spectrum matching strategy that assigns equal weight to the level of the spectrum at all frequencies (Bladon 1982) or a broad spectral integration strategy (e.g., Chistovich 1984) would tend to incorporate noise into the spectral estimation process and thus be more susceptible to error. For example, a narrow band of noise adjacent to a formant peak could substantially alter the spectral center of gravity without changing the frequency of the peak itself.

While it is generally agreed that vowel quality is determined primarily by the frequencies of the two or three lowest formants (Pols et al. 1969; Rosner and Pickering 1994), there is considerable controversy over the mechanisms underlying the perception of these formants in vowel identification. Theories generally fall into one of two main classes—those that assert that the identity of a vowel is determined by a distributed analysis of the shape of the entire spectrum (e.g., Pols et al. 1969; Bakkum et al. 1993; Zahorian and Jagharghi 1993), and those that assume an intermediate stage in which spectral features in localized frequency regions are extracted (e.g., Chistovich 1984; Carlson et al. 1974). Consistent with the first approach is the finding that listeners rely primarily on the two most prominent harmonics near the first-formant peak in perceptual judgments involving front vowels (e.g., [i] and [ɛ]), which have a large separation of the lowest formants, F_1 and F_2. For example, listeners rely only on the most prominent harmonics in the region of the formant peak to distinguish changes in F_1 center frequency (Sommers and Kewley-Port 1996) as well as to match vowel quality as a function of F_1 frequency (Assmann and Nearey 1986; Dissard and Darwin 2000) and identify vowels along a phonetic continuum (Carlson et al. 1974; Darwin 1984; Assmann and Nearey 1986).

A different pattern of sensitivity is found when listeners judge the phonetic quality of back vowels (e.g., [u] and [o]), where F_1 and F_2 are close together in frequency. In this instance, harmonics remote from the F_1 peak can make a contribution, and additional aspects of spectral shape (such as the center of spectral gravity in the region of the formant peaks or the relative amplitude of the formants) are taken into account (Chistovich and Lublinskaya 1979; Beddor and Hawkins 1990; Assmann 1991; Fahey et al. 1996).

The presence of competing sounds is a problem for models of formant estimation. Extraneous sounds in the F_1 region might change the apparent amplitudes of resolved harmonics and so alter the phonetic quality of the vowel. Roberts and Moore (1990, 1991a) demonstrated that this effect can occur. They found that additional components in the F_1 region of a vowel as well as narrow bands of noise could alter its phonetic quality. The shift in vowel quality was measured in terms of changes in the phonetic segment boundary along a continuum ranging from [I] to [ɛ] (Darwin 1984). Roberts and Moore hypothesized that the boundary shift was the result of excitation from the additional component being included in the perceptual estimate of the amplitudes of harmonics close to the first formant of the vowel.

How do listeners avoid integrating evidence from other sounds when making vowel quality judgments? Darwin (1984, 1992; Darwin and Carlyon 1995) proposed that the perception of speech is guided by perceptual grouping principles that exclude the contribution of sounds that originate from different sources. For example, Darwin (1984) showed that the influence of a harmonic component on the phoneme boundary was reduced when that harmonic started earlier or later than the remaining harmonics of the vowel. The perceptual exclusion of the asynchronous component is consistent with the operation of a perceptual grouping mechanism that segregates concurrent sounds on the basis of onset or offset synchrony. Roberts and Moore (1991a) extended these results by showing that segregation also occurs with inharmonic components in the region of F_1.

Roberts and Moore (1991b) suggested that the perceptual segregation of components in the F_1 region of vowels might benefit from the operation of a harmonic sieve (Duifhuis et al. 1982). The harmonic sieve is a hypothetical mechanism that excludes components whose frequencies do not correspond to integer multiples of a given fundamental. It accounts for the finding that a component of a tonal complex contributes less to its pitch when its frequency is progressively mistuned from its harmonic frequency (Moore et al. 1985). Analogously, a mistuned component near the F_1 peak makes a smaller contribution to its phonetic quality than that of its harmonic counterparts (Darwin and Gardner 1986).

The harmonic sieve utilizes a "place" analysis to group together components belonging to the same harmonic series, and thereby excludes inharmonic components. This idea has proved to have considerable explanatory power. However, it has not always been found to offer the most accurate account of the perceptual data. For example, computational models based on the harmonic sieve have not generated accurate predictions of listeners' identification of concurrent pairs of vowels with different f_0s (Scheffers

1983; Assmann and Summerfield 1990). The excitation patterns of "double vowels" often contain insufficient evidence of concurrent f_0s to allow for their segregation using a harmonic sieve. Alternative mechanisms, based on a temporal (or place-time) analysis, have been shown to make more accurate predictions of the pattern associated with listeners' identification responses (Assmann and Summerfield 1990; Meddis and Hewitt 1992).

Meddis and Hewitt (1991, 1992) describe a computational model that

1. carries out a frequency analysis of the signal using a bank of bandpass filters,
2. compresses the filtered waveforms using a model of mechanical-to-neural transduction,
3. performs a temporal analysis using autocorrelation functions (ACFs), and
4. sums the ACFs across the frequency channels to derive a summary autocorrelogram.

The patterning of peaks in the summary autocorrelogram is in accord with many of the classic findings of pitch perception (Meddis and Hewitt 1991). The patterning can also yield accurate estimates of the f_0s of concurrent vowels (Assmann and Summerfield 1990). Meddis and Hewitt (1992) segregated pairs of concurrent vowels by combining the ACFs across channels with a common periodicity to provide evidence of the first vowel, and then grouping the remaining ACFs to reconstruct the second segment. They showed that the portion of the summary autocorrelogram with short time lags (<4.5 ms) could be used to predict the phonetic identities of the vowels with reasonable accuracy.

The harmonic sieve and autocorrelogram embody different solutions to the problem of segregating a vowel from interfering sounds (including a second competing vowel). It can be complicated to compare models of vowel identification that incorporate these mechanisms because the models may differ not only in the technique used to represent the spectrum (or temporal pattern) of a vowel, but also in the approach to classifying the spectrum. Most models of categorization assume that the pattern to be classified is compared with a set of templates, and that the pattern is characterized as belonging to the set defined by the template to which it is most similar. "Similarity" is usually measured by an implicit representation of perceptual distance. The choice of distance metric can have a substantial effect on the accuracy with which a model predicts the pattern of vocalic identification made by a listener. Table 5.1 summarizes the results of several studies that have evaluated the efficacy of such perceptual distance metrics for vowels.

Three conclusions emerge from these comparisons. First, no single metric is optimal across all conditions. Different metrics appear to be best suited for different tasks. Second, metrics that highlight spectral peaks [and possibly also spectral "shoulders" (Assmann and Summerfield 1989; Lea and

TABLE 5.1. Sample of perceptual distance metrics for vowels

Factor	Metric	Reference
Vowel similarity judgments	Dimensions derived from principal components analysis (PCA) of one-third-octave spectra	Pols et al. (1969)
Speaker quality judgments for normal and profoundly hearing-impaired talkers	One-third-octave spectra + PCA	Bakkum et al. (1993)
Talker normalization	Excitation patterns	Suomi (1984)
Vowel quality matching	Loudness density patterns	Bladon and Lindblom (1981)
Prediction of vowel systems	Loudness density patterns	Lindblom (1986)
Vowel similarity judgments	Weighted spectral slope metric	Carlson et al. (1979); Nocerino et al. (1985)
Vowel identification by hearing impaired listeners	Weighted spectral slope metric	Turner and Henn (1989)
Concurrent vowel identification	Negative second differential of excitation pattern; peak metric	Assmann and Summerfield (1989)
Discrimination of vowel formant frequencies (F_1 and F_2)	Peak-weighted excitation pattern, specific loudness difference	Sommers and Kewley-Port (1996); Kewley-Port and Zheng (1998)

Summerfield 1994)] perform best when the task is to phonetically identify vowels. Third, metrics that convey information about the entire shape of the spectrum are more appropriate when the task is to discriminate vowels acoustically, that is, on the basis of timbre rather than using differences in phonetic quality (Klatt 1982).

The fact that no single metric is optimal for all vowel tasks and that the sensitivity of perceptual distance metrics to distortion and noise is so highly variable suggests that a simple template-matching approach with fixed frequency weights is inappropriate for vowel perception. Similar conclusions have been reached in recent reviews of speech-recognition research (Gong 1994; Lippmann 1996a; see Morgan et al., Chapter 6). To a much greater extent than humans, most existing speech recognizers are adversely affected by transmission-channel distortion, noise, and reverberation. A major difficulty is that these types of distortion can obscure or mask weak formants and other aspects of spectral shape, resulting in the problem of "missing data" (Cooke et al. 1996; Cooke and Ellis 2001). They can introduce "spurious" peaks and alter the shape of the spectrum, resulting in greater than predicted perceptual distances. Adult listeners with normal hearing possess remarkable abilities to compensate for such distortions. Unlike machine-based speech recognizers, they do so without the need for explicit practice or "recalibration" (Watkins 1991; Buuren et al. 1996; Lippmann 1996a).

The effects of two different types of noise on the spectrum of a vowel is illustrated in Figure 5.3. Panel A shows the Fourier amplitude spectrum of a 102.4-ms segment of the vowel in the word "head," spoken by an adult female speaker in a sound-attenuated recording chamber. Panel B shows the same signal combined with white noise at an SNR of 0dB. The envelope of the spectrum [obtained by linear predictive coding (LPC) analysis (Markel and Gray 1976)] shows that the spectral contrast is greatly diminished and that the peaks generated by the higher formants (F_2, F_3, F_4) are no longer distinct. The harmonicity of the vowel is not discernible in the upper formants, but remains evident in the F_1 region.

In natural listening environments steady-state broadband noise with a flat spectrum is uncommon. A more common form of noise is created when several individuals talk at once, creating multispeaker babble. Panel C shows the amplitude spectrum of a 102.4-ms segment of such babble,

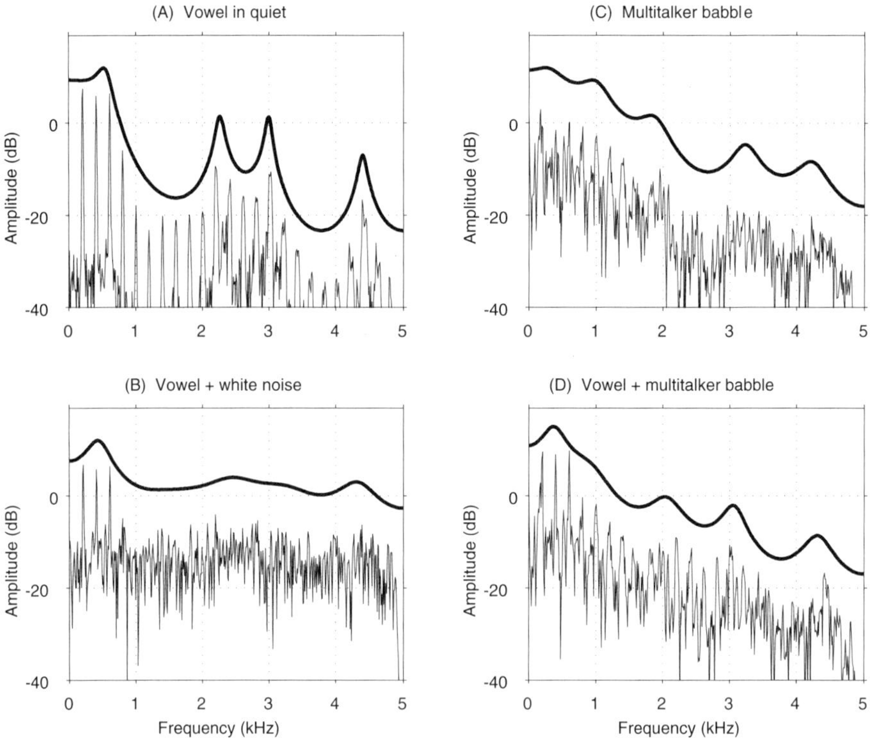

FIGURE 5.3. Effects of noise on formant peaks. A: The Fourier amplitude spectrum of a vowel similar to [ɛ]. The solid line shows the spectrum envelope estimated by linear predictive coding (LPC) analysis. B: White noise has been superimposed at an SNR of 0dB. C: The spectrum of a sample of multitalker babble. D: The spectrum of the vowel mixed with the babble at an SNR of 0dB.

created by mixing speech from four different speakers (two adult males, one adult female, and a child) at comparable intensities. In panel D the speech babble is combined with the vowel shown in panel A at an SNR of 0dB. Compared with panel A, there is a reduction in the degree of spectral contrast and there are changes in the shape of the spectrum. There are additional spectral peaks introduced by the competing voices, and there are small shifts in the frequency locations of spectral peaks that correspond to formants of the vowel. The harmonicity of the vowel is maintained in the low-frequency region, and is preserved to some degree in the second and third formant regions. These examples indicate that noise can distort the shape of the spectrum, change its slope, and reduce the contrast between peaks and adjacent valleys. However, the frequency locations of the formant peaks of the vowel are preserved reasonably accurately in the LPC analysis in panel D, despite the fact that other aspects of spectral shape, such as spectral tilt and the relative amplitudes of the formants, are lost.

Figure 5.3 also illustrates some of the reasons why formant tracking is such a difficult engineering problem, especially in background noise (e.g., Deng and Kheirallah 1993). An example of the practical difficulties of locating particular formants is found in the design of speech processors for cochlear implants.[4] Explicit formant tracking was implemented in the processor developed by Cochlear PTY Ltd. during the 1980s, but was subsequently abandoned in favor of an approach that seeks only to locate spectral peaks without assigning them explicitly to a specific formant. The latter strategy yields improved speech intelligibility, particularly in noise (McKay et al. 1994; Skinner et al. 1994).

Listeners with normal hearing have little difficulty understanding speech in broadband noise at SNRs of 0dB or greater. Environmental noise typically exhibits a sloping spectrum, more like the multispeaker babble of panels C and D than the white noise of panel B. For such noises, a subset of formants (F_1, F_2, and F_3) is often resolved, even at an SNR of 0dB, and generates distinct peaks in the spectrum envelope. However, spectral contrast (the difference in dB between the peaks and their adjacent valleys) is reduced by the presence of noise in the valleys between formants. As a result, finer frequency selectivity is required to locate the peaks. Listeners with sensorineural hearing loss generally have difficulty understanding speech under such conditions. Their difficulties are likely to stem, at least in part, from reduced frequency selectivity (Simpson et al. 1990; Baer et al. 1993). This hypothesis has been tested by the application of digital signal processing techniques to natural speech designed to either (1) reduce the

[4] Cochlear implants provide a useful means of conveying auditory sensation to the profoundly hearing impaired by bypassing the malfunctioning parts of the peripheral auditory system and stimulating auditory-nerve fibers directly with electrical signals through an array of electrodes implanted within the cochlea (cf. Clark, Chapter 8).

spectral contrast by smearing the spectral envelope (Keurs et al. 1992, 1993a,b; Baer and Moore 1994) or (2) enhance the contrast by sharpening the formant peaks (Veen and Houtgast 1985; Simpson et al. 1990; Baer et al. 1993). Spectral smearing results in a degradation of speech intelligibility, particularly for vowels, as well as an elevation in the speech reception threshold (SRT) in noise (Plomp and Mimpen 1979). However, the magnitude of the reduction in spectral contrast is not closely linked to measures of frequency selectivity (Keurs 1993a,b). Conversely, attempts to enhance intelligibility by increasing spectral contrast have shown a modest improvement for listeners with cochlear hearing impairment [corresponding to an increase in SNR of up to about 4dB (Baer et al. 1993)]. These results are consistent with the hypothesis that the difficulties experienced by the hearing-impaired when listening to speech in noise are at least partially due to the reduced ability to resolve formant peaks (cf. Edwards, Chapter 7).

2.3 Periodicity of Voiced Speech

The regularity with which the vocal folds open and close during voicing is one of the most distinctive attributes of speech—its periodicity (in the time domain) and corresponding harmonicity (in the frequency domain). This pattern of glottal pulsing produces periodicity in the waveform at rates between about 70 and 500Hz. Such vocal fold vibrations are responsible for the perception of voice pitch and provide the basis for segmental distinctions between voiced and unvoiced sounds (such as [b] and [p]), as well as distinctions of lexical tone in many languages. At the suprasegmental level, voice pitch plays a primary role in conveying different patterns of intonation and prosody.

Evidence of voicing is broadly distributed across frequency and time, and is therefore a robust property of speech. Figure 5.4 illustrates the effects of background noise on the periodicity of speech. The left panel shows the waveforms generated by a set of gammatone filters in response to the syllable [ga] in quiet. In this example, the speaker closed her vocal tract about 30ms after the syllable's onset and then released the closure 50ms later. The frequency channels below 1kHz are dominated by the fundamental frequency and the auditorily resolved, low-order harmonics. In the higher-frequency channels, filter bandwidths are broader than the frequency separation of the harmonics, and hence several harmonics interact in the passband of the filter to create amplitude modulation (AM) at the period of the fundamental. The presence of energy in the lowest-frequency channel during the stop closure provides evidence that the consonant is voiced rather than voiceless.

The panel on the right shows that periodicity cues are preserved to some extent in background noise at an SNR of +6dB. The noise has largely obliterated the silent interval created by the stop consonant and has masked the burst. However, there is continued domination of the output of the low-

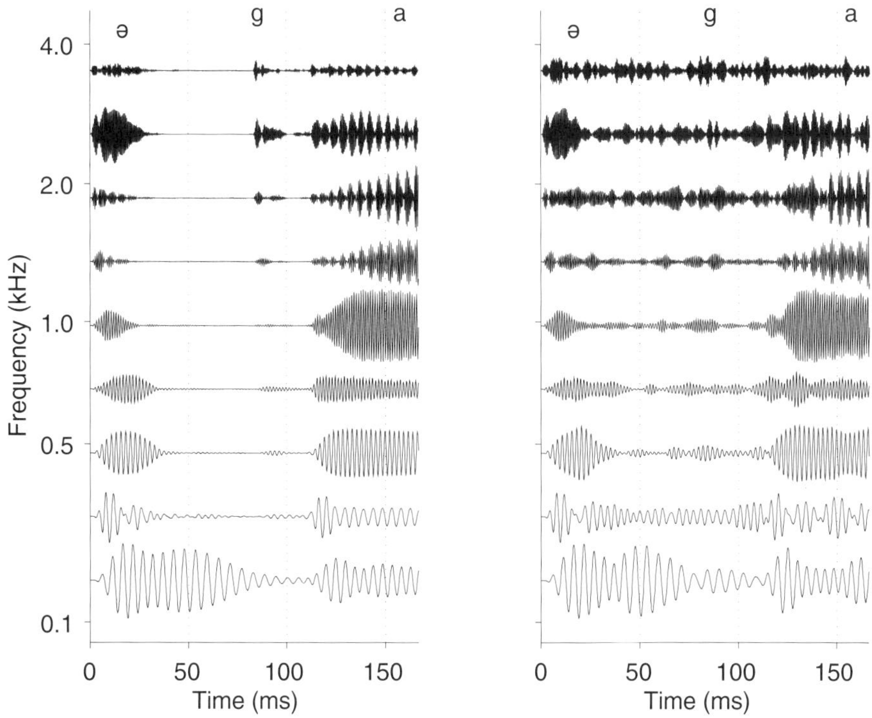

FIGURE 5.4. Effects of background noise on voicing periodicity. The left panel shows the results of a gammatone filter bank analysis (Patterson et al. 1992) of the voiced syllable [ga] spoken by an adult female talker. Filter center frequencies and bandwidths were chosen to match auditory filters measured psychophysically (Moore and Glasberg 1987) across the 0.1–4.0 kHz range. The panel on the right is an analysis of the same syllable combined with broadband (pink) noise at +6 dB SNR.

frequency channels by the individual harmonics and the modulation at the period of the fundamental remains in several of the higher-frequency channels.

It has been suggested that the presence of voicing underlies speech's robustness to noise. One source of evidence comes from a comparison of voiced and whispered speech. In the latter, the periodic glottal pulses are replaced with aperiodic turbulent noise, which has a continuous, rather than harmonic spectrum. Whispered speech is intelligible under quiet listening situations and is generally reserved for short-range communication, but can be less intelligible than voiced speech under certain conditions (Tartter 1991).

Periodicity cues in voiced speech may contribute to noise robustness via auditory grouping processes (Darwin and Carlyon 1995). A common periodicity across frequency provides a basis for associating speech components originating from the same larynx and vocal tract (Scheffers 1983; Assmann and Summerfield 1990; Bregman 1990; Darwin 1992; Langner 1992; Meddis and Hewitt 1992). Compatible with this idea, Brokx and Nooteboom (1982), Bird and Darwin (1998), and Assmann (1999) have shown that synthesized target sentences are easier to understand in the presence of a continuous speech masker if targets and maskers are synthesized with different f_0s, than with the same f_0. Similarly, when pairs of synthesized vowels are presented concurrently, listeners are able to identify them more accurately if they are synthesized with different fundamental frequencies, compared to the case where

1. both have the same fundamental (Scheffers 1983; Chalikia and Bregman 1989; Assmann and Summerfield 1990),
2. one is voiced and the other is noise-excited (Scheffers 1983), or
3. both are noise-excited (Scheffers 1983; Lea 1992).[5]

A further source of evidence for a contribution of voicing periodicity to speech intelligibility comes from studies of sine-wave speech (Remez et al. 1981). Sine-wave speech uses frequency-modulated sinusoids to model the movements of F_1, F_2, and F_3 from a natural speech signal, and thus lacks harmonic structure. Despite this spectral reduction, it can be understood, to a certain extent, under ideal listening conditions, though not in background noise. Carrell and Opie (1992), however, have shown that sine-wave speech is easier to understand when it is amplitude modulated at a rate similar to that imposed by the vocal folds during voicing. Thus, common, coherent AM may help listeners to group the three sinusoidal formant together to distinguish them from background noise.

2.4 Rapid Spectral Changes

Stevens (1980, 1983) has emphasized that consonants are differentiated from vowels and other vocalic segments (glides, liquids) by their rate of change in the short-time spectrum. The gestures accompanying consonantal closure and release result in rapid spectral changes (associated with bursts and formant transitions) serving as landmarks or pointers to regions of the signal where acoustic evidence for place, manner, and voicing are concentrated (Liu 1996). Stevens proposed that the information density in speech is highest during periods when the vocal tract produces this sort of rapid opening or closing gestures associated with consonants.

[5] However, if one vowel is voiced and the other is noise-excited, listeners can identify the noise-excited (or even an inharmonic) vowel at lower SNRs than its voiced counterpart (Lea 1992). Similar results are obtained using inharmonic vowels whose frequency components are randomly displaced in frequency (Cheveigné et al. 1995). These findings suggest that harmonicity or periodicity may provide a basis for "subtracting" interfering sounds, rather than selecting or enhancing target signals.

Stop consonants are less robust than vowels in noise and more vulnerable to distortion. Compared to vowels, they are brief in duration and low in intensity, making them particularly susceptible to masking by noise (e.g., Miller and Nicely 1955), temporal smearing via reverberation (e.g., Gelfand and Silman 1979), and attenuation and masking in hearing impairment (e.g., Walden et al. 1981). Given their high susceptibility to distortion, it is surprising that consonant segments contribute more to overall intelligibility than vowels, particularly in view of the fact that the latter are more intense, longer in duration, and less susceptible to masking. In natural environments, however, there are several adaptations that serve to offset, or at least partially alleviate, these problems. One is a form of auditory enhancement resulting from peripheral or central adaptation, which increases the prominence of spectral components with sudden onsets (e.g., Delgutte 1980, 1996; Summerfield et al. 1984, 1987; Summerfield and Assmann 1987; Watkins 1988; Darwin et al. 1989). A second factor is the contribution of lipreading, that is, the ability to use visually apparent articulatory gestures to supplement and/or complement the information provided by the acoustic signal (Summerfield 1983, 1987; Grant et al. 1991, 1994). Many speech gestures associated with rapid spectral changes provide visual cues that make an important contribution to intelligibility when the SNR is low.

2.5 *Temporal Envelope Modulations*

Although the majority of speech perception studies have focused on acoustic cues identified in the short-time Fourier spectrum, an alternative (and informative) way to describe speech is in terms of temporal modulations of spectral amplitude (Plomp 1983; Haggard 1985). The speech waveform is considered as the sum of amplitude-modulated signals contained within a set of narrow frequency channels distributed across the spectrum. The output of each channel is described as a carrier signal that specifies the waveform fine structure and a modulating signal that specifies its temporal envelope. The carrier signals span the audible frequency range between about 0.5 and 8 kHz, while the modulating signals represent fluctuations in the speech signal that occur at slower rates between 5 and 50 events per second—too low to evoke a distinctive sensation of pitch (Hartmann 1996) though they convey vital information for segmental and suprasegmental distinctions in speech.

Rosen (1992) summarized these ideas by proposing that the temporal structure of speech could be partitioned into three distinct levels based on their dominant fluctuation rates:

1. Envelope cues correspond to the slow modulations (at rates below 50 Hz) that are associated with changes in syllabic and phonetic-segment constituents.

2. Periodicity cues, at rates between about 70 and 500 Hz, are created by the opening and closing of the vocal folds during voiced speech.
3. Fine-structure cues correspond to the rapid modulations (above 250 Hz) that convey information about the formant pattern.

Envelope cues contribute to segmental (phonetic) distinctions that rely on temporal patterning (such as voicing and manner of articulation in consonants), as well as suprasegmental information for stress assignment, syllabification, word onsets and offsets, speaking rate, and prosody. Periodicity cues are responsible for the perception of voice pitch, and fine-structure cues are responsible for the perception of phonetic quality (or timbre).

One advantage of analyzing speech in this way is that the reduction in intelligibility caused by distortions such as additive, broadband noise, and reverberation can be modeled in terms of the corresponding reduction in temporal envelope modulations (Houtgast and Steeneken 1985). The capacity of a communication channel to transmit modulations in the energy envelope of speech is referred to as the temporal modulation transfer function (TMTF), which tends to follow a low-pass characteristic, with greatest sensitivity to modulations below about 20 Hz (Viemeister 1979; Festen and Plomp 1981).

Because the frequency components in speech are constantly changing, the modulation pattern of the *broadband* speech signal underestimates the information carried by spectrotemporal changes. Steeneken and Houtgast (1980) estimated that 20 bands are required to adequately represent variation in the formant pattern over time. They obtained the modulation (temporal envelope) spectrum of speech by

1. filtering the speech waveform into octave bands whose center frequencies range between 0.25 and 8 kHz;
2. squaring and low-pass-filtering the output (30-Hz cutoff); and
3. analyzing the resulting intensity envelope with a set of one-third octave, bandpass filters with center frequencies ranging between 0.63 and 12.5 Hz.

The output in each filter was divided by the long-term average of the intensity envelope and multiplied by $\sqrt{2}$ to obtain the modulation index. The modulation spectrum (modulation index as a function of modulation frequency) showed a peak around 3 to 4 Hz, reflecting the variational frequency of individual syllables in speech, as well as a gradual decline in magnitude at higher frequencies.

The modulation spectrum is sensitive to the effects of noise, filtering, nonlinear distortion (such as peak clipping), as well as time-domain distortions (such as those introduced by reverberation) imposed on the speech signal (Houtgast and Steeneken 1973, 1985; Steeneken and Houtgast 2002). Reverberation tends to attenuate the rapid modulations of speech by filling in the less-intense portions of the waveform. It has a low-pass filtering effect on the

TMTF.[6] Noise, on the other hand, attenuates all modulation frequencies to approximately the same degree. Houtgast and Steeneken showed that the extent to which modulations are preserved by a communication channel can be expressed by the TMTF and summarized using a numerical index of transmission fidelity, the speech transmission index (STI).

The STI measures the overall reduction in modulations present in the intensity envelope of speech and is obtained by a band-weighting method similar to that used in computing the AI. The input is either a test signal (sinusoidal intensity-modulated noise) or any complex modulated signal such as speech. The degree to which modulations are preserved by the communication channel is determined by analyzing the signal with 7 one-octave band filters whose center frequencies range between 0.125 and 8 kHz. Within each band, the modulation index is computed for 14 modulation frequencies between 0.63 and 12.5 Hz. Each index is transformed into an SNR, truncated to a 30-dB range, and averaged across the 14 modulation frequencies. Next, the octave bands are combined into a single measure, the STI, using band weightings in a manner similar to that used in computing the AI. The STI assumes a value of 1.0 when all modulations are preserved and 0 when they are observed no longer.

Houtgast and Steeneken showed that the reduction in intelligibility caused by reverberation, noise, and other distortions could be predicted accurately by the reduction of the TMTF expressed as the STI. As a result, the technique has been applied to characterizing the intelligibility of a wide range of communication channels ranging from telephone systems to individual seating positions in auditoria. The STI accounts for the effects of nonlinear signal processing in a way that makes it a useful alternative to the AI (which works best for linear distortions and normal hearing listeners). However, both methods operate on the long-term average properties of speech, and therefore do not account for effects of channel distortion on individual speech sounds or predict the pattern of confusion errors.

Figure 5.5 shows the analysis of a short declarative sentence, "The watchdog gave a warning growl." The waveform is shown at the top. The four traces below show the amplitude envelopes (left panels) and modulation spectra (right panels) in four frequency channels centered at 0.5, 1, 2, and 4 kHz. Distinctions in segmental and syllable structure are revealed by the modulation patterns in different frequency bands. For example, the affricate [č] generates a peak in the amplitude envelope of the 2- and 4-kHz channels, but not in the lower channels. The sentence contains eight syllables, with an average duration of about 200 ms, but only five give rise to distinct peaks in the amplitude envelope in the 1-kHz channel.

---

[6] In addition to suppressing modulations at low frequencies (less than 4 Hz), room reverberation may introduce spurious energy into the modulation spectrum at frequencies above 16 Hz as a result of harmonics and formants rapidly crossing the room resonances (Haggard 1985).

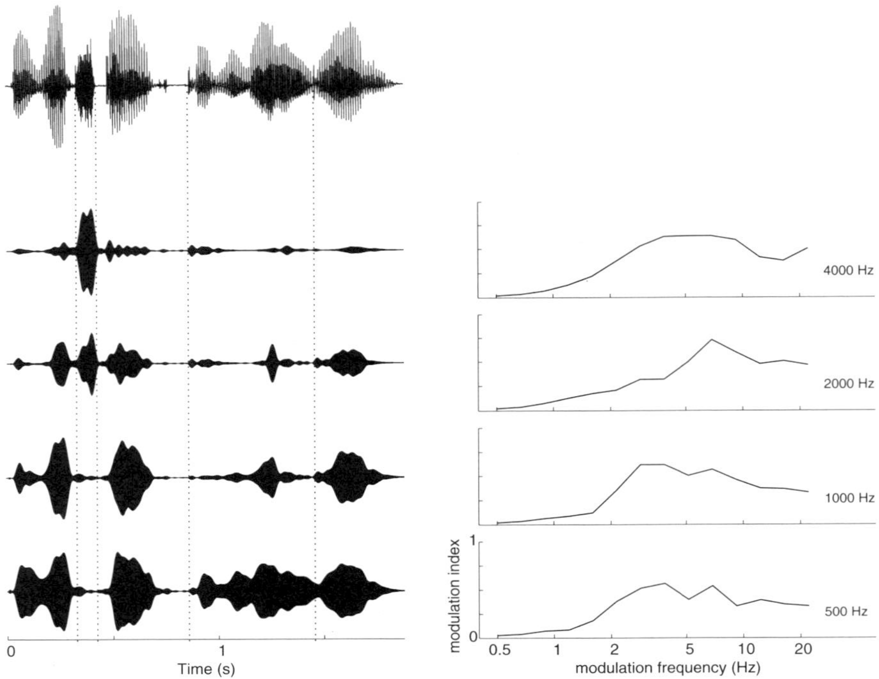

FIGURE 5.5. The upper trace shows the waveform of the sentence, "The watchdog gave a warning growl," spoken by an adult male. The lower traces on the left show the amplitude envelopes in four one-octave frequency bands centered at 0.5, 1, 2, and 4 kHz. The envelopes were obtained by (1) bandpass filtering the speech waveform (elliptical filters; one-octave bandwidth, 80 dB/oct slopes), (2) half-wave rectifying the output, and (3) low-pass filtering (elliptical filters; 80 dB/oct slopes, 30-Hz cutoff). On the right are envelope spectra (modulation index as a function of modulation frequency) corresponding to the four filter channels. Envelope spectra were obtained by (1) filtering the waveforms on the left with a set of bandpass filters at modulation frequencies between 0.5 and 22 Hz (one-third-octave bandwidth, 60 dB/oct slopes), and (2) computing the normalized root-mean-square (rms) energy in each filter band.

A powerful demonstration of the perceptual contribution of temporal envelope modulations to the robustness of speech perception was provided by Shannon et al. (1995). They showed that the rich spectral structure of speech recorded in quiet could be replaced by four bands of random noise that retained only the temporal modulations of the signal, eliminating all evidence of voicing and details of spectral shape. Nonetheless, intelligibility was reduced only marginally, both for sentences and for individual vowels and consonants in an [ɑCɑ] context (where C = any consonant). Subsequent studies showed that the precise corner frequencies and degree of overlap among the filter bands had a relatively minor effect on intelligibility (Shannon et al. 1998). Similar results were obtained when the noise carrier signals were replaced by amplitude-modulated sinusoids with fixed frequencies, equal to the center frequency of each filter band (Dorman et al. 1997). These findings illustrate the importance of the temporal modulation structure of speech and draw attention to the high degree of redundancy in the spectral fine-structure cues that have traditionally been regarded as providing essential information for phonetic identification. The results indicate that listeners can achieve high intelligibility scores when speech has been processed to remove much of the spectral fine structure, provided that the temporal envelope structure is preserved in a small number of broad frequency channels.

It is important to note that these results were obtained for materials recorded in quiet. Studies have shown greater susceptibility to noise masking for processed than unprocessed speech (Dorman et al. 1998; Fu et al. 1998). While performance in quiet reaches an asymptote with four or five bands, the decline in intelligibility as a function of decreasing SNR can be offset, to some degree, by increasing the number of spectral bands up to 12 or 16. Informal listening suggests that there is a radical loss of intelligibility when the speech is mixed with competing sounds. The absence of the spectrotemporal detail denies listeners access to cues such as voicing periodicity, which they would otherwise use to separate the sounds produced by different sources.

Although some global spectral shape information is retained in a four-band approximation to the spectrum, the precise locations of the formant peaks are generally not discernible. Reconstruction of the speech spectrum from just four frequency bands can be viewed as an extreme example of smearing the spectrum envelope. In several studies it has been demonstrated that spectral smearing over half an octave or more (thus exceeding the ear's critical bandwidth) results in an elevation of the SRT for sentences in noise (Keurs et al. 1992, 1993a,b; Baer and Moore 1993). These results are consistent with the notion that the spectral fine structure of speech plays a significant role in resisting distortion and noise.

Some investigators have studied the contribution of temporal envelope modulations in speech processed through what amounts to a *single*-channel, wideband version of the processor described by Shannon and colleagues.

These studies were motivated, in part, by the observation that temporal envelope cues are well preserved in the stimulation pattern of some present-day cochlear implants. Signal-correlated noise is created when noise is modulated by the temporal envelope of the wideband speech signal. It is striking that even under conditions where all spectral cues are removed, listeners can still recover some information for speech intelligibility. Grant et al. (1985, 1991) showed that this type of modulated noise could be an effective supplement to lip-reading for hearing-impaired listeners. Van Tasell et al. (1987b) generated signal-correlated noise versions of syllables in [aCa] context and obtained consonant identification scores from listeners with normal hearing. The temporal patterning preserved in the stimuli was sufficient for the determination of voicing, burst, and amplitude cues, although overall identification accuracy was low. Turner et al. (1995) created two-channel, signal-correlated noise by summing two noise bands, one modulated by low-frequency components, the other by high frequencies (the cutoff frequency was 1500 Hz). They found that two bands were more intelligible for normal listeners (40% correct syllable identification) than a single band (25% correct). This result is consistent with the findings of Shannon et al. (1995), who showed a progressive improvement in intelligibility as the number of processed channels increased from one to four (four bands yielding intelligibility comparable to unprocessed natural speech). Turner et al. (1995) reported similar abilities of normal and sensorineural hearing-impaired listeners to exploit the temporal information in the signal, provided that the reduced audibility of the signal for the hearing impaired was adequately compensated for. Taken together, these studies indicate that temporal envelope cues contribute strongly to intelligibility, but their contribution must be combined across a number of distinct frequency channels.

An alternative approach to studying the role of the temporal properties of speech was adopted in a series of studies by Drullman et al. (1994a,b). They filtered the modulation spectrum of speech to ascertain the contribution of different modulation frequencies to speech intelligibility. The speech waveform was processed with a bank of bandpass filters whose center frequencies ranged between 0.1 and 6.4 kHz. The amplitude envelope in each band (obtained by means of the Hilbert transform) was then low-pass filtered with cutoff frequencies between 0 and 64 Hz. The original carrier signal (waveform fine structure in each filter) was modulated by the modified envelope function. All of the processed waveforms were then summed using appropriate gain to reconstruct the wideband speech signal.

Drullman et al. found that low-pass filtering the temporal envelope of speech with cutoff frequencies below 8 Hz led to a substantial reduction in intelligibility. Low-pass filtering with cutoff frequencies above 8 Hz or high-pass filtering below 4 Hz did not lead to substantially altered SRTs for sentences in noise, compared to unprocessed speech. The intermediate range of modulation frequencies (4–16 Hz) made a substantial contribution to speech intelligibility, however. Removing high-frequency modulations in this range resulted in higher SRTs for sentences in noise and increased errors in phoneme identification, especially for stop consonants. Removing the low-frequency modulations led to poorer consonant identification, but stops (which are characterized by more rapid modulations) were well preserved, compared to other consonant types. Place of articulation was affected more than manner of articulation. Diphthongs were misclassified as monophthongs. Confusions between long and short vowels (in Dutch) were more prevalent when the temporal envelope was high-pass filtered.

The bandwidth of the analyzing filter had little effect on the results, except with filter cutoffs below 4 Hz. Listeners had considerable difficulty understanding speech from which all but the lowest modulation frequencies (0–2 Hz) had been removed. For these stimuli, the effect of temporal smearing was less deleterious when the bandwidths of the filters were larger (one octave rather than one-quarter octave). Drullman et al. interpreted this outcome in terms of a greater reliance on within-channel processes for low modulation rates. At higher modulation rates listeners may rely to a greater extent on across-channel processes. The cutoff was around 4 Hz, close to the mean rate of syllable and word alternation. If the analysis of temporal modulations relies on across-channel coupling, this would lead to the prediction that phase-shifting the carrier bands would disrupt the coupling and also result in lower intelligibility. However, this does not seem to be the case: Greenberg and colleagues (Greenberg 1996; Arai and Greenberg 1998; Greenberg and Arai 1998) reported that temporal desynchronization of frequency bands by up to 120 ms had relatively little effect on intelligibility on connected speech. Instead, the temporal modulation structure appears to be processed independently in different frequency bands (as predicted by the STI). In comparison, spectral transformations that involve frequency shifts (i.e., applying the temporal modulations to bands with different center frequencies) are extremely disruptive (Blesser 1972). One implication of the result is the importance of achieving the correct relationship between frequency and place within the cochlea when tuning multichannel cochlear-implant systems (Dorman et al. 1997; Shannon et al. 1998).

A further aspect of the temporal structure of speech was investigated by Greenberg et al. (1998). They partitioned the spectrum of spoken English sentences into one-third-octave bands and carried out intelligibility tests on these bands, alone and in combination. Even with just three bands, intelligibility remained high (up to 83% of the words were identified correctly). However, performance was severely degraded when these bands were desynchronized by more than 25 ms and the signal was limited to a small number of narrow bands. In contrast, previous findings by Arai and Greenberg (1998), as well as Greenberg and Arai (1998), show that listeners are relatively insensitive to temporal asynchrony when a larger number (19) of one-quarter-octave bands is presented in combination to create an approximation to (temporally desynchronized) full-bandwidth speech. This suggests that listeners rely on across-channel integration of the temporal structure to improve their recognition accuracy. Greenberg et al. suggested that listeners are sensitive to the phase properties of the modulation spectrum of speech, and that this sensitivity is revealed most clearly when the spectral information in speech is limited to a small number of narrow bands.

When speech is presented in a noisy background, it undergoes a reduction in intelligibility, in part because the noise reduces the modulations in the temporal envelope. However, the decline in intelligibility may also result from distortion of the temporal fine structure and the introduction of spurious envelope modulations (Drullman 1995a,b; Noordhoek and Drullman 1997). A limitation of the TMTF and STI methods is that they do not consider degradations in speech quality resulting from the introduction of spurious modulations absent from the input (Ludvigsen et al. 1990). These modulations can obscure or mask the modulation pattern of speech, and obliterate some of the cues for identification. Drullman's work suggests that the loss of intelligibility is mainly due to noise present in the temporal envelope troughs (envelope minima) rather than at the peaks (envelope maxima). Drullman (1995b) found that removing the noise from the speech peaks (by transmitting only the speech when the amplitude envelope in each band exceeded a threshold) had little effect on intelligibility. In comparison, removing the noise from the troughs (transmitting speech alone when the envelope fell below the threshold) led to a 2-dB elevation of the SRT.

In combination, these studies show that:

1. an analysis of the temporal structure of speech can make a valuable contribution to describing the perception of speech under adverse conditions;
2. the pattern of temporal amplitude modulation within a few frequency bands provides sufficient information for speech perception; and
3. a qualitative description of the extent to which temporal amplitude modulation is lost in a communication channel (but also, in the case of noise and reverberation, augmented by spurious modulations) is an informative way of predicting the loss of intelligibility that occurs when speech passes through that channel.

2.6 Speaker Adaptations Designed to Resist Noise and Distortion

The previous section considered several built-in properties of speech that help shield against interference and distortion. In addition, speakers actively adjust the parameters of their speech to offset reductions in intelligibility due to masking and distortion. In the current section, we consider specific strategies adopted by speakers under adverse conditions to promote successful communication under adverse conditions. These include the so-called Lombard effect, the use of distinct speaking styles such as "clear" and "shouted" speech, styles used to address hearing-impaired listeners and foreigners, as well as speech produced under high cognitive workload.

When speakers are given explicit instructions to "speak as clearly as possible," their speech differs in several respects from normal conversational speech. Clear speech is produced with higher overall amplitude, a higher mean f_0, and longer segmental durations (Picheny et al. 1985, 1986; Payton et al. 1994; Uchanski et al. 1994). Clear speech is more intelligible than conversational speech under a variety of conditions, including noise, reverberation, and hearing, impairment. Clear speech and conversational speech have similar long-term spectra, but differ with respect to their spectrotemporal patterning, which produces different TMTFs (Payton et al. 1994).

In long-distance communication, speakers often raise the overall amplitude of their voice by shouting. Shouted speech is produced with a reduced spectral tilt, higher mean f_0, and longer vocalic durations (Rostolland 1982). Despite its effectiveness in long-range communication, shouted speech is less intelligible than conversational speech at the same SNR (Pickett 1956; Pollack and Pickett 1958; Rostolland 1985).

When speech communication takes place in noisy backgrounds, such as crowded rooms, speakers modify their vocal output in several ways. The most obvious change is an increase in loudness, but there are a number of additional changes. Collectively these changes are referred to as the Lombard reflex (Lombard 1911).

The conditions that result in Lombard speech have been used to investigate the role of auditory feedback in speech production. Ladefoged (1967) used intense noise designed to mask both airborne and bone-conducted sounds from the speaker's own voice. His informal observations suggested that elimination of auditory feedback and its replacement by intense random noise have a disruptive effect, giving rise to inappropriate nasalization, distorted vowel quality, more variable segment durations, a narrower f_0 range, and an increased tendency to use simple falling intonation patterns. Dreher and O'Neill (1957) and Summers et al. (1988) have extended this work to show that speech produced under noisy conditions (flat-spectrum broadband noise) is more intelligible than speech produced in quiet conditions when presented at the same SNR. Thus, at SNRs where auditory feedback is not entirely eliminated, speakers adjust the parameters of their speech so as to preserve its intelligibility.

Table 5.2 summarizes the results of several studies comparing the production of speech in quiet and in noise. These studies have identified changes in a number of speech parameters. Taken together, the adjustments have two major effects: (1) improvement in SNR; and (2) a reduction in the information rate, allowing more time for decoding. Such additional time is needed, in view of demonstrations by Baer et al. (1993) that degradation of SNR leads to increases in the latency with which listeners make decisions about the linguistic content of a speech signal.

Researchers in the field of automatic speech recognition have sought to identify systematic properties of Lombard speech to improve the recogni-

TABLE 5.2. Summary of changes in the acoustic properties of speech produced in background noise (Lombard speech) compared to speech produced in quiet

Change	Reference
Increase in vocal intensity (about 5 dB increase in speech for every 10 dB increase in noise level)	Dreher and O'Neill (1957)
Decrease in speaking rate	Hanley and Steer (1949)
Increase in average f_0	Summers et al. (1988)
Increase in segment durations	Pisoni et al. (1985)
Reduction in spectral tilt (boost in high-frequency components)	Summers et al. (1988)
Increase in F_1 and F_2 frequency (inconsistent across talkers)	Summers et al. (1988); Junqua and Anglade (1990); Young et al. (1993)

tion accuracy of recognizers in noisy backgrounds (e.g., Hanson and Applebaum 1990; Gong 1994) given that Lombard speech is more intelligible than speech recorded in quiet (Summers et al. 1988). Lindblom (1990) has provided a qualitative account of the idea that speakers monitor their speech output and systematically adjust its acoustic parameters to maximize the likelihood of successful transmission to the listener. The hypospeech and hyperspeech (H & H) model assumes a close link between speech production and perception. The model maintains that speakers employ a variety of strategies to compensate for the demands created by the environment to ensure that their message will be accurately received and decoded. When the constraints are low (e.g., in quiet conditions), fewer resources are allocated to speech production, with the result that the articulators deviate less from their neutral positions and hypospeech is generated. When the demands are high (e.g., in noisy environments), speech production assumes a higher degree of flexibility, and speakers produce a form of speech known as hyperspeech.

Consistent with the H & H model, Lively et al. (1993) documented several changes in speech production under conditions of high cognitive work load (created by engaging the subject in a simultaneous task of visual information processing). Several changes in the acoustic correlates of speech were observed when the talker's attention was divided in this way, including increased amplitude, decreased spectral tilt, increased speaking rate, and more variable f_0. A small (2–5%) improvement in vowel identification was observed for syllables produced under such conditions. However, there were substantial differences across speakers, indicating that speaker adaptation under adverse conditions are idiosyncratic, and that it may be difficult to provide a quantitative account of their adjustments. Lively et al. did not inform the speakers that the intelligibility of their speech would be measured. The effects of work load might be greater in conditions where speakers are explicitly instructed to engage in conversation with listeners.

2.7 *Summary of Design Features*

In this section we have proposed that speech communication incorporates several types of shielding to protect the signal from distortion. The acoustic properties of speech suggest coding principles that contribute to noise reduction and compensation for communication channel distortion. These include the following:

1. The short-term amplitude spectrum dominated by low-frequency energy (i.e., lower than the region of maximum sensitivity of human hearing) and characterized by resonant peaks (formants) whose frequencies change gradually and coherently across time;
2. Periodicity in the waveform at rates between 50 and 500 Hz (along with corresponding harmonicity in the frequency domain) due to vocal fold vibration, combined with the slow fluctuations in the repetition rate that are a primary correlate of prosody;
3. Slow variations in waveform amplitude resulting from the alternation of vowels and consonants at a rate of roughly 3–5 Hz;
4. Rapid spectral changes that signal the presence of consonants.

To this list we can add two additional properties:

1. Differences in relative intensity and time of arrival at the two ears of a target voice and interfering sounds, which provide a basis for the spatial segregation of voices;

2. The visual cues provided by lip-reading provide temporal synchronization between the acoustic signal and the visible movements of the articulators (lips, tongue, teeth, and jaw); Cross-modal integration of acoustic and visual information can improve the effective by about 6 dB (MacLeod and Summerfield 1987).

Finally, the studies reviewed in section 2.6 suggest yet a different form of adaptation; under adverse conditions, speakers actively monitor the SNR and adjust the parameters of their speech to offset the effects of noise and distortion, thereby partially compensating for the reduction of intelligibility. The most salient modifications include an increase in overall amplitude and segmental duration, as well as a reduction in spectral tilt.

3. Speech Intelligibility Under Adverse Conditions

3.1 *Background Noise*

Speech communication nearly always takes place under conditions where some form of background noise is present. Traffic noise, competing voices, and the noise of fans in air conditioners and computers are common forms of interference. Early research on the effects of noise demonstrated that listeners with normal hearing can understand speech in the presence of white noise even when the SNR is as low as 0 dB (Fletcher 1953). However, under natural conditions the distribution of noise across time and frequency is rarely uniform. Studies of speech perception in noise can be grouped according to the type of noise maskers used. These include tones and narrowband noise, broadband noise, interrupted noise, speech-shaped noise, multispeaker babble, and competing voices. Each type of noise has a somewhat different effect on speech intelligibility, depending on its acoustic form and information content, and therefore each is reviewed separately.

The effects of different types of noise on speech perception have been compared in several ways. The majority of studies conducted in the 1950s and 1960s compared overall identification accuracy in quiet and under several different levels of noise (e.g., Miller et al. 1951). This approach is time-consuming, because it requires separate measurements of intelligibility for different levels of speech and noise. Statistical comparisons of conditions can be problematic if the mean identification level approaches either 0% or 100% correct in any condition. An alternative method, developed by Plomp and colleagues (e.g., Plomp and Mimpen 1979) avoids these difficulties by measuring the SRT. The SRT is a masked identification threshold, defined as the SNR at which a certain percentage (typically 50%) of the syllables, words, or sentences presented can be reliably identified. The degree of interference produced by a particular noise can be expressed in terms of the difference in dB between the SRT in quiet and in noise. Additional studies have compared the effects of different noises by conducting closed-set phonetic identification tasks and analyzing confusion matrices. The focus of this approach is phonetic perception rather than overall intelligibility, and its primary objective is to identify those factors responsible for the pattern of errors observed within and between different phonetic classes (e.g., Miller and Nicely 1955; Wang and Bilger 1973).

3.2 Narrowband Noise and Tones

A primary factor in determining whether a sound will be an effective masker is its frequency content and the extent of spectral overlap between masker and speech signal. In general, low-frequency noise (20–250 Hz) is more pervasive in the environment, propagates more efficiently, and is more disruptive than high-frequency interference (Berglund et al. 1996). At high intensities, noise with frequencies as low as 20 Hz can reduce the intelligibility of speech (Pickett 1957). Speech energy is concentrated between 0.1 and 6 kHz (cf. section 2.1), and noise with spectral components in this region is the most effective masker of speech. Within this spectral range, lower-frequency interference produces more masking than their higher-frequency counterparts (Miller 1947).

When speech is masked by narrowband maskers, such as pure tones and narrowband noise, low frequencies (<500 Hz) are more disruptive than higher frequencies (Stevens et al. 1946). As the sound pressure level increases, there is a progressive shift toward lower frequencies (300 Hz), presumably as the result of upward spread of masking by low frequencies (Miller 1947). Complex tonal maskers equated for sound pressure level (square and rectangular waves) are more effective maskers than sinusoids of comparable frequency, with little variation in masking effectiveness as a function of f_0 in the low-frequency (80–400 Hz) range. For frequencies above 1 kHz, neither pure tones nor square waves are effective maskers of speech (Stevens et al. 1946).

Licklider and Guttman (1957) varied the number and frequency spacing of sinusoidal components in a complex tonal masker, holding the overall power constant. Maskers, whose spectral energy is distributed across frequency in accordance with the "equal importance function" (proportional to the critical bandwidth), are more effective speech maskers than those with energy uniformly distributed. Masking effectiveness increased as the number of components was increased from 4 to 40, but there was little further change as the number of components increased beyond 40. Even with 256 components, the masking effectiveness of the complex was about 3 dB less than pink noise with the same frequency range and power.

3.3 Broadband Noise

When speech is masked by broadband noise with a uniform spectrum, its intelligibility is a linear function of SNR as long as the sound pressure level of the noise is greater than about 40 dB (Kryter 1946; Hawkins and Stevens 1950). For listeners with normal hearing, speech communication remains unhampered, unless the SNR is less than +6 dB. Performance remains above chance, even when the SNR is as low as −18 dB (Licklider and Miller 1951). The relationship between SNR and speech intelligibility is affected by context (e.g., whether the stimuli are nonsense syllables, isolated words, or words in sentences), by the size of the response set, and by the entropy associated with the speech items to be identified (Miller et al. 1951). In closed-set identification, the larger the response set the greater the susceptibility to noise. In open-set tasks the predictability of words within the sentence is a significant factor. Individual words in low-predictability sentences are more easily masked than those in high-predictability or neutral sentences (Kalikow et al. 1977; Elliot 1995).

Miller and Nicely (1955) examined the effects of broadband (white) noise on the identification of consonants in CV (consonant-vowel) syllables. They classified consonants in terms of such phonetic features as voicing, nasality, affrication, duration, and place of articulation. For each subgroup they examined overall error rates and confusion patterns, as well as a measure of the amount of information transmitted. Their analysis revealed that noise had the greatest effect on place of articulation. Duration and frication were somewhat more resistant to noise masking. Voicing and nasality were transmitted fairly successfully, and preserved to some extent, even at an SNR of

−12 dB. The effects of noise masking were similar to those of low-pass filtering, but did not resemble high-pass filtering, which resulted in a more random pattern of errors. They attributed the similarity in effects of low-pass filtering and noise to the sloping long-term spectrum of speech, which tends to make the high-frequency portion of the spectrum more susceptible to noise masking.

Pickett (1957) and Nooteboom (1968) examined the effects of broadband noise on the perception of vowels. Pickett suggested that vowel identification errors might result when phonetically distinct vowels exhibited similar formant patterns. An analysis of confusion matrices for different noise conditions revealed that listeners frequently confused front vowels (such as [i], with a high second formant) with a corresponding back vowel (e.g., [u], with a low F_2). When the F_2 peak is masked, the vowel is identified as a back vowel with a similar F_1. This error pattern supports the hypothesis that listeners rely primarily on the frequencies of formant peaks to identify vowels (rather than the entire shape of the spectrum), and are predicted by a formant-template model of vowel perception (Scheffers 1983). Scheffers (1983) found that the identification thresholds for synthesized vowels masked by pink noise could be predicted fairly well by the SNR in the region of the second formant. Scheffers found that unvoiced (whispered) vowels had lower thresholds than voiced vowels. He also showed that vowels were easier to identify when the noise was on continuously, or was turned on 20 to 30 ms before the onset of the vowel, compared with a condition where vowels and noise began together.

Pickett (1957) reported that duration cues (differences between long and short vowels) had a greater influence on identification responses when one or more of the formant peaks was masked by noise. This finding serves as an example of the exploitation of signal redundancy to overcome the deleterious effects of spectral masking. It has not been resolved whether results like these reflect a "re-weighting" of importance in favor of temporal over spectral cues or whether the apparent importance of cue B automatically increases when cue A cannot be detected.

3.4 Interrupted Speech and Noise

Miller and Licklider (1950) observed that under some condition the speech signal could be turned on and off periodically without substantial loss of intelligibility. Two factors, the interruption rate and the speech-time fraction, were found to be important. Figure 5.6 shows that intelligibility was lowest for interruption rates below 2 Hz (and a speech-time fraction of 50%), where large fragments of each word are omitted. If the interruption rate was higher (between 10 and 100 interruptions per second), listeners identified more than 80% of the monosyllabic words correctly. Regular, aperiodic, or random interruptions produced similar results, as long as the same constant average interruption rate and speech-time fraction were

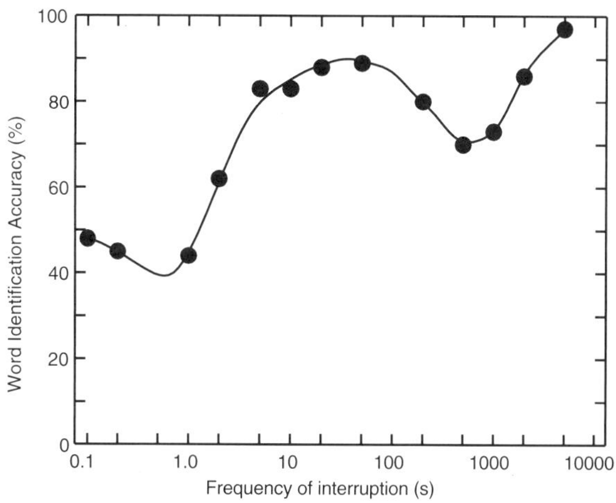

FIGURE 5.6. Word identification accuracy as a function of the rate of interruption for a speech-time fraction of 50%. (After Miller and Licklider 1950.)

maintained. The high intelligibility of interrupted speech is remarkable, considering that near-perfect identification is obtained in conditions where close to half of the power in the speech signal has been omitted. At the optimal (i.e., most intelligible) interruption rate (about 10 interruptions per second), listeners were able to understand conversational speech. Miller and Licklider suggested that listeners were able to do this by "patching together" successive glimpses of signal to reconstruct the intended message. For the speech materials in their sample (phonetically balanced monosyllabic words), listeners were able to obtain, on average, one "glimpse" per phonetic segment (although phonetic segments are not uniformly spaced in time). Miller and Licklider's findings were replicated and extended by Huggins (1975), who confirmed that the optimum interruption rate was around 10 Hz (100 ms) and demonstrated that the effect was at least partially independent of speaking rate. Huggins interpreted the effect in terms of a "gap-bridging" process that contributes to the perception of speech in noise.

Miller and Licklider (1950) also investigated the masking of speech by interrupted noise. They found that intermittent broadband noise maskers interfered less with speech intelligibility than did continuous maskers. An interruption rate of around 15 noise bursts per second produced the greatest release from masking. Powers and Wilcox (1977) have shown that the greatest benefit is observed when the interleaved noise and speech are comparable in loudness.

Howard-Jones and Rosen (1993) examined the possibility that the release from masking by interrupted noise might benefit from an independent analysis of masker alternations in different frequency regions. They proposed that listeners might benefit from a process of "un-comodulated" glimpsing in which glimpses are patched together across different frequency regions at different times. To test this idea they used a "checkerboard" masking noise. The noise was divided into 2, 4, 8, or 16 frequency bands of equal power. The noise bands were switched on and off at a rate of 10 Hz, either synchronously in time ("comodulated" interruptions) or asynchronously, with alternating odd and even bands ("un-comodulated" interruptions) to create a masker whose spectrogram resembled a checkerboard. Evidence for a contribution of un-comodulated glimpsing was obtained when the masker was divided into either two or four bands, resulting in a release from masking of 16 and 6 dB, respectively (compared to 23 dB for fully comodulated bands). The conclusion from this study is that listeners can benefit from un-comodulated glimpsing to integrate speech cues from different frequency bands at different times in the signal.

When speech is interrupted periodically by inserting silent gaps, it assumes a harsh, unnatural quality, and its intelligibility is reduced. Miller and Licklider (1950), using monosyllabic words as stimuli, noted that this harsh quality could be eliminated by filling the gaps with broadband noise. Although adding noise restored the natural quality of the speech, it did not improve intelligibility. Subsequent studies with connected speech found both greater naturalness and higher intelligibility when the silent portions of interrupted speech were filled with noise (Cherry and Wiley 1967; Warren et al. 1997). One explanation is that noise-filled gaps more effectively engage the listener's ability to exploit contextual cues provided by syntactic and semantic continuity (Warren and Obusek 1971; Bashford et al. 1992; Warren 1996).

3.5 Competing Speech

While early studies of the effects of noise on speech intelligibility often used white noise (e.g., Hawkins and Stevens 1950), later studies were interested in exploring more complex forms of noise that are more representative of noisy environments such as cafeterias and cocktail parties (e.g., Duquesnoy 1983; Festen and Plomp 1990; Darwin 1990; Festen 1993; Howard-Jones and Rosen 1993; Bronkhorst 2000; Brungart 2001; Brungart et al. 2001). Research on the perceptual separation of speech from competing spoken material has received particular attention because

1. the acoustic structure of the target and masker are similar,
2. listeners with normal hearing perform the separation of voices successfully and with little apparent effort, and
3. listeners with sensorineural hearing impairments find competing speech to be a major impediment to speech communication.

Accounting for the ability of listeners to segregate a mixture of voices and attend selectively to one of them has been described as the "cocktail party problem" by Cherry (1953). This ability is regarded as a prime example of auditory selective attention (Broadbent 1958; Bregman 1990).

The interfering effect of competing speech is strongly influenced by the number of competing voices present. Figure 5.7 illustrates the effects of competing voices on syllable identification with data from Miller (1947). Miller obtained articulation functions (percent correct identification of monosyllabic words as a function of intensity of the masker) in the presence of one, two, four, six, or eight competing voices. The target voice was always male, while the interfering voices were composed of equal numbers of males and females. A single competing (male) voice was substantially less effective as a masker than two competing voices (one male and one female). Two voices were less effective than four, but there was little subsequent change in masking effectiveness as the number was increased to six and eight voices. When a single competing voice is used as a masker, variation in its overall amplitude creates dips or gaps in the waveform that enable the listener to hear out segments of the target voice. When several voices are present, the masker becomes more nearly continuous in overall amplitude and the opportunity for "glimpsing" the target voice no longer arises.

When speech and nonspeech broadband maskers were compared in terms of their masking effect, competing speech maskers from a single speaker and amplitude-modulated noise were found to produce less masking than steady-state broadband noise (Speaks et al. 1967; Carhart et al. 1969; Gustafsson and Arlinger 1994). The advantage for speech over nonspeech maskers disappeared when several speakers were combined. Mixing

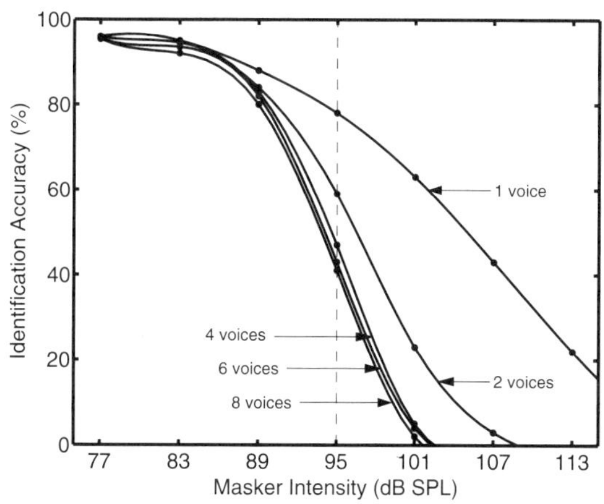

FIGURE 5.7. Syllable identification accuracy as a function of the number of competing voices. The level of the target speech (monosyllabic nonsense words) was held constant at 95 dB. (After Miller 1947.)

the sounds of several speakers produces a signal with a more continuous amplitude envelope and a more uniform spectrum (cf. Fig. 5.1). The masking effect of a mixture of speakers, or a mixture of samples of recorded music (Miller 1947), was similar to that of broadband noise.

A competing voice may interfere with speech perception for at least two reasons. First, it may result in spectral and temporal overlap, which leads to auditory masking. Second, it may interrupt the linguistic processing of the target speech. Brungart and colleagues (Brungart 2001; Brungart et al. 2001) measured the intelligibility of a target phrase masked by one, two, or three competing talkers as a function of SNR and masker type. Performance was generally worse with a single competing talker than with temporally modulated noise with the same long-term average spectrum as the speech. Brungart et al. suggested that part of the interference produced by a masking voice is due to informational masking, distinct from energetic masking caused by spectral and temporal overlap of the signals.

In contrast with these results, Dirks and Bower (1969) and Hygge et al. (1992) obtained similar results for speech maskers played forward or backward. In these studies there was little evidence that the masking effect was enhanced by semantic or syntactic interference of the masker. Their results suggest that the interfering effects of speech maskers can be partially alleviated by temporal dips in the masker that permit the listener to "glimpse" the acoustic structure of the target voice.

Support for the idea that listeners with normal hearing can exploit the temporal modulations associated with a single competing voice comes from studies that compared speech maskers with steady-state and amplitude-modulated noise maskers. There is a large difference in masking effect of a steady-state noise and a modulated noise (or a single interfering voice), as measured by the SRT. Up to 8 dB of masking release is provided by an amplitude-modulated noise masker, compared to the steady-state masker (Duquesnoy 1983; Festen and Plomp 1990; Gustafsson and Arlinger 1994). Speech-spectrum–shaped noise is a more effective masker than a competing voice (Festen and Plomp 1990; Festen 1993; Peters et al. 1998). Speech reception thresholds for sentences in modulated noise are 4 to 6 dB lower than comparable sentences in unmodulated noise. For sentences masked by a competing voice, the masking difference increased to 6 to 8 dB. However, masker modulation does not appear to play a significant role in masking of isolated nonsense syllables or spondee words (Carhart et al. 1969), and hence may be related to the syllable structure of connected speech.

For hearing-impaired listeners the benefits of modulation are reduced, and both types of maskers are equally disruptive (Dirks et al. 1969; Festen and Plomp 1990; Gustafsson and Arlinger 1994). This result is attributable to reduced temporal resolution associated with sensorineural hearing loss (Festen and Plomp 1990). Festen and Plomp (1990) suggested two possible bases for the effect: (1) listening in the temporal dips of the masker, providing a locally favorable SNR; and (2) comodulation masking release

(CMR). Festen (1993) described experiments in which across-frequency coherence of masker fluctuations was disrupted. He concluded that across-frequency processing of masker fluctuations (CMR) makes only a small (about 1.3 dB) contribution to the effect. The effect of masker fluctuation is level-dependent, in a manner consistent with an alternative explanation based on forward masking (the modulation is expected to produce less masking release at low sensation levels because the decay in forward masking is more gradual near threshold).

When listening to a mixture of two voices, listeners with normal hearing have exceptional abilities to hear out components of the composite signal that stem from the same larynx and vocal tract. For example, when a target sentence is combined with an interfering sentence spoken by a different speaker, listeners can correctly identify 70% to 80% of the target words at an SNR of 0 dB (Stubbs and Summerfield 1991). One factor that contributes to intelligibility is auditory grouping and segregation on the basis of f_0 (Brokx and Nooteboom 1982; Scheffers 1983; Assmann and Summerfield 1990, 1994; Darwin and Carlyon 1995; Bird and Darwin 1998). Summerfield and Culling (1992) demonstrated that listeners can exploit simultaneous differences in f_0 to segregate competing voices even at disadvantageous SNRs when the formants of a target voice do not appear as distinct peaks in the composite spectrum. They determined masked identification thresholds for target vowels in the presence of vowel-like maskers. Thresholds were about 15 dB lower when the masker and target differed in f_0 by two semitones (about 12%). At threshold, the formants of the target did not appear as distinct peaks in the composite spectrum envelope but rather as small bumps or "shoulders." An autocorrelation analysis, based on Meddis and Hewitt's (1992) model, revealed that the periodicity of the masker was stronger than that of the target in the majority of frequency channels. Summerfield and Culling proposed that the identity of the target vowel was determined on the basis of the disruption it produced in the periodicity of the masker, rather than on the basis of its own periodicity. This explanation is consistent with models of source segregation that remove the evidence of an interfering voice on the basis of its periodicity (Meddis and Hewitt 1992; Cheveigné 1997).

During voiced speech the pulsing of the vocal folds gives rise to a consistent pattern of periodicity in the waveform and harmonicity in the spectrum. In a mixture of two voices, the periodicity or harmonicity associated with the target voice provides a basis for grouping together signal components with the same f_0. Time-varying changes in f_0 also provide a basis for tracking properties of the voice over time.

Brokx and Nooteboom (1982) demonstrated benefits of differences in average f_0 using LPC-resynthesized speech. Brokx and Nooteboom analyzed natural speech using an LPC vocoder to artificially modify the characteristics of the excitation source and create synthesized, monotone versions of a set of 96 semantically anomalous sentences. They then varied the difference in fundamental frequency between the target sentence and a continuous speech masker. Identification accuracy was lowest when the target and masker had the same f_0, and gradually improved as a function of increasing difference in f_0. Identification accuracy was lower when the two voices were exactly an octave apart, a condition where every second harmonic of the higher-pitched voice overlaps with a harmonic of the lower f_0. These results were replicated and extended by Bird and Darwin (1998) who used monotone versions of short declarative sentences consisting of entirely voiced sounds. They presented the sentences concurrently in pairs, with one long masker sentence and a short target sentence in each pair. They found an improvement in intelligibility with differences in f_0 between ±2 and ±8 semitones. Using a similar method, Assmann (1999) confirmed the benefits of f_0 difference using both monotone sentence pairs (in which f_0 was held constant) and sentence pairs with natural intonation (in which the natural variation in f_0 was preserved in each sentence, but shifted up or down along the frequency scale to produce the corresponding mean difference in f_0). An unexpected result was that sentences with natural intonation were not significantly more intelligible than monotone sentences, suggesting that f_0 differences are more important for segregating competing speech sounds than time-varying changes in f_0.

In natural environments, competing voices typically originate from different locations in space. A number of laboratory studies have confirmed that a difference in spatial separation can aid the perceptual segregation of competing voices (e.g., Cherry 1953). Yost et al. (1996) presented speech (words, letters, or numbers) to listeners with normal hearing in three listening conditions. In one condition the listener was seated in a sound-deadened room and signals were presented over loudspeakers arranged in a circle around the listener. In a second condition, speech was presented in the free field as in the first condition, but was recorded using a stationary KEMAR manikin and delivered binaurally over headphones to a listener in a remote room. In the third condition, a single microphone was used and the sounds presented monaurally. Sounds were presented individually, in pairs, or in triplets from different randomly chosen subsets of the loudspeakers. Identification scores were highest when the free-field conditions were comparable to the monaural. Intermediate scores were observed under conditions where the binaural recordings were made with the KEMAR manikin. Differences among the conditions were reduced substantially when only two, rather than three, utterances were presented simultaneously, suggesting that listening with two ears in free field is most effective when more than two concurrent sound sources are present.

3.6 Binaural Processing and Noise

When a sound source is located directly in front of an observer in the free field, the acoustic signals reaching the two ears are nearly identical. When the source is displaced to one side or the other, each ear receives a slightly different signal. Interaural level differences (ILDs) in sound pressure level, which are due to head shadow, and interaural time differences (ITDs) in the time of arrival provide cues for sound localization and can also contribute to the intelligibility of speech, especially under noisy conditions. When speech and noise come from different locations in space, interaural disparities can improve the SRT by up to 10 dB (Carhart 1965; Levitt and Rabiner 1967; Dirks and Wilson 1969; Plomp and Mimpen 1981). Some benefit is derived from ILDs and ITDs, even when listening under monaural conditions (Plomp 1976). This benefit is probably a result of the improved SNR at the ear ipsilateral to the signal.

Bronkhorst and Plomp (1988) investigated the separate contributions of ILDs and ITDs using free-field recordings obtained with a KEMAR manikin. Speech was recorded directly in front of the manikin, and noise with the same long-term spectrum as the speech was recorded at seven different angles in the azimuthal plane, ranging from 0 to 180 degrees in 30-degree steps. Noise samples were processed to contain only ITD or only ILD cues. The binaural benefit was greater for ILDs (about 7 dB) than for ITDs (about 5 dB). In concert, ILDs and ITDs yielded a 10-dB binaural gain, comparable to that observed in earlier studies.

The binaural advantage is frequency dependent (Kuhn 1977; Blauert 1996). Low frequencies are diffracted around the head with relatively little attenuation (a consequence of the wavelength of such signals being appreciably longer than the diameter of the head), while high frequencies (>4 kHz for human listeners) are attenuated to a much greater extent (thus providing a reliable cue based on ILDs in the upper portion of the spectrum). The encoding of ITDs is based on neural phase-locking, which declines appreciably above 1500 Hz (in the upper auditory brain stem). Thus, ITD cues are generally not useful for frequencies above this limit, except when high-frequency carrier signals are modulated by low frequencies. Analysis of the pattern of speech errors in noise suggests that binaural listening may provide greater benefits at low frequencies. For example, in binaural conditions listeners made fewer errors involving manner-of-articulation features, which rely predominantly on low-frequency cues, and they were better able to identify stop consonants with substantial low-frequency energy, such as the velar stops [k] and [g] (Helfer 1994).

3.7 Effects of Reverberation

When speech is spoken in a reverberant environment, the signal emanating from the mouth is combined with reflections that are time-delayed, scaled versions of the original. The sound reaching the listener is a mixture of direct and reflected energy, resulting in temporal "smearing" of the speech signal. Echoes tend to fill the dips in the temporal envelope of speech and increase the prominence of low-frequency energy that masks the speech spectrum. Sounds with time-invariant cues, such as steady-state vowels, suffer little distortion, but the majority of speech sounds are characterized by changing formant patterns. For speech sounds with time-varying spectra, reverberation leads to a blurring of spectral detail. Hence, speech sounds with rapidly changing spectra (such as stop consonants) are more likely to suffer deleterious effects of reverberation than segments with more stationary formants. Factors that affect speech intelligibility include volume of the enclosure, reverberation time, ambient noise level, the speaker's vocal output level, as well as the distance between speaker and listener. Hearing-impaired listeners are more susceptible to the effects of reverberation than listeners with normal hearing (Finitzo-Hieber and Tillman 1978; Duquesnoy and Plomp 1983; Humes et al. 1986).

An illustration of the effects of reverberation on the speech spectrogram is shown in Figure 5.8. Overall, the most visible effect is the transformation of dynamic features of the spectrogram into more static features. Differences between the spectrogram of the utterance in quiet and in reverberation include:

1. Reverberation fills the gaps and silent intervals associated with vocal-tract closure in stop consonants. For example, the rapid alternation of noise

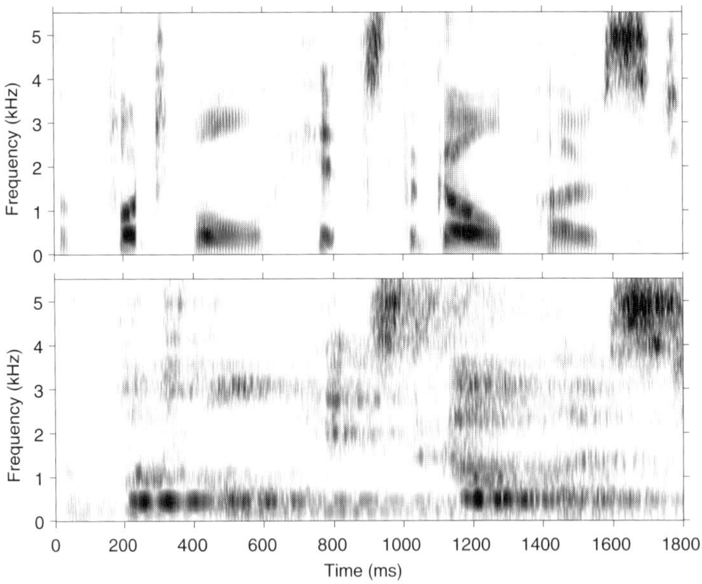

FIGURE 5.8. The upper panel displays the spectrogram of a wideband sentence, "The football hit the goal post," spoken by an adult male. The lower panel shows the spectrogram of a version of the sentence in simulated reverberation, modeling the effect of a highly reverberant enclosure with a reverberation time of 1.4 seconds at a location 2 m from the source.

and silence surrounding the [t] burst in "football" (occurring at about the 300-ms frame on the spectrogram) is blurred under reverberant conditions (lower spectrogram).

2. Both onsets and offsets of syllables tend to be blurred, but the offsets are more adversely affected.

3. Noise bursts (fricatives, affricates, stop bursts) are extended in duration. This is most evident in the [t] burst of the word "hit" (cf. the 900-ms frame in the upper spectrogram).

4. Reverberation blurs the relationship between temporal events, such as the voice onset time (VOT), the time interval between stop release and the onset of voicing. Temporal offsets are blurred, making it harder to determine the durations of individual speech segments, such as the [U] in "football" (at approximately the 200-ms point in the upper spectrogram).

5. Formant transitions are flattened, causing diphthongs and glides to appear as monophthongs, such as the [o^w] in "goal" (cf. the 1100-ms frame).

6. Amplitude modulations associated with f_0 are reduced, smearing the vertical striation pattern in the spectrogram during the vocalic portions of the utterance (e.g., during the word "goal").

In a reverberant sound field, sound waves reach the ears from many directions simultaneously and hence their sound pressure levels and phases vary as a function of time and location of both the source and receiver. Plomp and Steeneken (1978) estimated the standard deviation in the levels of individual harmonics of complex tones and steady-state vowels to be about 5.6 dB, while the phase pattern was effectively random in a diffuse sound field (a large concert hall with a reverberation time of 2.2 seconds). This variation is smaller than that associated with phonetic differences between pairs of vowels, and is similar in magnitude to differences in pronunciations of the same vowel by different speakers of the same age and gender (Plomp 1983). Plomp and Steeneken showed that the effects of reverberation on timbre are well predicted by differences between pairs of amplitude spectra, measured in terms of the output levels of a bank of one-third-octave filters. Subsequent studies have confirmed that the intelligibility of spoken vowels is not substantially reduced in a moderately reverberant environment for listeners with normal hearing (Nábělek and Letowski 1985).

Nábělek (1988) suggested two reasons why vowels are typically well preserved in reverberant environments. First, the spectral peaks associated with formants are generally well defined in relation to adjacent spectral troughs (Leek et al. 1987). Second, the time trajectory of the formant pattern is relatively stationary (Nábělek and Letowski 1988; Nábělek and Dagenais 1986). While reverberation has only a minor effect on steady-state speech segments and monophthongal vowels, diphthongs are affected more dramatically (as illustrated in Fig. 5.8). Nábelek et al. (1994) noted that reverberation often results in confusions among diphthongs such as [ai] and [au]. Frequently, diphthongs are identified as monophthongs whose onset formant pattern is similar to the original diphthong (e.g., [ai] and [a]). Nábelek et al. proposed that the spectral changes occurring over the final portion of the diphthong are obscured in reverberant conditions by a temporal-smearing process they refer to as "reverberant self-masking." Errors can also result from "reverberant overlap-masking," which occurs when the energy originating from a preceding segment overlaps a following segment. This form of distortion often leads to errors in judging the identity of a syllable-final consonant preceded by a relatively intense vowel, but rarely causes errors in vowel identification per se (Nábelek et al. 1989).

Reverberation tends to "smear" and prolong spectral-change cues, such as formant transitions, smooth out the waveform envelope, and increase the prominence of low-frequency energy capable of masking higher frequencies. Stop consonants are more susceptible to distortion than other consonants, particularly in syllable-final position (Nábelek and Pickett 1974; Gelfand and Silman 1979). When reverberation is combined with background noise, final consonants are misidentified more frequently than initial consonants. Stop consonants, in particular, may undergo "filling in" of the silent gap during stop closure (Helfer 1994). Reverberation tends to obscure cues that specify rate of spectral change (Nábelek 1988), and hence can create ambiguity between stop consonants and semivowels (Liberman et al. 1956). Reverberation results in "perseveration" of formant transitions, and formant transitions tend to be dominated by their onset frequencies. Relational cues, such as the frequency slope of the second formant from syllable onset to vocalic midpoint (Sussman et al. 1991), may be distorted by reverberation, and this distortion may contribute to place-of-articulation errors.

When listening in the free field, reverberation diminishes the interaural coherence of speech because of echoes reaching the listener from directions other than the direct path. Reverberation also reduces the interaural coherence of sound sources and tends to randomize the pattern of ILDs and ITDs. The advantage of binaural listening under noisy conditions is reduced, but not eliminated in reverberant environments (Moncur and Dirks 1967; Nábelek and Pickett 1974). Plomp (1976) asked listeners to adjust the intensity of a passage of read speech until it was just intelligible in the presence of a second passage from a competing speaker. Compared to the case where both speakers were located directly in front of the listener, spatial separation of the two sources produced a 6-dB advantage in SNR. This advantage dropped to about 1 dB in a room with a reverberation time of 2.3 seconds. The echo suppression process responsible for this binaural advantage is referred to as binaural squelching of reverberation and is particularly pronounced at low frequencies (Bronkhorst and Plomp 1988).

The deterioration of spatial cues in reverberant environments may be one reason why listeners do not use across-frequency grouping by common ITD to separate sounds located at different positions in the azimuthal plane. Culling and Summerfield (1995a) found no evidence that listeners could exploit the pattern of ITDs across frequency for the purpose of grouping vocalic formants that share the same ITD as a means of segregating them from formants with different ITDs. Their results were corroborated by experiments showing that listeners were unable to segregate a harmonic from a vowel when the remaining harmonics were assigned a different ITD (Darwin and Hukin 1997). Some segregation was obtained when ITDs were combined with other cues (e.g., differences in f_0 and onset asynchrony), but the results suggest that ITDs exert their influence by drawing attention to sounds that occupy a specific location in space, rather than by grouping frequency components that share a common pattern of ITD (Darwin 1997; Darwin and Hukin 1998).

Binaural processes that minimize the effects of reverberation are supplemented by monaural processes to offset the deleterious effects of reverberation (Watkins 1988, 1991; Darwin 1990). In natural environments high frequencies are often attenuated by obstructions, and the longer wavelengths of low-frequency signals allow this portion of the spectrum to effectively bend around corners. Darwin et al. (1989) examined the effects of imposing different spectral slopes on speech signals to simulate such effects of reverberant transmission channels. A continuum of vowels was synthesized, ranging from [ɪ] to [ɛ] within the context of various [b__t] words, and the vowels filtered in such a manner as to impose progressively steeper slopes in the region of the first formant. When the filtered signals were presented in isolation, the phonemic boundary between the vocalic categories shifted in accordance with the apparent shift in the F_1 peak. However, when the filtered speech was presented after a short carrier phrase filtered in comparable fashion, the magnitude of the boundary shift was reduced. This result is consistent with the idea that listeners perceptually compensate for spectral tilt. However, this compensation may occur only under extreme conditions, since it was present only with extreme filtering (30-dB change in spectral slope) and did not completely eliminate the effects of filtering.

Watkins (1991) used an LPC vocoder to determine the ability of listeners to compensate for distortions of the spectrum envelope. He synthesized a set of syllables along a perceptual continuum ranging from [Ič] ("itch") to [ɛč] ("etch") by varying the F_1 frequency of the vowel and processing each segment with a filter whose transfer function specified the *difference* between the spectral envelopes of the two end-point vowels (i.e., [ɪ] minus [ɛ], as well as its inverse). The two filtering operations resulted in shifts of the phonemic boundary associated with F_1 toward a higher apparent formant peak when the first form of subtractive filter was used, and toward a lower apparent peak for the second type of filter. However, when the signals were embedded in a short carrier phrase processed in a comparable manner, the magnitude of the shift was reduced, suggesting that listeners are capable of compensating for the effects of filtering if given sufficiently long material with which to adapt. The shifts were not entirely eliminated by presenting the carrier phrase and test signals to the opposing ears or by using different apparent localization directions (by varying the ITD). Subsequent studies (Watkins and Makin 1994, 1996) showed that perceptual compensation was based on the characteristics of the following, as well as those of the preceding, signals. The results indicate that perceptual compensation does not reflect peripheral adaptation directly, but is based on some form of central auditory process.

When harmonically rich signals, such as vowels and other voiced segments, are presented in reverberation, echoes can alter the sound pressure level of individual harmonics and scramble the original phase pattern, but the magnitude of these changes is generally small relative to naturally occurring differences among vocalic segments (Plomp 1983). However, when the f_0 is nonstationary, the echoes originating from earlier time points overlap with later portions of the waveform. This process serves to diffuse cues relating to harmonicity, and could therefore reduce the effectiveness of f_0 differences to segregate competing voices. Culling et al. (1994) confirmed this supposition by simulating the passage of speech from a speaker to the ears of a listener in a reverberant room. They measured the benefit afforded by f_0 differences under reverberant conditions sufficient to counteract the effects of spatial separation (produced by a 60-degree difference in azimuth). They showed that this degree of reverberation reduces the ability of listeners to use f_0 differences in segregating pairs of concurrent vowels under conditions where f_0 is changing, but not in the condition where both masker and target had stationary f_0s. When f_0 is modulated by an amount equal to or greater than the difference in f_0 between target and masker, the benefits of using a difference in f_0 are no longer present.

The effects of reverberation on speech intelligibility are complex and not well described by a spectral-based approach such as the AI. This is illustrated in Figure 5.8, which shows that reverberation radically changes the appearance of the speech spectrogram and eliminates or distorts many traditional speech cues such as formant transitions, bursts, and silent intervals. Reverberation thus provides an illustration of perceptual constancy in speech perception. Perceptual compensation for such distortions is based on a number of different monaural and binaural "dereverberation" processes acting in concert. Some of these processes operate on a local (syllable-internal) basis (e.g., Nábelek et al. 1989), while others require prior exposure to longer stretches of speech (e.g., Watkins 1988; Darwin et al. 1989).

A more quantitatively predictive means of studying the impact of reverberation is afforded by the TMTF, and an accurate index of the effects of reverberation in intelligibility is provided by the STI (Steeneken and Houtgast 1980; Humes et al. 1987). Such effects can be modeled as a low-pass filtering of the modulation spectrum. Although the STI is a good predictor of overall intelligibility, it does not attempt to model processes underlying perceptual compensation. In effect, the STI transforms the effects of reverberation into an equivalent change in SNR. However, several properties of speech intelligibility are not well described by this approach. First, investigators have noted that the pattern of confusion errors is not the same for noise and reverberation. The combined effect of reverberation and noise is more harmful than noise alone (Nábelek et al. 1989; Takata and Nábelek 1990; Helfer 1994). Second, some studies suggest there may be large individual subject differences in susceptibility to the effects of reverberation (Nábelek and Letowski 1985; Helfer 1994). Third, children are affected more by reverberation than adults, and such differences are observed up to age 13, suggesting that acquired perceptual strategies contribute to the ability of compensating for reverberation (Finitzo-Hieber and Tillman 1978; Nábelek and Robinson 1982; Neuman and Hochberg 1983). Fourth, elderly listeners, with normal sensitivity, are more adversely affected by reverberation than younger listeners, suggesting that aging may lead to a diminished ability to compensate for reverberation (Gordon-Salant and Fitzgibbons 1995; Helfer 1992).

3.8 Frequency Response of the Communication Channel

In speech perception, the vocal-tract resonances provide phonetic and lexical information, as well as information about the source, such as personal identity, gender, age, and dialect of the speaker (Kreiman 1997). However, under everyday listening conditions the spectral envelope is frequently distorted by properties of the transmission channel. Indoors, sound waves are reflected and scattered by various surfaces (e.g., furniture and people), while room resonances and antiresonances introduce peaks and valleys into the spectrum. Outdoor listening environments contain potential obstructions such as buildings and trees, and exhibit atmospheric distortions due to wind and water vapor. For this reason damping is not uniform as a function of frequency. In general, high-frequency components tend to be absorbed more rapidly than their low-frequency counterparts. As a result of the need to communicate efficiently in all of these conditions, listeners compensate for a variety of distortions of the communication channel rapidly and without conscious effort.

3.8.1 Low-Pass and High-Pass Filtering

Fletcher (1953) studied the effects of low-pass (LP) and high-pass (HP) filtering on the intelligibility of nonsense syllables. His objective was to measure the independent contribution of the low- and high-frequency channels. Eliminating the high frequencies reduced the articulation scores of consonants more than vowels, while eliminating the low-frequency portion of the spectrum had the opposite effect. Fletcher noted that the articulation scores for both the LP and HP conditions did not actually sum to the full-band score. He developed a model, the AI (Fletcher and Galt 1950), as a means of transforming the partial articulation scores (Allen 1994) into an additive form (cf. section 2.1). Accurate predictions of phone and syllable articulation were obtained using a model that assumed that (1) spectral information is processed independently in each frequency band and (2) is combined in an "optimal" way to derive recognition probabilities. As discussed in section 2.1, the AI generates accurate and reliable estimates of the intelligibility of filtered speech based on the proportion of energy within the band exceeding the threshold of audibility and the width of the band. One implication of the model is that speech "features" (e.g., formant peaks) are extracted from each frequency band independently, a strategy that may contribute to noise robustness (Allen 1994).

Figure 5.9 illustrates the effects of LP and HP filtering on speech intelligibility. Identification of monosyllabic nonsense words remains high when LP-filtered at a cutoff frequency of 3 kHz or greater, or HP-filtered at a cutoff frequency of 1 kHz or lower. For a filter cutoff around 2 kHz, the effects of LP and HP filtering are similar, resulting in intelligibility of around 68% (for nonsense syllables).

When two voices are presented concurrently it is possible to improve the SNR by restricting the bandwidth of one of the voices. Egan et al. (1954) found that HP-filtering either voice with a cutoff frequency of 500 Hz led to improved articulation scores. Spieth and Webster (1955) confirmed that

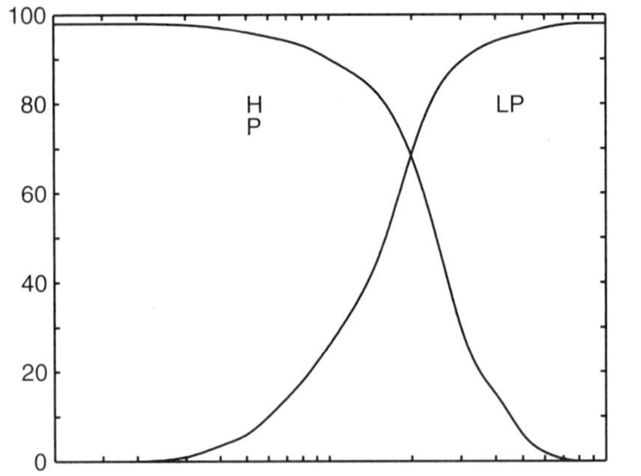

FIGURE 5.9. Effects of high-pass and low-pass filtering on the identification of monosyllabic nonsense words. (After French and Steinberg 1947.)

differential filtering led to improved scores whenever one of the two voices was filtered, regardless of whether such filtering was imposed on the target or interfering voice. Intelligibility was higher when one voice was LP-filtered and the other HP-filtered, compared to the case where both voices were unfiltered. The effectiveness of the filtering did not depend substantially on the filter-cutoff frequency (565, 800, or 1130Hz for the HP filter, and 800, 1130, and 1600Hz for the LP filter). Egan et al. (1954) found that intensity differences among the voices could be beneficial. Slight attenuation of the target voice provided a small benefit, offset, in part, by the increased amount of masking exerted by the competing voice. Presumably, such benefits of attenuation are a consequence of perceptual grouping processes sensitive to common amplitude modulation. Webster (1983) suggested that any change in the signal that gives one of the voices a "distinctive sound" could lead to improved intelligibility.

3.8.2 Bandpass and Bandstop Filtering

Several studies have examined the effects of narrow, bandpass (ca. one-third octave) filtering on the identification of vowels (Lehiste and Peterson 1959; Carterette and Møller 1962; Castle 1964). Two conclusions emerge from these studies. First, vowel identification is substantially reduced, but remains above chance, when the signals are subjected to such bandpass filtering. Second, the pattern of errors is not uniform but varies as a function of the intended vowel category—a conclusion not in accord with template theories of vowel perception.[7] For example, when the filter is centered near the first formant, a front vowel may be confused for a back vowel with similar F_1 (e.g., American English [e] is heard as [o]), consistent with the observation that back vowels (e.g., [o]) can be approximated using only a single formant, while front vowels (e.g., [e]) cannot (Delattre et al. 1952).

The studies of LP and HP filtering, reviewed in section 3.8.1, indicate that speech intelligibility is not substantially reduced by removing that portion of the spectrum below 1kHz or above 3kHz. In addition, speech that is band limited between 0.3 and 3.4kHz (i.e., telephone bandwidth) is only marginally less intelligible than full-spectrum speech. These findings suggest that frequencies between 0.3 and 3.4kHz provide the bulk of the information in speech. However, several studies have shown that speech can withstand disruption of the midfrequency region without substantial loss of intelligibility. Lippmann (1996b) filtered CVC nonsense syllables to remove the frequency band between 0.8 and 3kHz and found that speech intelligi-

[7] Note that this applies equally to "whole-spectrum" and feature-based models that classify vowels on the basis of template matching using the frequencies of the two or three lowest formants.

bility was not substantially reduced (better than 90% correct consonant identification from a 16-item set). Warren et al. (1995) reported high intelligibility for everyday English sentences that had been filtered using narrow bandpass filters, a condition they described as "listening through narrow spectral slits." With one-third-octave filter bandwidths, about 95% of the words could be understood in sentences filtered at center frequencies of 1100, 1500, and 2100Hz. Even when the bandwidth was reduced to 1/20th of an octave, intelligibility was about 77% for the 1500-Hz band.

The high intelligibility of spectrally limited sentences can be attributed, in part, to the ability of listeners to exploit the linguistic redundancy in everyday English sentences. Stickney and Assmann (2001) replicated Warren et al.'s findings using gammatone filters (Patterson et al. 1992) with bandwidths chosen to match psychophysical estimates of auditory filter bandwidth (Moore and Glasberg 1987). Listeners identified the final keywords in high-predictability sentences from the Speech Perception in Noise (SPIN) test (Kalikow et al. 1977) at rates similar to those reported by Warren et al. (between 82% and 98% correct for bands centered at 1500, 2100, and 3000Hz). However, performance dropped by about 20% when low-predictability sentences were used, and by a further 23% when the filtered, final keywords were presented in isolation. These findings highlight the importance of linguistic redundancy (provided both within each sentence, and in the context of the experiment where reliable expectations about the prosody, syntactic form, and semantic content of the sentences are established). Context helps to sustain a high level of intelligibility even when the acoustic evidence for individual speech sounds is extremely sparse.

4. Perceptual Strategies for Retrieving Information from Distorted Speech

The foregoing examples demonstrate that speech communication is a remarkably robust process. Its resistance to distortion can be attributed to many factors. Section 2 described acoustic properties of speech that contribute to its robustness and discussed several strategies used by speakers to improve intelligibility under adverse listening conditions. Section 3 reviewed the spectral and temporal effects of distortions that arise naturally in everyday environments and discussed their perceptual consequences. The overall conclusion is that the information in speech is shielded from distortion in several ways. First, peaks in the envelope of the spectrum provide robust cues for the identification of vowels and consonants even when the spectral valleys are obscured by noise. Second, periodicity in the waveform reflects the fundamental frequency of voicing, allowing listeners to group together components that stem from the same voice across frequency and time in order to segregate them from competing signals (Brokx and Nooteboom 1982; Bird and Darwin 1998). Third, at disadvantageous SNRs, the formants of voiced sounds can exert their influence by disrupting the periodicity of competing harmonic signals or by disrupting the interaural correlation of a masking noise (Summerfield and Culling 1992; Culling and Summerfield 1995a). Fourth, the amplitude modulation pattern across frequency bands can serve to highlight informative portions of the speech signal, such as prosodically stressed syllables. These temporal modulation patterns are redundantly specified in time and frequency, making it possible to remove large amounts of the signal via gating in the time domain (e.g., Miller and Licklider 1950) or filtering in the frequency domain (e.g., Warren et al. 1995). Even when the spectral details and periodicity of voiced speech are eliminated, intelligibility remains high if the temporal modulation structure is preserved in a small number of bands (Shannon et al. 1995). However, speech processed in this manner is more susceptible to interference by other signals.

In this section we consider the perceptual and cognitive strategies used by listeners to facilitate the extraction of information from speech signals corrupted by noise and other distortions of the communication channel. Background noise and distortion generally lead to a reduction in SNR, as portions of the signal are rendered inaudible or are masked by other signals. Masking, the inability to resolve auditory events closely spaced in time and frequency, is a consequence of the fact that the auditory system has limited frequency selectivity and temporal resolution (Moore 1995). The processes described below can be thought of as strategies used by listeners to overcome these limitations.

In sections 4.1 and 4.2 we consider the role of two complementary strategies for recovering information from distorted speech: glimpsing and tracking. Glimpsing exploits moment-to-moment fluctuations in SNR to focus auditory attention on temporal regions of the composite signal where the target voice is best defined. Tracking processes exploit moment-to-moment correlations in fundamental frequency, amplitude envelope, and formant pattern to group together components of the signal originating from the same voice.

Glimpsing and tracking are low-level perceptual processes that require an ongoing analysis of the signal within a brief temporal window, and both can be regarded as *sequential* processes. Perceptual grouping also involves *simultaneous* processes (Bregman 1990), as when a target voice is separated from background signals on the basis of either a static difference in fundamental frequency (Scheffers 1983), or differences in interaural timing and level (Summerfield and Culling 1995).

In the final subsections of the chapter we consider additional processes (both auditory and linguistic) that help listeners to compensate for distortions of the communication channel. In section 4.3 we examine the role of perceptual grouping and adaptation in the enhancement of signal onsets. In section 4.4 we review evidence for the existence of central processes that compensate for deformations in the frequency responses of communication channels, and we consider their time course. Finally, in section 4.5, we briefly consider how linguistic and pragmatic context helps to resolve the ambiguities created by gaps and discontinuities in the signal, and thereby contributes to the intelligibility of speech under adverse acoustic conditions.

4.1 Glimpsing

In vision, glimpsing occurs when an observer perceives an object based on fragmentary evidence (i.e., when the object is partly obscured from view). It is most effective when the object is highly familiar (e.g., the face of a friend) and when it serves as the focus of attention. Visual objects can be glimpsed from a static scene (e.g., a two-dimensional image). Likewise, auditory glimpsing involves taking a brief "snapshot" from an ongoing temporal sequence. It is the process by which distinct regions of the signal, separated in time, are linked together when intermediate regions are masked or deleted. Empirical evidence for the use of a glimpsing strategy comes from a variety of studies in psychoacoustics and speech perception. The following discussion offers some examples and then considers the mechanism that underlies glimpsing in speech perception.

In comodulation masking release, the masked threshold of a tone is lower in the presence of an amplitude-modulated masker (with correlated amplitude envelopes across different and widely separated auditory channels) compared to the case where the modulation envelopes at different frequencies are uncorrelated (Hall et al. 1984). Buus (1985) proposed a model of CMR that implements the strategy of "listening in the valleys" created by the masker envelope. The optimum time to listen for the signal is when the envelope modulations reach a minimum. Consistent with this model is the finding that CMR is found only during periods of low masker energy, that is, in the valleys where the SNR is highest (Hall and Grose 1991).

Glimpsing has been proposed as an explanation for the finding that modulated maskers produce less masking of connected speech than unmodulated maskers. Section 3.5 reviewed studies showing that listeners with normal hearing can take advantage of the silent gaps and amplitude minima in a masking voice to improve their identification of words spoken by a target voice. The amplitude modulation pattern associated with the alternation of syllable peaks in a competing sentence occur at rates between 4 and 8 Hz (see section 2.5). During amplitude minima of the masker, entire syllables or words of the target voice can be glimpsed.

Additional evidence for glimpsing comes from studies of the identification of concurrent vowel pairs. When two vowels are presented concurrently, they are identified more accurately if they differ in f_0 (Scheffers 1983). When the difference in f_0 is small (less than one semitone, 6%), corresponding low-frequency harmonics from the two vowels occupy the same auditory filter and beat together, alternately attenuating and then reinforcing one another. As a result, there can be segments of the signal where the harmonics defining the F_1 of one vowel are of high amplitude and hence are well defined, while those of the competing vowel are poorly defined. The variation in identification accuracy as a function of segment duration suggests that listeners can select these moments to identify the vowels (Culling and Darwin 1993a, 1994; Assmann and Summerfield 1994).

Supporting evidence for glimpsing comes from a model proposed by Culling and Darwin (1994). They applied a sliding temporal window across the vowel pair, and assessed the strength of the evidence favoring each of the permitted response alternatives for each position of the window. Because the window isolated those brief segments where beating resulted in a particularly favorable representation of the two F_1s, strong evidence favoring the vowels with those F_1s was obtained. In effect, their model was a computational implementation of glimpsing. Subsequently, their model was extended to account for the improvement in identification of a target vowel when the competing vowel is preceded or followed by formant transitions (Assmann 1995, 1996). These empirical studies and modeling results suggest that glimpsing may account for several aspects of concurrent vowel perception.

The ability to benefit from glimpsing depends on two separate processes. First, the auditory system must perform an analysis of the signal with a sliding time window to search for regions where the property of the signal being sought is most evident. Second, the listener must have some basis for distinguishing target from masker. In the case of speech, this requires some prior knowledge of the structure of the signal and the masker (e.g., knowledge that the target voice is female and the masker voice is male). Further research is required to clarify whether glimpsing is the consequence of a unitary mechanism or a set of loosely related strategies. For example, the time intervals available for glimpsing are considerably smaller for the identification of concurrent vowel pairs (on the order of tens of milliseconds) compared to pairs of sentences, where variation in SNR provides intervals of 100 ms or longer during which glimpsing could provide benefits.

4.2 Tracking

Bregman (1990) proposed that the perception of speech includes an early stage of auditory scene analysis in which the components of a sound mixture are grouped together according to their sources. He suggested that listeners make use of gestalt grouping principles such as proximity, good continuation, and common fate to link together the components of signals and segregate them from other signals. Simultaneous grouping processes make use of co-occuring properties of signals, such as the frequency spacing of harmonics and the shape of the spectrum envelope, in order to group together components of sound that emanate from the same source. Sequential grouping is used to forge links over time with the aid of tracking processes. Tracking exploits correlations in signal characteristics across time and frequency to group together acoustic components originating from the same larynx and vocal tract.

Two properties of speech provide a potential basis for tracking. First, changes in the rate of vocal-fold vibration during voiced speech tend to be graded, giving rise to finely granulated variations in pitch. Voiced signals have a rich harmonic structure, and hence changes in f_0 generate a pattern of correlated changes across the frequency spectrum. Second, the shape of the vocal tract tends to change slowly and continuously during connected speech, causing the trajectories of formant peaks to vary smoothly in time and frequency. When the trajectories of the formants and f_0 are partially obscured by background noise and other forms of distortion, the perceptual system is capable of recovering information from the distorted segments by a process of tracking (or trajectory extrapolation).

4.2.1 Fundamental Frequency Tracking

Despite the intuitive appeal of the idea that listeners track a voice through background noise, the empirical support for such a tracking mechanism, sensitive to f_0 modulation, is weak (Darwin and Carlyon 1995). Modulation of f_0 in a target vowel can increase its prominence relative to a steady-state masker vowel (McAdams 1989; Marin and McAdams 1991; Summerfield and Culling 1992; Culling and Summerfield 1995b). However, there is little evidence that listeners can detect the coherent (across-frequency) changes produced by f_0 modulation (Carlyon 1994). Gardner et al. (1989) were able to induce alternative perceptual groupings of subsets of formants by synthesizing them with different stationary f_0s, but not with different patterns of f_0 modulation. Culling and Summerfield (1995b) found that coherent f_0 modulation improved the identification of a target vowel presented in a background of an unmodulated masker vowel. However, the improvement occurred both for coherent (sine phase) and incoherent (random phase) sinusoidal modulation of the target. Overall, these results suggest that f_0 modulation can affect the perceptual prominence of a vowel but does not provide any benefit for sound segregation. In continuous speech, the benefits of f_0 modulation may have more to do with momentary differences in instantaneous f_0 between two voices (providing opportunities for simultaneous grouping processes and glimpsing) than with correlated changes in different frequency regions. One reason why f_0 modulation does not help may be that the harmonicity in voiced speech obviates the need for a computationally expensive operation of tracking changes in the frequencies of harmonics (Summerfield 1992). A further reason is that in enclosed environments, reverberation tends to blur the pattern of modulation created by changes in the frequencies of the harmonics, making f_0 modulation an unreliable source of information (Gardner et al. 1989; Culling et al. 1994).

4.2.2 Formant Tracking

Bregman (1990) suggested that listeners might exploit the trajectories of formant peaks to track the components of a voice through background noise. In section 2.1 it was suggested that peaks in the spectrum envelope provide robust cues because they are relatively impervious to the effects of background noise, as well as to modest changes in the frequency response of communication channels and deterioration in the frequency selectivity of the listener. A complicating factor is that the trajectories of different formants are often uncorrelated (Bregman 1990). For example, during the transition from the consonant to the vowel in the syllable [da], the frequency of the first formant increases while the second formant decreases. Moreover, in voiced speech the individual harmonics also generate peaks in the fine structure of the spectrum. Changes in the formant patterns are independent of changes in the frequencies of harmonics, and thus listeners need to distinguish among different types of spectral peaks in order to track formants over time. The process is further complicated by the limited frequency selectivity in hearing [i.e., the low-order harmonics of vowels and other voiced signals are individually resolved, while the higher harmonics are not (Moore and Glasberg 1987)].

Despite the intuitive plausibility of the idea that listeners track formants through background noise, there is little direct evidence to support its role in perceptual grouping. Assmann (1995) presented pairs of concurrent vowels in which one member of the pair had initial or final flanking formant transitions that specified a [w], [j], or [l] consonant. He found that the addition of formant transitions helped listeners identify the competing vowel, but did not help identify the vowel to which they were linked. The results are not consistent with a formant-tracking process, but instead support an alternative hypothesis: formant transitions provide a time window over which the formant pattern of a competing vowel can be glimpsed.

Indirect support for perceptual extrapolation of frequency trajectories comes from studies of frequency-modulated tones that lie on a smooth temporal trajectory. When a frequency-modulated sinusoid is interrupted by noise or a silent gap, listeners hear a continuous gliding pitch (Ciocca and Bregman 1987; Kluender and Jenison 1992). This illusion of continuity is also obtained when continuous speech is interrupted by brief silent gaps or noise segments (Warren et al. 1972). In natural environments speech is often interrupted by extraneous impulsive noise, such as slamming doors, barking dogs, and traffic noise, that masks portions of the speech signal. Warren et al. describe a perceptual compensatory mechanism that appears to "fill in," or restore, the masked portions of the original signal. This process is called auditory induction and is thought to occur at an unconscious level since listeners are unaware that the perceptually restored sound is actually missing.

Evidence for auditory induction comes from a number of studies that have examined the effect of speech interruptions (Verschuure and Brocaar 1983; Bashford et al. 1992; Warren et al. 1997). These studies show that the intelligibility of interrupted speech is higher when the temporal gaps are filled with broadband noise. Adding noise provides benefits for conditions with high-predictability sentences, as well as for low-predictability sentences, but not with isolated nonsense syllables (Miller and Licklider 1950; Bashford et al. 1992; Warren et al. 1997). Warren and colleagues (Warren 1996; Warren et al. 1997) attributed these benefits of noise to a "spectral restoration" process that allows the listener to "bridge" noisy or degraded portions of the speech signal. Spectral restoration is an unconscious and automatic process that takes advantage of the redundancy of speech to minimize the interfering effects of extraneous signals. It is likely that spectral restoration involves the evocation of familiar or overlearned patterns from long-term memory (or schemas; Bregman 1990) rather than the operation of tracking processes or trajectory extrapolation.

4.3 Role of Adaptation and Grouping in Enhancing Onsets

A great deal of information is conveyed in temporal regions where the speech spectrum is changing rapidly (Stevens 1980). The auditory system is particularly responsive to such changes, especially when they occur at the onsets of signals (Palmer 1995; see Palmer and Shamma, Chapter 4). For example, auditory-nerve fibers show increased rates of firing at the onsets of syllables and during transient events such as stop consonant bursts (Delgutte 1996). Such "adaptation" is associated with a decline in discharge rate observed over a period of prolonged stimulation and is believed to arise because of the depletion of neurotransmitter in the synaptic junction between inner hair cells and the auditory nerve (Smith 1979). The result is a sharp increase in firing rate at the onset of each pitch pulse, syllable or word, followed by a gradual decline to steady-state levels. It has been suggested that adaptation plays an important role in enhancing the spectral contrast between successive signals, and increases the salience of a stimulus immediately following its onset (Delgutte and Kiang 1984; Delgutte 1996).

Adaptation has also been suggested as an explanation for the phenomenon of psychophysical enhancement. *Enhancement* is the term used to describe the increase in perceived salience of a frequency component omitted from a broadband sound when it is subsequently reintroduced (Viemeister 1980; Viemeister and Bacon 1982). Its relevance for speech was demonstrated by Summerfield et al. (1984, 1987), who presented a sound whose spectral envelope was the "complement" of a vowel (i.e., formant peaks were replaced by valleys and vice versa) followed by a tone complex with a flat amplitude spectrum. The flat-spectrum sound was perceived as having a timbral quality similar to the vowel whose complement had preceded it. Summerfield and Assmann (1989) showed that the identification of a target vowel in the presence of a competing masker vowel was substantially improved if the vowel pair was preceded by a precursor with the same spectral envelope as the masker. By providing prior exposure to the spectral peaks of the masker vowel, the precursor served to enhance the spectral peaks in the target vowel. These demonstrations collectively illustrate the operation of an auditory mechanism that enhances the prominence of spectral components subjected to sudden changes in amplitude. It may also play an important role in compensating for distortions of the communication channel by emphasizing frequency regions containing newly arriving energy relative to background components (Summerfield et al. 1984, 1987). Enhancement is thus potentially attributable to the reduction in discharge rate in auditory-nerve fibers whose characteristic frequencies (CFs) are close to spectral peaks of the precursor. Less adaptation will appear in fibers whose CFs occur in the spectral valleys. Hence, newly arriving sounds generate higher discharge rates when their spectral components stimulate unadapted fibers (tuned to the spectral valleys of the precursor) than when they stimulate adapted fibers (tuned to the spectral peaks). In this way the neural response to newly arriving signals could be greater than the response to preexisting components.

An alternative explanation for enhancement assumes that this perceptual phenomenon is the result of central grouping processes that link auditory channels with similar amplitude envelopes (Darwin 1984; Carlyon 1989). According to this grouping account, the central auditory system selectively enhances the neural response in channels that display abrupt increases in level. Central grouping processes have been invoked to overcome several problems faced by the peripheral adaptation account (or a related account based on the adaptation of suppression; Viemeister and Bacon 1982). First, under some circumstances, enhancement has been found to persist for as long as 20 seconds, a longer time period than the recovery time constants for adaptation of nerve fibers in the auditory periphery (Viemeister 1980). Second, while adaptation is expected to be strongly level-dependent, Carlyon (1989) demonstrated a form of enhancement whose magnitude was *not* dependent on the level of the enhancing stimulus (but cf. Hicks and Bacon 1992). Finally, in physiological recordings from peripheral auditory-nerve fibers, Palmer et al. (1995) found no evidence for an increased gain in the neural responses of fibers tuned to stimulus components that evoke enhancement. The conclusion is that peripheral adaptation contributes to the enhancement effect, but does not provide a complete explanation for the observed effects. This raises the question, What is grouping and how does it relate to peripheral adaptation? A possible answer is that adaptation in peripheral analysis highlights frequency channels in which abrupt increments in spectral level have occurred

(Palmer et al. 1995). Central grouping processes must then establish whether these increments have occurred concurrently in different frequency channels. If so, the "internal gain" in those channels is elevated relative to other channels.

4.4 Compensation for Communication Channel Distortions

Nonuniformities in the frequency response of a communication channel can distort the properties of the spectrum envelope of speech, yet unintelligibility is relatively unaffected by manipulations of spectral tilt (Klatt 1989) or by the introduction of a broad peak into the frequency response of a hearing aid (Buuren et al. 1996). A form of perceptual compensation for spectral-envelope distortion was demonstrated by Watkins and colleagues (Watkins 1991; Watkins and Makin 1994, 1996), who found that listeners compensate for complex changes in the frequency response of a communication channel when identifying a target word embedded in a brief carrier sentence. They synthesized a continuum of sounds whose end points defined one of two test words. They showed that the phoneme boundary shifted when the test words followed a short carrier phrase that was filtered using the inverse of the spectral envelope of the vowel to simulate a transmission channel with a complex frequency response. The shift in perceived quality was interpreted as a form of perceptual compensation for the distortion introduced by the filter. Watkins and Makin showed that the effect persists, in reduced form, in conditions where the carrier phrase follows the test sound, when the carrier is presented to the opposite ear, and when a different pattern of interaural timing is applied. For these reasons they attributed the perceptual shifts to a central (as opposed to peripheral) auditory (rather than speech-specific) process that compensates for distortions in the frequency responses of communication channels.

The effects described by Watkins and colleagues operate within a very brief time window, one or two syllables at most. There are indications of more gradual forms of compensation for changes in the communication channel. Perceptual acclimatization is a term often used to describe the long-term process of adjustment to a hearing aid (Horwitz and Turner 1997). Evidence for perceptual acclimatization comes from informal observations of hearing-aid users who report that the benefits of amplification are greater after a period of adjustment, which can last up to several weeks in duration. Gatehouse (1992, 1993) found that some listeners fitted with a single hearing aid understand speech more effectively with their aided ear at high presentation levels, but perform better with their unaided ear at low sound pressure levels. He proposed that each ear performs best when receiving a pattern of stimulation most like the one it commonly receives. The internal representation of the spectrum is assumed to change in a frequency-dependent way to adapt to alterations of the stimulation pattern. Such changes have been observed to take place over periods as long as 6 to 12 weeks. In elderly listeners, this may involve a process of relearning the phonetic interpretation of (high-frequency) speech cues that were previously inaudible. Reviews of the contribution of perceptual acclimatization have concluded, however, that the average increase in hearing-aid benefit over time is small at best (Turner and Bentler 1998); the generality of this phenomenon bears further study.

Sensorineural hearing loss is often associated with elevated thresholds in the high-frequency region. It has been suggested (Moore 1995) that there may be a remapping of acoustic cues in speech perception by hearing-impaired listeners, with greater perceptual weight placed on the lower frequencies and on the temporal structure of speech. An extreme form of this remapping is seen with cochlear-implant users, for whom the spectral fine structure and tonotopic organization of speech is greatly reduced (Rosen 1992; see Clark, Chapter 8). For such listeners, temporal cues may play an enhanced role. Most cochlear implant users show a gradual process of adjustment to the device, accompanied by improved speech recognition performance. This suggests that acclimatization processes may shift the perceptual weight assigned to different aspects of the temporal structure of speech preserved by the implant.

Shannon et al. (1995) showed that listeners with normal hearing could achieve a high degree of success in understanding speech that retained only the temporal information in four broad frequency channels and lacked both voicing information and spectral fine structure (see section 2.5). Rosen et al. (1998) used a similar processor to explore the effects of shifting the bands so that each temporal envelope stimulated a frequency band between 1.3 and 2.9 octaves higher in frequency than the one from which it was originally obtained. Similar shifts may be experienced by multichannel cochlear implants when the apical edge of the electrode reaches only part of the way down the cochlea. Consistent with other studies (Dorman et al. 1997; Shannon et al. 1998), Rosen et al. found a sharp decline in intelligibility of frequency-shifted speech presented to listeners with normal hearing. However, over the course of a 3-hour training period, performance improved significantly, indicating that some form of perceptual reorganization had taken place. Their findings suggest that (1) a coarse temporal representation may, under some circumstances, provide sufficient cues for understanding speech with little or no need for training; and (2) a period of perceptual adjustment may be needed when the bands are shifted from their expected locations along the tonotopic array.

4.5 Use of Linguistic Context

The successful recovery of information from distorted speech depends on properties of the signal. Nonuniformities in the distribution of energy across time and frequency enable listeners to glimpse the target voice, while regularities in time and frequency allow for the operation of perceptual grouping principles. Intelligibility is also determined by the ability of the listener to exploit various aspects of linguistic and pragmatic context, especially when the signal is degraded (Treisman 1960, 1964; Warren 1996). For example, word recognition performance in background noise is strongly affected by such factors as the size of the response set (Miller et al. 1951), lexical status, familiarity of the stimulus materials and word frequency (Howes 1957; Pollack et al. 1959; Auer and Bernstein 1997), and lexical neighborhood similarity (Luce et al. 1990; Luce and Pisoni 1998).

Miller (1947) reported that conversational babble in an unfamiliar language was neither more nor less interfering than babble in the native language of the listeners (English). He concluded that the spectrum of a masking signal is the crucial factor, while the linguistic content is of secondary importance. A different conclusion was reached by Treisman (1964), who used a shadowing task to show that the linguistic content of an interfering message was an important determinant of its capacity to interfere with the processing of a target message. Most disruptive was a competing message in the same language and similar in content, followed by a foreign language familiar to the listeners, followed by reversed speech in the native language, followed by an unfamiliar foreign language. Differences in task demands (the use of speech babble or a single competing voice masker), the amount of training, as well as instructions to the subjects may underlie the difference between Triesman's and Miller's results. The importance of native-language experience was demonstrated by Gat and Keith (1978) and Mayo et al. (1997). They found that native English listeners could understand monosyllabic words or sentences of American English at lower SNRs than could nonnative students who spoke English as a second language. In addition, Mayo et al. found greater benefits of linguistic context for native speakers of English and for those who learned English as a second language before the age of 6, compared to bilinguals who learned English as a second language in adulthood. Other studies have confirmed that word recognition by nonnative listeners can be severely reduced in conditions where fine phonetic discrimination is required and background noise is present (Bradlow and Pisoni 1999).

When words are presented in sentences, the presence of semantic context restricts the range of plausible possibilities. This leads to higher intelligibility and greater resistance to distortion (Kalikow et al. 1977; Boothroyd and Nittrouer 1988; Elliot 1995). The SPIN test (Kalikow et al. 1977) provides a clinical measure of the ability of a listener to take advantage of context to identify the final keyword in sentences, which are either of low or high predictability.

Boothroyd and Nittrouer (1988) presented a model that assumes that the effects of context are equivalent to providing additional, statistically independent channels of sensory information. First, they showed that the probability of correct recognition of speech units (phones or words) in context (p_c) could be predicted from their identification without context (p_i) from the following relationship:

$$p_c = 1 - (1 - p_i)^k \tag{3}$$

The factor k is a constant that measures the use of contextual information. It is computed from the ratio of the logarithms of the error probabilities:

$$k = \frac{\log(1 - p_c)}{\log(1 - p_i)} \tag{4}$$

Boothroyd and Nittrouer extended this model to show that the recognition of complex speech units (e.g., words) could be predicted from the identification of their component parts (phones). Their model was based on earlier work by Fletcher (1953) showing that the probability of correct identification of individual consonants and vowels within CVC nonsense syllables could be accurately predicted by assuming that the recognition of the whole depends on prior recognition of the component parts, and that the probabilities of recognizing the parts are statistically independent. According to this model, the probability of recognizing the whole (p_w) depends on the probability of identifying the component parts (p_p):

$$p_w = p_p^j \tag{5}$$

where $1 \leqslant j \leqslant n$, and n is the number of parts. The factor j is computed from the ratio of the logarithms of the recognition probabilities:

$$j = \frac{\log(1 - p_w)}{\log(1 - p_p)} \tag{6}$$

The value of j ranges between 1 (in situations where context plays a large role) and n (where context has no effect on recognition). For nonsense syllables and nonmeaningful sentences, the value of j is assumed to be equal to n, the number of component parts.

Boothroyd and Nittrouer applied these models to predict context effects in CVC syllables and in sentences. They included high-predictability and low-predictability sentences differing in the degree of semantic context, as well as zero-predictability sentences in which the words were presented randomly so that neither semantic nor syntactic context was available. They found values of k ranging between 1.3 for CVCs and 2.7 for high-predictability sentences and values of j ranging from 2.5 in nonsense CVC syllables to 1.6 in four-word, high-predictability sentences. The derived j and k factors were constant across a range of probabilities, supporting the assumption that these measures provide good quantitative measures of the effects of linguistic context.

Another modeling approach was used by Rooij and Plomp (1991), who characterized the effects of linguistic context on sentence perception in terms of linguistic entropy, a measure derived from information theory (Shannon and Weaver 1949). The entropy, H, of an information source (in bits) represents the degree of uncertainty in receiving a given item from a vocabulary of potential elements, and is defined as

$$H = -\sum_{i=1}^{n} \log p_i \tag{7}$$

where p_i is the probability of selecting item i from a set of N independent items. The entropy increases as a function of the number of items in the set and is dependent on the relative probabilities of the individual items in the set. The degree of linguistic redundancy is inversely proportional to its entropy. Rooij and Plomp estimated the linguistic entropy of a set of sentences (originally chosen to be as similar as possible in overall redundancy) by means of a visual letter-guessing procedure proposed by Shannon (1951). They estimated the entropy in bits per character (for individual letters in sentences) from the probability of correct guesses made by subjects who were given successive fragments of each sentence, presented one letter at a time. After each guess the subject was told the identity of the current letter and all those that preceded it. Rooij and Plomp showed that estimates of the linguistic entropy of a set of sentences (presented auditorily) could predict the susceptibility of the sentences to masking by speech-shaped noise. Differences in linguistic entropy had an effect of about 4 dB on the SRT and followed a linear relationship for listeners with normal hearing. Despite the limitations of this approach (e.g., the assumption that individual letters are equi-probable, and the use of an orthographic measure of linguistic entropy, rather than one based on phonological, morphological, or lexical units), this study illustrates the importance of linguistic factors in accounting for speech perception abilities in noise. The model has been extended to predict speech recognition in noise for native and nonnative listeners (van Wijngaarden et al. 2002).

Listeners exploit their knowledge of linguistic constraints to restrict the potential interpretations that can be placed on the acoustic signal. The process involves the active generation of possible interpretations, combined with a method for filtering or restricting lexical candidates (Klatt 1989; Marslen-Wilson 1989). When speech is perceived under adverse conditions, the process of restricting the set of possible interpretations requires a measure of quality or "goodness of fit" between the candidate and its acoustical support. The process of evaluating and assessing the reliability of incoming acoustic properties depends both on the signal properties (including some measure of distortion) and the strength of the linguistic hypotheses that are evoked. If the acoustic evidence is weak, then the linguistic hypotheses play a stronger role. If the signal provides clear and unambiguous evidence for a given phonetic sequence, then linguistic plausibility makes little or no contribution (Warren et al. 1972). One challenge for future research is to describe how this attentional switching is achieved on-line by the central nervous system.

5. Summary

The overall conclusion from this review is that the information in speech is shielded from distortion in several ways. First, peaks in the envelope of the spectrum provide robust cues for the identification of vowels and consonants, even when the spectral valleys are obscured by noise. Second, periodicity in the waveform reflects the fundamental frequency of voicing, allowing listeners to group together components that stem from the same voice across frequency and time in order to segregate them from competing signals (Brokx and Nooteboom 1982; Bird and Darwin 1998). Third, at disadvantageous SNRs, the formants of voiced sounds can exert their influence by disrupting the periodicity of competing harmonic signals or by disrupting the interaural correlation of a masking noise (Summerfield and Culling 1992; Culling and Summerfield 1995a). Fourth, the amplitude-modulation pattern across frequency bands can serve to highlight informative portions of the speech signal, such as prosodically stressed syllables. These temporal modulation patterns are redundantly specified in time and frequency, making it possible to remove large amounts of the signal via gating in the time domain (e.g., Miller and Licklider 1950) or filtering in the frequency domain (e.g., Warren et al. 1995). Even when the spectral details and periodicity of voiced speech are eliminated, intelligibility remains high if the temporal modulation structure is preserved in a small number of bands (Shannon et al. 1995). However, speech processed in this manner is more susceptible to interference by other signals (Fu et al. 1998).

Competing signals, noise, reverberation, and other imperfections of the communication channel can eliminate, mask, or distort the information-providing segments of the speech signal. Listeners with normal hearing rely on a range of perceptual and linguistic strategies to overcome these effects and bridge the gaps that appear in the time-frequency distribution of the distorted signal. Time-varying changes in the SNR allow listeners to focus their attention on temporal and spectral regions where the target voice is best defined, a process described as glimpsing. Together with complementary processes such as perceptual grouping and tracking, listeners use their knowledge of linguistic contraints to fill in the gaps in the signal and arrive at the most plausible interpretations of the distorted signal.

Glimpsing and tracking depend on an analysis of the signal within a sliding temporal window, and provide effective strategies when the distortion is intermittent. When the form of distortion is relatively stationary (e.g., a continuous, broadband noise masker, or the nonuniform frequency response of a large room), other short-term processes such as adaptation and perceptual grouping can be beneficial. Adaptation serves to emphasize newly arriving components of the signal, enhancing syllable onsets and regions of the signal undergoing rapid spectrotemporal change. Perceptual grouping processes link together acoustic components that emanate from the same sound source. Listeners may also benefit from central auditory processes that compensate for distortions of the frequency response of the channel. The nature and time course of such adaptations remain topics of current interest and controversy.

List of Abbreviations

ACF	autocorrelation function
AI	articulation index
AM	amplitude modulation
CF	characteristic frequency
CMR	comodulation masking release
f_0	fundamental frequency
F_1	first formant
F_2	second formant
F_3	third formant
HP	high pass
ILD	interaural level difference
ITD	interaural time difference
LP	low pass
LPC	linear predictive coding
LTASS	long-term average speech spectrum
rms	root mean square
SNR	signal-to-noise ratio
SPIN	speech perception in noise test
SRT	speech reception threshold
STI	speech transmission index
TMTF	temporal modulation transfer function
VOT	voice onset time

References

Allen JB (1994) How do humans process and recognize speech? IEEE Trans Speech Audio Proc 2:567–577.

ANSI (1969) Methods for the calculation of the articulation index. ANSI S3.5-1969. New York: American National Standards Institute.

ANSI (1997) Methods for the calculation of the articulation index. ANSI S3.5-1997. New York: American National Standards Institute.

Arai T, Greenberg S (1998) Speech intelligibility in the presence of cross-channel spectral asynchrony, IEEE Int Conf Acoust Speech Signal Proc, pp. 933–936.

Assmann PF (1991) Perception of back vowels: center of gravity hypothesis. Q J Exp Psychol 43A:423–448.

Assmann PF (1995) The role of formant transitions in the perception of concurrent vowels. J Acoust Soc Am 97:575–584.

Assmann PF (1996) Modeling the perception of concurrent vowels: role of formant transitions. J Acoust Soc Am 100:1141–1152.

Assmann PF (1999) Fundamental frequency and the intelligibility of competing voices. Proceedings of the 14th International Congress of Phonetic Sciences, pp. 179–182.

Assmann PF, Katz WF (2000) Time-varying spectral change in the vowels of children and adults. J Acoust Soc Am 108:1856–1866.

Assmann PF, Nearey TM (1986) Perception of front vowels: the role of harmonics in the first formant region. J Acoust Soc Am 81:520–534.

Assmann PF, Summerfield AQ (1989) Modeling the perception of concurrent vowels: vowels with the same fundamental frequency. J Acoust Soc Am 85: 327–338.

Assmann PF, Summerfield AQ (1990) Modeling the perception of concurrent vowels: vowels with different fundamental frequencies. J Acoust Soc Am 88: 680–697.

Assmann PF, Summerfield Q (1994) The contribution of waveform interactions to the perception of concurrent vowels. J Acoust Soc Am 95:471–484.

Auer ET Jr, Bernstein LE (1997) Speechreading and the structure of the lexicon: computationally modeling the effects of reduced phonetic distinctiveness on lexical uniqueness. J Acoust Soc Am 102:3704–3710.

Baer T, Moore BCJ (1993) Effects of spectral smearing on the intelligibility of sentences in noise. J Acoust Soc Am 94:1229–1241.

Baer T, Moore BCJ (1994) Effects of spectral smearing on the intelligibility of sentences in the presence of interfering speech [letter]. J Acoust Soc Am 95: 2277–2280.

Baer T, Moore BCJ, Gatehouse S (1993) Spectral contrast enhancement of speech in noise for listeners with sensorineural hearing impairment: effects on intelligibility, quality, and response times. J Rehabil Res Dev 30:49–72.

Bakkum MJ, Plomp R, Pols LCW (1993) Objective analysis versus subjective assessment of vowels pronounced by native, non-native, and deaf male speakers of Dutch. J Acoust Soc Am 94:1983–1988.

Bashford JA, Reiner KR, Warren RM (1992) Increasing the intelligibility of speech through multiple phonemic restorations. Percept Psychophys 51:211–217.

Beddor PS, Hawkins S (1990) The influence of spectral prominence on perceived vowel quality. J Acoust Soc Am 87:2684–2704.

Beranek LL (1947) The design of speech communication systems. Proc Inst Radio Engineers 35:880–890.

Berglund B, Hassmen P, Job RF (1996) Sources and effects of low-frequency noise. J Acoust Soc Am 99:2985–3002.

Bird J, Darwin CJ (1998) Effects of a difference in fundamental frequency in separating two sentences. In: Palmer A, Rees A, Summerfield Q, Meddis R (eds) Psychophysical and physiological advances in hearing. London: Whurr.

Bladon RAW (1982) Arguments against formants in the auditory representation of speech In: Carlson R, Granstrom B (eds) The Representation of Speech in the Peripheral Auditory System, Amsterdam Elsevier Biomedical Press, pp. 95–102.

Bladon RAW, Lindblom B (1981) Modeling the judgement of vowel quality differences. J Acoust Soc Am 69:1414–1422.

Blauert J (1996) Spatial Hearing: The Psychophysics of Human Sound Localization, 2nd ed. Cambridge, MA: MIT Press.
Blesser B (1972) Speech perception under conditions of spectral transformation. I. Phonetic characteristics. J Speech Hear Res 15:5–41.
Boothroyd A, Nittrouer S (1988) Mathematical treatment of context effects in phoneme and word recognition. J Acoust Soc Am 84:101–114.
Bradlow AR, Pisoni DB (1999). Recognition of spoken words by native and non-native listeners: talker-, listener-, and item-related factors. J Acoust Soc Am 106:2074–2085.
Bregman AS (1990) Auditory Scene Analysis. Cambridge, MA: MIT Press.
Broadbent DE (1958) Perception and Communication. Oxford: Pergamon Press.
Brokx JPL, Nooteboom SG (1982) Intonation and the perception of simultaneous voices. J Phonetics 10:23–26.
Bronkhorst AW (2000) The cocktail party phenomenon: a review of research on speech intelligibility in multiple-talker conditions. Acustica 86:117–128.
Bronkhorst AW, Plomp R (1988) The effect of head-induced interaural time and level differences on speech intelligibility in noise. J Acoust Soc Am 83:1508–1516.
Brungart DS (2001) Informational and energetic masking effects in the perception of two simultaneous talkers. J Acoust Soc Am 109:1101–1109.
Brungart DS, Simpson BD, Ericson MA, Scott KR (2001) Informational and energetic masking effects in the perception of multiple simultaneous talkers. J Acoust Soc Am 110:2527–2538.
Buuren RA van, Festen JM, Houtgast T (1996) Peaks in the frequency response of hearing aids: evaluation of the effects on speech intelligibility and sound quality. J Speech Hear Res 39:239–250.
Buus S (1985) Release from masking caused by envelope fluctuations. J Acoust Soc Am 78:1958–1965.
Byrne D, Dillon H, Tran K, et al. (1994) An international comparison of long-term average speech spectra. J Acoust Soc Am 96:2108–2120.
Carhart R (1965) Monaural and binaural discrimination against competing sentences. Int Audiol 4:5–10.
Carhart R, Tillman TW, Greetis ES (1969) Perceptual masking in multiple sound background. J Acoust Soc Am 45:411–418.
Cariani PA, Delgutte B (1996a) Neural correlates of the pitch of complex tones. I. Pitch and pitch salience. J Neurophys 76:1698–1716.
Cariani PA, Delgutte B (1996b) Neural correlates of the pitch of complex tones. II. Pitch shift, pitch ambiguity, phase invariance, pitch circularity, rate pitch, and the dominance region for pitch. J Neurophys 76:1717–1734.
Carlson R, Fant G, Granstrom B (1974) Two-formant models, pitch, and vowel perception. Acustica 31:360–362.
Carlson R, Granstrom B, Klatt D (1979) Vowel perception: the relative perceptual salience of selected acoustic manipulations. Speech Transmission Laboratories (Stockholm) Quarterly Progress Report SR 3–4, pp. 73–83.
Carlyon RP (1989) Changes in the masked thresholds of brief tones produced by prior bursts of noise. Hear Res 41:223–236.
Carlyon RP (1994) Further evidence against an across-frequency mechanism specific to the detection of FM incoherence between resolved frequency components. J Acoust Soc Am 95:949–961.

Carney LH, Yin TCT (1988) Temporal coding of resonances by low-frequency auditory nerve fibers: single-fiber responses and a population model. J Neurophys 60:1653–1677.

Carrell TD, Opie JM (1992) The effect of amplitude comodulation on auditory object formation in sentence perception. Percept Psychophys 52:437–445.

Carterette EC, Møller A (1962) The perception of real and synthetic vowels after very sharp filtering. Speech Transmission Laboratories (Stockholm) Quarterly Progress Report SR 3, pp. 30–35.

Castle WE (1964) The Effect of Narrow Band Filtering on the Perception of Certain English Vowels. The Hague: Mouton.

Chalikia M, Bregman A (1989) The perceptual segregation of simultaneous auditory signals: pulse train segregation and vowel segregation. Percept Psychophys 46:487–496.

Cherry C (1953) Some experiments on the recognition of speech, with one and two ears. J Acoust Soc Am 25:975–979.

Cherry C, Wiley R (1967) Speech communication in very noisy environments. Nature 214:1164.

Cheveigné A de (1997) Concurrent vowel identification. III: A neural model of harmonic interference cancellation. J Acoust Soc Am 101:2857–2865.

Cheveigné A de, McAdams S, Laroche J, Rosenberg M (1995) Identification of concurrent harmonic and inharmonic vowels: a test of the theory of harmonic cancellation and enhancement. J Acoust Soc Am 97:3736–3748.

Chistovich LA (1984) Central auditory processing of peripheral vowel spectra. J Acoust Soc Am 77:789–805.

Chistovich LA, Lublinskaya VV (1979) The "center of gravity" effect in vowel spectra and critical distance between the formants: psychoacoustic study of the perception of vowel-like stimuli. Hear Res 1:185–195.

Ciocca V, Bregman AS (1987) Perceived continuity of gliding and steady-state tones through interrupting noise. Percept Psychophys 42:476–484.

Coker CH, Umeda N (1974) Speech as an error correcting process. Speech Communication Seminar, SCS-74, Stockholm, Aug. 1–3, pp. 349–364.

Cooke MP, Ellis DPW (2001) The auditory organization of speech and other sources in listeners and computational models. Speech Commun 35:141–177.

Cooke MP, Morris A, Green PD (1996) Recognising occluded speech. In: Greenberg S, Ainsworth WA (eds) Proceedings of the ESCA Workshop on the Auditory Basis of Speech Perception, pp. 297–300.

Culling JE, Darwin CJ (1993a) Perceptual separation of simultaneous vowels: within and across-formant grouping by f_0. J Acoust Soc Am 93:3454–3467.

Culling JE, Darwin CJ (1994) Perceptual and computational separation of simultaneous vowels: cues arising from low frequency beating. J Acoust Soc Am 95: 1559–1569.

Culling JF, Summerfield Q (1995a) Perceptual separation of concurrent speech sounds: absence of across-frequency grouping by common interaural delay. J Acoust Soc Am 98:785–797.

Culling JF, Summerfield Q (1995b) The role of frequency modulation in the perceptual segregation of concurrent vowels. J Acoust Soc Am 98:837–846.

Culling JF, Summerfield Q, Marshall DH (1994) Effects of simulated reverberation on the use of binaural cues and fundamental-frequency differences for separating concurrent vowels. Speech Commun 14:71–95.

Darwin CJ (1984) Perceiving vowels in the presence of another sound: constraints on formant perception. J Acoust Soc Am 76:1636–1647.

Darwin CJ (1990) Environmental influences on speech perception. In: Advances in Speech, Hearing and Language Processing, vol. 1. London: JAI Press, pp. 219–241.

Darwin CJ (1992) Listening to two things at once. In: Schouten MEH (ed) The Auditory Processing of Speech: From Sounds to Words. Berlin: Mouton de Gruyter, pp. 133–147.

Darwin CJ (1997) Auditory Grouping. Trends in Cognitive Science 1:327–333.

Darwin CJ, Carlyon RP (1995) Auditory Grouping. In: Moore BCJ (ed) The Handbook of Perception and Cognition, vol. 6, Hearing. London: Academic Press.

Darwin CJ, Gardner RB (1986) Mistuning a harmonic of a vowel: grouping and phase effects on vowel quality. J Acoust Soc Am 79:838–845.

Darwin CJ, Hukin RW (1997) Perceptual segregation of a harmonic from a vowel by interaural time difference and frequency proximity. J Acoust Soc Am 102: 2316–2324.

Darwin CJ, Hukin RW (1998) Perceptual segregation of a harmonic from a vowel by interaural time difference in conjunction with mistuning and onset asynchrony. J Acoust Soc Am 103:1080–1084.

Darwin CJ, McKeown JD, Kirby D (1989) Compensation for transmission channel and speaker effects on vowel quality. Speech Commun 8:221–234.

Delattre P, Liberman AM, Cooper FS, Gerstman LJ (1952) An experimental study of the acoustic determinants of vowel color: observations on one- and two-formant vowels synthesized from spectrographic patterns. Word 8:195–201.

Delgutte B (1980) Representation of speech-like sounds in the discharge patterns of auditory-nerve fibers. J Acoust Soc Am 68:843–857.

Delgutte B (1996) Auditory neural processing of speech. In: Hardcastle WJ, Laver J (eds) The Handbook of Phonetic Sciences. Oxford: Blackwell.

Delgutte B, Kiang NYS (1984) Speech coding in the auditory nerve: IV. Sounds with consonant-like dynamic characteristics. J Acoust Soc Am 75:897–907.

Deng L, Kheirallah I (1993) Dynamic formant tracking of noisy speech using temporal analysis on outputs from a nonlinear cochlear model. IEEE Trans Biomed Eng 40:456–467.

Dirks DD, Bower DR (1969) Masking effects of speech competing messages. J Speech Hear Res 12:229–245.

Dirks DD, Wilson RH (1969) The effect of spatially separated sound sources on speech intelligibility. J Speech Hear Res 12:5–38.

Dirks DD, Wilson RH, Bower DR (1969) Effects of pulsed masking on selected speech materials. J Acoust Soc Am 46:898–906.

Dissard P, Darwin CJ (2000) Extracting spectral envelopes: formant frequency matching between sounds on different and modulated fundamental frequencies. J Acoust Soc Am 107:960–969.

Dorman MF, Loizou PC, Rainey D (1997). Speech intelligibility as a function of the number of channels of stimulation for signal processors using sine-wave and noise outputs. J Acoust Soc Am 102:2403–2411.

Dorman MF, Loizou PC, Fitzke J, Tu Z (1998). The recognition of sentences in noise by normal- ring listeners using simulations of cochlear-implant signal processors with 6–20 channels. J Acoust Soc Am 104:3583–3585.

Dreher JJ, O'Neill JJ (1957) Effects of ambient noise on speaker intelligibility for words and phrases. J Acoust Soc Am 29:1320–1323.
Drullman R (1995a) Temporal envelope and fine structure cues for speech intelligibility. J Acoust Soc Am 97:585–592.
Drullman R (1995b) Speech intelligibility in noise: relative contribution of speech elements above and below the noise level. J Acoust Soc Am 98:1796–1798.
Drullman R, Festen JM, Plomp R (1994a) Effect of reducing slow temporal modulations on speech reception. J Acoust Soc Am 95:2670–2680.
Drullman R, Festen JM, Plomp R (1994b) Effect of temporal envelope smearing on speech reception. J Acoust Soc Am 95:1053–1064.
Dubno J, Ahlstrom JB (1995) Growth of low-pass masking of pure tones and speech for hearing-impaired and normal-hearing listeners. J Acoust Soc Am 98:3113–3124.
Duifhuis H, Willems LF, Sluyter RJ (1982) Measurement of pitch on speech: an implementation of Goldstein's theory of pitch perception. J Acoust Soc Am 71:1568–1580.
Dunn HK, White SD (1940) Statistical measurements on conversational speech. J Acoust Soc Am 11:278–288.
Duquesnoy AJ (1983) Effect of a single interfering noise or speech source upon the binaural sentence intelligibility of aged persons. J Acoust Soc Am 74:739–743.
Duquesnoy AJ, Plomp R (1983) The effect of a hearing aid on the speech-reception threshold of a hearing-impaired listener in quiet and in noise. J Acoust Soc Am 73:2166–2173.
Egan JP, Wiener FM (1946) On the intelligibility of bands of speech in noise. J Acoust Soc Am 18:435–441.
Egan JP, Carterette EC, Thwing EJ (1954) Some factors affecting multi-channel listening. J Acoust Soc Am 26:774–782.
Elliot LL (1995) Verbal auditory closure and the Speech Perception in Noise (SPIN) test. J Speech Hear Res 38:1363–1376.
Fahey RP, Diehl RL, Traunmuller H (1996) Perception of back vowels: effects of varying F_1–f_0 Bark distance. J Acoust Soc Am 99:2350–2357.
Fant G (1960) Acoustic Theory of Speech Production. Mouton: The Hague.
Festen JM (1993) Contributions of comodulation masking release and temporal resolution to the speech-reception threshold masked by an interfering voice. J Acoust Soc Am 94:1295–1300.
Festen JM, Plomp R (1981) Relations between auditory functions in normal hearing. J Acoust Soc Am 70:356–369.
Festen JM, Plomp R (1990) Effects of fluctuating noise and interfering speech on the speech-reception threshold for impaired and normal hearing. J Acoust Soc Am 88:1725–1736.
Finitzo-Hieber T, Tillman TW (1978) Room acoustics effects on monosyllabic word discrimination ability for normal and hearing impaired children. J Speech Hear Res 21:440–458.
Fletcher H (1952) The perception of sounds by deafened persons. J Acoust Soc Am 24:490–497.
Fletcher H (1953) Speech and Hearing in Communication. New York: Van Nostrand (reprinted by the Acoustical Society of America, 1995).
Fletcher H, Galt RH (1950) The perception of speech and its relation to telephony. J Acoust Soc Am 22:89–151.

French NR, Steinberg JC (1947) Factors governing the intelligibility of speech sounds. J Acoust Soc Am 19:90–119.

Fu Q-J, Shannon RV, Wang X (1998) Effects of noise and spectral resolution on vowel and consonant recognition: acoustic and electric hearing. J Acoust Soc Am 104:3586–3596.

Gardner RB, Gaskill SA, Darwin CJ (1989) Perceptual grouping of formants with static and dynamic differences in fundamental frequency. J Acoust Soc Am 85:1329–1337.

Gat IB, Keith RW (1978) An effect of linguistic experience. Auditory word discrimination by native and non-native speakers of English. Audiology 17:339–345.

Gatehouse S (1992) The time course and magnitude of perceptual acclimitization to frequency responses: evidence from monaural fitting of hearing aids. J Acoust Soc Am 92:1258–1268.

Gatehouse S (1993) Role of perceptual acclimitization to frequency responses: evidence from monaural fitting of hearing aids. J Am Acad Audiol 4:296–306.

Gelfand SA, Silman S (1979) Effects of small room reverberation on the recognition of some consonant features. J Acoust Soc Am 66:22–29.

Glasberg BR, Moore BCJ (1986) Auditory filter shapes in subjects with unilateral and bilateral cochlear impairments. J Acoust Soc Am 79:1020–1033.

Gong Y (1994) Speech recognition in noisy environments: a survey. Speech Commun 16:261–291.

Gordon-Salant S, Fitzgibbons PJ (1995) Recognition of multiply degraded speech by young and elderly listeners. J Speech Hear Res 38:1150–1156.

Grant KW, Ardell LH, Kuhl PK, Sparks DW (1985) The contribution of fundamental frequency, amplitude envelope, and voicing duration cues to speechreading in normal-hearing subjects. J Acoust Soc Am 77:671–677.

Grant KW, Braida LD, Renn RJ (1991) Single band amplitude envelope cues as an aid to speechreading. Q J Exp Psychol 43A:621–645.

Grant KW, Braida LD, Renn RJ (1994) Auditory supplements to speechreading: combining amplitude envelope cues from different spectral regions of speech. J Acoust Soc Am 95:1065–1073.

Greenberg S (1995) Auditory processing of speech. In: Lass NJ (ed) Principles of Experimental Phonetics. St. Louis: Mosby-Year Book, pp. 362–407.

Greenberg S (1996) Understanding speech understanding: Towards a unified theory of speech perception. In: Greenberg S, Ainsworth WA (eds) Proceedings of the ESCA Workshop on the Auditory Basis of Speech Perception, pp. 1–8.

Greenberg S, Arai T (1998) Speech intelligibility is highly tolerant of cross-channel spectral asynchrony. Proceedings of the Joint Meeting of the Acoustical Society of America and the International Congress on Acoustics, pp. 2677–2678.

Greenberg S, Arai T, Silipo R (1998) Speech intelligibility derived from exceedingly sparse spectral information. Proceedings of the International Conference on Spoken Language Processing, Sydney, pp. 74–77.

Gustafsson HA, Arlinger SD (1994) Masking of speech by amplitude-modulated noise. J Acoust Soc Am 95:518–529.

Haggard MP (1985) Temporal patterning in speech: the implications of temporal resolution and signal processing. In: Michelson A (ed) Time Resolution in Auditory Systems. Berlin: Springer-Verlag, pp. 217–237.

Hall JW, Grose JH (1991) Relative contributions of envelope maxima and minima to comodulation masking release. Q J Exp Psychol 43A:349–372.

Hall JW, Haggard MP, Fernandez MA (1984) Detection in noise by spectro-temporal analysis. J Acoust Soc Am 76:50–56.

Hanky TD, Steer MD (1949) Effect of level of distracting noise upon speaking rate, duration and intensity. J Speech Hear Dis 14:363–368.

Hanson BA, Applebaum TH (1990) Robust speaker-independent word recognition using static, dynamic and acceleration features: experiments with Lombard and noisy speech. Proc Int Conf Acoust Speech Signal Processing 90:857–860.

Hartmann WM (1996) Pitch, periodicity, and auditory organization. J Acoust Soc Am 100:3491–3502.

Hawkins JE Jr, Stevens SS (1950) The masking of pure tones and of speech by white noise. J Acoust Soc Am 22:6–13.

Helfer KS (1992) Aging and the binaural advantage in reverberation and noise. J Speech Hear Res 35:1394–1401.

Helfer KS (1994) Binaural cues and consonant perception in reverberation and noise. J Speech Hear Res 37:429–438.

Hicks ML, Bacon SP (1992) Factors influencing temporal effects with notched-noise maskers. Hear Res 64:123–132.

Hillenbrand JM, Nearey TM (1999) Identification of resynthesized /hVd/ utterances: effects of formant contour. J Acoust Soc Am 105:3509–3523.

Hillenbrand JM, Getty LA, Clark MJ, Wheeler K (1995) Acoustic characteristics of American English vowels. J Acoust Soc Am 97:3099–3111.

Hockett CF (1955) A Manual of Phonology. Bloomington, IN: Indiana University Press.

Horwitz AR, Turner CW (1997) The time course of hearing aid benefit. Ear Hear 18:1–11.

Houtgast T, Steeneken HJM (1973) The modulation transfer function in room acoustics as a predictor of speech intelligibility. Acustica 28:66–73.

Houtgast T, Steeneken HJM (1985) A review of the MTF concept in room acoustics and its use for estimating speech intelligibility in auditoria. J Acoust Soc Am 77:1069–1077.

Howard-Jones PA, Rosen S (1993) Uncomodulated glimpsing in "checkerboard" noise. J Acoust Soc Am 93:2915–2922.

Howes D (1957) On the relation between the intelligibility and frequency of occurrence of English words. J Acoust Soc Am 29:296–303.

Huggins AWF (1975) Temporally segmented speech. Percept Psychophys 18:149–157.

Hukin RW, Darwin CJ (1995) Comparison of the effect of onset asynchrony on auditory grouping in pitch matching and vowel identification. Percept Psychophys 57:191–196.

Humes LE, Dirks DD, Bell TS, Ahlstrom C, Kincaid GE (1986) Application of the articulation index and the speech transmission index to the recognition of speech by normal-hearing and hearing-impaired listeners. J Speech Hear Res 29:447–462.

Humes LE, Boney S, Loven F (1987) Further validation of the speech transmission index (STI). J Speech Hear Res 30:403–410.

Hygge S, Rönnberg J, Larsby B, Arlinger S (1992) Normal-hearing and hearing-impaired subjects' ability to just follow conversation in competing speech, reversed speech, and noise backgrounds. J Speech Hear Res 35:208–215.

Joris PX, Yin TC (1995) Envelope coding in the lateral superior olive. I. Sensitivity to interaural time differences. J Neurophys 73:1043–1062.

Junqua JC, Anglade Y (1990) Acoustic and perceptual studies of Lombard speech: application to isolated words automatic speech recognition. Proc Int Conf Acoust Speech Signal Processing 90:841–844.

Kalikow DN, Stevens KN, Elliot LL (1977) Development of a test of speech intelligibility in noise using sentence materials with controlled word predictability. J Acoust Soc Am 61:1337–1351.

Kates JM (1987) The short-time articulation index. J Rehabil Res Dev 24:271–276.

Keurs M ter, Festen JM, Plomp R (1992) Effect of spectral envelope smearing on speech reception. I. J Acoust Soc Am 91:2872–2880.

Keurs M ter, Festen JM, Plomp R (1993a) Effect of spectral envelope smearing on speech reception. II. J Acoust Soc Am 93:1547–1552.

Keurs M ter, Festen JM, Plomp R (1993b) Limited resolution of spectral contrast and hearing loss for speech in noise. J Acoust Soc Am 94:1307–1314.

Kewley-Port D, Zheng Y (1998) Auditory models of formant frequency discrimination for isolated vowels. J Acoust Soc Am 103:1654–1666

Klatt DH (1982) Speech processing strategies based on auditory models. In: Carlson R, Granstrom B (eds) The Representation of Speech in the Peripheral Auditory System. Amsterdam: Elsevier.

Klatt DH (1989) Review of selected models of speech perception. In: Marslen-Wilson W (ed) Lexical Representation and Process. Cambridge, MA: MIT Press, pp.169–226.

Kluender KR, Jenison RL (1992) Effects of glide slope, noise intensity, and noise duration in the extrapolation of FM glides through noise. Percept Psychophys 51:231–238.

Kreiman J (1997) Listening to voices: theory and practice in voice perception research. In: Johnson K, Mullenix J (eds) Talker Variability in Speech Processing. San Diego: Academic Press.

Kryter KD (1946) Effects of ear protective devices on the intelligibility of speech in noise. J Acoust Soc Am 18:413–417.

Kryter KD (1962) Methods for the calculation and use of the articulation index. J Acoust Soc Am 34:1689–1697.

Kryter D (1985) The Effects of Noise on Man, 2nd ed. London: Academic Press.

Kuhn GF (1977) Model for the interaural time differences in the azimuthal plane. J Acoust Soc Am 62:157–167.

Ladefoged P (1967) Three Areas of Experimental Phonetics. Oxford: Oxford University Press, pp. 162–165.

Lane H, Tranel B (1971) The Lombard sign and the role of hearing in speech. J Speech Hear Res 14:677–709.

Langner G (1992) Periodicity coding in the auditory system. Hear Res 60:115–142.

Lea AP (1992) Auditory modeling of vowel perception. PhD thesis, University of Nottingham.

Lea AP, Summerfield Q (1994) Minimal spectral contrast of formant peaks for vowel recognition as a function of spectral slope. Percept Psychophys 56:379–391.

Leek MR, Dorman MF, Summerfield, Q (1987) Minimum spectral contrast for vowel identification by normal-hearing and hearing-impaired listeners. J Acoust Soc Am 81:148–154.

Lehiste I, Peterson GE (1959) The identification of filtered vowels. Phonetica 4:161–177.

Levitt H, Rabiner LR (1967) Predicting binaural gain in intelligibility and release from masking for speech. J Acoust Soc Am 42:820–829.

Liberman AM, Delattre PC, Gerstman LJ, Cooper FS (1956) Tempo of frequency change as a cue for distinguishing classes of speech sounds. J Exp Psychol 52:127–137.

Liberman AM, Cooper FS, Shankweiler DP, Studdert-Kennedy M (1967) Perception of the speech code. Psychol Rev 74:431–461.

Licklider JCR, Guttman N (1957) Masking of speech by line-spectrum interference. J Acoust Soc Am 29:287–296.

Licklider JCR, Miller GA (1951) The perception of speech. In: Stevens SS (ed) Handbook of Experimental Psychology. New York: John Wiley, pp. 1040–1074.

Lindblom B (1986) Phonetic universals in vowel systems. In Ohala JJ, Jaeger JJ, (eds.) Experimental Phonology. New York: Academic Press, pp. 13–44.

Lindblom B (1990) Explaining phonetic variation: a sketch of the H&H theory. In: Hardcastle WJ, Marshall A (eds) Speech Production and Speech Modelling. Dordrecht: Kluwer Academic, pp. 403–439.

Lippmann R (1996a) Speech perception by humans and machines. In: Greenberg S, Ainsworth WA (eds) Proceedings of the ESCA Workshop on the Auditory Basis of Speech Perception. pp. 309–316.

Lippmann R (1996b) Accurate consonant perception without mid-frequency speech energy. IEEE Trans Speech Audio Proc 4:66–69.

Liu SA (1996) Landmark detection for distinctive feature-based speech recognition. J Acoust Soc Am 100:3417–3426.

Lively SE, Pisoni DB, Van Summers W, Bernacki RH (1993) Effects of cognitive workload on speech production: acoustic analyses and perceptual consequences. J Acoust Soc Am 93:2962–2973.

Lombard E (1911) Le signe de l'élévation de la voix. Ann Malad l'Oreille Larynx Nez Pharynx 37:101–119.

Luce PA, Pisoni DB (1998) Recognizing spoken words: the neighborhood activation model. Ear Hear 19:1–36.

Luce PA, Pisoni DB, Goldinger SD (1990) Similarity neighborhoods of spoken words. In: Altmann GTM (ed) Cognitive Models of Speech Processing. Cambridge: MIT Press, pp. 122–147.

Ludvigsen C (1987) Prediction of speech intelligibility for normal-hearing and cochlearly hearing impaired listeners. J Acoust Soc Am 82:1162–1171.

Ludvigsen C, Elberling C, Keidser G, Poulsen T (1990) Prediction of intelligibility for nonlinearly processed speech. Acta Otolaryngol Suppl 469:190–195.

MacLeod A, Summerfield Q (1987) Quantifying the contribution of vision to speech perception in noise. Br J Audiol 21:131–141.

Marin CMH, McAdams SE (1991) Segregation of concurrent sounds. II: Effects of spectral-envelope tracing, frequency modulation coherence and frequency modulation width. J Acoust Soc Am 89:341–351.

Markel JD, Gray AH (1976) Linear Prediction of Speech. New York: Springer-Verlag.

Marslen-Wilson W (1989) Access and integration: projecting sound onto meaning. In: Marslen-Wilson W (ed) Lexical Representation and Process. Cambridge: MIT Press, pp. 3–24.

Mayo LH, Florentine M, Buus S (1997) Age of second-language acquisition and perception of speech in noise. J Speech Lang Hear Res 40:686–693.

McAdams SE (1989) Segregation of concurrent sounds: effects of frequency-modulation coherence and a fixed resonance structure. J Acoust Soc Am 85:2148–2159.

McKay CM, Vandali AE, McDermott HJ, Clark GM (1994) Speech processing for multichannel cochlear implants: variations of the Spectral Maxima Sound Processor strategy. Acta Otolaryngol 114:52–58.

Meddis R, Hewitt M (1991) Virtual pitch and phase sensitivity of a computer model of the auditory periphery. I: Pitch identification. J Acoust Soc Am 89: 2866–2882.

Meddis R, Hewitt M (1992) Modelling the identification of concurrent vowels with different fundamental frequencies. J Acoust Soc Am 91:233–245.

Miller GA (1947) The masking of speech. Psychol Bull 44:105–129.

Miller GA, Licklider JCR (1950) The intelligibility of interrupted speech. J Acoust Soc Am 22:167–173.

Miller GA, Nicely PE (1955) An analysis of perceptual confusions among some English consonants. J Acoust Soc Am 27:338–352.

Miller GA, Heise GA, Lichten W (1951) The intelligibility of speech as a function of the context of the test materials. J Exp Psychol 41:329–335.

Moncur JP, Dirks D (1967) Binaural and monaural speech intelligibility in reverberation. J Speech Hear Res 10:186–195.

Moore BCJ (1995) Perceptual Consequences of Cochlear Hearing Impairment. London: Academic Press.

Moore BCJ, Glasberg BR (1983) Suggested formulae for calculating auditory-filter shapes and excitation patterns. J Acoust Soc Am 74:750–753.

Moore BCJ, Glasberg BR (1987) Formulae describing frequency selectivity as a function of frequency and level, and their use in calculating excitation patterns. Hear Res 28:209–225.

Moore BCJ, Glasberg BR, Peters RW (1985) Relative dominance of individual partials in determining the pitch of complex tones. J Acoust Soc Am 77:1861–1867.

Müsch H, Buus S (2001a). Using statistical decision theory to predict speech intelligibility. I. Model structure. J Acoust Soc Am 109:2896–2909.

Müsch H, Buus S (2001b). Using statistical decision theory to predict speech intelligibility. II. Measurement and prediction of consonant-discrimination performance. J Acoust Soc Am 109:2910–2920.

Nábělek AK (1988) Identification of vowels in quiet, noise, and reverberation: relationships with age and hearing loss. J Acoust Soc Am 84:476–484.

Nábělek AK, Dagenais PA (1986) Vowel errors in noise and in reverberation by hearing-impaired listeners. J Acoust Soc Am 80:741–748.

Nábělek AK, Letowski TR (1985) Vowel confusions of hearing-impaired listeners under reverberant and non-reverberant conditions. J Speech Hear Disord 50:126–131.

Nábělek AK, Letowski TR (1988) Similarities of vowels in nonreverberant and reverberant fields. J Acoust Soc Am 83:1891–1899.

Nábělek AK, Pickett JM (1974) Monaural and binaural speech perception through hearing aids under noise and reverberation with normal and hearing-impaired listeners. J Speech Hear Res 17:724–739.

Nábělek AK, Robinson PK (1982) Monaural and binaural speech perception in reverberation in listeners of various ages. J Acoust Soc Am 71:1242–1248.

Nábělek AK, Letowski TR, Tucker FM (1989) Reverberant overlap- and self-masking in consonant identification. J Acoust Soc Am 86:1259–1265.

Nábělek AK, Czyzewski Z, Crowley H (1994) Cues for perception of the diphthong [ai] in either noise or reverberation: I. Duration of the transition. J Acoust Soc Am 95:2681–2693.
Nearey TM (1989) Static, dynamic, and relational properties in vowel perception. J Acoust Soc Am 85:2088–2113.
Neuman AC, Hochberg I (1983) Children's perception of speech in reverberation. J Acoust Soc Am 73:2145–2149.
Nocerino N, Soong FK, Rabiner LR, Klatt DH (1985) Comparative study of several distortion measures for speech recognition. Speech Commun 4:317–331.
Noordhoek IM, Drullman R (1997) Effect of reducing temporal intensity modulations on sentence intelligibility. J Acoust Soc Am 101:498–502.
Nooteboom SG (1968) Perceptual confusions among Dutch vowels presented in noise. IPO Ann Prog Rep 3:68–71.
Palmer AR (1995) Neural signal processing. In: Moore BCJ (ed) The Handbook of Perception and Cognition, vol. 6, Hearing. London: Academic Press.
Palmer AR, Summerfield Q, Fantini DA (1995) Responses of auditory-nerve fibers to stimuli producing psychophysical enhancement. J Acoust Soc Am 97: 1786–1799.
Patterson RD, Moore BCJ (1986) Auditory filters and excitation patterns as representations of auditory frequency selectivity. In: Moore BCJ (ed) Frequency Selectivity in Hearing. London: Academic Press.
Patterson RD, Robinson K, Holdsworth J, McKeown D, Zhang C, Allerhand MH (1992) Complex sounds and auditory images. In: Cazals Y, Demany L, Horner K (eds) Auditory Physiology and Perception. Oxford: Pergamon Press, pp. 429–446.
Pavlovic CV (1987) Derivation of primary parameters and procedures for use in speech intelligibility predictions. J Acoust Soc Am 82:413–422.
Pavlovic CV, Studebaker GA (1984) An evaluation of some assumptions underlying the articulation index. J Acoust Soc Am 75:1606–1612.
Pavlovic CV, Studebaker GA, Sherbecoe RL (1986) An articulation index based procedure for predicting the speech recognition performance of hearing-impaired individuals. J Acoust Soc Am 80:50–57.
Payton KL, Uchanski RM, Braida LD (1994) Intelligibility of conversational and clear speech in noise and reverberation for listeners with normal and impaired hearing. J Acoust Soc Am 95:1581–1592.
Peters RW, Moore BCJ, Baer T (1998) Speech reception thresholds in noise with and without spectral and temporal dips for hearing-impaired and normally hearing people. J Acoust Soc Am 103:577–587.
Peterson GE, Barney HL (1952) Control methods used in a study of vowels. J Acoust Soc Am 24:175–184.
Picheny M, Durlach N, Braida L (1985) Speaking clearly for the hard of hearing I: Intelligibility differences between clear and conversational speech. J Speech Hear Res 28:96–103.
Picheny M, Durlach N, Braida L (1986) Speaking clearly for the hard of hearing II: Acoustic characteristics of clear and conversational speech. J Speech Hear Res 29:434–446.
Pickett JM (1956) Effects of vocal force on the intelligibility of speech sounds. J Acoust Soc Am 28:902–905.
Pickett JM (1957) Perception of vowels heard in noises of various spectra. J Acoust Soc Am 29:613–620.

Pisoni DB, Bernacki RH, Nusbaum HC, Yuchtman M (1985) Some acoustic-phonetic correlates of speech produced in noise. Proc Int Conf Acoust Speech Signal Proc, pp. 1581–1584.
Plomp R (1976) Binaural and monaural speech intelligibility of connected discourse in reverberation as a function of azimuth of a single competing sound source (speech or noise). Acustica 24:200–211.
Plomp R (1983) The role of modulation in hearing. In: Klinke R (ed) Hearing: Physiological Bases and Psychophysics. Heidelberg: Springer-Verlag, pp. 270–275.
Plomp R, Mimpen AM (1979) Improving the reliability of testing the speech reception threshold for sentences. Audiology 18:43–52.
Plomp R, Mimpen AM (1981) Effect of the orientation of the speaker's head and the azimuth of a sound source on the speech reception threshold for sentences. Acustica 48:325–328.
Plomp R, Steeneken HJM (1978) Place dependence of timbre in reverberant sound fields. Acustica 28:50–59.
Pollack I, Pickett JM (1958) Masking of speech by noise at high sound levels. J Acoust Soc Am 30:127–130.
Pollack I, Rubenstein H, Decker L (1959) Intelligibility of known and unknown message sets. J Acoust Soc Am 31:273–279.
Pols L, Kamp L van der, Plomp R (1969) Perceptual and physical space of vowel sounds. J Acoust Soc Am 46:458–467.
Powers GL, Wilcox JC (1977) Intelligibility of temporally interrupted speech with and without intervening noise. J Acoust Soc Am 61:195–199.
Rankovic CM (1995) An application of the articulation index to hearing aid fitting. J Speech Hear Res 34:391–402.
Rankovic CM (1998) Factors governing speech reception benefits of adaptive linear filtering for listeners with sensorineural hearing loss. J Acoust Soc Am 103: 1043–1057.
Remez RE, Rubin PE, Pisoni DB, Carrell TD (1981) Speech perception without traditional speech cues. Science 212:947–950.
Roberts B Moore BCJ (1990) The influence of extraneous sounds on the perceptual estimation of first-formant frequency in vowels. J Acoust Soc Am 88:2571–2583.
Roberts B, Moore BCJ (1991a) The influence of extraneous sounds on the perceptual estimation of first-formant frequency in vowels under conditions of asynchrony. J Acoust Soc Am 89:2922–2932.
Roberts B, Moore BCJ (1991b) Modeling the effects of extraneous sounds on the perceptual estimation of first-formant frequency in vowels. J Acoust Soc Am 89:2933–2951.
Rooij JC van, Plomp R (1991) The effect of linguistic entropy on speech perception in noise in young and elderly listeners. J Acoust Soc Am 90:2985–2991.
Rosen S (1992) Temporal information in speech: acoustic, auditory and linguistic aspects. In: Carlyon RP, Darwin CJ, Russell IJ (eds) Processing of Complex Sounds by the Auditory System. Oxford: Oxford University Press, pp. 73–80.
Rosen S, Faulkner A, Wilkinson L (1998) Perceptual adaptation by normal listeners to upward shifts of spectral information in speech and its relevance for users of cochlear implants. Abstracts of the 1998 Midwinter Meeting of the Association for Research in Otolaryngology.
Rosner BS, Pickering JB (1994) Vowel Perception and Production. Oxford: Oxford University Press.

Rostolland D (1982) Acoustic features of shouted voice. Acustica 50:118–125.

Rostolland D (1985) Intelligibility of shouted voice. Acustica 57:103–121.

Scheffers MTM (1983) Sifting Vowels: Auditory Pitch Analysis and Sound Segregation. PhD thesis, Rijksuniversiteit te Groningen, The Netherlands.

Shannon CE (1951) Prediction and entropy of printed English. Bell Sys Tech J 30:50–64.

Shannon CE, Weaver W (1949) A Mathematical Theory of Communication. Urbana, IL: University of Illinois Press.

Shannon RV, Zeng F-G, Kamath V, Wygonski J, Ekelid M (1995) Speech recognition with primarily temporal cues. Science 270:303–304.

Shannon RV, Zeng F-G, Wygonski J (1998). Speech recognition with altered spectral distribution of envelope cues. J Acoust Soc Am 104:2467–2476.

Simpson AM, Moore BCJ, Glasberg BR (1990) Spectral enhancement to improve the intelligibility of speech in noise for hearing-impaired listeners. Acta Otolaryngol Suppl 469:101–107.

Skinner MW, Clark GM, Whitford LA, et al. (1994) Evaluation of a new spectral peak coding strategy for the Nucleus 22 Channel Cochlear Implant System. Am J Otol 15 (suppl 2):15–27.

Smith RL (1979) Adaptation, saturation, and physiological masking in single auditory-nerve fibers. J Acoust Soc Am 65:166–178.

Sommers M, Kewley-Port D (1996) Modeling formant frequency discrimination of female vowels J Acoust Soc Am 99:3770–3781.

Speaks C, Karmen JL, Benitez L (1967) Effect of a competing message on synthetic sentence identification. J Speech Hear Res 10:390–395.

Spieth W, Webster JC (1955) Listening to differentially filtered competing messages. J Acoust Soc Am 27:866–871.

Steeneken HJM, Houtgast T (1980) A physical method for measuring speech-transmission quality. J Acoust Soc Am 67:318–326.

Steeneken HJM, Houtgast T (2002) Validation of the revised STI_r method. Speech Commun 38:413–425.

Stevens KN (1980) Acoustic correlates of some phonetic categories. J Acoust Soc Am 68:836–842.

Stevens KN (1983) Acoustic properties used for the identification of speech sounds. In: Parkins CW, Anderson SW (eds) Cochlear Prostheses: An International Symposium Ann NY Acad Sci 403:2–17.

Stevens SS, Miller GA, Truscott I (1946) The masking of speech by sine waves, square waves, and regular and modulated pulses. J Acoust Soc Am 18:418–424.

Stickney GS, Assmann PF (2001) Acoustic and linguistic factors in the perception of bandpass-filtered speech. J Acoust Soc Am 109:1157–1165.

Stubbs RJ, Summerfield AQ (1991) Effects of signal-to-noise ratio, signal periodicity, and degree of hearing impairment on the performance of voice-separation algorithms. J Acoust Soc Am 89:1383–1393.

Studebaker GA, Sherbecoe RL (2002) Intensity-importance functions for band-limited monosyllabic words. J Acoust Soc Am 111:1422–1436.

Studebaker GA, Pavlovic CV, Sherbecoe RL (1987) A frequency importance function for continuous discourse. J Acoust Soc Am 81:1130–1138.

Summerfield Q (1983) Audio-visual speech perception, lipreading, and artificial stimulation. In: Lutman ME, Haggard MP (eds) Hearing Science and Hearing Disorders. London: Academic Press, pp. 131–182.

Summerfield Q (1987) Speech perception in normal and impaired hearing. Br Med Bull 43:909–925.

Summerfield Q (1992) Role of harmonicity and coherent frequency modulation in auditory grouping. In: Schouten, MEH (ed) The Auditory Processing of Speech. Berlin: Mouton de Gruyter.

Summerfield Q, Assmann PF (1987) Auditory enhancement in speech perception. In: Schouten MEH (ed) The Psychophysics of Speech Perception. Dordrecht: Martinus Nijhoff, pp. 140–150.

Summerfield Q, Assmann PF (1989) Auditory enhancement and the perception of concurrent vowels. Percept Psychophys 45:529–536.

Summerfield Q, Culling JF (1992) Auditory segregation of competing voices: absence of effects of FM or AM coherence. Philos Trans R Soc Lond B 336:357–366.

Summerfield Q, Culling JF (1995) Auditory computations which separate speech from competing sounds: a comparison of binarual and monaural processes. In: Keller E (ed) Speech Synthesis and Speech Recognition. London: John Wiley.

Summerfield Q, Haggard MP, Foster JR, Gray S (1984) Perceiving vowels from uniform spectra: phonetic exploration of an auditory aftereffect. Percept Psychophys 35:203–213.

Summerfield Q, Sidwell A, Nelson T (1987) Auditory enhancement of changes in spectral amplitude. J Acoust Soc Am 81:700–708.

Summers WV, Pisoni DB, Bernacki RH, Pedlow RI, Stokes MA (1988) Effects of noise on speech production: acoustic and perceptual analyses. J Acoust Soc Am 84:917–928.

Suomi K (1984) On talker and phoneme information conveyed by vowels: A whole spectrum approach to the normalization problem. Speech Common 3:199–209.

Sussman HM, McCaffrey HA, Matthews SA (1991) An investigation of locus equations as a source of relational invariance for stop place categorization. J Acoust Soc Am 90:1309–1325.

Takata Y, Nábelek AK (1990) English consonant recognition in noise and in reverberation by Japanese and American listeners. J Acoust Soc Am 88:663–666.

Tartter VC (1991) Identifiability of vowels and speakers from whispered syllables. Percept Psychophys 49:365–372.

Trees DA, Turner CC (1986) Spread of masking in normal and high-frequency hearing-loss subjects. Audiology 25:70–83.

Treisman AM (1960) Contextual cues in selective listening. Q J Exp Psychol 12:242–248.

Treisman AM (1964) Verbal cues, language, and meaning in selective attention. Am J Psychol 77:206–219.

Turner CW, Bentler RA (1998) Does hearing aid benefit increase over time? J Acoust Soc Am 104:3673–3674.

Turner CW, Henn CC (1989) The relation between frequency selectivity and the recognition of vowels. J Speech Hear Res 32:49–58.

Turner CW, Souza PE, Forget LN (1995) Use of temporal envelope cues in speech recognition by normal and hearing-impaired listeners. J Acoust Soc Am 97:2568–2576.

Uchanski RM, Choi SS, Braida LD, Reed CM, Durlach NI (1994) Speaking clearly for the hard of hearing. IV: Further studies on speaking rate. J Speech Hear Res 39:494–509.

Van Tasell DJ, Fabry DA, Thibodeau LM (1987a) Vowel identification and vowel masking patterns of hearing-impaired listeners. J Acoust Soc Am 81:1586–1597.
Van Tasell DJ, Soli SD, Kirby VM, Widin GP (1987b) Temporal cues for consonant recognition: training, talker generalization, and use in evaluation in cochlear implants. J Acoust Soc Am 82:1247–1257.
Van Wijngaarden SJ, Steeneken HJM, Houtgast T (2002) Quantifying the intelligibility of speech in noise for non-native listeners. J Acoust Soc Am 111:1906–1916.
Veen TM, Houtgast T (1985) Spectral sharpness and vowel dissimilarity. J Acoust Soc Am 77:628–634.
Verschuure J, Brocaar MP (1983) Intelligibility of interrupted meaningful and nonsense speech with and without intervening noise. Percept Psychophys 33:232–240.
Viemeister N (1979) Temporal modulation transfer functions based upon modulation transfer functions. J Acoust Soc Am 66:1364–1380.
Viemeister NF (1980) Adaptation of masking. In: Brink G van der, Bilsen FA (eds) Psychophysical, Physiological and Behavioural Studies in Hearing. Delft: Delft University Press.
Viemeister NF, Bacon S (1982) Forward masking by enhanced components in harmonic complexes. J Acoust Soc Am 71:1502–1507.
Walden BE, Schwartz DM, Montgomery AA, Prosek RA (1981) A comparison of the effects of hearing impairment and acoustic filtering on consonant recognition. J Speech Hear Res 24:32–43.
Wang MD, Bilger RC (1973) Consonant confusions in noise: a study of perceptual features. J Acoust Soc Am 54:1248–1266.
Warren RM (1996) Auditory illusions and the perceptual processing of speech. In: Lass NJ (ed) Principles of Experimental Phonetics. St Louis: Mosby-Year Book.
Warren RM, Obusek CJ (1971) Speech perception and perceptual restorations. Percept Psychophys 9:358–362.
Warren RM, Obusek CJ, Ackroff JM (1972) Auditory induction: perceptual synthesis of absent sounds. Science 176:1149–1151.
Warren RM, Riener KR, Bashford Jr JA, Brubaker BS (1995) Spectral redundancy: intelligibility of sentences heard through narrow spectral slits. Percept Psychophys 57:175–182.
Warren RM, Hainsworth KR, Brubaker BS, Bashford A Jr, Healy EW (1997) Spectral restoration of speech: intelligibility is increased by inserting noise in spectral gaps. Percept Psychophys 59:275–283.
Watkins AJ (1988) Spectral transitions and perceptual compensation for effects of transmission channels. Proceedings of Speech '88: 7th FASE Symposium, Institute of Acoustics, pp. 711–718.
Watkins AJ (1991) Central, auditory mechanisms of perceptual compensation for spectral-envelope distortion. J Acoust Soc Am 90:2942–2955.
Watkins AJ, Makin SJ (1994) Perceptual compensation for speaker differences and for spectral-envelope distortion. J Acoust Soc Am 96:1263–1282.
Watkins AJ, Makin SJ (1996) Effects of spectral contrast on perceptual compensation for spectral-envelope distortion J Acoust Soc Am 99:3749–3757.
Webster JC (1983) Applied research on competing messages. In: Tobias JV, Schubert ED (eds) Hearing Research and Theory, vol. 2. New York: Academic Press, pp. 93–123.
Wegel RL, Lane CL (1924) The auditory masking of one pure tone by another and its probable relation to the dynamics of the inner ear. Phys Rev 23:266–285.

Young K, Sackin S, Howell P (1993) The effects of noise on connected speech: a consideration for automatic processing. In: Cooke M, Beet S, Crawford M (eds) Visual Representations of Speech. Chichester: John Wiley.
Yost WA, Dye RH, Sheft S (1996) A simulated "cocktail party" with up to three sound sources. Percept Psychophys 58:1026–1036.
Zahorian SA, Jagharghi AJ (1993) Spectral-shape features versus formants as acoustic correlates for vowels. J Acoust Soc Am 94:1966–1982.

6
Automatic Speech Recognition: An Auditory Perspective

Nelson Morgan, Hervé Bourlard, and Hynek Hermansky

1. Overview

Automatic speech recognition (ASR) systems have been designed by engineers for nearly 50 years. Their performance has improved dramatically over this period of time, and as a result ASR systems have been deployed in numerous real-world tasks. For example, AT&T developed a system that can reliably distinguish among five different words (such as "collect" and "operator") spoken by a broad range of different speakers. Companies such as Dragon and IBM marketed PC-based voice-dictation systems that can be trained by a single speaker to perform well even for speech spoken in a relatively fluent manner. Although the performance of such contemporary systems is impressive, their capabilities are still quite primitive relative to what human listeners are capable of doing under comparable conditions. Even state-of-the-art ASR systems still perform poorly under adverse acoustic conditions (such as background noise and reverberation) that present little challenge to human listeners (see Assmann and Summerfield, Chapter 5). For this reason the robust quality of human speech recognition provides a potentially important benchmark for the evaluation of automatic systems, as well as a fertile source of inspiration for developing effective algorithms for future-generation ASR systems.

2. Introduction

2.1 Motivations

The speaking conditions under which ASR systems currently perform well are not typical of spontaneous speech but are rather reflective of more formal conditions such as carefully read text, recorded under optimum conditions (typically using a noise-canceling microphone placed close to the speaker's mouth). Even under such "ideal" circumstances there are a number of acoustic factors, such as the frequency response of the microphone, as well as various characteristics of the individual speaker, such as dialect, speaking style, and gender, that can negatively affect ASR performance. These characteristics of speech communication, taken for granted by human listeners, can significantly obscure linguistically relevant information, particularly when the ASR system has been trained to expect an input that does not include such sources of variability.

It is possible to train a system with powerful statistical algorithms and training data that incorporates many (if not all) of the degradations anticipated in real-world situations (e.g., inside a passenger car). In this fashion the system "knows" what each word is supposed to sound like in a particular acoustic environment and uses this information to compensate for background noise when distinguishing among candidate words during recognition. However, the vast spectrum of potential degradations rules out the possibility of collecting such data for all possible background conditions. A statistical system can be trained with "clean data" (i.e., recorded under pristine acoustic conditions) that can later be adjusted during an adaptation phase incorporating a representative sample of novel, nonlinguistic factors (e.g., speaker identity). However, this latter strategy does not ensure higher recognition performance, since current approaches often require a significant amount of additional training data for the system to properly adapt. For this reason many forms of degradation cannot simply be compensated for by using algorithms currently in practice.

Because speech communication among humans is extremely stable despite great variability in the signal, designers of ASR systems have often turned to human mechanisms as a source of algorithmic inspiration. However, such humanly inspired algorithms must be applied with caution and care since the conditions under which ASR systems operate differ substantially from those characteristic of speech communication in general. The algorithms need to be customized to the specific applications for which they are designed or recognition performance is likely to suffer. For this reason it is often only the most general principles of human speech communication that form the basis of auditory approaches to automatic speech recognition. This chapter describes some of the auditory-inspired algorithms used in current-generation ASR systems.

2.2 Nonlinguistic Sources of Variance for Speech

There are many sources of acoustic variance in the speech signal not directly associated with the linguistic message, including the following:

A. Acoustic degradations
   1. Constant or slowly varying additive noise (e.g., fan noise)
   2. Impulsive, additive noise (e.g., door slam)
   3. Microphone frequency response
   4. Talker or microphone movement

5. Nonlinearities within the microphone
6. Segment-specific distortion attributable to the microphone (e.g., distortion resulting from high-energy signals)
7. Analog channel effects (such as delays and crosstalk)
8. Room reverberation

B. Speech production variations
1. Accent and dialect
2. Speaking style (reflected in speaking rate, vocal effort, and intonation)
3. Changes in speech production in response to the background acoustics (e.g., increase in vocal effort due to noise—the "Lombard" effect)
4. Acoustic variability due to specific states of health and mental state

2.3 Performance of Speech Recognition Systems with Respect to Human Hearing

Human speech communication degrades gracefully under progressively deteriorating acoustic conditions for the vast majority of circumstances associated with spoken language, even when the signal deviates appreciably from the "ideal" (i.e., carefully enunciated speech spoken in a relatively unreverberant environment). The ASR systems can be quite sensitive to such unanticipated changes to the signal or background. For the purposes of the present discussion, ASR performance will be evaluated in terms of the proportion of words incorrectly identified (word-error rate, WER), since this is the most common metric used. The WER is generally defined as the sum of word deletions, substitutions, and insertions. This score can, in principle, exceed 100%. Although this metric has its limitations with respect to characterizing the usability of an ASR system (whose performance also depends on such capabilities as rejecting out-of-vocabulary speech and error-recovery strategies), it can serve as a useful means for comparing the effectiveness of very different systems.

Recognition performance is quite high for corpora consisting of read text (e.g., *Wall Street Journal*), which has a 6% to 9% WER, relative to that associated with more spontaneous material (e.g., Switchboard, a corpus of recorded telephone conversations on a range of topics), which has a 20% to 30% WER. The large difference in performance appears to be at least partly a consequence of the variability in speaking style. Text is typically read in a more formal, precisely enunciated fashion than spontaneous speech, even when the text is a transcript of a spontaneous dialog, resulting in far fewer recognition errors (Weintraub et al. 1996).

When the performance of ASR systems is compared to that of human listeners, the machines come up short. Thus, humans have no trouble understanding most of the Switchboard dialog material or for *Wall Street Journal* material read under a wide range of signal-to-noise ratios (Lippmann 1997).

2.4 Auditory Perspectives and ASR

Since human speech perception is stable over a range of sources of variability (as discussed above), it seems reasonable to attempt to incorporate some of the principles of human hearing in our machine systems. While simple mimicry is unlikely to be effective (since the underlying human and machine mechanisms are so different), functional modeling of some of the human subsystems provide a plausible direction for research. In fact, many significant advances in speech recognition can be attributed to models of some of the gross properties of human speech perception. Examples include:

1. Computing measures of the short-term (ca. 25 ms) power spectrum, de-emphasizing the phase spectrum over this interval of time;
2. Bark or Mel-scale warping of the frequency axis, providing approximately linear spacing in the spectrum at low frequencies and logarithmic resolution above 1 kHz;
3. Spectral smoothing to minimize the influence of harmonic structure on a segment's phonetic identification;
4. Spectral normalization that reduces sensitivity to constant or slowly varying spectral colorations;
5. Multiple processing strategies and sources of knowledge.

The logic underlying the incorporation of auditory-like principles into ASR algorithms is straightforward—a "strong," robust model can use a limited amount of training data to maximum effect. Given the broad spectrum of acoustic variability inherent in the speech signal, it is unrealistic to train an ASR system on the full panoply of speech it is likely to encounter under real-world conditions. If such nonlinguistic variability can be represented by a low-dimensional model (using auditory-derived principles), it may be possible for an ASR system to "learn" all likely combinations of linguistic and nonlinguistic factors using a relatively limited amount of training data. One example of this approach models the effect of vocal-tract length by compression or expansion of the frequency axis (e.g., Kamm et al. 1995), resulting in a significant improvement of ASR performance (on the Switchboard corpus) despite the relative simplicity of the algorithm.

Analogous algorithms are likely to be developed over the next few years that will appreciably improve the performance of ASR systems. This chapter describes some of the recent research along these lines pertaining to the development of auditory-like processing for speech recognition.

3. Speech Recognition System Overview

To understand the auditory-inspired algorithms currently being used in speech recognition, it is useful to describe the basic structure of a typical ASR system.

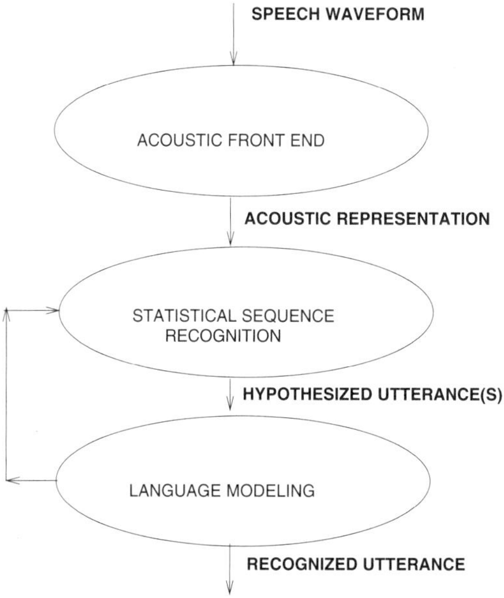

FIGURE 6.1. Generic block diagram for a speech recognition system.

3.1 A Typical System

Figure 6.1 illustrates the basic structure of such a system. Its primary components are the acoustic front end, statistical sequence recognition, and language modeling.

3.1.1 The Acoustic Front End

The input to this initial component is a digital representation of the signal waveform (typically sampled at a rate ranging from 8 to 16 kHz). This signal may be recorded by a microphone close to the speaker's mouth, from a telephone, or even in a relatively open acoustic environment. The output of this component is a sequence of variables computed to represent the speech signal in a fashion that facilitates the task of recognition. Current ASR systems typically compute some variant of a local spectrum (or its simple transformation into a cepstrum[1]). This signal processing is performed in order to emphasize those properties of the speech signal most directly associated with the linguistic message and to minimize the contribution of

[1]The cepstrum is the Fourier transform of the log of the magnitude spectrum. This is equivalent to the coefficients of the cosine components of the Fourier series of the log magnitude spectrum, since the magnitude spectrum is of even (symmetric) order. See Avendaño et al., Chapter 2, for a more detailed description of the cepstrum. Filtering of the log spectrum is called "liftering". It is often implemented by multiplication in the cepstral domain.

acoustic effects (such as reverberation, voice quality, and regional dialect) unrelated to the phonetic composition of the speech signal. It is at this stage of processing that most of the auditory-like algorithms are currently applied. Further details concerning the mathematical foundations underlying this stage of processing can be found in Avendaño et al., Chapter 2, as well as in Rabiner and Schaefer (1978).

3.1.2 Statistical Sequence Recognition

Once the acoustic representation of the speech signal has been computed, it is necessary to determine the local "costs" (corresponding to 10-ms frames of the acoustic analysis) for each "hypothesis" of a specific speech class (e.g., a phone). These local costs are then integrated into more global costs for hypothesizing an entire sequence of linguistic units (e.g., words). The cost functions are typically generated by a statistical system (i.e., a set of acoustic models for which statistics have been computed on a preexisting corpus of training data). To make this process tractable, many simplifying assumptions about the statistical distributions of the data are often made. For example, the data may be assumed to be distributed according to a Gaussian distribution. However, even if such an assumption were correct, the statistics derived from the training data may differ appreciably from what is observed during recognition (e.g., the presence of background noise not included in the training portion of the corpus). The major components of this processing stage are (1) distance or probability estimation, (2) hypothesis generation, and (3) hypothesis testing (search).

3.1.3 Language Modeling

This stage of the system is designed to constrain the recognition of sequences performed in the statistical modeling described above. Given a hypothesized sequence of words, this module generates either a list of allowable words or a graded lexical list with associated probabilities of occurrence. Simple statistical models currently dominate language modeling for most ASR systems due to their relative simplicity and robustness. Sometimes more highly structured models of natural language are used at a subsequent stage of processing to more accurately ascertain the desired flow of action. Improving the integration of such linguistic structure within the recognition process constitutes one of the major challenges for future development of ASR systems.

3.2 Auditory-Based Approaches to the Systems

The following sections describe some of the auditory-based approaches that have been used to improve the performance of the stages concerned with the acoustic front end and the statistical sequence recognition. The role played by auditory models is different for each of these components (cf. Fig. 6.1):

1. Acoustic front end: ASR systems perform feature extraction using signal processing techniques to compensate for variability in speaker characteristics as well as the acoustic environment. Nearly all attempts to integrate auditory strategies have focused primarily on this stage.

2. Statistical sequence recognition: The specific function of this module is highly dependent on the characteristics of the acoustic front end. Thus, a change in the front-end representation could degrade recognition performance if the statistical module has not been adapted to accommodate the new features. For example, models of forward temporal masking create additional contextual dependencies that need to be incorporated within the statistical models of phonetic sequences.

4. Acoustic Analysis in ASR

This section describes the acoustic front-end block of Figure 6.1, with an emphasis on those properties most directly associated with auditory properties.

4.1 Auditory-Like Processing and ASR

The earliest ASR systems utilized the power spectrum (thus ignoring short-term phase) for representing the speech signal at the output of the front-end stage in an effort to emphasize spectral peaks (formants) in the signal because of their significance for vocalic identification and speech synthesis. One of the first systems for recognizing speech automatically in this fashion was developed in the early 1950s by Davis et al. (1952). The system attempted to recognize 10 isolated digits spoken by a single speaker. The features extracted from the speech signal were two different types of zero-crossing quantities, each updated every 150 to 200 ms. One quantity was obtained from the lower frequency (below 900 Hz) band, the second from the upper band. The goal was to derive an estimate of the first two dominant formants in the signal. Despite this crude feature representation, the recognizer achieved 97% to 99% recognition accuracy.[2]

Subsequent ASR systems introduced finer spectral analysis employing a number of bandpass filters computing the instantaneous energy on several different frequency channels across the spectrum. In the earliest systems the filters were spaced linearly, while later implementations incorporated filter spacing inspired by models of the auditory filter bank, using either a Bark or Mel-frequency scale (Bridle and Brown 1974; Davis and Mermelstein

---

[2] It should be kept in mind that this was a highly constrained task. Results were achieved for a single speaker for whom the system had been trained, recorded under nearly ideal acoustic conditions, with an extremely limited vocabulary of isolated words.

1980). Other properties of the acoustic front end derived from models of hearing are

1. spectral amplitude compression (Lim 1979; Hermansky 1990);
2. decreasing sensitivity of hearing at lower frequencies (equal-loudness curves) (Itahashi and Yokoyama 1976; Hermansky 1990); and
3. large spectral integration (Fant 1970; Chistovich 1985) by principal component analysis (Pols 1971), either by cepstral truncation (Mermelstein 1976), or by low-order autoregressive modeling (Hermansky 1990).

Such algorithms are now commonly used in ASR systems in the form of either *Mel cepstral analysis* (Davis and Mermelstein 1980) or *perceptual linear prediction* (PLP) (Hermansky 1990). Figure 6.2 illustrates the basic steps in these analyses. Each of the major blocks in the diagram is associated with a generic module. To the side of the block is an annotation describing how the module is implemented in each technique. The initial preemphasis of the signal is accomplished via high-pass filtering. Such filtering removes any direct current (DC) offset[3] contained in the signal. The high-pass filtering also flattens the spectral envelope, effectively compensating for the 6-dB roll-off of the acoustic spectrum (cf. Avendaño et al., Chapter 2).[4] This simplified filter characteristic, implemented in Mel cepstral analysis with a first-order, high-pass filter, substantially improves the robustness of the ASR system. Perceptual linear prediction uses a somewhat more detailed weighting function, corresponding to an equal loudness curve at 40 dB sound pressure level (SPL) (cf. Fletcher and Munson 1933).[5]

The second block in Figure 6.2 refers to the short-term spectral analysis performed. This analysis is typically implemented via a fast Fourier transform (FFT) because of its computational speed and efficiency, but is equivalent in many respects to a filter-bank analysis. The FFT is computed for a predefined temporal interval (usually a 20- to 32-ms "window") using a specific [typically a Hamming (raised cosine)] function that is multiplied by the data. Each analysis window is stepped forward in time by 50% of the window length (i.e., 10–16 ms) or less (cf. Avendaño et al., Chapter 2, for a more detailed description of spectral analysis).

---

[3] The DC level can be of significant concern for engineering systems, since the spectral splatter resulting from the analysis window can transform the DC component into energy associated with other parts of the spectrum.

[4] The frequency-dependent sensitivity of the human auditory system performs an analogous equalization for frequencies up to about 4 kHz. Of course, in the human case, this dependence is much more complicated, being amplitude dependent and also having reduced sensitivity at still higher frequencies, as demonstrated by researchers such as Fletcher and Munson (1933) (see a more extended description in Moore 1989).

[5] As originally implemented, this function did not eliminate the DC offset. However, a recent modification of PLP incorporates a high-order, high-pass filter that acts to remove the DC content.

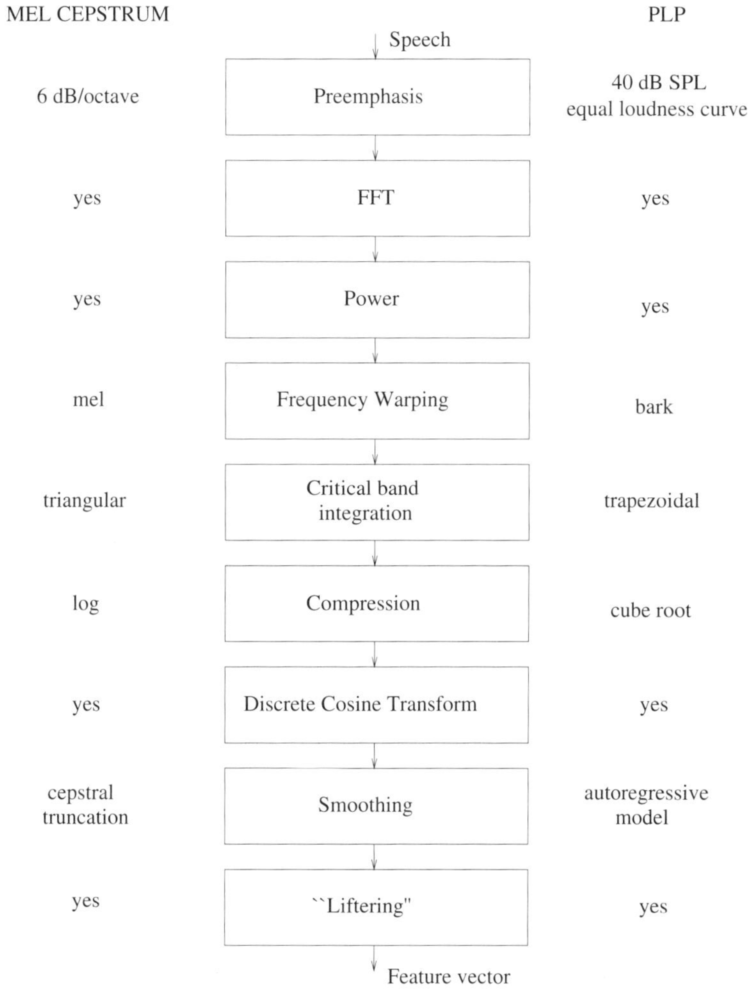

FIGURE 6.2. Generic block diagram for a speech recognition front end.

Only the short-term power spectrum is estimated on the assumption that the phase component of the spectrum can be disregarded over this short time interval. The power spectrum is computed by squaring the magnitude of the complex FFT coefficients.

In the fourth block, the frequency axis is warped in nonlinear fashion to be concordant with an auditory spatial-frequency coordinate system. Perceptual linear prediction uses the Bark scale and is derived from Schroeder (1977):

$$\Omega(\omega) = 6\ln\left(\frac{\omega}{1200\pi} + \left[\left(\frac{\omega}{1200\pi}\right)^2 + 1\right]^{0.5}\right) \tag{1}$$

where ω is the frequency in radians/second. A corresponding approximation used for the warping the frequency axis for Mel-cepstral processing (O'Shaughnessy 1987) is

$$\Omega(\omega) = 2595 \log_{10}\left[1 + \frac{\omega}{1400\pi}\right] \tag{2}$$

The effect of each transformation is to create a scaling that is quasi-linear below 1 kHz and logarithmic above this limit.

In the fifth component of the system, the power in the signal is computed for each critical-band channel, according to a specific weighting formula. In Mel-cepstral analysis, triangular filter characteristics are used, while in PLP the filters are trapezoidal in shape. This trapezoidal window is an approximation to the power spectrum of the critical band masking curve (Fletcher 1953) and is used to model the asymmetric properties of auditory filters (25 dB/Bark on the high-frequency slope and 10 dB/Bark on the lower slope).

In the compression module, the amplitude differential among spectral peaks is further reduced by computing a nonlinear transform of the critical band power spectrum. In PLP, this function is a cube root [based on Stevens's power law relating intensity to loudness (Stevens 1957)]. In Mel-cepstral analysis, a comparable compression is achieved by computing the logarithm of the critical-band power spectrum.

The discrete cosine transform module transforms the auditory-like spectrum into (solely real) coefficients specifying the amplitude of the cosine terms in a decomposition of the compressed spectrum. In the case of the Mel cepstrum, the output is in the form of cepstral coefficients (or amplitudes of the Fourier components of the log spectrum). For PLP this stage results in an output that is similar to an autocorrelation corresponding to the compressed power spectrum of the previous stage.

The penultimate module performs spectral smoothing. Although the critical band spectrum suppresses a certain proportion of the spectral fine structure, a separate level of integration is often useful for reducing the effects of nonlinguistic information on the speech signal. In Mel-cepstral processing, this step is accomplished by cepstral truncation—typically the lower 12 or 14 components are computed from 20 or more filter magnitudes. Thus, the higher Fourier components in the compressed spectrum are ignored and the resulting representation corresponds to a highly smoothed spectrum. In the case of PLP, an autoregressive model (derived by solving linear equations constructed from the autocorrelation of the previous step) is used to smooth the compressed critical-band spectrum. The resulting (highly smoothed) spectrum typically matches the peaks of the spectrum far better than the valleys. This property of the smoothing provides a more robust representation in the presence of additive noise. In the case of PLP, the autoregressive coefficients are converted to cepstral variables. In both instances, the end result is an implicit spectral integration that is somewhat broader than a critical-band (cf. Klatt 1982).

The final module multiplies the cepstral parameters by a simple function (such as n^{α}, where n is the cepstral index and α is a parameter between 0 and 1). The purpose of this liftering module is to modify the computed distances so as to adjust the sensitivity of the dynamic range of the peaks in the spectrum. In many modern systems the optimum weighting associated with each cepstral feature is automatically determined from the statistics of the training set, thus alleviating the need to compute this simple function.

4.2 Dynamic Features in ASR

Although spectral smoothing has significantly improved the performance of ASR systems, it has proven insufficient, by itself, to achieve the sort of robustness required by real-world applications. The following dynamic features, incorporating information about the temporal dependence among successive frames, have increased the robustness of ASR systems even further:

1. Delta (and delta-delta) features: The introduction of these features by Furui (1981) was the first successful attempt to model the acoustic dynamics of the speech signal, and these features were used to characterize the time trajectories of a variety of acoustic parameters. The delta features correspond to the slope (or velocity) and the delta-delta features to the curvature (or acceleration) associated with a specific parameter trajectory. The trajectories were based on cepstral coefficients in Furui's original implementation, but have since been applied by others to other spectral representations (including PLP). Such dynamic features are used in virtually all state-of-the-art ASR systems as an important extension of frame-by-frame short-term features. Their success may be at least partially attributed to the fact that dynamic features contribute new information pertaining to the context of each frame that was unavailable to the pattern classification component of an ASR system with purely static features.
2. RASTA processing: Analogous to delta computations, RASTA processing (Hermansky and Morgan 1994) filters the time trajectories of the acoustic features of the speech signal. However, it differs from dynamic ("delta") feature calculation in two respects: First, RASTA typically incorporates a nonlinearity applied between the compression and discrete cosine transform stages (cf. Fig. 6.2), followed by a bandpass filter, and an (approximately) inverse nonlinearity. For the logarithmic version of RASTA ("log RASTA"), the nonlinearity consists of a log operation and the inverse is an exponentiation. This latter variant of RASTA is optimal for reducing the effects of spectral coloration (that are either constant or varying very slowly with time). Second, RASTA typically uses a rather broad bandpass filter with a relatively flat pass-band (typically 3 to 10 Hz, with a gradual attenuation above 10 Hz and a spectral zero at 0 Hz), which preserves much of the phonetically important information in the feature representation. Although

the development of RASTA was initially motivated by the requirement of some form of auditory normalization (differentiation followed by integration), subsequent work has demonstrated that it bears some relation to both frequency modulation and temporal masking.

3. Cepstral mean subtraction: RASTA processing has been shown to provide greater robustness to very slowly varying acoustic properties of the signal, such as spectral coloration imposed by a particular input channel spectrum. When it is practical to compute an long-term average cepstrum of the signal this quantity can be subtracted from cepstra computed for short-time frames of the signal (cf. Stern and Acero 1989). The resulting cepstrum is essentially normalized in such a manner as to have an average log spectrum of zero. This simple technique is used very commonly in current ASR systems.

What each of these dynamic techniques has in common is a focus on information spanning a length of time greater than 20 to 30 ms. In each instance, this sort of longer time analysis results in an improvement in recognition performance, particularly with respect to neutralizing the potentially deleterious effects of nonlinguistic properties in the signal. The potential relevance of temporal auditory phenomena to these techniques will be examined in section 7.1.

4.3 Caveats About Auditory-Like Analysis in ASR

While a priori knowledge concerning the speech signal can be useful in guiding the design of an ASR system, it is important to exclude representational details not directly germane to the linguistic message. Because of this potential pitfall, incorporating highly detailed models has met with mixed success as a consequence of the following conditions:

1. *Testing on tasks that fail to expose the weaknesses inherent in conventional feature extraction techniques*: Speech recognizers generally work well on clean, well-controlled laboratory-collected data. However, successful application of auditory models requires demonstrable improvements in recognition performance under realistic acoustic conditions where conventional ASR techniques often fail.

2. *Failure to adapt the entire suite of recognition algorithms to the newly introduced feature extraction derived from auditory models*: Novel algorithms are often tested on a well-established task using a system finely tuned to some other set of techniques. In a complex ASR system there are many things that can go wrong, and usually at least one of them does when the new technique is interchanged with an older one.

3. *Evaluation by visual inspection*: The human visual cognitive system is very different from current ASR systems and has far greater (and somewhat different) powers of generalization. It is possible for a given representation to look better in a visual display than it functions in an ASR

system. The only current valid test of a representation's efficacy for recognition is to use its features as input to the ASR system and evaluate its performance in terms of error rate (at the level of the word, phone, frame, or some other predefined unit).

4. *Certain properties of auditory function may play only a tangential role in human speech communication*: For example, properties of auditory function characteristic of the system near threshold may be of limited relevance when applied to conversational levels (typically 40 to 70 dB above threshold). Therefore, it is useful to model the hearing system for conditions typical of real-world speech communication (with the appropriate levels of background noise and reverberation).

Clearly, listeners do not act on all of the acoustic information available. Human hearing has its limits, and due to such limits, certain sounds are perceptually less prominent than others. What might be more important for ASR is not so much what human hearing can detect, but rather what it does (and does not) focus on in the acoustic speech signal. Thus, if the goal of speech analysis in ASR is to filter out certain details from the signal, a reasonable constraint would be to either eliminate what human listeners do not hear, or at least reduce the importance of those signal properties of limited utility for speech recognition. This objective may be of greater importance in the long run (for ASR) than improving the fidelity of the auditory models.

4.4 Conclusions and Discussion

Certain properties of auditory function are currently being used for the extraction of acoustic features in ASR systems. Potentially important processing constraints may be derived in this way since properties of human speech recognition ultimately determine which components of the signal are useful for decoding the linguistic message and how to perform the decoding. The most popular analysis techniques currently used in ASR (e.g., Mel-cepstral analysis and PLP) use a nonlinear frequency scale, spectral amplitude compression, and an equal loudness curve. Longer-term temporal information, which is also incorporated in auditory perception, is just starting to be exploited. However, most ASR algorithms have been designed primarily from an engineering perspective without consideration of the actual ways in which the human auditory system functions.

5. Statistical Sequence Recognition

Despite the importance of acoustic analysis for ASR, statistical techniques are critically important for representing the variability inherent in these acoustic parameters. For this reason statistical models have been developed for learning and inference used in ASR.

This section discusses the mechanisms shown in the second block of Figure 6.1. Once the acoustic representation has been computed, statistical models are formulated to recognize specific sequences of speech units. Higher level knowledge, as incorporated in a language model (the third block of Fig. 6.1) is not addressed in this chapter for sake of brevity. A discussion of this important topic can be found in Rabiner and Juang (1993) and Jelinek (1998).

5.1 Hidden Markov Models

Although current speech recognition models are structurally simple, they can be difficult to understand. These models use rigorous mathematics and explicit optimization criteria, but the math is often "bent" to make the processing computationally tractable. For this reason, certain assumptions made by the computational models may not always be valid within the ASR domain. For instance, it is often necessary to exponentiate the language and/or acoustic model probabilities before they are combined as a means of weighting the relative importance of different probabilistic contributions to the classification decision.

Currently, the most effective family of statistical techniques for modeling the variability of feature vector sequences is based on a structure referred to as a *hidden Markov model* (HMM) (cf. section 4.1). An HMM consists of states, the transitions between them, and some associated parameters. It is used to represent the statistical properties of a speech unit (such as a phone, phoneme, syllable, or word) so that hypothetical state sequences associated with the model can have associated probabilities. The word "hidden" refers to the fact that the state sequence is not actually observed, but is hypothesized by choosing a series of transitions through the model. The term "Markov" refers to the fact that the mathematics does not involve statistical dependence on states earlier than the immediately previous one. An example of a simple HMM is shown in Figure 6.3. Consider it to represent the model of a brief word that is assumed to consist of three stationary parts. Parameters for these models are trained using sequences of feature vectors computed from the acoustic analysis of all available speech utterances. The resulting models are used as reference points for recognition, much as in earlier systems examples of sound units (e.g., words) were stored as reference patterns for comparison with unknown speech during recognition. The advantage of using statistical models for this purpose rather than literal examples of speech sounds (or their associated feature vectors) is that the statistics derived from a large number of examples often generalize much better to new data (for instance, using the means and variances of the acoustic features).

The sequences of state categories and observation vectors are each viewed as stochastic (statistical) processes. They are interrelated through "emission" probabilities, so called because the model views each feature

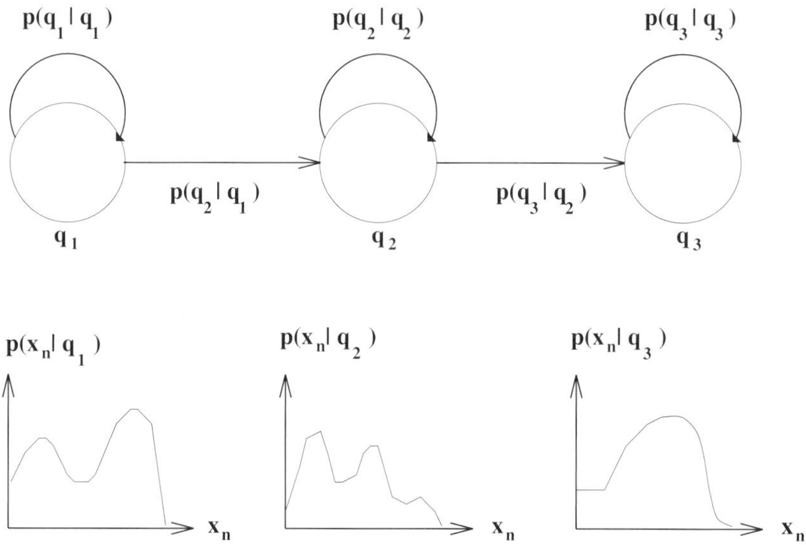

FIGURE 6.3. A three-state hidden Markov model (HMM). An HMM is a stochastic, finite-state machine, consisting of a set of states and corresponding transitions between states. The HMMs are commonly specified by a set of states q_i, an emission probability density $p(x_n|q_i)$ associated with each state, and transition probabilities $P(q_j \mid q_i)$ for each permissible transition from state q_i to q_j.

vector as having been generated or "emitted" on a transition to a specific state. The emission probabilities are actually probability densities for acoustic vectors conditioned on a choice for another probabilistic quantity, the state.

As will be discussed in section 5.4, HMMs are not derived from either auditory models or acoustic-phonetic knowledge per se, but are simply models that enable statistical mechanisms to deal with nonstationary time series. However, HMMs incorporate certain strong assumptions and descriptions of the data in practice, some of which may not be well matched to auditory models. This mismatch must be taken into account when auditory features or algorithms are incorporated in an HMM-based system.

The basic ideas and assumptions underlying HMMs can be summarized as follows:

1. Although speech is a nonstationary process, the sequence of feature vectors is viewed as a *piecewise stationary* process, or one in which there are regions of the sequence (each "piece") for which the statistics are the same. In this way, words and sentences can be modeled in terms of piecewise stationary segments. In other words, it is assumed that for each distinct state the probability density for the feature vectors will be the same for any time in the sequence associated with that state. This limitation is tran-

scended to some degree when models for the duration of HMMs are incorporated. However, the densities are assumed to instantaneously change when a transition is taken to a state associated with a different piecewise segment.

2. When this piecewise-stationarity assumption is made, it is necessary to estimate the statistical distribution underlying each of these stationary segments. Although the formalism of the model is very simple, HMMs currently require detailed statistical distributions to model each of the possible stationary classes. All observations associated with a single state are typically assumed to be conditionally independent and identically distributed, an assumption that may not particularly correspond to auditory representations.

3. When using HMMs, lexical units can be represented in terms of the statistical classes associated with the states. Ideally, there should be one HMM for every possible word or sentence in the recognition task. Since this is often infeasible, a hierarchical scheme is usually adopted in which sentences are modeled as a sequence of words, and words are modeled as sequences of subword units (usually phones). In this case, each subword unit is represented by its own HMM built up from some elementary states, where the topology of the HMMs is usually defined by some other means (for instance from phonological knowledge). However, this choice is typically unrelated to any auditory considerations. In principle, any subunit could be chosen as long as (a) it can be represented in terms of sequences of stationary states, and (b) one knows how to use it to represent the words in the lexicon. However, it is possible that choices made for HMM categories may be partially motivated by both lexical and auditory considerations. It is also necessary to restrict the number of subword units so that the number of parameters remains tractable.

5.2 *HMMs for ASR*

Hidden Markov models are the statistical basis for nearly all current ASR systems. Theory and methodology pertaining to HMMs are described in many sources, including Rabiner (1989) and Jelinek (1998). The starting point for the application of HMMs to ASR is to establish the optimality criterion for the task. For each observed sequence of acoustic vectors, X, the fundamental goal is to find the statistical model, M, that is most probable. This is most often reexpressed using Bayes's rule as

$$P(M|X) = \frac{P(X|M)P(M)}{P(X)} \tag{3}$$

in which $P(M|X)$ is the posterior probability of the hypothesized Markov model, M (i.e., associated with a specific sequence of words), given the acoustic vector sequence, X. Since it is generally infeasible to compute the left-hand side of the equation directly, this relation is usually used to split this posterior probability into a likelihood, $P(X|M)$, which represents the contribution of the *acoustic model*, and a prior probability, $P(M)$, which represents the contribution of the *language model*. $P(X)$ (in Eq. 3) is independent of the model used for recognition. Acoustic training derives parameters of the estimator for $P(X|M)$ in order to maximize this value for each example of the model. The language model, which will generate $P(M)$ during recognition, is optimized separately.

Once the acoustic parameters are determined during training, the resulting estimators will generate values for the "local" (per frame) emission and transition probabilities during recognition (see Fig. 6.3). These can then be combined to produce an approximation to the "global" (i.e., per utterance) acoustic probability $P(X|M)$ (assuming statistical independence). This multiplication is performed during a dedicated search phase, typically using some form of dynamic programming (DP) algorithm. In this search, state sequences are implicitly hypothesized. As a practical matter, the global probability values are computed in the logarithmic domain, to restrict the arithmetic range required.

5.3 Probability Estimators

Although many approaches have been considered for the estimation of acoustic probability densities, three are most commonly used in ASR systems:

1. Discrete estimators: In this approach feature vectors are quantized to the nearest entry in a codebook table. The number of times that an index co-occurs with a state label (e.g., the number of frames for which each index has a particular phonetic label) is used to generate a table of joint probabilities. This table can be turned into a set of conditional probabilities through normalizing by the number of frames used for each state label. Thus, training can be done by counting, and this estimation during recognition only requires a simple table lookup.
2. Mixtures of Gaussians: The most common approach is to use iterative estimation of means, variances, and mixture weights for a combination of Gaussian probabilities. The basic idea underlying this technique is that a potentially multimodal distribution can be represented by a sum of weighted Gaussian probability functions. Mixtures of Gaussian functions are often estimated using only diagonal elements in each covariance matrix (implicitly assuming no correlation among features). In practice this often provides a more effective use of parameters than using a full covariance matrix for a single Gaussian density.
3. Neural networks: Some researchers have utilized artificial neural networks (ANNs) to estimate HMM probabilities for ASR (Morgan and Bourlard 1995). The ANNs provide a simple mechanism for handling

acoustic context, local correlation of feature vectors, and diverse feature types (such as continuous and discrete measures). Probability distributions are all represented by the same set of parameters, providing a parsimonious computational structure for the ASR system. The resulting system is referred to as an HMM/ANN hybrid.

5.4 HMMs and Auditory Perspectives

Hidden Markov models are very useful for temporal pattern matching in the presence of temporal and spectral variability. However, it is apparent that as statistical representations they incorporate very little of the character of spoken language, since exactly the same HMMs may be used for other tasks, such as handwriting recognition, merely by changing the nature of the feature extraction performed.

This has been demonstrated for cursive handwriting recognition for the *Wall Street Journal* corpus (Starner et al. 1994). In this instance the same HMMs (retrained with handwriting features) were used to represent the written form of the corpus as was done for the spoken examples and the approach worked quite well. Thus, there was little specialization for the auditory modality beyond the feature-extraction stage.

In the most common implementations, the HMM formalism, as used for ASR, has no relation to the properties of speech or to auditory processing. An HMM is just a mathematical model, based on the assumptions described above, to compute the likelihood of a time series. It may ultimately be necessary to incorporate more of the characteristics of speech in these statistical models to more closely approach the robustness and accuracy of human speech recognition. Some of the assumptions that are incorporated in sequence recognition may also be incompatible with choices made in the signal-processing design. For instance, using dynamic auditory features may emphasize contextual effects that are poorly modeled by HMMs that incorporate strong assumptions about statistical independence across time. For example, RASTA processing can impair recognition performance unless explicit context dependence is incorporated within the HMMs (e.g., using different phone models for different phonetic contexts).

A critical issue is the choice of the units to be modeled. When there are a sufficient number of examples to learn from, word models have often been used for tasks such as digit recognition. For larger vocabulary tasks, models are most often developed from phone or subphone states. It is also possible to use transition-based units such as the diphone in place of the phone. Intermediate units such as the syllable or demisyllable have appealing properties, both in terms of providing commonality across many words and in terms of some of their relation to certain temporal properties of hearing. It may be desirable to develop statistical representations for portions of the spectrum, particularly to accommodate asynchrony among disparate spectral bands in realistic environments (cf. section 7.3). More generally, it may be advantageous to more thoroughly integrate the statistical modeling and the acoustic analysis, providing mechanisms for integrating multiple auditory perspectives by merging statistical models that are associated with each perspective.

Finally, the probability estimation may need to be quite flexible to incorporate a range of new acoustical analyses into the statistical framework. Section 7 describes an approach that provides some of this flexibility.

6. Capabilities and Limitations of Current ASR Systems

Thus far, this chapter has focused on the basic engineering tools used in current ASR systems. Using such computational methods, it has been possible to develop highly effective large-vocabulary, speaker-independent, continuous speech recognition systems based on HMMs or hybrid HMM/ANNs, particularly when these systems incorporate some form of individual speaker adaptation (as many commercial systems currently do). However, for the system to work well, it is necessary to use a close-talking microphone in a reasonably quiet acoustic environment. And it helps if the user is sufficiently well motivated as to speak in a manner readily recognized by the system. The system need not be as highly controlled for smaller-scale vocabulary tasks such as those used in many query systems over the public telephone network.

All of these commercial systems rely on HMMs. Such models represent much of the temporal and spectral variability of speech and benefit from powerful and efficient algorithms that enable training to be performed on large corpora of both continuous and isolated-word material. Flexible decoding strategies have also been developed for such models, in some cases resulting in systems capable of performing large vocabulary recognition in real time on a Pentium-based computer.

For training an HMM, explicit segmentation of the speech stream is not required as long as the identity and order of the representational units (typically phones and words) are provided. Given their very flexible topological structure, HMMs can easily be enhanced to include statistical information pertaining to either phonological or syntactic rules.

However, ASR systems still commit many errors in the presence of additive noise or spectral colorations absent from the training data. Other factors that present potential challenges to robust ASR include reverberation, rapid speaking rate, and speech babble in the background, conditions that rarely affect the recognition capabilities of human listeners (but cf. Assmann and Summerfield, Chapter 5; Edwards, Chapter 7).

There have been many attempts to generalize the application of HMMs to ASR in an attempt to provide a robust statistical framework for optimizing recognition performance. This is usually accomplished by increasing the variety of representational units of the speech signal (typically by using context-dependent phone models) and the number of parameters to describe each model (e.g., using mixtures of Gaussian probability density functions). This approach has led to significant improvements in recognition performance for large-vocabulary corpora. Unfortunately, such improvements have not been transferred, as yet, to a wide range of spontaneous speech material due to such factors as variation in speaking rate, vocal effort, pronunciation, and disfluencies.

7. Future Directions

Thus far, this chapter has focused on approaches used in current-generation ASR systems. Such systems are capable of impressive performance under ideal speaking conditions in highly controlled acoustic environments but do not perform as well under many real-world conditions. This section discusses some promising approaches, based on auditory models, that may be able to ultimately improve recognition performance under a wide range of circumstances that currently foil even the best ASR systems.

7.1 Temporal Modeling

Temporal properties often serve to distinguish between speech and nonspeech signals and can also be used to separate different sources of speech. Various lines of evidence suggest that the "short-term memory" of the auditory system far exceeds the 20- to 30-ms interval that is conventionally used for ASR analysis. For example, studies of forward masking (Zwicker 1975; Moore 1989), adaptation of neural firing rates at various levels of the auditory pathway (Aitkin et al. 1966), and the buildup of loudness (e.g., Stevens and Hall 1966) all suggest that a time window of 100 to 250ms may be required to model many of the auditory mechanisms germane to speech processing.

What temporal properties of speech are likely to serve as important cues for recognition? A consideration of human perceptual capabilities is likely to provide some insight. Among the perceptual properties of interest are the following:

1. Forward masking: If one stimulus (masker) is followed closely in time by another (probe), the detectability of the latter is diminished. This process is highly nonlinear since, independently of the masker amplitude, masking is not evident after about 100 to 200ms (e.g., Moore 1989).
2. Perception of modulation components: Since the early experiments of Riesz (1928), it has often been noted that the sensitivity of human hearing to both amplitude and frequency modulation is highest for modulation frequencies between 3 and 6Hz. Thus, the perception of modulated signals appears to be governed by a bandpass characteristic. This is matched by the

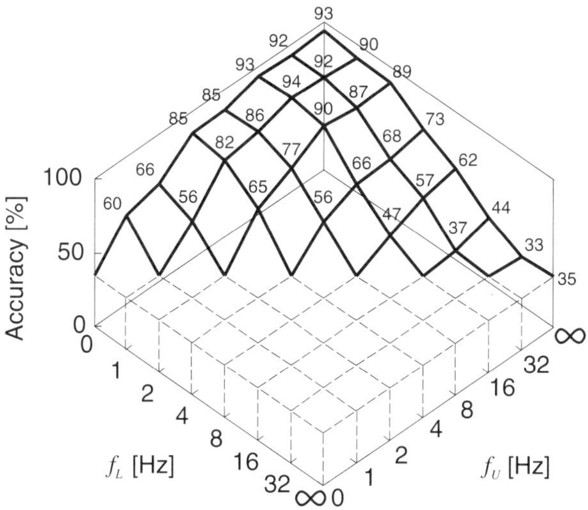

FIGURE 6.4. Intelligibility of a subset of Japanese monosyllables as a function of temporal filtering of spectral envelopes of speech. The modified speech is highly intelligible when $f_L \leq 1$ Hz and $f_U \geq 16$ Hz. The data points show the average over 124 trials. The largest standard error of a binomial distribution with the same number of trials is less than 5%.

properties of the speech signal whose dominant modulation frequencies lie in the same range (Houtgast and Steeneken 1985). Drullman et al. (1994a,b) showed that low-pass filtering of envelope information at frequencies higher than 16 Hz or high-pass filtering at frequencies lower than 2 Hz causes virtually no reduction in the intelligibility of speech. Thus, the bulk of linguistically relevant information appears to lie in the region between 2 and 16 Hz, with dominant contributions made by components between 4 and 6 Hz. Arai et al. (1996) conducted experiments similar to Drullman et al.'s, but using a somewhat different signal processing paradigm [based on a residual-excited linear predictive coding (LPC) vocoder, and aiming for bandpass processing of trajectories of spectral envelope] and speech materials (Japanese monosyllables rather than Dutch sentences or words). The results of this experiment are illustrated in Figure 6.4.

3. Focusing on transitions: Those portions of the signal where there is significant spectral change appear to be particularly important for encoding phonetic information (e.g., Furui 1986). Such regions, often associated with phonetic transitions, may also be of critical importance in ensuring the robustness of human listening to adverse acoustic conditions. Certain signal proccssing techniques, such as RΛSTΛ or delta features, tend to place greater emphasis on regions of spectral change. Other methods,

such as dropping frames of quasi- stationary spectral information (Tappert and Das 1978) or using diphones as speech units (Ghitza and Sondhi 1993), implicitly emphasize transitions as well. Although emphasizing spectral dynamics may be beneficial from a signal processing perspective, such an approach may conflict with the underlying structural formalism of HMMs. For example, when similar frames are dropped, there is a potential reduction in correlation among successive observation vectors. However, this frame-reduction process may violate certain assumptions inherent in the design of HMMs, as well as in their training and decoding procedures. Frames that remain may correspond to regions in the signal that exhibit a significant amount of spectral movement. A sequence of frames consisting entirely of such rapidly varying spectral features may be difficult to model with standard HMM techniques that usually assume some local or "quasi" stationarity property of the representational units and could thus require major changes to the basic HMM architecture (cf. section 7.2 and Morgan et al. 1995).

Certain other properties of auditory processing have been modeled and successfully applied to recognition (e.g., Cohen 1989); however, much of this work has emphasized relatively short-term (<50 ms) phenomena. Current efforts in applying auditory models to ASR are focused on relatively long time constants, motivated in part by the desirability of incorporating sufficiently long time spans as to distinguish between the actual speech spectrum and slowly varying (or static) degradations, such as additive stationary noise. Thus, it would be a mistake to apply either RASTA filtering or cepstral mean subtraction on speech segments too short to encompass phonetic context associated with neighboring phones. It is important for the processing to "see" enough of the signal to derive relativistic acoustic information. The length of time required is on the order of a demi- or whole syllable (i.e., 100–250 ms).

Using subsequences of input vectors (centered on the current frame) as a supplement to the current acoustic vector generally improves recognition performance. Different approaches have been taken. Among the most popular are:

1. Within the context of Gaussian-mixture-based HMMs, linear discriminant analysis (LDA) is occasionally applied over a span of several adjacent frames to reduce the dimensionality of the acoustic features while concurrently minimizing the intraclass and maximizing the interclass variance (Brown 1987).
2. It has often been shown within the context of an hybrid HMM/ANN system that both frame classification and word recognition performance improves when multiple frames are provided at the input of the network in order to compute local probabilities for the HMM (Morgan and Bourlard 1995).

The multivector input technique relies on a classifier to deduce the relative importance of temporally asynchronous analysis vectors for classifying a specific span of speech.

The strength of such methods probably lies in their computational expansion of the "local" perspective of the speech signal to one that is inherently multiscale in time and complexity (i.e., consisting of segmental blocks ranging from phonetic segments to syllables). These longer spans of time can be useful for deriving features for the subsequent classification (cf. Hermansky 1995).

In summary, many techniques have been developed for the incorporation of temporal properties of the speech signal over longer time regions than are traditionally used for acoustic analysis in ASR systems. Although RASTA processing can be viewed as a crude approximation to the effects of forward masking, more detailed models of forward masking have yet to be fully incorporated into ASR systems. Techniques are only beginning to be developed for coupling the longer-term features with statistical models that are associated with longer-term units (e.g., the syllable), and for combining multiple sources of information using different time ranges. This approach has been successfully applied to speech recognition under reverberant conditions (Wu et al. 1998). It can also be useful to analyze much longer time regions (e.g., 1 to 2 seconds) to estimate speaking rate, a variable that has a strong effect on spectral content, phonetic durations, and pronunciations (Morgan et al. 1997). It may also be advantageous to consider statistical units that are more focused on phonetic transitions and acoustic trajectories rather than on piecewise stationary segments.

7.2 Matching the Statistical Engine

Incorporating auditory models into ASR systems does not always result in improved recognition performance. Changing the acoustic front end while keeping the remaining components of the system unchanged may not provide an optimal solution. The HMM formalism is very powerful and can easily be extended in different ways to accommodate factors such as context-dependent classes and higher-order models. However, HMMs are also severely constrained by the assumptions they make about the signal.

One attempt to handle the inconsistency of the piecewise stationary assumption of conventional HMMs was the proposal of Ghitza and Sondhi (1993) to use the diphone as a basic unit in HMM recognition based on a hand-segmented training corpus. Another approach has been proposed by Deng et al. (1994) using a so-called nonstationary HMM. This form of HMM represents a phoneme not only by its static properties but also by its first-order (and sometimes higher-order) dynamics. Yet a separate approach, stochastic perceptual auditory-event–based modeling (SPAM), was been explored by Morgan et al. (1995). In SPAM speech is modeled as a succession of auditory events or "avents," separated by relatively stationary periods (ca. 50–150 ms). Avents correspond to points of decision concerning a phonetic segment. Such avents are associated with times when the signal spectrum and amplitude are rapidly changing. The stationary periods are mapped to a single tied state, so that modeling power is focused on regions of significant change. This approach appears to provide additional robustness to additive noise when used in combination with a more traditional, phone-based system.

Other researchers have explored using explicit segmentation where all of the segments are treated as stochastic variables. Although reliable automatic phonetic segmentation is difficult to achieve, it is possible to choose a likely segmentation pattern from many hypothetical segmentations based on a variety of representational and statistical criteria. If such a strategy is adopted, it is then easier to work at the segment level and to adapt the analysis strategies based on specific hypotheses (cf. Zue et al. 1989 for a successful example of such approaches). Another alternative, referred to as "stochastic segment modeling" (Ostendorf et al. 1992), relies on standard HMMs to generate multiple segmentations by using a search strategy that incorporates N hypothesized word sequences (N-best) rather than the single best. Phonetic segments can thus be considered as stochastic variables for which models can be built and later used to rescore the list of candidate sentences. Segmentation and modeling at the syllabic level may also provide an alternative strategy for overcoming some of the inherent complexity of the segmentation process (cf. Hunt et al. 1980; Wu et al. 1998).

7.3 Sub-Band Analysis

Virtually all ASR systems estimate state probability densities from a 16- to 32-ms "slice" of the speech signal distributed across the spectrum. However, it is also possible to compute phone probabilities based on sub-bands of the spectrum and then combine these probability estimations at a subsequent stage of processing. This perspective has been motivated by the articulation theory of Harvey Fletcher and colleagues (Fletcher 1953; Allen 1994). Fletcher suggested that the decoding of the linguistic message is based on decisions made within relatively circumscribed frequency channels processed independently of one another. Listeners recombine decisions from these frequency bands so that the global error rate is equal to the product of "band-limited" error rates within independent frequency channels. This reasoning implies that if any of the frequency bands yield zero error rate, the resulting error rate should also be zero, independent of the error rates in the remaining bands. While the details of this model have often been challenged, there are several reasons for designing systems that combine decisions (or probabilities) derived from independently computed channels:

1. Better robustness to band-limited noise;
2. Asynchrony across channels;
3. Different strategies for decoding the linguistic message may be used in different frequency bands.

Preliminary experiments (Bourlard and Dupont 1996; Hermansky et al. 1996) have shown that this multiband technique could lead to ASR systems that are more robust to band-limited noise. Improvement in recognition has also been observed when a sub-band system is used in conjunction with a full-band system (http://www.clsp.jhu.edu/ris/results-report.html). Furthermore, the multiband approach may be able to better model the asynchrony across frequency channels (Tomlinson et al. 1997). The variance of sub-band phonetic boundaries has been observed to be highly dependent on the speaking rate and the amount of acoustic reverberation (Mirghafori and Morgan 1998), suggesting that sub-band asynchrony is indeed a source of variability that could potentially be compensated for in a sub-band–based system that estimates acoustic probabilities over time spans longer than a phone.

7.4 Multiple Experts for ASR

Sub-band analysis is only one of several potential approaches for combining multiple perspectives of the speech signal. Acoustic analyses can be performed separately on different spectral bands, and combined after some preliminary soft decision process. A number of experiments have shown that combining probability estimates from streams with differing temporal characteristics can also improve performance (Dupont et al. 1997; Wu et al. 1998). More generally, different acoustic, prosodic, and linguistic experts can combine their partial decisions at some higher level. Multimodal input (for instance, from lip reading and acoustics) is also being explored (e.g., Wolff et al. 1994).

7.5 Auditory Scene Analysis

Auditory processes have recently begun to be examined from the standpoint of acoustic object recognition and scene analysis (cf. Assmann and Summerfield, Chapter 5; Bregman 1990). This perspective, referred to as "auditory scene analysis," may provide a useful complement to current ASR strategies (cf. Cooke et al. 1994). Much of the current work is focused on separating sound "streams," but researchers in this area are also exploring these perspectives from the standpoint of robustness to "holes" in the spectrum that arise from noise or extreme spectral coloration (e.g., Hermansky et al. 1996; Cooke et al. 1997).

8. Summary

This chapter has described some of the techniques used in automatic speech recognition, as well as their potential relation to auditory mechanisms. Human speech perception is far more robust than ASR. However, it is still unclear how to incorporate auditory models into ASR systems as a means of increasing their performance and resilience to environmental degradation.

Researchers will need to experiment with radical changes in the current paradigm, although such changes may need to be made in a stepwise fashion so that their impact can be quantified and therefore better understood. It is likely that any radical changes will lead initially to increases in the error rate (Bourlard et al. 1996) due to problems integrating novel algorithms into a system tuned for more conventional types of processing.

As noted in a commentary written three decades ago (Pierce 1969), speech recognition research is often more like tinkering than science, and an atmosphere that encourages scientific exploration will permit the development of new methods that will ultimately be more stable under real-world conditions. Researchers and funding agencies will therefore need to have the patience and perseverance to pursue approaches that have a sound theoretical and methodological basis but that do not improve performance immediately.

While the pursuit of such basic knowledge is crucial, ASR researchers must also retain their perspective as engineers. While modeling is worthwhile in its own right, application of auditory-based strategies to ASR requires a sense of perspective—Will particular features potentially affect performance? What problems do they solve? Since ASR systems are not able to duplicate the complexity and functionality of the human brain, researchers need to consider the systemwide effects of a change in one part of the system. For example, generation of highly correlated features in the acoustic front end can easily hurt the performance of a recognizer whose statistical engine assumes uncorrelated features, unless the statistical engine is modified to account for this (or, alternatively, the features are decorrelated).

Although there are many weaknesses in current-generation systems, the past several decades have witnessed development of powerful algorithms for learning and statistical pattern recognition. These techniques have worked very well in many contexts and it would be counterproductive to entirely discard such approaches when, in fact, no alternative mathematical structure currently exists. Thus, the mathematics applied to dynamic systems has no comparably powerful learning techniques for application to fundamentally nonstationary phenomena. On the other hand, it may be necessary to change current statistical sequence recognition approaches to improve their applicability to models and strategies based on the deep structure of the phenomenon (e.g., production or perception of speech), to better integrate the different levels of representation (e.g., acoustics and language), or to remove or reduce the inaccurate assumptions that are used in the practical application of these methods.

The application of auditory strategies to ASR may help in developing auditory models. Although reduction of machine word-error rate does not in any way prove that a particular strategy is employed by humans, the failure of an approach to handle a specific signal degradation can occasionally rule out specific hypotheses. Both ASR researchers and auditory modelers must face the fundamental quandaries of dealing with partial information and signal transformations during recognition that are not well represented in the data used to train the statistical system.

List of Abbreviations

ANN	artificial neural networks
ASR	automatic speech recognition
DC	direct current (ØHz)
DP	dynamic programming
FFT	fast Fourier transform
HMM	hidden Markov model
LDA	linear discriminant analysis
LPC	linear predictive coding
PLP	perceptual linear prediction
SPAM	stochastic perceptual auditory-event–based modeling
SPL	sound pressure level
WER	word-error rate

References

Aitkin L, Dunlop C, Webster W (1966) Click-evoked response patterns of single units in the medial geniculate body of the cat. J Neurophysiol 29:109–123.

Allen JB (1994) How do humans process and recognize speech? IEEE Trans Speech Audiol Proc 2:567–577.

Arai T, Pavel M, Hermansky H, Avendano C (1996) Intelligibility of speech with filtered time trajectories of spectral envelopes. Proc Int Conf Spoken Lang Proc, pp. 2490–2493.

Bourlard H, Dupont S (1996) A new ASR approach based on independent processing and recombination of partial frequency bands. Proc Int Conf Spoken Lang Proc, pp. 426–429.

Bourlard H, Hermansky H, Morgan N (1996) Towards increasing speech recognition error rates. Speech Commun 18:205–231.

Bregman AS (1990) Auditory Scene Analysis. Cambridge: MIT Press.

Bridle JS, Brown MD (1974) An experimental automatic word recognition system. JSRU Report No. 1003. Ruislip, England: Joint Speech Research Unit.

Brown P (1987) The Acoustic-Modeling Problem in Automatic Speech Recognition. Ph.D. thesis, Computer Science Department, Carnegie Mellon University, Pittsburgh.

Chistovich LA (1985) Central auditory processing of peripheral vowel spectra. J Acoust Soc Am 77:789–805.

Cohen JR (1989) Application of an auditory model to speech recognition. J Acoust Soc Am 85:2623–2629.

Cooke MP, Green PD, Crawford MD (1994) Handling missing data in speech recognition. Proc Int Conf Spoken Lang Proc, pp. 1555–1558.

Cooke MP, Morris A, Green PD (1997) Missing data techniques for robust speech recognition. Proc IEEE Int Conf Acoust Speech Signal Proc, pp. 863–866.

Davis KH, Biddulph R, Balashek S (1952) Automatic recognition of digits. J Acoust Soc Am 24:637–642.

Davis S, Mermelstein P (1980) Comparison of parametric representations of monosyllabic word recognition in continuously spoken sentences. IEEE Trans Acoust Speech Signal Proc 28:357–366.

Deng L, Aksmanovic M, Sun X, Wu C (1994) Speech recognition using hidden Markov models with polynomial regression functions as nonstationary states. IEEE Trans Speech Audiol Proc 2:507–520.

Drullman R, Festen JM, Plomp R (1994a) Effect of temporal envelope smearing on speech reception. J Acoust Soc Am 95:1053–1064.

Drullman R, Festen JM, Plomp R (1994b) Effect of reducing slow temporal modulations on speech reception. J Acoust Soc Am 95:2670–2680.

Dupont S, Bourlard H, Ris C (1997) Using multiple time scales in a multi-stream speech recognition system. Proc Euro Speech Tech Comm (Eurospeech-97), pp. 3–6.

Fant G (1970) Acoustic Theory of Speech Production. Mouton: The Hague.

Fletcher H (1953) Speech and Hearing in Communication. New York: Van Nostrand.

Fletcher H, Munson W (1933) Loudness, its definition, measurement, and calculation. J Acoust Soc Am 5:82–108.

Furui S (1981) Cepstral analysis technique for automatic speaker verification. IEEE Trans Acoust Speech Signal Proc 29:254–272.

Furui S (1986) On the role of spectral transition for speech perception. J Acoust Soc Am 80:1016–1025.

Ghitza O, Sondhi MM (1993) Hidden Markov models with templates as nonstationary states: an application to speech recognition. Comput Speech Lang 2:101–119.

Hermansky H (1990) Perceptual linear predictive (PLP) analysis of speech. J Acoust Soc Am 87:1738–1752.

Hermansky H (1995) Modulation Spectrum in Speech Processing, in "Signal Analysis and Prediction", Prochazka, Uhlin, Royner, Kingshury, Eds. Boston: Birkhauser.

Hermansky H, Morgan N (1994) RASTA processing of speech. IEEE Trans Speech Audiol Proc 2:578–589.

Hermansky H, Tibrewala S, Pavel M (1996) Toward ASR on partially corrupted speech. Proc Int Conf Spoken Lang Proc, pp. 462–465.

Houtgast T, Steeneken HJM (1985) A review of the MTF concept in room acoustics and its use for estimating speech intelligibility in auditoria. J Acoust Soc Am 77: 1069–1077.

http://www.clsp.jhu.edu/ws96/ris/results-report.html (1996) WWW page for Johns Hopkins Switchboard Workshop 96, speech data group results page.

Hunt M, Lennig M, Mermelstein P (1980) Experiments in syllable-based recognition of continuous speech. Proc IEEE Int Conf Acoust Speech Signal Proc, pp. 880–883.

Itahashi S, Yokoyama S (1976) Automatic formant extraction utilizing mel scale and equal loudness contour. Proceedings of the IEEE International Conference on Acoustics, Speech, and Signal Processing, Philadelphia, PA, pp. 310–313.

Jelinek F (1998) Statistical Methods for Speech Recognition. Cambrdige: MIT Press.

Kamm T, Andreou A, Cohen J (1995) Vocal tract normalization in speech recognition: compensating for systematic speaker variability. Proc 15th Ann Speech Res Symp, Baltimore, MD, pp. 175–178.

Klatt DH (1982) Speech processing strategies based on auditory models. In: Carlson R, Granstrom B (eds) The Representation of Speech in the Peripheral Auditory System. Amsterdam: Elsevier, pp. 181–202.

Lim JS (1979) Spectral root homomorphic deconvolution system. IEEE Trans Acoust Speech Signal Proc 27:223–233.

Lippmann RP (1997) Speech recognition by machines and humans. Speech Commun 22:1–16.

Mermelstein P (1976) Distance measures for speech recognition, psychological and instrumental. In: Chen RCH (ed) Pattern Recognition and Artificial Intelligence. New York: Academic Press, pp. 374–388.

Mirghafori N, Morgan, N (1998) Transmissions and transitions: a study of two common assumptions in multi-band ASR. Proc IEEE Int Conf Acoust Speech Signal Proc, pp. 713–716.

Moore B (1989) An Introduction to the Psychology of Hearing. London: Academic Press.

Morgan N, Bourlard H (1995) Continuous speech recognition: an introduction to the hybrid HMM/connectionist approach. Signal Processing Magazine 25–42.

Morgan N, Bourlard H, Greenberg S, Hermansky H (1995) Stochastic perceptual models of speech. Proc IEEE Int Conf Acoust Speech Signal Proc, pp. 397–400.

Morgan N, Fosler E, Mirghafori N (1997) Speech recognition using on-line estimation of speaking rate. Proc Euro Conf Speech Tech Comm (Eurospeech-97), pp. 1951–1954.

O'Shaughnessy D (1987) Speech Communication, Human and Machine. Reading, MA: Addison-Wesley.

Ostendorf M, Bechwati I, Kimball O (1992) Context modeling with the stochastic segment model. Proc IEEE Int Conf Acoust Speech Signal Proc, pp. 389–392.

Pierce JR (1969) Whither speech recognition? J Acoust Soc Am 46:1049–1051.

Pols LCW (1971) Real-time recognition of spoken words. IEEE Trans Comput 20:972–978.

Rabiner LR (1989) A tutorial on hidden Markov Models and selected applications in speech recognition. Proc IEEE 77:257–285.

Rabiner LR, Juang BH (1993) Fundamentals of Speech Recognition. Englewood Cliffs, NJ: Prentice Hall.

Rabiner LR, Schaefer RW (1978) Digital Processing of Speech Signals. Englewood Cliffs, NJ: Prentice Hall.

Riesz RR (1928) Differential intensity sensitivity of the ear for pure tones. Phys Rev 31:867–875.

Schroeder M (1977) Recognition of Complex Signals. In: Bullock TH (ed) Life Sciences Research Report 5. Berlin: Abakon Verlag, p. 324.
Starner T, Makhoul J, Schwartz R, Chou G (1994) On-line cursive handwritten recognition using speech recognition methods. Proc IEEE Int Conf Acoust Speech Signal Proc, pp. V-125–128.
Stern R, Acero A (1989) Acoustical pre-processing for robust speech recognition. Proc Speech Nat Lang Workshop, pp. 311–318.
Stevens SS (1957) On the psychophysical law. Psychol Rev 64:153–181.
Stevens JC, Hall JW (1966) Brightness and loudness as function of stimulus duration. Perception and Psychophysis, pp. 319–327.
Tappert CC, Das SK (1978) Memory and time improvements in a dynamic programming algorithm fpr matching speech pattersn. IEEE Trans Acoust Speech Signal Proc 26:583–586.
Tomlinson MJ, Russell MJ, Moore RK, Buckland AP, Fawley MA (1997) Modeling asynchrony in speech using elementary single-signal decomposition. Proc IEEE Int Conf Acoust Speech Signal Proc, pp. 1247–1250.
Weinstraub M, Taussig K, Hunicke-Smith K, Snodgrass A (1996) Effect of speaking style on LVCSR performance. Proc Int Conf Spoken Lang Proc, pp. S1–4.
Wolff G, Prasad K, Stork D, Hennecke M (1994) Lipreading by neural networks: visual preprocessing, learning, and sensory integration. In: Cowan J, Tesauro G, Alspector J (eds) Advances in Neural Information Processing 6, San Francisco: Morgan-Kaufmann, pp. 1027–1034.
Wu SL, Kingsbury B, Morgan N, Greenberg S (1998) Performance improvements through combining phone and syllable-scale information in automatic speech recognition. Proc IEEE Int Conf Acoust Speech Signal Proc, pp. 854–857.
Zue V, Glass J, Phillips M, Seneff S (1989) Acoustic segmentation and phonetic classification in the Summit system. Proceedings of the IEEE International Conference on Acoustics, Speech, and Signal Processing, Glasgow, Scotland, pp. 389–392.
Zwicker E (1975) Scaling. In: Keidel D, Neff W (eds) Handbook of Sensory Physiology. Berlin: Springer-Verlag, 3:401–448.

7 Hearing Aids and Hearing Impairment

Brent Edwards

1. Introduction

Over 25 million people in the United States have some form of hearing loss, yet less than 25% of them wear a hearing aid. Several reasons have been cited for the unpopularity of the hearing aid:

1. the stigma associated with wearing an aid,
2. the perception that one's hearing loss is milder than it really is,
3. speech understanding is satisfactory without one,
4. cost, and
5. one's hearing has not been tested (Kochkin 1993).

One compelling reason may be an awareness that today's hearing aids do not adequately correct the hearing loss of the user.

The performance of hearing aids has been limited by several practical technical constraints. The circuitry must be small enough to fit behind the pinna or inside the ear canal. The required power must be sufficiently low that the aid can run on a low-voltage (1.3 V) battery for several consecutive days. Until recently, the signal processing required had to be confined to analog technology, precluding the use of more powerful signal-processing algorithms that can only effectively be implemented on digital chips.

A more important factor has been the absence of a scientific consensus on precisely what a hearing aid should do to properly compensate for a hearing loss. For the better part of the 20th century, research pertaining to this specific issue had been stagnant (reasons for this circumstance are discussed by Studebaker 1980). But over the past 25 years there has been a trend toward increasing sophistication of the processing performed by a hearing aid, as well as an attempt to match the aid to specific properties of an individual's hearing loss. The 1980s produced commercially successful hearing aids using nonlinear processing based on the perceptual and physiological consequences of damage to the outer hair cells (the primary, and most common, cause of hearing loss). In 1995, two highly successful models of hearing aid were introduced that process the acoustic signal in the digital domain. Until the introduction of these digital aids the limiting factor on what could be done to ameliorate the hearing loss was the technology used in the hearing aids. Nowadays the limiting factor is our basic knowledge pertaining to the functional requirements of what a hearing aid should actually do.

This chapter discusses basic scientific and engineering issues. Because the majority of hearing-impaired individuals experience mild-to-moderate levels of sensorineural hearing loss, the discussion is limited to impairment of sensorineural origin and the processing that has been proposed for its amelioration. The physiological and perceptual consequences of a profound hearing loss frequently differ from those associated with less severe losses, and therefore require different sorts of ameliorative strategies than would be effective with only a mild degree of impairment. Because profound loss is less commonly observed among the general population, current hearing-aid design has focused on mild-to- moderate loss (cf. Clark, Chapter 8, for a discussion of prosthetic strategies for the profoundly hearing impaired). Conductive loss (due to damage of the middle or outer ear) is also not addressed here for similar reasons.

2. Amplification Strategies

Amplification strategies for amelioration of hearing loss have tended to use either linear or syllabic compression processing. Linear compression has received the most attention, while syllabic compression remains highly controversial as a hearing-aid processing strategy. Dillon (1996) and Hickson (1994) provide excellent overviews of other forms of compression used in hearing aids.

2.1 Recruitment and Damaged Outer Hair Cells

The majority of hearing aid users have mild-to-moderate sensorineural hearing loss resulting from damage to the outer hair cells. The most prominent perceptual consequence of this damage is a decrease in auditory sensitivity a hypersensitive response to changes (particularly increases) in sound pressure level (SPL). This specific property of loudness coding in the hearing impaired is known as *recruitment* (Fowler 1936). The growth of loudness as a function of SPL is far steeper (and hence abnormal) than for a healthy ear.

A typical pattern of recruitment is illustrated in Figure 7.1, which shows the loudness functions for a normal-hearing and a hearing-impaired listener obtained using loudness scaling with half- octave bands of noise (Allen et al. 1990). The functions shown differ by greater than 25 dB at low loudness levels; however, this level difference decreases as the loudness increases until being nearly identical to the normal function at very high SPLs.

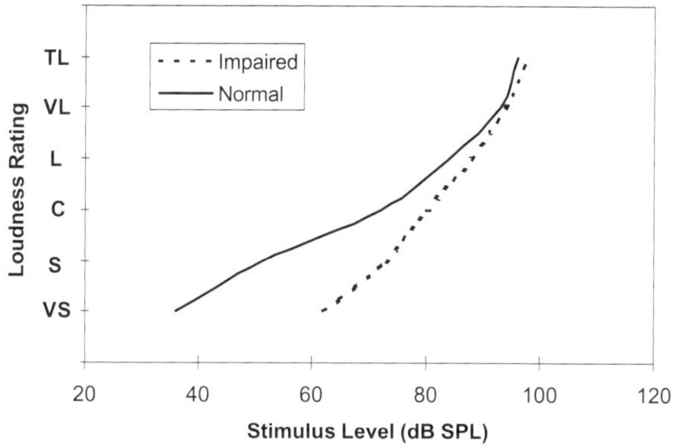

FIGURE 7.1. Typical loudness growth functions for a normal-hearing person (solid line) and a hearing-impaired person (dashed line). The abscissa is the sound pressure level of a narrowband sound and the ordinate is the loudness category applied to the signal. VL, very soft; S, soft; C, comfortable; L, loud; VL, very loud; TL, too loud.

The increased growth of loudness illustrated in Figure 7.1 is caused by the loss of the compressive nonlinearity in the transducer properties of the outer hair cells. The active biological amplification resident in the outer hair cells provides the sensitivity to low-intensity sounds and also sharpens the tuning of the basilar membrane. Measurements of auditory-nerve (Evans and Harrison 1976) and basilar-membrane (Sellicket et al. 1982) tuning curves show that damage to the outer hair cells eliminates the tip component of the tuning curve, elevating a fiber's tip threshold by over 40dB and thereby broadening the filtering. More importantly for hearing aids, outer hair cells append a compressive nonlinearity to the basilar membrane response. Input- output (I/O) functions associated with basilar-membrane velocity as a function of SPL exhibit a compressive nonlinearity between about 30 and 90dB SPL. When the outer hair cells are damaged, this function not only becomes less sensitive but tends to become more linear as well. Figure 7.2 replots two curves from Ruggero and Rich (1991), showing that sensitivity is significantly reduced for low-SPL signals but remains near normal for high-SPL sounds. This I/O pattern is remarkably similar in form to the psychoacoustic characterization of hearing loss illustrated in Figure 7.1. Based on these data, Killion (1996) has estimated the compression ratio provided by the outer hair cells to be approximately 2.3 to 1 (others studies have estimated it to be as much as 3 to 1). Loss of outer hair cell functioning thus decreases sensitivity by as much as 40 to 60dB and removes the compressive nonlinearity associated with the transduction mechanism.

For the majority of hearing-impaired listeners, the inner hair cells remain undamaged and thus the information-carrying capacity of the system remains, in principle, unaffected. This leaves open the possibility of reintroducing function provided by the outer hair cells at a separate stage of

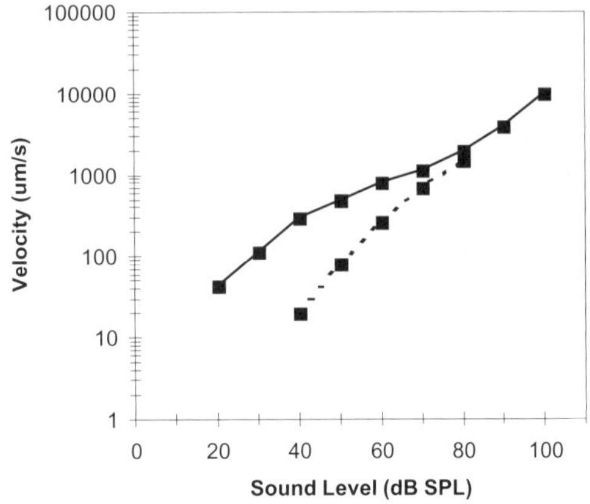

FIGURE 7.2. The response of a healthy basilar membrane (solid line) and one with deadened outer hair cells (dashed line) to best-frequency tone at different sound pressure levels (replotted from Ruggero and Rich 1991). The slope reduction in the mid-level region of the solid line indicates compression; this compression is lost in the response of the damaged cochlea.

the transduction process (via a hearing aid). Once damage to the inner hair cells occurs, however, auditory-nerve fibers lose their primary input, thereby reducing the effective channel capacity of the system (cf. Clark, Chapter 8 for a discussion of direct electrical stimulation of the auditory nerve as a potential means of compensating for such damage). It is generally agreed that a hearing loss of less than 60 dB (at any given frequency) is primarily the consequence of outer hair cell damage, and thus the strategy of amplification in the region of hearing loss is to increase the effective level of the signal transmitted to the relevant portion of the auditory nerve. Ideally, the amplification provided should compensate perfectly for the outer hair cell damage, thereby providing a signal identical to the one typical of the normal ear. When inner hair cell does damage occur, no amount of amplification will result in normal stimulation of fibers innervating the affected region of the cochlea. Under such circumstances the amplification strategy needs to be modified. In the present discussion it is assumed that the primary locus of cochlear damage resides in the outer hair cells. (Clark Chapter 8) discusses the prosthetic strategies used for individuals with damage primarily to the inner hair cells.

2.2 Linear Amplification

Until recently, hearing aids attempted to compensate for a hearing loss by application of linear gain (i.e., gain that is invariant with level for any given frequency). If sufficient gain is provided to lower the user's threshold to

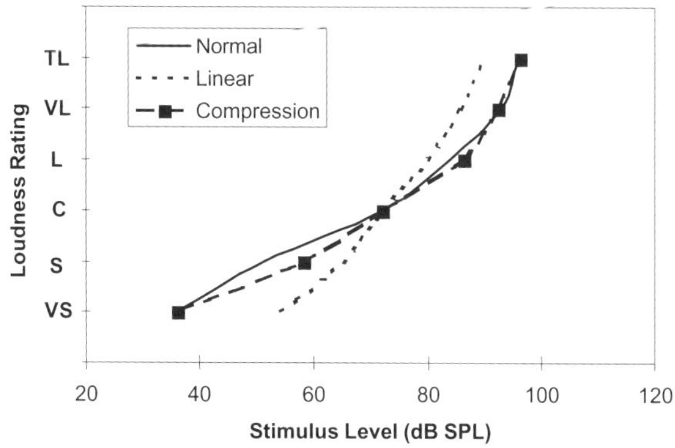

FIGURE 7.3. Loudness growth functions for a normal-hearing listener (solid line), a hearing-impaired listener wearing a linear hearing aid (short dashed line), and a hearing-impaired listener wearing a compression hearing aid (long dashed line with symbol).

normal, high-intensity sounds are perceived as being louder than normal, and may even exceed the threshold of pain. The dashed line in Figure 7.3 ilustrates a compromise strategy, incorporating certain functional properties of linear amplification, but where the comfortable loudness regions have been equated. This strategy often provides too little gain for low-intensity signals and too much gain for high-level signals. This occurs because the aid's loudness function remains steeper than normal. Because of this phenomenon, linear amplification requires some form of limiting so that signals are not presented at an uncomfortable or painful level. Compression limiting is usually preferred over clipping from the standpoint of both quality and intelligibility (Dreschler 1989).

For several decades the same gain function was thought to be acceptable to all hearing aid wearers, regardless of the form of the individual's hearing loss. This notion resulted from an interpretation of the Harvard Report (Davis et al. 1947) that a slope of 6dB/octave was an optimal hearing aid response for all hearing loss configurations, and this idea was not seriously challenged until the 1960s (Studebaker 1980). Since then, for what are now obvious reasons, the amount of gain prescribed at a given frequency typically increases with the hearing loss at that frequency and different gain functions are provided for different hearing losses. Since most hearing losses are high frequency in nature, the gain provided by modern prescriptions still increases with frequency, but usually not with a straight 6dB/octave slope.

Several researchers have suggested formuli for determining the gain function based on audiometric thresholds (Barfod 1972; Pascoe 1975). The most popular of these is the National Acoustic Laboratory (NAL) procedure (Byrne and Dillon 1986), which is based on theoretical considerations and empirical evidence. Studebaker (1992) has shown that the NAL prescription is indeed near optimal over a certain range of stimulus levels since it maximizes the articulation index (AI) given a constraint on loudness.

Because of their different slopes, the aided loudness function of a linear hearing aid wearer matches the normal loudness function at only one stimulus level. A volume control is usually provided with linear aids, which allows wearers to adjust the gain as the average level of the environment changes, effectively shifting their aided loudness curve along the dimension representing level in order to achieve normal loudness at the current level of their surroundings.

From the perspective of speech intelligibility, the frequency response of a linear aid should provide sufficient gain to place the information-carrying dynamic range of speech above the audibility threshold of listeners while keeping the speech signal below their threshold of discomfort. The slope of the frequency response can change considerably and not affect intelligibility as long as speech remains between the thresholds of audibility and discomfort (Lippman et al. 1981; van Dijkhuizen et al. 1987), although a negative slope may result in a deterioration of intelligibility due to upward spread of masking (van Dijkhuizen et al. 1989).

2.3 Compressive Amplification

The dynamic range of speech that carries useful information for intelligibilty is 30 dB, and some have estimated this range to be even wider (Studebaker et al. 1997). With the additional variability in average speaker level of 35 dB (Pearson et al. 1977), the overall dynamic range of speech that a hearing aid wearer can expect to encounter is over 60 dB. Given that linear aids with a fixed frequency response can provide normal loudness levels over only a limited stimulus range, and given that many hearing-impaired listeners have dynamic ranges of less than 60 dB, linear aids and their corresponding gain prescriptions are clearly a compromise solution to hearing loss compensation from the perspective of preserving speech audibility under all conditions.

As previously stated, a natural goal for a hearing aid should be to process the acoustic stimuli such that the signal reaching the inner hair cells is as close as possible to the signal that would be presented to the inner hair cells by a healthy auditory system. The hearing aid should perform the functioning that the damaged outer hair cells can no longer perform. With respect to loudness, then, the hearing aid should compress the stimuli in the same manner as a properly functioning outer hair cell, providing less gain in a frequency region as the stimulus level in that region increases. With a perfect hearing aid, every inner hair cell would receive the same stimulation that would have been received if the outer hair cells were not damaged. If this were achieved in an aid, the audibility of the wide dynamic range of speech would automatically be maintained, at least as far as it is maintained for normal listeners.

As Allen (1996) has pointed out, the strategy of using compression to compensate for the expansive characteristic of loudness perception in impaired ears was suggested (Steinberg and Gardner 1937) only a year after the phenomenon of loudness recruitment was reported, yet it would be decades before such a hearing aid was made available, due to the technical challenges involved. The concept of compression would also take as long, if not longer, to be accepted as a legitimate hearing loss compensation technique due to difficulties validating it with speech intelligibility tests, and its merits are still being debated.

Simply put, compression is implemented in a hearing aid by continuously estimating the level of a signal and varying the gain with level as specified by an I/O curve. A typical I/O curve is shown in Figure 7.4. The slope at any point on the I/O curve is equal to the inverse of the compression ratio. Here, the slope between 45 and 85 dB is 0.5, so the compression ratio is 2, and a 2-dB increase in the input level will result in a 1-dB increase in the output level. The compressive action of the outer hair cells does not appear to operate at high and low levels, so compression hearing aids should return to linear processing at low and high levels as well. While some form of high-level limiting is usually implemented to prevent the overloading of the hearing aid circuit or receiver, there is no perceptual reason for providing such limiting for users whose hearing returns to normal at high levels (Killion 1996). The solid line with symbols in Figure 7.3 shows the result of

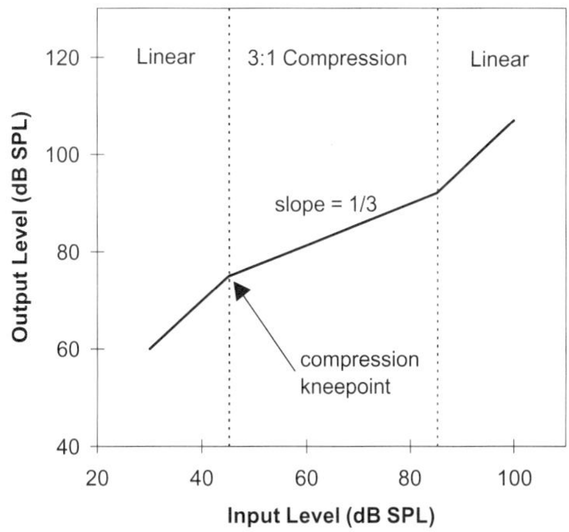

FIGURE 7.4. Typical input-output function of a compression hearing aid measured with a pure tone stimulus at multiple levels. The function depicted shows linear operation at low and high input levels, and 3:1 compression at mid-levels. Different compression hearing aids have different compression ratios and different levels over which compression occurs.

applying this I/O function to the recruiting loudness curve. The resulting aided loudness function matches the normal function almost exactly.

The gain of a hearing aid must be able to adjust to sound environments of different levels: the gain required for understanding someone's soft speech across the table at an elegant restaurant is quite different from the gain required to understand someone shouting at you at a noisy bar. A survey by the *San Francisco Chronicle* of the typical background noise experienced in San Francisco restaurants found a difference of nearly 30 dB between the most elegant French restaurant and the current trendiest restaurant. A person with linear hearing aids fit to accommodate the former environment would be better off removing the hearing aids in the latter environment. While a volume control can address this in a crude sense—assuming that the user doesn't mind frequently adjusting the level of the aid—the frequency response of the gain should change with level as well to provide maximum intelligibility (Skinner 1980; Rankovic 1997). A person with a sloping high-frequency hearing loss may require gain with a steep frequency response at low levels, where one's equal-loudness contours significantly differ from normal as frequency increases. At high levels, where their equal loudness contours are nearer to normal, the frequency response of the gain needed is significantly shallower in slope. The speed with which the gain should be allowed to change is still being debated: on the order of tens of milliseconds to adjust to phonemic-rate level variations (fast acting), hundreds of milliseconds for word- and speaker-rate variations (slow acting), or longer to accommodate changes in the acoustic environment (very slow acting).

With respect to fast-acting compression, it is generally accepted that syllabic compression should have attack times as short as possible (say, <5 ms), and recommendations for acceptable ranges of release times vary from between 60 and 150 ms (Walker and Dillon 1982), less than 100 ms (Jerlvall and Lindblad 1978), and between 30 and 90 ms (Nabelek 1983). Release times should be short enough that the gain can sufficiently adapt to the level variations of different phonemes, particularly low-amplitude consonants that carry much of the speech information (Miller 1951). Recommendations for slow-acting compression (more commonly referred to as slow-acting automatic gain control (AGC) to eliminate confusion with fast-acting compression) typically specify attack and release times of around 500 ms (Plomp 1988; Festen et al. 1990; Moore et al. 1991). This time scale is too long to be able to adjust the gain for each syllable, which has a mean duration of 200 ms in spontaneous conversational speech (Greenberg 1997), but could follow the word level variations. Neuman et al. (1995) suggest that release time preference may vary with listener and with noise level.

The type of compression that this chapter focuses on is fast-acting compression since it is the most debated and complex form of compression and has many perceptual consequences. It also represents the most likely candidate for mimicking the lost compressive action of the outer hair cells.

2.4 Time Constants

From the perspective of mimicking outer hair cells, the gain in a compression system should adjust almost instantaneously since there appears to be little lag in the compressive action of the outer hair cells. The perceptual consequences of fast gain adjustments are discussed later. Since compression is a nonlinear process, a compression hearing aid creates harmonic and intermodulation distortions that strengthen as the time constant of the compressive action shortens. In order that these distortion components are minimized, the action of the compressor must be slow enough that it does not act upon the fine structure of the signal but only on the envelope.

The speed with which a compressor responds to a change in stimulus level is characterized by the attack and release times. The attack time is defined by the time it takes the gain to adjust to within 3 dB of its final value when the level of the stimulus increases from 55 to 90 dB SPL. The release time is defined by the time it takes the gain to adjust to within 4 dB of its final value when the stimulus changes from 90 to 55 dB SPL (ANSI 1996). Both this definition and that of previous ANSI specification (ANSI 1987) result in time constant values that are dependent on the compression ratio and other factors, causing some difficulty when comparing results from different studies that used different compression configurations. Kates (1993) has suggested a technique that might address this problem, but it has not yet been adopted by the hearing aid industry or by subsequent researchers investigating fast-acting compression.

For practical reasons, the gain should reduce quickly when the level of the stimulus increases suddenly, to prevent the presentation of a painfully loud sound to the listener. This requires an extremely short attack time, with a correspondingly longer release time to prevent the distortions discussed above. Since compressed speech is equal in loudness to uncompressed speech if both signals have equal 90% cumulative amplitude distributions (Bakke et al. 1974; Levitt and Neuman 1991), and the loudness of modulated stimuli seems to be determined by the peaks of signals more than the root mean square (rms) of signals (Zhang and Zeng 1997), using fast attack times to ensure that the peaks are placed at normal loudness levels should ensure loudness normalization of speech.

Fast-acting compressors designed to normalize loudness from phoneme to adjacent phoneme have release times short enough such that low-level phonemes that follow high-level phonemes, such as a consonant following a vowel, are presented at an audible level. A stop consonant, which can be 20 to 30 dB lower in level than the preceding vowel (Fletcher 1953), could be underamplified by 10 to 15 dB with 2:1 compression if the release time is too slow. Jerlvall and Lindblad (1978), for example, found that confusions among the unvoiced final consonants in a consonant-vowel-consonant (CVC) sequence increased significantly when the release time increased from 10 to 1000 ms, most likely due to an insufficiently quick increase in gain following the vowel with the longer release time. Considering that one phonetic transcription of spontaneous conversational speech found that most phonetic classes had a median duration of 60 to 100 ms (Greenberg et al. 1996), the recovery time should be less than 60 ms in order to adjust to each phoneme properly.

Attack times from 1 to 5 ms and release times of 20 to 70 ms are typical of fast-acting compressors, which are sometimes called syllabic compressors since the gain acts quickly enough to provide different gains to different syllables. The ringing of narrow-bandwidth filters that precede the compressors can provide a lower limit on realizable attack and release times (e.g., Lippman et al. 1981).

2.5 Overshoot and Undershoot

The dynamic behavior of a compressor is demonstrated in Figure 7.5. The top panel shows the level of a sinusoid that increases by 25 dB before returning to its original level 100 ms later. The middle panel shows the gain trajectory resulting from a level estimation of the signal using a 5-ms attack and 50-ms release time constants and an I/O curve with 3:1 compression.

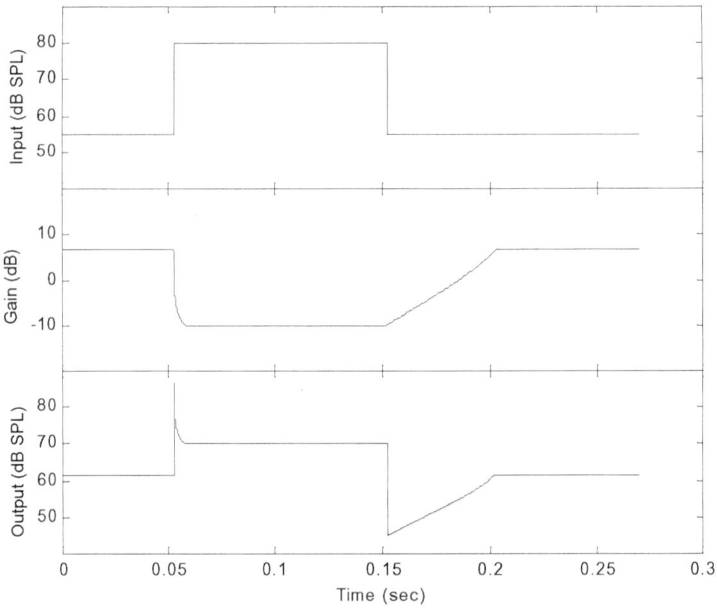

FIGURE 7.5. A demonstration of the dynamic behavior of a compressor. Top: Level of the input signal Middle: Gain that will be applied to the input signal for 3:1 compression, incorporating the dynamics of the attack and release time constants. Bottom: The level of the output signal, demonstrating overshoot (at 0.05 second) and undershoot (at 0.15 second).

The bottom panel shows the level of the signal at the output of the compressor where the effects of the attack and release time lag are clearly evident. These distortions are known as overshoot and undershoot.

Because of forward masking, the effect of the undershoot is probably not significant as long as the undershoot recovers quickly enough to provide enough gain to any significant low-level information. If the level drops below audibility, however, then the resulting silent interval could be mistaken for the pressure buildup before release in a stop consonant and cause poorer consonant identification. Overshoot may affect the quality of the sound and would have a more significant perceptual effect with hearing-impaired listeners because of recruitment, providing an unnatural sharpness to sounds if too severe. Verschuure et al. (1993) have argued that overshoot may cause some consonants to be falsely identified as plosives, and thus speech recognition could be improved if overshoot were eliminated. Nabelek (1983) clipped the overshoot resulting from compression and found a significant improvement in intelligibility. Robinson and Huntington (1973) introduced a delay to the stimulus before the gain was applied such that the increase in stimulus level and corresponding decrease in gain were more closely synchronized, resulting in a reduction in overshoot, as illustrated in Figure 7.6. Because of the noncausal effect of this delay (the gain appears to adjust before the stimulus level change occurs), a small overshoot may result at the release stage. This can be reduced with

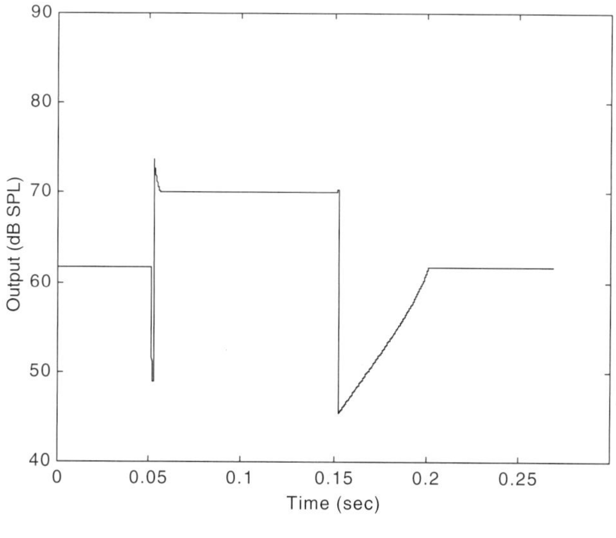

FIGURE 7.6. The level of the output signal resulting the same input level and gain calculation as in Figure 7.5, but with a delay to the stimulus before gain application. This delay results in a reduction in the overshoot, as seen by the lower peak level at 0.05 second.

a simple hold circuit (Verschuure et al. 1993). Verschuure et al. (1993, 1994, 1996) found that the intelligibility of embedded CVCs improved when the overshoots were smoothed with this technique. Additionally, compression with this overshoot reduction produced significantly better intelligibility than linear processing, but compression without the delay was not signficantly better than linear processing. The authors suggested that previous studies showing no benefit for compression over linear processing may have been due to overshoot distortions in the compressed signals, perhaps affecting the perception of plosives and nasals, which are highly correlated with amplitude cues (Summerfield 1993). Indeed, other studies that used this delay-and-hold technique either showed positive or at least failed to show negative results over linear processing (Yund and Buckles 1995a,b,c).

It should be noted that the overshoot and undershoot that results from compression is not related to nor does it "reintroduce" the normal overshoot and undershoot phenomenon found at the level of the auditory nerve and measured psychoacoustically (Zwicker 1965), both of which are a result of neural adaptation (Green 1969). Hearing-impaired subjects, in fact, show the same overshoot effect with a masking task as found with normals (Carlyon and Sloan 1987) and no effect of sensorineural impairment on overshoot at the level of auditory nerve exists (Gorga and Abbas 1981a). Overshoot, then, cannot be viewed as anything but unwanted distortion of the signal's envelope.

2.6 Wideband Compression

A topic of numerous studies and of considerable debate is the effect of the number of bands in a compression system. With multiband compression, the signal is filtered into a number of contiguous frequency channels. The gain applied to each bandpass signal is dependent on the level of that signal as determined by the I/O function for that band. The postcompression bandpass signals are then linearly summed to create a broadband stimuli. Figure 7.6 depicts this processing. Multiband compression allows the amount of compression to vary with frequency, and separates the dependencies such that the signal level in one frequency region does not affect the compressive action in another frequency region. This is not the only way in which the processing can achieve frequency-dependent compression; the summation of a linear system and a high-pass compressive system, the use of principal components (Bustamante and Braida 1987b), and the similar use of orthogonal polynomials (Levitt and Neuman 1991) all produce this effect.

The simplest form of compression is single band or wideband: the gain is controlled by the overall level of the signal, and the gain is adjusted equally across all frequencies. This has the effect of preserving the spectral shape of a signal over short time scales, apart from any separate frequency-dependent linear gain that is applied to the signal pre- or postcompression.

It has been argued that this preservation is necessary to maintain any speech cues that may rely on spectral shape.

Dreschler (1988a, 1989) compared the consonant confusions obtained with wideband compression to confusions obtained with linear processing. Consonant identification is of particular importance under wideband compression since temporal distortions have a large effect on consonant intelligibility while vowel intelligibility is more susceptible to spectral distortions (Verschuure et al. 1994). Using multidimensional scaling, Dreschler found that the presence of compression increased the weighting for plosiveness and decreased the weighting of both frication and nasality relative to linear processing. Dreschler attributed the increased importance of plosiveness to the reduction in forward masking caused by compression. The perception of the silent interval before the plosive burst was more salient, due to the reduced gain of the preceding sound, consistent with the fact that temporal gap detection is improved by compression with hearing-impaired listeners (Moore 1991).

In an attempt to relate speech perception in hearing-impaired listeners to their loss of high-frequency audibility, Wang et al. (1978) investigated consonant confusions for filtered speech with normal-hearing listeners. They found that decreasing the low-pass cutoff from 2800 to 1400 Hz increased the importance of both nasality and voice, while it also reduced the importance of sibilance, high anterior place of articulation, and, to a lesser extent, duration. For the consonant in the vowel-consonant (VC) position, frication also increased in weight. Consonant confusion patterns for low-pass filtered speech presented to normal-hearing listeners were similar to those found for subjects with high-frequency hearing loss, indicating that reduced audibility accounts for most of the error patterns that hearing-impaired listeners produce. Bosman and Smoorenberg (1987) also showed that nasality and voicing are the primary consonant features shared by normal-hearing and hearing-impaired listeners. Since different hearing aid signal-processing algorithms may produce similar intelligibility scores while presenting significantly different representations of speech to the hearing aid listener, one measure of the success of a specific signal-processing strategy could be the extent to which the resulting confusion patterns are similar to those found in normal-hearing listeners. On this basis, wideband compression better transmits speech information compared to linear processing (Dreschler 1988a, 1989), since the consonant confusion patterns are more similar to those found with normal-hearing subjects.

One significant drawback of wideband compression, however, is that a narrowband signal in a region of normal hearing will be compressed the same amount as a narrowband signal in a region of hearing loss. Even if the configuration of a listener's hearing loss required the same compression ratio at all frequencies, however, wideband compression would still not properly compensate for the damaged outer hair cells since the gain applied to all frequencies would be determined primarily by the frequency region with the highest level. If speech were a narrowband signal with only one spectral peak at any given time, then this might be more appropriate for processing speech in quiet, but speech is wideband with information-bearing components at different levels across the spectrum. Under wideband compression, the gain applied to a formant at 3kHz could be set by a simultaneous higher-level formant at 700Hz, providing inadequate gain to the lower-level spectral regions. Gain would not be independently applied to each of the formants in a vowel that would ensure their proper audibility. Wideband compression, then, is inadequate from the viewpoint of providing the functioning of healthy outer hair cells, which operate in localized frequency regions. This is particularly important for speech perception in the presence of a strong background noise that is spectrally dissimilar to speech, when a constant high-level noise in one frequency region could hold the compressor to a low-gain state for the whole spectrum.

Additionally, the gain in frequency regions with low levels could fluctuate even when the level in those regions remains constant because a strong fluctuating signal in another frequency region could control the gain. One can imagine music where stringed instruments are maintaining a constant level in the upper frequency region, but the pounding of a kettle drum causes the gain applied to the strings to increase and decrease to the beat. It is easy to demonstrate with such stimuli for normal listeners that single-band compression introduces disturbing artifacts that multiband compression excludes (Schmidt and Rutledge 1995). This perceptual artifact would remain even in a recruiting ear where the unnatural fluctuations would in fact be perceptually enhanced in regions of recruitment.

Schmidt and Rutledge (1996) calculated the peak/rms ratio of jazz music within 28 one-quarter-octave bands, and also calculated the peak/rms ratio in each band after the signal had been processed by either a wideband or multiband compressor. Figure 7.7 shows the effective compression ratio measured in each band for the two different types of compressors that have been calculated from the change to the peak/rms caused by the compressors. The open symbols plot the effective compression ratio calculated from the change to the peak/rms ratio of the broadband signal. Even though the wideband compressor shows significantly greater compression than the multiband compressor when considering the effect in the broadband signal, the wideband compressor produces significantly less compression than the multiband compressor when examining the effect in localized frequency regions. The wideband compressor even expands the signal in the higher frequency regions, the opposite of what it should be doing. Additionally, the multiband processor provides more consistent compression across frequency. Wideband compression is thus a poor method for providing compression in localized frequency regions, as healthy outer hair cells do.

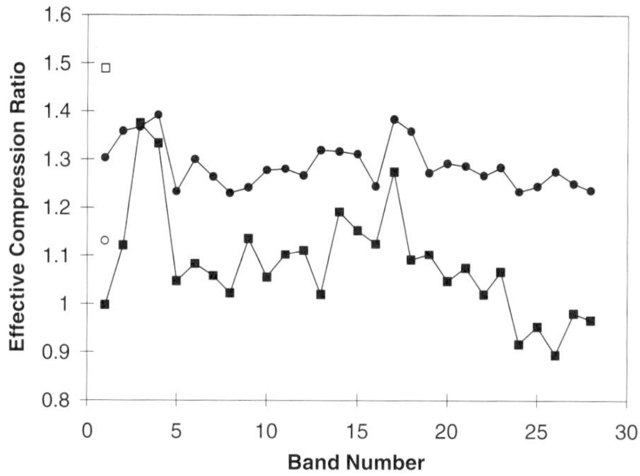

FIGURE 7.7. Amount of compression applied to music by a wideband compressor (squares) and a multiband compressor (circles). The compression was measured by comparing the peak/root mean square (rms) ratio of the music into and out of the compressor over different frequency regions. The open symbols on the left show the compression ratio calculated from the change to the broadband peak/rms ratio. The filled symbols show the change to the peak/rms ratio in localized frequency regions.

2.7 Multiband Compression

For the above reasons, compressors are typically designed such that the gain is adjusted independently in at least two separate frequency regions. Of particular importance is the uncoupling of the gain in the upper frequency region, where high information-carrying consonants reside, from the lower-frequency region, where most of the energy in vowels and most environmental noises reside (Klumpp and Webster 1963; Kryter 1970; Ono et al. 1983). Multiband compression produces this effect: the input to the hearing aid is passed through a bank of bandpass filters, compression is independently applied to the output of each band, and the processed bandpass signals are summed into a single broadband signal, which is the hearing aid output.

The dynamics of the gain is favorably affected by the multiband architecture. Under wideband compression with a 50-ms release time, the gain could require 50 ms to increase the signal amplitude to the proper level for a low-level, high-frequency sound that immediately followed a high-level low-frequency sound, such as would occur when a consonant follows a vowel. If the duration of the low-level sound is short, then the gain may not increase in time for the sound to be presented at an audible level. Given a multiband compressor for which the two sounds fall in separate bands, the gain in the high-frequency region will not be reduced during the presence of the low-frequency signal and thus can adjust to the high-frequency signal with the speed of the much shorter attack time.

2.8 Reduced Spectral Contrast by Multiband Compression

The spectral contrast of a compressed signal is reduced as the number of independent bands increases. If two bands have the same compressive gain I/O function, then the difference in the signal level in each band will be reduced by a factor equal to the inverse of the compression ratio. This results in a reduction in the contrast between spectral peaks and valleys. Under 3:1 compression, for example, a 12-dB peak in the signal relative to the level in a different compression band would become a 4-dB peak after compression. Since the spectral contrast within a band is preserved (the same gain is applied across all frequencies within a band), only contrast across bands is changed, so this effect becomes more prominent as the number of bands increase.

This degradation of the spectral contrast might degrade speech cues that are encoded by spectral shape, such as place of articulation (Stevens and Blumstein 1978). On the surface, this seems a specious argument against multiband compression. If a multiband compressor perfectly replaced the lost compression of the outer hair cells, then any distortion of the acoustic spectral shape would be inconsequential since the perceptual spectral shape would be preserved. The validity of this argument against multiband compression results from the perceptual consequences of hearing impairment, which are not characterized by the abnormal growth of loudness but which cause the perceptual spectral contrast to be reduced, such as changes to lateral suppression, auditory bandwidths, and other effects. Smoorenburg (1990) has hypothesized that broadened auditory filter bandwidths are the cause of 10dB excess masking found with hearing-impaired listeners in the frequency region where noise-masked thresholds intersect auditory thresholds. The upper spread of masking can be worse in some hearing-impaired listeners due to lowered tails in their tuning curves (Gorga and Abbas 1981b). Restoring the loudness levels within localized frequency regions will not restore normal spectral contrast when these other effects occur, indicating that compression may not be enough to restore the perceptual signal to normal. Ideally, though, the point remains that the change to the characteristic of a sound by compression must be judged not by the effect on the acoustic signal but by its perceptual consequences.

2.9 Compression and the Speech Transmission Index

Plomp (1988, 1994) has argued effectively that any fast-acting compression is detrimental to speech intelligibility due to its effect on the modulation spectrum as quantified by the speech transmission index (STI). The STI is a measure relating the modulation transfer function (MTF) of a system to speech understanding (Houtgast and Steeneken 1973, 1985). It was originally developed for characterizing room acoustics, and it accurately predicts the effect of noise and reverberation on speech intelligibility using the change to the modulations patterns of speech. Noise reduces the level of all frequencies in the modulation spectrum of speech by filling in the envelope valleys and reducing the peak-to-trough range, while reverberation attenuates the higher modulation frequencies more than the lower modulation frequencies. The STI predicts that any reduction in the modulation spectrum between approximately 0.5 and 16 Hz will have a detrimental effect on intelligibility.

Since compression reduces the temporal fluctuations of a signal, the modulation spectrum is reduced. The effect of compression on the modulation spectrum is like a high-pass filter, with the knee point dependent on the time constants of the compressor. Plomp (1988) has shown that, for attack and release times of 8 ms each, the MTF of compression with a speech signal is close to 0.5 for modulation frequencies below 12 Hz, meaning that the modulation spectrum of speech is reduced by almost half at these frequencies. Since the modulation frequencies that affect speech intelligibility are approximately between 0.5 and 16 Hz, an important portion of the modulation spectrum is affected by compression, which may have a detrimental effect on speech understanding.

As equal-valued attack and release times of a multiband 2:1 compressor are reduced from 1 second to 10 ms, transitioning from near-linear processing to fast-acting compression, the STI reduces from a value of 1 to almost 0.5 (Festen et al. 1990). Since speech in noise at a 0-dB signal-to-noise ratio (SNR) also has an STI of 0.5 (Duquesnoy and Plomp 1980; Plomp 1988), many have concluded that fast-acting compression has the same detrimental effect on the intelligibility of speech in quiet as the addition of noise at a 0-dB SNR. One useful measure of speech understanding is the speech reception threshold (SRT), which estimates the SNR necessary for 50% sentence correct scores (Plomp and Mimpen 1979). Since most hearing-impaired listeners have an SRT greater than 0 dB (Plomp 1988), the fact that the STI equates fast-acting 2:1 compression with a 0-dB SNR means that such compression should result in sentence correct scores of less than 50%. No such detrimental effect results from compression on speech in quiet.

Noordhoek and Drullman (1997) measured an average SRT of −4.3 dB with 12 normal-hearing subjects. Since noise at −4.3 dB SNR reduces the modulations in speech by a factor of 0.27, the STI predicts that compression that produces the same reduction in modulation would also result in a 50% sentence correct score. They found with the same subjects that modulations had to be reduced by a factor of 0.11 to reduce the sentence score to 50%, compressing the envelopes of speech with 24 independent bands. Conversely, since a modulation transfer function of 0.11 was necessary to reduce sentence scores to 50% correct, one would expect the measured SRT to be –9 dB, using the equations given in Duquesnoy and Plomp (1980), instead of the actual measured value of –4.3 dB. This 5-dB discrepancy indicates that compression is not distorting speech intelligibility as much as the STI calculations indicate. Other reseachers have found that fast-acting compression produced SRT scores that were as good as or better than those produced with linear processing (Laurence et al. 1983; Moore and Glasberg 1988; Moore et al. 1992), although these results were with two-band compression, which does not affect the STI as much as compression with more bands since the modulations in narrow bands are not as compressed as effectively. Thus, the reduction in intelligibility by noise cannot be attributed to reductions in modulation alone, and the STI cannot be used to predict the effect of compression since it was derived from the modulation changes caused by reverberation and noise.

This apparent discrepancy in the relation between the STI and speech intelligibility for compression is most likely due to the effects of noise and reverberation on speech that are not introduced by compression and not characterized by the modulation spectrum. Both noise and reverberation distort the phase of signals besides adding energy in local temporal-frequency regions where it didn't previously exist, deteriorating phase-locking cues (noise), and adding ambiguity to the temporal place of spectral cues (reverberation). Compression simply adds more gain to low-level signals, albeit in a time-varying manner, such that no energy is added in a time-frequency region where it didn't already exist. More importantly, the fine structure is preserved by compression while severely disturbed by both noise and reverberation. Slaney and Lyon (1993) have argued that the temporal representation of sound is important to auditory perception by using the correlogram to represent features such as formants and pitch percepts that can be useful for source separation. Synchronous onsets and offsets across frequency can allow listeners to group the sound from one source in the presence of competing sources to improve speech identification (Darwin 1981, 1984). The preservation and use of these features encoded in the fine temporal structure of the signal are fundamental to cognitive functions such as those demonstrated through auditory scene analysis (Bregman 1990) and are preserved under both linear processing and compression but not with the addition of noise and reverberation. Similar STIs, then, do not necessarily imply similar perceptual consequences of different changes to the speech signal.

Investigating the effect of compression on speech in noise, Noordhoek and Drullman (1997) found that modulation reduction, or compression, had a significant effect on the SRT; a modulation reduction of 0.5 (2:1 compression) increased the SRT from –4.3 dB to –0.8 dB, though the noise and speech were summed after individual compression instead of compressing after summation. These results indicate that compression with a large number of bands may affect speech perception more drastically for normal-hearing listeners in noise than in quiet. This is most likely due to the reduction of spectral contrast that accompanies multiband compression and to the fact that spectral contrast may be a less salient percept in noise and thus more easily distorted by compression. Drullman et al. (1996) investigated the correlation between modulation reduction and spectral contrast reduction under multiband compression. They found that spectral contrasts, measured as a spectral modulation function in units of cycles/octave, were reduced by the same factor as the reduction in temporal modulation, confirming the high correlation between reduced temporal contrast and reduced spectral contrast with multiband compression.

2.10 Compression Versus Linear Processing

Several studies have shown that speech recognition performance deteriorates as the number of compression bands increases. Plomp (1994) presented data that demonstrated that sentence correct scores reduced to 0% as the number of bands increased from 1 to 16 for both normal-hearing and hearing-impaired listeners with infinite compression. These results and others must be tempered by the way in which compression is applied. In these studies, the same compression ratio is applied to each band regardless of the hearing loss of the subject. For example, 4:1 compression would be applied to all bands, even though hearing loss, and thus the amount of compression needed for loudness normalization, most likely varied with frequency (Plomp 1994; Festen 1996). This fact was pointed out by Crain and Yund (1995), who investigated the effect of the number of bands both with identical compression ratios applied to each band and with compression ratios that varied with bands as a function of the hearing loss within the frequency region of each band. They found that intelligibility decreased as the number of bands increased when the same compression ratio was applied in each band; intelligibility was not affected as the number of bands increased for the condition with band-varying compression. These results are consistent with the the data presented by Plomp (1994), who showed that performance deteriorates as the number of bands increases when the same compression ratio is applied to each band. Verschuure et al. (1993, 1994), using a technique in which the compression ratio increased with frequency, found that compression was not worse than linear as long as the compression ratio was less than 4:1; in all cases, performance was worse for compression ratios of 8:1 compared to 2:1. This seems logical since loss of outer hair cell functioning results in the loss of at most 3:1 compression processing. Thus, negative results found for compression ratios greater than this are not as meaningful as for smaller compression ratios since proper compensation for damaged outer hair cells should not need ratios larger that 3:1. Both Verschuure et al. and Crain and Yund used the delay-and-hold technique previously described to reduce overshoot.

Lippman et al. (1981) fit compression in a 16-band processor to the loss profile of the individual patients and found poorer performance with compression than with linear processing. Several researchers have pointed out that the linear gain response provided in this study and in others showing no benefit for compression were usually optimized for the speech stimuli used in the test, making as much of the speech audible while maintaining the speech level below the level of discomfort for the listener (Lippman et al. 1981; Villchur 1987; Yund and Buckles 1995b). Villchur (1989) has suggested that these tests are not representative of real-world situations since the hearing aid will not always be adjusted to the optimal gain setting for different levels of speech as is the case in the experiments, and some means is necessary to do so in a real hearing aid. Villchur (1996) has also pointed out that the level variation of speech materials used to test speech understanding is smaller than that found in everyday conversational speech, eliminating one of the factors that might cause compression to show a benefit over linear processing. This is in addition to the absence in most speech tests of level variations encountered in different environments and in the same environment but with different speakers. Lippman et al. (1981) have called these speech materials "precompressed," and noted that with a subset of their stimuli for which the last key word in the sentence was 11 dB below the level of the first key word, performance under compression was 12% better compared to linear for the last two key words but only 2% better for the first two key words. The compressor appeared to increase the gain for the final word compared to linear, making the speech cues for that word more audible. While Lippman et al. (1981) found that 16-band compression was slightly worse than linear processing when the linear gain put the speech stimuli at the most comfortable level of each individual listener, they also found that compression performed increasingly better than linear when the level of the speech stimuli was reduced while the gain characteristics of the linear and compressive processing were unadjusted. Compression increases the gain as the speech level decreases, maintaining more of the speech signal above the threshold of the listener. This simulated the benefit that compression would provide when real-world level variations occur.

This observation was forecast by Caraway and Carhart (1967, p. 1433):

> One must be careful not to restrict his thinking on compressor systems to comparisons that are equated only at one level. . . . One may select a gain adjustment of a compressor which causes high input levels to produce high sensation levels of output. Such an adjustment then allows greater reductions of input without radical drops in intelligibility than is allowed by a system without compression. . . . It must also be remembered, however, that an improper gain setting of the compressor system can have an opposite [detrimental] result.

This last point foreshadows the results later obtained by those who demonstrated the negative effects of compression with high compression ratios or a compression function not fit to the hearing loss of each subject. These negative reports on fast-acting compression might have been modified if Caraway and Carhart's comment had been paid more attention. Of course, the converse is also true, and several studies that demonstrated the benefit of compression over linear processing did not shape the linear as precisely as possible to the hearing loss of the subjects (Villchur 1973; Yanick 1976; Yanick and Drucker 1976).

2.11 Consonant Perception

In an extensive study, Yund and Buckles (1995a,b) found that speech discrimination performance improved as the number of bands in a compressor increased from 1 to 8 and was no worse at 16 bands when the compression was fit to the hearing loss of each individual. When increasing the number of bands from 4 to 16, the most significant difference in consonant confusions was an improved perception of the duration feature and, for voiceless consonants, the stop feature. Yund and Buckles relate this to more gain provided in the high-frequency region due to a gain control split between more bands. If a single band controls the gain above 2kHz, for example, then less gain than necessary will be provided at 4kHz in order to prevent too much gain being provided at 2kHz. The authors also explain better voiceless stop perception due to finer resolution being provided in the gain function to the low-level spectral features.

Compared to linear processing, Yund and Buckles (1995b) found that information on manner and voiceless duration were better transmitted by a 16-band compressor than by linear processing, with the gain in both processors shaped to the hearing loss of the individual subjects. Place information was better transmitted by the linear processing for voiced consonants, but poorer for voiceless consonants. Hearing-impaired listeners in general have difficulty identifying place of articulation for stop consonants (Owens et al. 1972), which Turner et al. (1997) suggest is due to difficulty with their perception of rapid formant transitions. These results are generally consistent with those found by other researchers (Lippmann et al. 1981; De Gennaro et al. 1986; CHABA Working Group 1991), although Lippmann et al. found duration to be more poorly transmited by compression.

When all spectral cues are removed from speech, the identification of consonant place is more adversely affected than the identification of consonant manner (Boothroyd et al. 1996). Since multiband compression reduces spectral contrast relative to linear processing, place cues can be expected to be the most detrimentally affected. Lippmann et al. (1981) noted that poor place percept under compression may be due to the listener's unfamiliarity with the new place cue created by the spectral shaping of multiband compression. This is supported by Yund and Buckles (1995c), who found that more place and duration information was transmitted as listeners gained more experience with the processing of multiband compression. Difference in experience might be why Yund and Buckles (1995b) found that compression increased the responses for middle stops and middle fricatives, while Nabelek (1983) found the opposite. Compression, however, increases the audibility of low-level spectral cues, which can improve the perception of frication even for normal-hearing subjects (Hickson and Byrne 1997). Perception of voice onset time also seems to be negatively affected by frequency-varying compression (Yund and Buckles 1995c), which Yund and Buckles suggest is due to the envelope being affected differently across frequency, athough other envelope effects were not noted in their study.

2.12 Vowel Perception

Vowel perception is generally not a problem with hearing-impaired listeners since the level of vowels is much greater than the level of consonants. Large spectral contrasts exist with vowels in quiet (Leek et al. 1987) and most of the vowel information is contained in the highest 10dB of the dynamic range of speech (van Harten-de Bruijn et al. 1997). Raised auditory thresholds will affect consonant perception much more severely than vowel perception (Dreschler 1980), and only the most severely hearing impaired have difficulty with vowel recognition (Owens et al. 1968; Pickett et al. 1970; Hack and Erber 1982; de Gennaro et al. 1986). Since vowels are most significantly identified by their relative formant frequency locations (Klatt 1982; Syrdal and Gopal 1986), the effects of compression on relative formant amplitude, peak-valley differences, and spectral shape are not as important as long as the frequency of the formants is identifiable. Additionally, vowels in continuous speech have significant dynamic cues other than their steady-state spectral shape that can be used for identification, such as duration of on-glides and off-glides, formant trajectory shape, and steady-state duration (Peterson and Lehiste 1960). Hearing aid processing must ensure that such coarticulatory cues are audible to the wearer and not distorted.

Since vowel discrimination is based on formant differences of spectral ripples up to 2 cycles/octave (Van Veen and Houtgast 1985), formants will be individually amplified if compression bands are less than one-half-octave wide. Similar implications result from the study of Boothroyd et al. (1996), where vowel identification performance was reduced from 100% to 75% when the spectral envelope was smeared with a bandwidth of 700Hz. If more than one formant falls within a single compression band, the lower-level formant may be inaudible since the higher-level formant controls the gain of the compressor, causing the gain to lower in that band. If the formants fall in separate bands, then lower-level formants will receive higher gain, and the reduced gain applied to the higher-level, lower-frequency formant may reduce masking at high levels (Summers and Leek 1995).

Crain and Yund (1995) also found that vowel discrimination deteriorated as the number of bands increased when each band was set to the same compression ratio, but performance didn't change when the compression ratios were fit to the hearing loss of the subjects.

2.13 Restoration of Loudness Summation

One overlooked benefit of multiband compression aids that is unrelated to speech recognition is the reintroduction of loudness summation, which is lost in regions of damaged outer hair cells. As the bandwidth of a signal increases from narrowband to broadband while maintaining the same overall energy, the loudness level increases as the bandwidth widens beyond a critical band. This effect is not seen in regions of hearing loss and can be explained by the loss of the outer hair cells' compressive nonlinearity (Allen 1996). Multiband compression can reintroduce this effect.

As the frequency separation between two tones increases from a small separation wherein they both fall within a critical band to one in which they fall into separate critical bands, the level of a single tone matched to their loudness increases by 7 dB for an undamaged auditory system (Zwicker et al. 1990). For a multiband compressor with 3:1 compression in each band, the gain applied to the two-tone complex increases by 2 dB when frequency separation places the two tones in separate multiband compressors, relative to the level when they both fall into the same compression band. Because of the 3:1 compression, the level of the matching tone would have to increase by 6 dB in order to match the 2-dB level increase of the two-tone complex with the wider separation. This 6-dB increase in the aided loudness level for the impaired listener under multiband compression is similar to the 7-dB increase experienced by normal-hearing listeners. Similarly, a four-tone complex increases in loudness level by 11 dB (Zwicker et al. 1957) for a normal-hearing listener while it would increase by 12 dB with 3:1 multiband compression for a hearing-impaired listener. Equating the effect of increasing the bandwidth of more complex broadband stimuli, such as noise or speech, is more difficult to analyze since it depends on the bandwidth of the compression filters and the amount of self-masking that is produced. Thus, multiband compression can reintroduce partially restored loudness summation, an effect not achievable by linear processing or wideband compression.

2.14 Overamplification from Band Processing

The nonlinear nature of multiband compression can also produce unwanted differences between narrowband and broadband processing. The slopes of the bandpass filters in multiband compression can introduce artifacts that result from the crossover region between the filters (Edwards and Struck 1996). Figure 7.8A shows the filters of a three-band compressor, designed

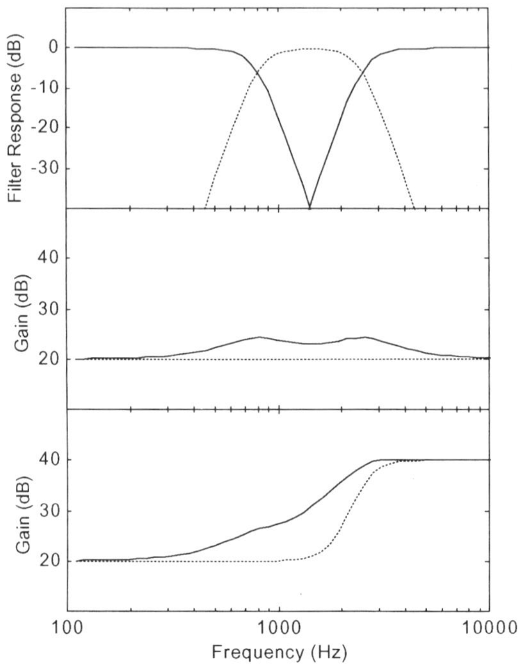

FIGURE 7.8. Top: The magnitude responses of three filters designed to produce a flat response when summed linearly. Middle: The gain applied to a 65 dB sound pressure level (SPL) pure-tone sweep (solid) and noise with 65 dB SPL in each band (dotted), indicating the effect of the filter slopes on the gain in a compression system. All bands are set for equal gain and compression ratios. Bottom: The same as the middle filter, but with gain in the highest band set 20 dB higher.

to give a flat response when equal gain is applied to each filter. The I/O function of each band is identical, with 3:1 compression that produces a 20-dB gain for a 65-dB SPL signal within the band. The dashed line in Figure 7.8B shows the frequency response of the compressor measured with broadband noise that has a 65-dB SPL level in each band. As expected, the response is 20 dB at all frequencies. The solid line shows the response measured with a 65-dB SPL tone swept across the spectrum. The gain is significantly higher than expected due to the skirts of the filters in the crossover region between filters. As the level of the tone within a filter decreases due to attenuation by the filter skirt, the compressor correspondingly increases the gain. One shouldn't expect the transfer functions measured with the noise and tone to be the same since this expectation comes from linear systems theory, and the system being measured is nonlinear. The increased gain to the narrowband signal is disconcerting, though, particularly since more gain is being applied to the narrowband stimuli than the broadband signal, the opposite of what one would want from the perspective of loudness summation. This can be particularly problematic for multiband hearing aids whose gain is set to match a target measured with tonal stimuli since less gain than required will be provided for more common broadband stimuli.

The effect is worse when the bands are programmed for different gain settings, as shown in Figure 7.8C where each band has 3:1 compression but the I/O function of the highest band is 20-dB higher than the band in the frequency region below, a more common configuration for high-frequency loss. Additional problems can occur with harmonic signals where the bandwidths of the multiband compressors are on the order of the frequency spacing of the harmonics. In addition, the amount of gain applied to each harmonic can depend on the number of harmonics that fall within any given band since the level passed by the filter will increase as the number of harmonics increases, resulting in decreased gain.

A solution to this was described by Lindemann (1997). Treating multiband compression as a sampling of spectral power across frequency, he noted that band-limiting the autocorrelation function of the spectral power will resolve these problems. The functional effect of this band-limiting is to increase the overlap of the bands such that the gain at any given frequency is controlled by more than two bands. This design is consistent with the functioning of the auditory system. Considering each inner hair cell to be a filter in a multiband compressor, then the human auditory system consists of 3500 highly overlapping bands. A similar solution was also proposed by White (1986) to lessen the reduction in spectral contrast caused by compression. Implementations that provide frequency-dependent compression but do not used multiple bands (Bustamante and Braida 1987b; Levitt and Neuman 1991) also avoid this problem.

3. Temporal Resolution

3.1 Speech Envelopes and Modulation Perception

Speech is a dynamic signal with the information-relevant energy levels in different frequency regions varying constantly over at least a 30-dB range. While it is important for hearing aids to ensure that speech is audible to the wearer, it may also be necessary to ensure that any speech information conveyed by the temporal structure of these level fluctuations is not distorted by the processing done in the aid. In addition, if the impaired auditory systems of hearing-impaired individuals distort the information transmitted by these dynamic cues, then one would want to introduce processing into the aid that would restore the normal perception of these fluctuations.

Temporal changes in the envelope of speech conveys information about consonants, stress, voicing, phoneme boundaries, syllable boundaries, and phrase boundaries (Erber 1979; Price and Simon 1984; Rosen et al. 1989). One way in which the information content of speech in envelopes has been investigated is by filtering speech into one or more bands, extracting the envelope from these filtered signals, and using the envelopes to modulate noise bands in the same frequency region from which the envelopes were extracted. Using this technique for a single band, the envelope of wideband speech has been shown to contain significant information for intelligibility (Erber 1972; Van Tasell et al. 1987a). Speech scores for normal-hearing subjects rose from 23% for speech reading alone to 87% for speech reading with the additional cue of envelopes extracted from two octave bands of the speech (Breeuwer and Plomp 1984). Shannon et al. (1995) found that the envelopes from four bands alone were sufficient for providing near-100% intelligibility. It should be emphasized that this technique eliminates fine spectral cues—only information about the changing level of speech in broad frequency regions is given to the listener.

These experiments indicate that envelope cues contain significant and perhaps sufficient information for the identification of speech. If hearing-impaired listeners for some reason possess poorer than normal temporal acuity, then they might not be able to take advantage of these cues to the same extent that normal listeners can. Poorer temporal resolution would cause the perceived envelopes to be "smeared," much in the same manner that poor frequency resolution smears perceived spectral information.

Psychoacoustic temporal resolution functions are measures of the auditory system's ability to follow the time-varying level fluctuations of a signal. Standard techniques for measuring temporal acuity have shown the auditory system of normal-hearing listeners to have a time-constant of approximately 2.5 ms (Viemeister and Plack 1993). For example, gaps in broadband noise of duration less than that are typically not detectable (Plomp 1964). These functions measure the sluggishness of auditory processing, the limit beyond which the auditory system can no longer follow changes in the envelope of a signal.

Temporal resolution performance in individuals has been shown to be correlated with their speech recognition scores. Tyler et al. (1982a) demonstrated a correlation between gap detection thresholds and SRTs in noise. Dreschler and Plomp (1980) also showed a relationship between the slopes of forward and backward masking and SRTs in quiet. Good temporal resolution, in general, is important for the recognition of consonants where fricatives and plosives are strongly identified by their time structure (Dreschler 1989; Verschuure et al. 1993). This is supported by the reduction in consonant recognition performance when reduced temporal resolution is simulated in normal subjects (Drullman et al. 1994; Hou and Pavlovic 1994).

As discussed in section 2.9, physical acoustic phenomena that reduce the fluctuations in the envelope of speech are known to reduce speech intelligibility. The reduction in speech intelligibility caused by noise or reverberation can be predicted by the resulting change in the modulation spectrum of the speech (Houtgast and Steeneken 1973; Nabelek and Robinson 1982; Dreschler and Leeuw 1990). If an impaired auditory system caused poorer temporal resolution, the perceived modulation spectrum would be altered. Thus, it becomes pertinent to determine whether hearing-impaired people have poorer temporal resolution than normal-hearing listeners since this would result in poorer speech intelligibility beyond the effect of reduced audibility. If evidence suggested poorer temporal resolution in people with damaged auditory systems, then a potential hearing aid solution would be to enhance the modulations of speech, knowing that the impaired listener would "smear" the envelope back to its original form and thus restore the perceptual modulation spectrum to normal.

3.2 Psychoacoustic Measures of Temporal Resolution

Contrary to the findings previously cited, there exists a large body of evidence that shows little or no correlation between temporal resolution measures and speech intelligibility (Festen and Plomp 1983; Dubno and Dirks 1990; van Rooij and Plomp 1990; Takahashi and Bacon 1992). The differences between these studies and the previously cited ones that do show a correlation may relate to the audibility of the signal, i.e., the reduced audible bandwidth of the stimuli due to hearing loss.

To address this issue, Bacon and Viemeister (1985) used temporal modulation transfer functions (TMTFs) to measure temporal acuity in hearing-impaired subjects. In this task, sinusoids were used to modulate broadband noise, and modulation detection thresholds were obtained as a function of modulation frequency. They found that TMTFs obtained from subjects with high-frequency hearing loss displayed the low-pass characteristic seen in TMTFs obtained with normal-hearing subjects, the sensitivity to modulation was reduced since thresholds were increased overall, the 3-dB point was lower in frequency, and the slope at high modulation frequencies was steeper. These characteristics indicate reduced temporal resolution in hearing-impaired subjects under unaided listening conditions.

Bacon and Viemeister then simulated in normal-hearing subjects reduced audibility by low-pass filtering the broadband noise carrier, also adding high-pass noise to prevent the use of the high-frequency region and then remeasured TMTFs. The results under these conditions were similar to those obtained with the impaired subjects. This indicates that the reduction in audible bandwidth of the damaged auditory system most likely reflects the reduced sensitivity to the modulation. This is supported by Bacon and Gleitman (1992) and Moore et al. (1992), who showed that TMTFs for hearing-impaired individuals were identical to those for normal-hearing listeners when the signals were presented at equal sensation level (SL). Derleth et al. (1996) showed similar results when signals were presented at the same loudness level. These findings were extended by Schroder et al.

(1994), who found normal modulation-depth discrimination thresholds in hearing-impaired subjects except where high levels of hearing loss completely eliminated regions of audibility that could not be overcome by increased stimulus intensity. It seems, then, that poor modulation perception is not due to poorer temporal acuity in the impaired auditory systems but is due to a reduced listening bandwidth caused by the hearing loss. This is consistent with the notion that temporal acuity is limiting by central processing and not by the auditory periphery.

A similar effect has been obtained with gap detection measures, where gap detection thresholds are typically larger for hearing-impaired subjects than for normal-hearing subjects (Fitzgibbons and Wightman 1982; Florentine and Buus 1984; Fitzgibbons and Gordon-Salant 1987; Glasberg et al. 1987). In normals, gap detection thresholds measured with noise reduce as the frequency of the noiseband increases (Shailer and Moore 1983). One would then expect people with high-frequency loss who are unable to use information in the high frequencies to manifest a poorer ability to perform temporal acuity tasks. Florentine and Buus (1984) have shown that gap detection thresholds with hearing-impaired subjects are equivalent to those with normal-hearing listeners when the stimuli are presented at the same SL. Plack and Moore (1991), who measured temporal resolution at suprathreshold levels in subjects with normal and impaired hearing, found equivalent rectangular durations (ERDs) for the two groups (ERDs are the temporal equivalent of equivalent rectangular bandwidths, ERBs). The similarity in results between gap detection and modulation detection is consistent with the finding of Eddins (1993) that the two represent the same underlying phenomenon. Evidence suggests that better temporal resolution in both TMTF and gap detection is related to the increase in audible bandwidth of the stimuli and not to the added use of broader auditory filters found at high frequencies (Grose et al. 1989; Eddins et al. 1992; Strickland and Viemeister 1997), although Snell et al. (1994) have found a complex interaction between bandwidth and frequency region in gap detection. Were temporal resolution primarily affected by the bandwidth of auditory filters, then one would expect damaged outer hair cells to enhance auditory acuity because of the increased auditory bandwidth and corresponding reduction of filter ringing (Duifhuis 1973).

3.3 Recruitment and Envelopes

It might be expected that hearing-impaired individuals might exhibit better performance than by normal-hearing subjects in such tasks as modulation detection because of recruitment in the impaired ear. For a given modulation depth the instantaneous signal level varies between the envelope peak and trough. Because of the increased growth of loudness in the impaired ear (see Figure 7.1), the difference in loudness level between envelope maxima and minima is greater for the impaired ear than for the normal ear.

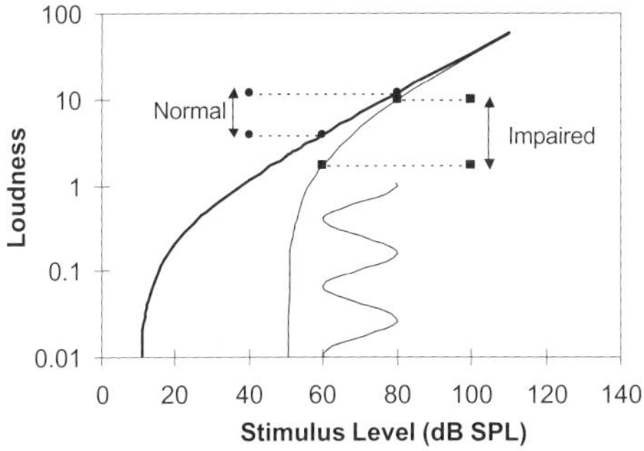

FIGURE 7.9. A demonstration of how perception of modulation strength is affected by an abnormal growth of loudness. Two loudness growth curves are shown: the thicker curve on the left represents that for a normal-hearing listener and the thinner one of the right represents that for a hearing-impaired listener. A stimulus with its level fluctuating between 60 and 80 dB SPL (vertical oscillation pattern) produces a larger loudness fluctuation for the hearing-impaired listener (shown by the separation between the filled squares) than for the normal-hearing listener (shown by the separation between the filled circles).

As shown in Figure 7.9, a 20-dB fluctuation in the envelope of the amplitude modulation (AM) signal corresponds to a larger variation in loudness level for the impaired ear compared to the loudness variation in normals. Thus, one would surmise that the perception of envelope fluctuations should be enhanced in recruitment. This hypothesis is supported by a study using a scaling technique with one unilateral subject (Wojtczak 1996) where magnitude estimates of AM were larger for the impaired ear than for the normal ear.

Further evidence is reported in a study by Moore et al. (1996) in which modulation matching tasks were performed by subjects with unilateral hearing loss. The AM depth in one ear was adjusted until the fluctuations were perceived to be equated in strength to the AM fluctuations in the other ear. Figure 7.10 replots the results for a single (but representative) subject. The fact that the equal-strength curve lies above the diagonal indicates that less modulation is necessary in the impaired than in the normal ear for the same perceived fluctuation strength. The enhancement is well accounted for by loudness recruitment in the impaired ear. The dashed line in Figure 7.10 (replotted from Moore et al. 1996) shows the predicted match given the differences in the slopes of the loudness functions for the normal and impaired ear. Thus, the perception of the strength of the envelope fluctuations seems to be enhanced by the loss of the compressive nonlinearity in the damaged ear. If the slope of the loudness matching function is 2:1, then the envelope fluctuations in the impaired ear are perceived as twice as strong as in the normal ear.

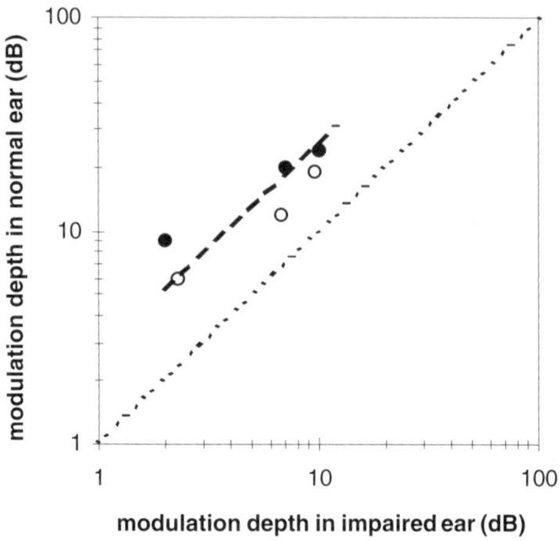

FIGURE 7.10. Modulation matching data from unilaterally hearing-impaired subjects (replotted from Moore et al. 1996). The circles plot the points of equally perceived modulation strength between the two ears. The dotted diagonal is the prediction from assuming that the impaired ear hears modulation as equally strong as in the normal ear. The dashed line is the prediction from the loudness growth curves measured for each ear.

These results seem to contradict the modulation results discussed previously, particularly those that show TMTFs for hearing-impaired listeners are no different from those of normal subjects when audibility is taken into account. If the perceptual strength of the fluctuations are being enhanced by the damaged cochlea, then one might expect the damaged auditory system to be more sensitive than a normal one to AM. Instead, listeners with hearing impairment are no more sensitive to AM than normals when the stimuli are loud enough to be above the level of hearing loss.

This line of reasoning, relating an expanded perceptual scale to just noticeable differences (jnds), is similar to Fechner's (1933) theory relating loudness perception and intensity discrimination. This theory states that steeper loudness functions should produce smaller intensity jnds. If a smaller than normal dB increment is required to produce the same loudness change, then the intensity discrimination threshold should also be smaller. In the same manner that Fechner's law does not hold for loudness and intensity discrimination, it also does not appear to hold for the relationship between perceived modulation strength and modulation discrimination. Consistent with this are the findings of Wojtczak and Viemeister (1997), who showed that AM jnds at low modulation frequencies are related to intensity jnds, as well as the findings of Moore et al. (1997), who showed that AM scaling is related to loudness scaling.

An alternate hypothesis to account for the lack of an increase in the modulation jnd with enhanced modulation perception was suggested by

Moore and Glasberg (1988). They suggest that the fluctuations inherent in the noise carrier are enhanced by recruitment, along with the modulation being detected. These enhanced noise fluctuations confound the detection of modulation and gaps in noise and thus thresholds are not better than normal. This theory is supported by Wojtczak (1996), who used spectrally triangular noise carriers instead of spectrally rectangular carriers to show that AM detection thresholds are in fact lower for hearing-impaired listeners than for normal-hearing listeners. The triangular carrier has significantly fewer envelope fluctuations, so the enhanced fluctuations of the noise carrier did not confound the detection of the modulation.

The general results of these psychoacoustic data suggest that as long as signals are made audible to listeners with hearing loss, their temporal resolution will be normal and no processing is necessary to enhance this aspect of their hearing ability. Any processing that is performed by the hearing aid should not reduce the temporal processing capability of the listener in order that recognition of speech information not be impaired. As noted, however, the perceived strength of envelope fluctuations are enhanced by the loss of compression in the impaired ear, and the amount of the enhancement is equal to the amount of loudness recruitment in that ear, indicating that a syllabic compression designed for loudness correction should also be able to correct the perceived envelope fluctuation strength.

3.4 The Effect of Compression on Modulation

Before discussing the implications of temporal resolution on hearing aid design, it is necessary to first discuss the effect that hearing aids have on the signal envelope. Since linear and slow-acting compression aids do not alter the phonemic-rate fluctuations of the speech envelopes, only the fast-acting compression hearing aids are discussed.

The effect of compression is straightforward when the attack and release times are very short (known as instantaneous compression). Under 3:1 compression, for example, a 12-dB peak-to-trough level difference reduces to a 4-dB difference—a significant reduction in the envelope fluctuation. Syllabic compression reduces the modulation sensitivity by preprocessing the signal such that the magnitude of the envelope fluctuations is reduced. It will be seen, however, that this effect is both modulation-frequency and modulation-level dependent. The effect of compression is less dramatic with more realistic time constants, such as release times of several tens of milliseconds.

The effect of instantaneous 3:1 compression on AM as a function of modulation depth is plotted with a dashed line in Figure 7.11. As can be seen, the result of the compression is to reduce the modulation depth of the output of the compressor by approximately 9.5 dB. Here, the decibel definition of modulation depth is $20 \log m$, where m is the depth of the sinusoidal modulator that modifies the envelope of the carrier by a factor of $[1 + m \sin(w_m t)]$,

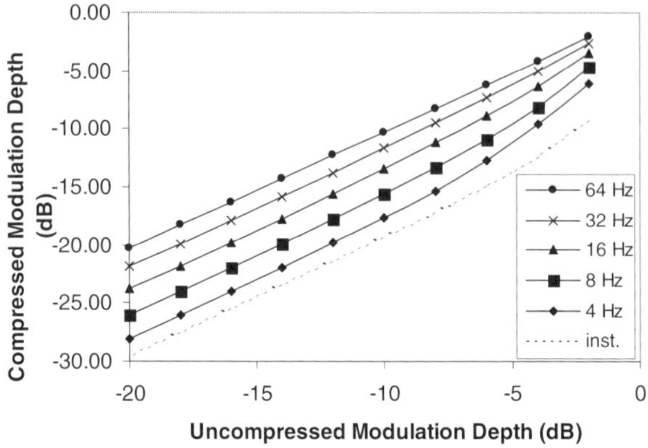

FIGURE 7.11. Effect of fast-acting compression on sinusoidal amplitude modulation depth. The abscissa is the modulation depth of a sinusoidal amplitude modulated tone at the input to a fast-acting 3:1 compressor; the ordinate is the corresponding modulation depth of the output of the compressor. The different curves correspond to different modulation frequencies. The dashed line is the effect on all modulation frequencies for instantaneous compression.

where w is the modulation frequency in radians and t is time. Compression as a front end to the auditory system therefore makes a listener 8-dB less sensitive to the fluctuations of the stimuli. Note that the effect of compression is reduced at the highest modulation depths. This is consistent with the findings of Stone and Moore (1992) and Verschuure et al. (1996), both of whom found that the effect of compression is constant as a function of modulation depth only for modulation depths of 0.43 (–7.3 dB) or lower.

More realistic implementations of compression, however, do not have quite as significant an effect on modulation. As pointed out in section 2.4, typical syllabic compression schemes have fast (~1 ms) attack times and slower (20–70 ms) release times. With the longer release time, the gain does not follow the level of the stimulus as accurately and thus does not reduce the modulation depth as effectively as instantaneous compression. Stone and Moore (1992) and Verschuure et al. (1996) describe this as a reduction in the effective compression ratio of the compressor, and they have shown that the effective compression ratio reduces as the modulation frequency increases. As the modulation frequency of the envelope increases, the sluggishness of the compressor prevents the gain from tracking the level as accurately, and the gain is not changed as much as the change in stimulus level dictates. Indeed, for modulation periods significantly shorter than the release time, the gain will hardly change at all and the modulation depth of the input to the compressor will be preserved at the output.

In the following analysis, a syllabic compressor with a compression ratio of 3:1, an attack time of 1 ms, and a release time of 50 ms is simulated. A 1-kHz tone is sinusoidally modulated at several modulation frequencies and depths, and the modulation depth of the compressed signal is measured. The effects of the compressor on modulation depth are shown in Figure 7.11, with modulation frequency as the parameter. The abscissa represents the modulation depth of the input signal to the compressor, and the ordinate represents the modulation depth of the compressor output. The results show that 64-Hz modulation is too fast for the compressor to follow and thus the modulation depth at this frequency and higher is unaffected. For lower modulation frequencies, the compressor reduces the modulation depth by approximately 2 dB for every octave-lowering of the modulation frequency.

The data for input modulation depths less than –8 dB are replotted with a dashed line in Figure 7.12 as a *compression modulation transfer function* (CMTF). The abscissa is the modulation depth into the compressor and the ordinate is the amount that the modulation depth is reduced at the output of the compressor, the same method that others have used to describe the effect of compression on envelopes (Verschuure et al. 1996; Viemeister et al. 1997). Since the effect of compression is relatively constant at these low modulation depths, the CMTF is essentially independent of input modula-

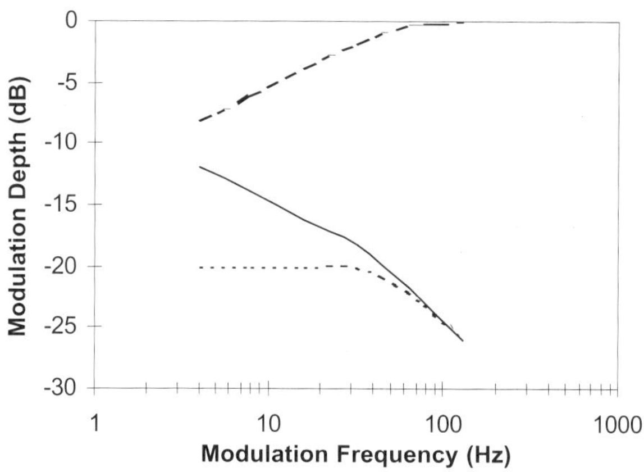

FIGURE 7.12. Effect of fast-acting compression on modulation perception. The lower dotted line is a typical temporal modulation transfer function (TMTF) for a normal-hearing listener. The solid line is the predicted aided TMTF of a hearing-impaired listener that is wearing a compression hearing aid. The upper dashed curve is the compression modulation transfer function (MTF), characterizing the effect of this compressor on modulation.

tion depth when –8 dB or less, but flattens out and approaches unity as input modulation depth approaches 0 dB. The form of the CMTF emphasizes how the compressor will reduce the sensitivity of the hearing aid wearer to envelope modulations.

3.5 Application to Hearing Aid Design

If the loudness-normalization approach to hearing-aid design is extended to an envelope-normalization approach, the I/O function of the hearing aid would be designed such that sinusoidal modulation of a given depth is perceived in an impaired ear as equally strong as it would be perceived in a normal ear. The results of the study by Moore et al. (1996) show that the perception of sinusoidal AM is enhanced in recruiting auditory systems, indicating the need for a fast-acting compression system that reduces the envelope fluctuations to normally perceived levels. Since the enhanced AM perception seems to be caused by the loudness growth in the impaired ear, a hearing aid that compresses the signal to compensate for loudness recruitment will automatically compensate for the abnormal sensitivity to AM. Approximately the same compression ratios are necessary to normalize modulation as to normalize loudness. As shown in Figure 7.11, however, not all modulation frequencies will be properly processed since the effective compression ratio reduces as modulation frequency increases (Stone and Moore 1992), and only modulations of the lowest modulation frequencies will be fully normalized. This points to one reason why the attack and release times of fast-acting compression should be as short as possible. If instantaneous compression were implemented, then all modulation frequencies would be properly normalized by the compressor. However, release times must have a minimum value in order to minimize harmonic distortion.

3.5.1 Aided Modulation Detection

The psychoacoustic results from the temporal resolution tasks with hearing-impaired subjects, however, point to a problem with this argument. Once signals have been made audible across the entire bandwidth, these subjects display normal modulation detection. Any reduction in magnitude of envelope fluctuations due to compression will make those fluctuations less perceptible and thus less detectable. The range between the envelope maxima and minima will be compressed, reducing the modulation depth and possibly the important temporal cues used to identify speech. Envelope differences will also be less discriminable, since the differences will be compressed, e.g., a 3-dB increase in an envelope peak could be reduced to a 1-dB increase. In adverse listening situations such as speech in noise with a low SNR, small differences in envelope fluctuations may provide important cues necessary for correct identification of speech. The flattening of these envelope fluctuations by fast-acting compression may make important envelope differences more difficult to discriminate, if not completely indiscriminable (cf. Plomp 1988, 1994).

The solid curve in Figure 7.12 plots the expected TMTF using a hearing aid with fast-acting 3:1 compression, derived from a normal TMTF (shown with the dotted curve and the CMTF shown with the dashed curve). The modulation detection threshold is significantly elevated for modulation frequencies below 64 Hz. Since hearing-impaired listeners appear to have normal modulation perception as long as signals are audible, fast-acting compression may have the effect of normalizing the perceived fluctuation strengths of envelopes even though the signal detection capabilities of the listener have been clearly impaired by compression to subnormal levels (cf. Plomp 1988, 1994).

Villchur (1989) counters this argument in the following way. Speech is rarely heard with threshold levels of modulation. Indeed, speech perception experiments are typically not concerned with whether speech is detectable, i.e., discriminating noise-alone from speech in noise, but with whether speech is identifiable, e.g., discriminating /ba/ in noise from /da/ in noise. The SRTs in noise for normal listeners are typically on the order of –6 dB (Plomp and Mimpen 1979). Given that speech peaks are about 12 dB above the rms level of speech, the resulting envelope of speech in noise at SRT will have approximately a 6-dB range, corresponding to an effective modulation depth of –9.6 [the relationship between dynamic range, DR, and modulation depth, m, is $DR = 20\log([1 + m]/[1 - m])$]. Since modulation detection thresholds are smaller than –20 dB, the amount of modulation in speech at 50% intelligibility is more than 10 dB above modulation detection threshold. Thus, the elevated thresholds resulting from compression (Figure 7.12) are not necessarily relevant to the effect of compression on the ability of a listener to use envelope cues. What is important is the effect of compression on the listener's ability to discriminate envelopes that are at suprathreshold depths.

3.5.2 Aided Modulation Discrimination

Consider the modulation-depth discrimination task, where thresholds are obtained for the jnd in modulation depth between two sinusoidal AM signals. This task is perhaps more relevant to speech recognition than modulation detection per se, because the standard from which the jnd is measured is not a steady-state tone or stationary noise but a signal with suprathreshold levels of envelope fluctuation. Arguments similar to those used for speech discrimination can be applied to modulation depth discrimination, anticipating that performance will be worsened by compression since modulation depth differences are compressed. To determine if indeed this is the case, the effect of compression on the more relevant measure of modulation discrimination is discussed below.

Since modulation thresholds are at a normal level once stimuli are completely audible, it is possible to analyze modulation discrimination data from normal subjects and assume that the results hold for impaired listeners as well. The dashed line in Figure 7.13 shows the modulation discrimination data for a representative normal-hearing subject. The task was to discriminate the modulation of the comparison signal from the modulation depth of the standard signal. The modulation depth of the standard is plotted along the abscissa as $20\log m_s$, where m_s is the modulation depth of the standard. Since Wakefield and Viemeister (1990) found that psychometric functions were parallel if discrimination performance is plotted as 10 $\log(m_c^2 - m_s^2)$, thresholds in Figure 7.13 are plotted with this measure, where m_c is the modulation depth of the comparison.

It is not entirely clear that 3:1 compression results in a threefold increase in peak-to-trough discrimination threshold. Compression will reduce the modulation depth in the standard and comparison alike. Since discrimination thresholds are lower at smaller modulation depths of the standard, compressing the envelope of the standard reduces the threshold for the task. The modulation discrimination data of Wakefield and Viemeister, combined with the level-dependent CMTF, can be used to determine the effect of compression on modulation discrimination. The CMTF provides the transfer function from pre- to postcompression modulation depth, and the discrimination data is applied to the postcompression modulation. The

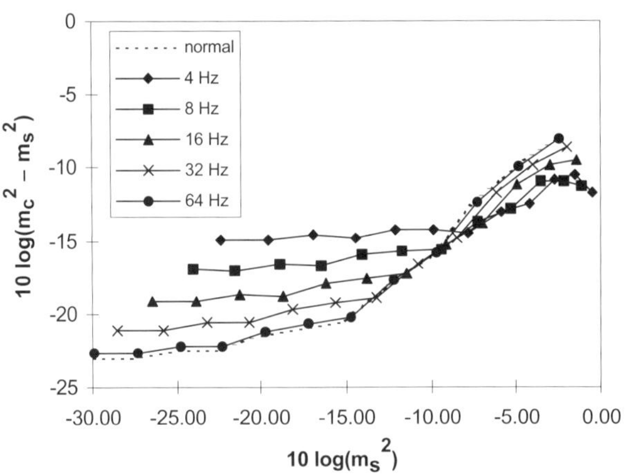

FIGURE 7.13. Effect of fast-acting compression on modulation discrimination. The dashed line shows modulation discrimination performance for a normal-hearing subject (from Wakefield and Viemeister 1990). The solid lines show modulation discrimination performance for different modulation frequencies, indicating that this compressor impairs modulation discrimination below approximately –8 dB but enhances modulation discrimination above –8 dB.

modulation discrimination thresholds under compression can be calculated as follows. Given the modulation depth of the standard at the input to the compressor, the modulation depth at the compressor output is calculated using the CMTF. From the discrimination data, the modulation depth of the compressed comparison required for adequate discrimination from the compressed standard is first determined. Then, the uncompressed modulation depth necessary to produce the depth of the compressed comparison is determined from the inverse of the CMTF. Essentially, known discrimination data are applied to the output of the compressor to calculate discrimination thresholds measured at the input to the compressor.

The results of this analysis are shown in Figure 7.13 using modulation frequency as the parameter. Modulation depths for the standard and comparison are shown, as measured at the input to the compressor. Discrimination thresholds for linear processing are plotted with a dashed line. The results show that for modulation depths smaller than −12 dB to −8 dB, thresholds are considerable higher under 3:1 compression than under linear processing. For example, with 8-Hz modulation and a precompression modulation depth of −15 dB, the precompression modulation depth necessary for discrimination is 4 dB greater than with linear processing. Compression has made envelope discrimination more difficult at small modulation depths. For a given depth of the standard, discrimination threshold increases as modulation frequency decreases due to the increased effectiveness of compression at lower frequencies, as shown in the CMTF in Figure 7.12.

Note that above a certain knee point in the curve, envelope discrimination with 3:1 compression is better than with linear processing (as shown by the solid lines below the dashed line in Fig. 7.13). This result indicates that compression improves envelope discrimination at high levels of fluctuation. It now remains to relate these modulation depth results to speech envelopes that are typically measured by peak-to-trough dynamic ranges.

3.5.3 Application to Speech in Noise

As far as modulation detection and discrimination can be related to the envelope cues of speech either in quiet or in the presence of noise, these results can be related to speech perception as follows. Since a modulation depth of −10 dB corresponds to a level difference between envelope maxima and minima of approximately 6 dB, the results in Figure 7.13 indicate that 3:1 compression will have a negative impact on the perception of envelope cues with less than 6 dB of dynamic range. Speech has a dynamic range of 30 dB, so compression should not have a negative impact on the discrimination of the envelope fluctuations of speech in quiet. Indeed, the results of Figure 7.13 indicate that compression may enhance the discriminability of these envelopes. Noise, however, reduces the dynamic range of speech by "filling in the gaps," effectively reducing the modulation in the envelope (Steeneken and Houtgast 1983). Because of the smaller modulation depth of speech in noise and the negative effect of compression itself at low modulation depths, compression could have a significant impact on the perception of speech envelopes in the presence of background noise. For a 0-dB SNR, the peaks in the speech signal are typically 12 dB above the level of the noise (Miller and Nicely 1955). For the modulation depth of the speech in noise to be less than –10 dB, the SNR must be less than –6 dB. It is for these conditions of extremely poor SNRs that compression can be expected to have a negative impact on the disciminability of speech envelope cues. Since SRTs for normals are –6 dB and even higher for hearing-impaired subjects, compression will only impair envelope cues when sentence correct scores are 50% or less. This analysis assumes that results from sinusoidal AM can translate to results with the more complex envelope of speech, an assumption that is not clearly valid since a compressor is not a linear system. The general effect, however, is still applicable.

The question remains as to whether the overall effect of compression has reduced the information-carrying ability of envelopes. This can be addressed by counting the number of modulation-depth jnds with and without compression to determine if compression has reduced the number of discriminable "states" that the envelope can encode with modulation depth. It is clear that compression cannot reduce the number of envelope jnds based on modulation depth because of the 1:1 mapping of precompressor modulation depth to postcompressor modulation depth. Thus, there is no reduction in the number of bits that an envelope can encode with depth of modulation. Most of the jnds under compression, however, occur at modulation depths of –8 dB and above. Thus, compression transmits more information about envelopes with dynamic ranges of 10 dB and above than does linear processing, while linear processing better transmits envelope information at smaller modulation depths.

In summary, these results do not support the assumption that compression will impair envelope perception and reduce envelope discriminability under all conditions. Compression that normalizes the perception of loudness will also normalize the perception of envelopes, while improving discrimination of envelopes with dynamic ranges greater than 10 dB and worsening discrimination of envelopes with dynamic ranges less than 10 dB. It should also be noted that the analysis performed was done for 3:1 compression, which is usually the maximum required for loudness compensation (Killion 1996). Negative effects for smaller compression ratios will be less than what is shown here.

4. Frequency Resolution

4.1 Psychoacoustic Measures

Frequency resolution is a measure of the auditory system's ability to encode sound based on its spectral characteristics, such as the ability to detect one frequency component in the presence of other frequency components. This is somewhat ambiguous according to Fourier theory, in which a change in the spectrum of a signal results in a corresponding change in the temporal structure of the signal (which might provide temporal cues for the detection task). Frequency resolution can be more accurately described as the ability of the auditory periphery to isolate a certain specific frequency component of a stimulus by filtering out stimulus components of other frequencies. It is directly proportional to the bandwidth of the auditory filters whose outputs stimulate inner hair cells. From a perceptual coding perspective, if the cue to a detectable change in the spectrum of a signal is a change in the shape of the excitation along the basilar membrane (as opposed to a change in the temporal pattern of excitation), then it is clear that the change in spectral content exceeds the listener's frequency resolution threshold. If no difference can be perceived between two sounds that differ in spectral shape, however, then the frequency resolution of the listener was not sufficiently fine to discriminate the change in spectral content.

Poorer than normal frequency-resolving ability of an impaired listener with sensorineural hearing might result in the loss of certain spectral cues used for speech identification. At one extreme, with no frequency resolution capability whatever, the spectrum of a signal would be irrelevant and the only information that would be coded by the auditory system would be the broadband instantaneous power of the signal. Partially degraded frequency resolution might affect the perception of speech by, for example, impairing the ability to distinguish between vowels that differ in the frequency of a formant. Poorer than normal frequency resolution would also result in greater masking of one frequency region by another, which again could eliminate spectral speech cues. If one considers the internal perceptual spectrum of a sound to be the output of a bank of auditory filters, then broadening those auditory filters is equivalent to smearing the signal's amplitude spectra (see, e.g., Horst 1987). Small but significant spectral detail could be lost by this spectral smearing. If hearing loss exerts a concomitant reduction in frequency resolving capabilities, then it is important to know the extent to which this occurs and what the effect is on speech intelligibility.

Both physiological and psychoacoustic research have shown that cochlear damage results in both raised auditory thresholds and poorer frequency resolution (Kiang et al. 1976; Wightman et al. 1977; Liberman and Kiang 1978; Florentine et al. 1980; Gorga and Abbas 1981b; Tyler et al. 1982a; Carney and Nelson 1983; Tyler 1986). Humes (1982) has suggested that the data showing poorer frequency resolution in hearing-impaired listeners simply reflects proper cochlear functioning at the high stimulus levels presented to those with hearing loss since auditory filter bandwidths in healthy cochleas naturally increase with stimulus level. Auditory filter bandwidths measured at high stimulus levels (in order to exceed the level of the hearing loss) may be mistakenly classified as broader than normal if compared to bandwidths measured in normal-hearing individuals at equal SL but lower SPL. Thus, care must be taken to control for the level of the stimuli when measuring the frequency resolution of hearing-impaired subjects.

4.2 Resolution with Equalized Audibility

Dubno and Schaefer (1991) equalized audibility among subjects, with and without hearing loss, by raising the hearing threshold of the latter with broadband background noise and then compared frequency resolution between the two groups. While they found no difference in the critical ratio of the two groups, frequency selectivity measured with tuning curves and narrowband noise-masked thresholds were poorer for the hearing-impaired subjects than for the noise-masked normal-hearing subjects. Dubno and Schaefer (1992) used a notch noise to show that auditory filters in subjects with hearing loss have broader ERBs than normal-hearing subjects with artificially elevated thresholds. Similar conclusions are drawn from the detection of ripples in the spectrum of broadband noise, where impaired listeners require greater spectral contrast than normal when the stimuli are presented at the same high level (Summers and Leek 1994). These results are perhaps not surprising given that Florentine and Buus (1984) have shown that noise maskers do not entirely simulate the reduced frequency resolution observed among the hearing impaired.

Not only are auditory filters broader in impaired listeners, but the low-frequency slope of the filters is shallower than normal, resulting in a greater degree of upward spread of masking (Glasberg and Moore 1986). Gagné (1983) compared frequency selectivity in normals and impaired listeners by measuring upward spread of masking at equal dB SPL levels of maskers. He found that those with hearing loss showed more masking than those with normal hearing and similar results were found by Trees and Turner (1986). Using noise maskers to simulate hearing loss in normals, it is clear that the increased upward spread in hearing-impaired listeners is not simply due to the loss of off-frequency listening resulting from hearing loss above the frequency of the signal (Dubno and Schaefer 1992; Dubno and Ahlstrom 1995).

There is overwhelming evidence that individuals with hearing impairment have poorer frequency resolution than normal, and that this cannot be accounted for entirely by the higher stimulus level necessary for audibility. Indeed, the high stimulus levels compound the problem since auditory filter bandwidths increase with stimulus level for both normal-hearing and hearing-impaired listeners. Thus, frequency-resolving capabilities are impaired not only by the damaged auditory system but also by the higher stimulus levels used to place the signal well above the listener's threshold of audibility.

4.3 The Effect of Reduced Frequency Resolution on Speech Perception

The short-term spectral shape of speech contains significant information that can help with the accurate identification of speech. The frequency location of the formant peaks are the primary cues for the identification of vowels (Delattre et al. 1952; Klatt 1980), as evidenced by Dreschler and Plomp (1980), who used multidimensional scaling to show that the F1 and F2 frequencies were the two most significant cues for vowel identification. Important cues for consonant place of articulation are also contained in the spectral shape of speech (Stevens and Blumstein 1978; Bustamante and Braida 1987a), both in the spectral contour of consonants and in the second formant transition of adjoining vowels (Pickett 1980). Poor frequency resolution might smear the details of these spectral features, making such speech cues more difficult to identify.

Several researchers have investigated this issue by measuring speech recognition in hearing-impaired subjects who have different auditory filters bandwidths, looking for evidence that poorer than normal performance is related to poorer than normal frequency resolution. The difficulty with this approach, and with many studies that have shown speech intelligibility to be correlated with frequency resolution (e.g., Bonding 1979; Dreschler and Plomp 1980), is that both measures are themselves correlated with hearing threshold. The poor speech scores obtained may be due to the inaudibility of part of the speech material and not due to the reduced frequency resolving ability of the listener.

In experiments where speech signals were presented to impaired listeners with less than the full dynamic range of speech above the listener's audibility threshold, poorer than normal speech recognition was correlated with both frequency resolution and threshold (Stelmachowicz et al. 1985; Lutman and Clark 1986). The correlation between speech intelligibility and frequency resolution was found to be significantly smaller when the effect of audibility was partialed out.

To investigate the effect of frequency resolution on speech intelligibility while eliminating threshold effects, speech tests must be performed at equally audible levels of speech across subjects with different auditory filter bandwidths. This can be achieved either by presenting the speech to the normal and impaired subjects at equal SLs such that equal amounts of the dynamic range of speech fall below the threshold of both groups, or by increasing the hearing threshold of normals to the level of the impaired subjects using masking noise and presenting speech at the same overall SPL to both groups. The latter technique ensures that level effects are the same for both groups. Background noise, however, may introduce artifacts for the normal-hearing subjects that do not occur for hearing-impaired subjects (such as the random level fluctuations inherent in the noise).

Dubno and Dirks (1989) equalized the audibility of their speech understanding paradigm by presenting speech stimuli at at levels that produced equal AI values to listeners with varying degrees of hearing loss. They found that stop-consonant recognition was not correlated with auditory filter bandwidth. While Turner and Robb (1987) did find a significant difference in performance between impaired and normal-hearing subjects when the stop-consonant recognition scores were equated for audibility, their results are somewhat ambiguous since they did not weight the spectral regions according to the AI as Dubno and Dirks did. Other research has found no difference in speech recognition between hearing-impaired listeners in quiet and normal-hearing listeners whose thresholds were raised to the level of the impaired listeners' by masking noise (Humes et al. 1987; Zurek and Delhorne 1987; Dubno and Schaefer 1992, 1995).

These results indicate that reduced frequency resolution does not impair the speech recognition abilities of impaired listeners in quiet environments. In general, hearing-impaired listeners have not been found to have a significantly more difficult time than normal-hearing listeners with understanding speech in quiet once the speech has been made audible by the appropriate application of gain (Plomp 1978). Under noisy conditions, however, those with impairment have a significantly more difficult time understanding speech compared to the performance of normals (Plomp and Mimpen 1979; Dirks et al. 1982). Several researchers have suggested that this difficulty in noise is due to the poorer frequency resolution caused by the damaged auditory system (Plomp 1978; Scharf 1978; Glasberg and Moore 1986; Leek and Summers 1993). Comparing speech recognition performance with psychoacoustic measures using multidimensional analysis, Festen and Plomp (1983) found that speech intelligibility in noise was related to frequency resolution, while speech intelligibility in quiet was determined by audibility thresholds. Horst (1987) found a similar correlation.

Using synthetic vowels to study the perception of spectral contrast, Leek et al. (1987) found that normal-hearing listeners required formant peaks to be 1 to 2 dB above the level of the other harmonics to be able to accurately identify different vowels. Normal listeners with thresholds raised by masking noise to simulate hearing loss needed 4-dB formant peaks, while impaired listeners needed 7-dB peaks in quiet. Thus, 3 dB of additional spectral contrast was needed with the impaired listeners because of reduced frequency resolution, while an additional 2 to 3 dB was needed because of the reduced audibility of the stimuli. Leek et al. (1987) determined that to obtain the same formant peak in the internal spectra or excitation patterns of both groups given their thresholds, the auditory filters of those with hearing loss needed bandwidths two to three times wider than normal auditory filters, consistent with results from other psychoacoustic (Glasberg and Moore 1986) and physiological (Pick et al. 1977) data comparing auditory bandwidth. These results are also consistent with the results of Summers and Leek (1994), who found that the hearing-impaired subjects required higher than normal spectral contrast to detect ripples in the amplitude spectra of noise, but calculated that the contrast of the internal rippled spectra were similar to the calculated internal contrasts of normal listeners when taking the broader auditory filters of impaired listeners into account. Consistent with equating the internal spectral of the impaired and normal subjects at threshold, Dubno and Ahlstrom (1995) found that the AI better predicted consonant recognition in hearing-impaired individuals when their increased upper spread of masking data was used to calculate the AI rather than using masking patterns found in normal-hearing subjects. In general, the information transmitted by a specific frequency region of the auditory periphery in the presence of noise is affected by the frequency resolution of that region since frequency-resolving capability affects the amount of masking that occurs at that frequency (Thibodeau and Van Tasell 1987).

Van Tasell et al. (1987a) attempted to measure the excitation pattern (or internal spectrum) of vowels directly by measuring the threshold of a brief probe tone that was directly preceded by a vowel, as a function of the frequency of the probe. They found that the vowel masking patterns (and by extension the internal spectrum) were smoother and exhibited less pronounced peaks and valleys due to broader auditory filters, although the level of the vowel was higher for the impaired subjects than for the normal subjects (which may have been part of the cause for the broader auditory filters). Figure 7.14 shows the transformed masking patterns of /λ/ for a

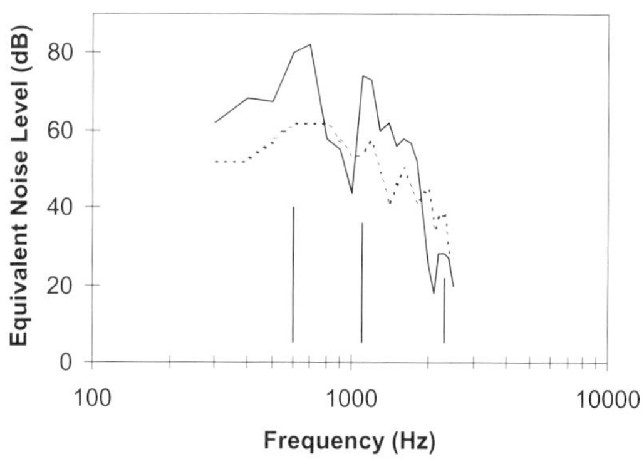

FIGURE 7.14. The vertical lines represent the first three formants of the vowel /ʌ/. The solid line plots the estimated masking pattern of the vowel for a normal-hearing listener. The dotted line shows the estimated masking pattern of the vowel for a hearing-impaired listener. (Data replotted from Van Tasell et al. 1987a.)

normal and an impaired listener, taken from Figures 7.3 and 7.5 of their paper (see their paper for how the masking patterns are calculated). Vowel identification was poorer for the subject with hearing loss, in keeping with the poorer representation of the spectral detail by the impaired auditory system. While the Van Tasell et al. study showed correlations of less than 0.5 between the masking patterns and the recognition scores, the authors note that this low degree of correlation is most likely due to the inappropriateness of using the mean-squared difference between the normal and impaired excitation patterns as the error metric for correlation analysis.

It has been assumed in the discussion so far that the hearing loss of the individual has been primarily, if not solely, due to outer hair cell damage. What the effect of inner hair cell damage is on frequency resolving abilities and coding of speech information is unclear. Vowel recognition in quiet is only impaired in listeners with profound hearing loss of greater than 100 dB (Owens et al. 1968; Pickett 1970; Hack and Erber 1982), who must have significant inner hair cell damage since outer hair cell loss only raises thresholds by at most 60 dB. Faulkner et al. (1992) have suggested that this group of individuals has little or no frequency resolving capability. This may indeed be the case since the benefit provided to normals by adding a broadband speech envelope cue when lipreading is the same as the benefit provided to severely hearing-impaired subjects by adding the complete speech signal when lipreading (Erber 1972). A continuum must exist, then, between normal listeners and the severely hearing impaired through which frequency resolution abilities get worse and the ability to use spectral information for speech recognition deteriorates even in quiet. For most hearing aid wearers, their moderate hearing loss results in the threshold of audibility limiting their speech in quiet performance while frequency resolution limits their speech in noise performance.

4.4 Implications for Hearing Aid Design

Reduced frequency selectivity in damaged auditory systems seems to impair speech understanding primarily in low SNR situations. The broader auditory filters and loss of lateral suppression smooth the internal spectral contrast and perhaps increase the SNR within each auditory filter (Leek and Summers 1996). In simulations of reduced spectral contrast, normal subjects have shown a reduction in the recognition of vowels and place-of-articulation information in consonants (Summerfield et al. 1985; Baer and Moore 1993; Boothroyd et al. 1996). One function of a hearing aid would be to sharpen the spectral contrast of the acoustic signal by increasing the level and narrowing the bandwidth of the spectral peaks, and decreasing the level of the spectral valleys. Ideally, the broader auditory filters in hearing-impaired listeners would smear the sharpened spectra to an internal level of contrast equivalent to that for a normal listener. This technique has found little success and may be due to the broad auditory filters overwhelming the sharpening technique (Horst 1987). Poor frequency resolution will smear a spectral peak a certain degree regardless of how narrow the peak in the signal is (Summerfield et al. 1985; Stone and Moore 1992; Baer and Moore 1993). A modest amount of success was obtained by Bunnell (1990), who applied spectral enhancement only to the midfrequency region, affecting the second and third formants. Bunnell's processing also had the consequence of reducing the level of the first formant, however, which has been shown to mask the second formant in hearing-impaired subjects (Danaher and Pickett 1975; Summers and Leek 1997), and this may have contributed to the improved performance.

Most of the studies that have investigated the relationship between frequency resolution and speech intelligibility did not amplify the speech with the sort of frequency-dependent gain found in hearing aids. Typically, the speech gain was applied equally across all frequencies. Since the most common forms of hearing loss increase with frequency, the gain that hearing aids provide also increases with frequency. This high-frequency emphasis significantly reduces the masking ability of low-frequency components on the high-frequency components and may therefore reduce the direct correlation found between frequency resolution and speech intelligibility. Upward spread of masking may still be a factor with some compression aids since the gain tends to flatten out as the stimulus level increases. The extent to which the masking results apply to speech perception under realistic hearing aid functioning is uncertain.

As discussed in section 2, Plomp (1988) has suggested that multiband compression is harmful to listeners with damaged outer hair cells since it reduces the spectral contrast of signals. The argument that loudness recruitment in an impaired ear compensates for this effect, or that the multiband compressor is simply performing the spectral compression that a healthy cochlea normally does, does not hold since recruitment and the abnormal growth of loudness does not take frequency resolution into account. Indeed, the spectral enhancement techniques that compensate for reduced frequency resolution described above produce expansion, the opposite of compression.

This reduction in spectral contrast occurs for a large number of independent compression bands. Wideband compression (i.e., a single band) does not affect spectral contrast since the AGC is affecting all frequencies equally. Two- or three-band compression preserves the spectral contrast in local frequency regions defined by the bandwidth of each filter. Multiband compression with a large number of bands can be designed to reduce the spectrum-flattening by correlating the AGC action in each band such that they are not completely independent. Such a stratagem sacrifices the compression ratio in each band somewhat but provides a better solution than simply reducing the compression ratio in each band. Bustamante and Braida (1987b) have also proposed a principal components solution to the conflicting demands of recruitment and frequency resolution. In this system, broadband compression occurs in conjunction with narrowband expansion, thereby enhancing the spectral peaks in local frequency regions, while presenting the average level of the signal along a loudness frequency contour appropriate to the hearing loss of the individual.

Finally, it has been noted that frequency resolution degrades and upward spread of masking increases as the SPL increases (Egan and Hake 1950) for both normal-hearing and hearing-impaired people. This suggests that loudness normalization may not be a desirable goal for small SNR environments when high overall levels are encountered, since the broadening of auditory bandwidths seems to have the largest effect on speech intelligibility at low SNRs. Under such situations, reducing the signal presentation level could actually improve frequency resolving capability and consequently improve speech intelligibility, as long as none of the speech is reduced below the listener's level of audibility. However, little benefit has been found with this technique so far (cf. section 5).

5. Noise Reduction

Hearing-impaired listeners have abnormal difficulty understanding speech in noise, even when the signal is completely audible. The first indication people usually have of their hearing loss is a reduced understanding of speech in noisy environments such as noisy restaurants or dinner parties. Highly reverberant environments (e.g., such as inside a church or lecture hall) also provide a more difficult listening environment for those with hearing loss. Difficulty with understanding speech in noise is a major complaint of hearing aid users (Plomp 1978; Tyler et al. 1982a), and one of the primary goals of hearing aids (after providing basic audibility) is to improve intelligibility in noise.

Tillman et al. (1970) have shown that normal listeners need an SNR of –5dB for 50% word recognition in the presence of 60dB SPL background noise, while impaired listeners under the same conditions require an SNR of 9dB. These results have been confirmed by several researchers (Plomp and Mimpen 1979; Dirks et al. 1982; Pekkarinen et al. 1990), each of whom has also found a higher SNR requirement for impaired listeners that was not accounted for by reduced audibility. Since many noisy situations have SNRs around 5 to 8dB (Pearsons et al. 1977), many listeners with hearing loss are operating in conditions with less than 50% word recognition ability. No change to the SNR is provided by standard hearing aids because the amplification increases the level of the noise and the speech equally. Difficulty with understanding speech in noisy situations remains.

The higher SNRs required by the hearing impaired to perform as well as normals is probably due to broader auditory filters and reduced suppression, resulting in poorer frequency resolution in the damaged auditory system. As discussed in section 4, reduced frequency resolution smears the internal auditory spectrum and makes phoneme recognition in background noise difficult. The fact that the AI successfully predicts the reduced speech intelligibility when the increased auditory filter bandwidths of hearing-impaired listeners are accounted for supports the hypothesis that reduced frequency resolution is the cause of higher SRTs.

Ideally, a hearing aid should compensate for the reduced frequency resolution of the wearers and thus improve their understanding of speech in noise. Unfortunately, spectral enhancement techniques that have attempted to compensate for poorer spectral resolution have been unsuccessful in improving intelligibility in noise. Thus, if a hearing aid is to improve speech understanding in noise beyond simply making all parts of the signal audible, the level of the noise must somehow be reduced in the physical stimulus.

Over the past several decades, signal-processing algorithms have been developed that reduce the level of noise in the presence of speech. These algorithms can be divided into two categories: single-microphone and multimicrophone systems.

5.1 Single-Microphone Techniques

In single-microphone systems, the desired signal and competing background noise are both transduced by the same microphone. The noise-reduction algorithm must take advantage of spectral and temporal mismatches between the target speech and background noise in order to attenuate the latter while preserving the former. In addition, the algorithm must frequently identify which part of the signal is the target speech and which part is background noise. This task is made even more difficult if the background contains speech from competing talkers. This is because a *priori* information about the structure of speech cannot easily be used to discriminate target speech from the background. Any practical implementation of a noise-reduction algorithm in a hearing aid requires that the processing be robust to a wide assortment of adverse environments: vehicular traffic, office machinery, speech conversation, as well as the combination of speech babble, music, and reverberation.

A significant problem with single-microphone noise-reduction algorithms is that improvement in the SNR does not necessarily imply a concurrent improvement in speech intelligibility. Indeed, Lim (1983) has pointed out that many techniques provide SNR improvements of up to 12 dB in wideband noise but none results in improved intelligibility. It has also been shown that noise-reduction algorithms can improve the intelligibility scores of automatic speech recognition systems while not improving intelligibility scores with human observers. Thus, it appears that quantifying algorithmic performance using physical measurements of the noise-reduced stimuli is not sufficient. The change in intelligibility with human observers must also be measured.

The most ambitious noise-reduction system to be applied to commercial hearing aids is that of Graupe et al. (1986, 1987), who used a multiparameter description of speech to estimate the SNR in different frequency regions. If the SNR was estimated to be low in a specific frequency region, then the gain was reduced in that channel. This system was implemented on a chip called the Zeta-Noise Blocker and used in several hearing aids. Stein and Dempsey-Hart (1984) and Wolinsky (1986) showed a significant improvement in intelligibility with this processing compared to linear processing alone. Further investigations, however, have not been able to demonstrate any increase in intelligibility with this processing (Van Tasell et al. 1988), and these negative results suggest that the original studies may not have had the proper control conditions since the subjects were allowed to adjust the volume. It was thus unclear whether the benefit found was due to the noise-reduction processing per se or to the increased audibility resulting from volume control adjustment. The company that sold the Zeta-Noise Blocker has since gone out of business.

5.1.1 Frequency-Specific Gain Reduction

In the late 1980s, several hearing aids were introduced possessing level-dependent high-pass filtering, a cruder version of the Graupe processing (Sigelman and Preves 1987; Staab and Nunley 1987). These systems were known at the time as automatic signal processing (ASP) and today are called BILL devices, for "bass increase at low levels." (A description that more accurately captures the noise reduction aspect of the design might be "bass decrease at high levels.")

This design, originally proposed by Lybarger (1947), is an attempt to reduce the masking of high-frequency speech cues by the low-frequency components of background noise. This seemed to be a natural strategy since speech babble has a maximum spectrum level of around 500 Hz and decreases above that at about 9 dB/octave. Moreover, most environmental noises are low frequency in nature (Klumpp and Webster 1963). In addition, upward spread of masking would seem to add to the difficulty of understanding speech in noise since most difficult noisy environments occur at high levels and thus more masking would be expected. This problem could extend to masking by the low-frequency components of speech on the high-frequency components of the same speech (Pollack 1948; Rosenthal et al. 1975). Finally, wider auditory bandwidths of hearing-impaired listeners could also cause greater masking of high frequencies by low. It seems, then, that reducing the gain in the low-frequency region when the average power in this region is high should result in improved perception of high-frequency speech cues and thus improve overall speech intelligibility.

In general, decreasing the low-frequency gain has not been found to improve intelligibility (Punch and Beck 1986; Neuman and Schwander 1987; Van Tasell et al. 1988; Tyler and Kuk 1989; Fabry and Van Tasell 1990;

Van Tasell and Crain 1992). Many, in fact, found that attenuating the low-frequency response of a hearing aid resulted in a decrement in speech intelligibility. This is perhaps not surprising since the low-frequency region contains significant information about consonant features such as voicing, sonorance, and nasality (Miller and Nicely 1955; Wang et al. 1978) and about vocalic features such as F_1. Consistent with this are the results of Gordon-Salant (1984), who showed that low-frequency amplification is important for consonant recognition by subjects with flat hearing losses. Fabry and Van Tasell (1990) suggested that any benefit obtained from a reduction in upward spread of masking is overwhelmed by the negative effect of reducing the audibility of the low-frequency speech signal. They calculated that the AI did not predict any benefit from high-pass filtering under high levels of noise, and if the attenuation of the low frequencies is sufficiently severe, then the lowest levels of speech in that region are inaudible and overall speech intelligibility is reduced. In addition to the poor objective measures of this processing, Neuman and Schwander (1987) found that the subjective quality of the high-pass filtered speech-in-noise was poorer than a flat 30-dB gain or a gain function, which placed the rms of the signal at the frequency response of the subjects' most-comfortable-level frequency contour. Punch and Beck (1980) also showed that hearing aid wearers actually prefer an extended low-frequency response rather than an attenuated one.

Some investigations into these attenuation-based noise-reduction techniques, however, have produced positive results. Fabry et al. (1993) showed both improved speech recognition and less upward spread of masking when high-pass filtering speech in noise, but found only a small ($r = 0.61$) correlation between the two results. Cook et al. (1997) found a significant improvement in speech intelligibility from high-pass filtering when the masking noise was low pass, but found no correlation between improved speech recognition scores and the measure of upward spread of masking and no improvement in speech intelligibility when the noise was speech-shaped.

Inasmuch as reducing the gain in regions of low SNR is desirable, Festen et al. (1990) proposed a technique for estimating the level of noise across different regions of the spectrum. They suggested that the envelope minima out of a bank of bandpass filters indicate the level of steady-state noise in each band. The gain in each band is then reduced to lower the envelope minima close to the listener's hearing threshold level, thereby making the noise less audible while preserving the SNR in each band. Festen et al. (1993) found that in cases where the level of a bandpass noise was extremely high, this technique improved the intelligibility of speech, presumably due to reduced masking of the frequency region above the frequency of the noise. For noise typical of difficult listening environments, however, no such improvement was obtained. Neither was an improvement found by Neuman and Schwander (1987), who investigated a similar type of processing.

A variation on this idea has been implemented in a recently developed commercial digital signal processing (DSP) hearing aid that adjusts the gain within several bands according to the dynamic range of the signal measured in each band. If the measured dynamic range is similar to that found in clean speech, then the noise reduction does nothing in that band. If the measured dynamic range in a band is found to be less than that for clean speech, then it is assumed that a certain amount of noise exists in the band and attenuation is applied in an amount inversely proportional to the measured dynamic range. This device has achieved enormous success in the marketplace, which is most likely due to the excitement of its being one of the first DSP aids on the market; the success indicates that, at the very least, the band-attenuation strategy is not being rejected by the hearing-impaired customers outright.

Levitt (1991) noted that reducing the gain in frequency regions with high noise levels is likely to be ineffective because the upward spread of masking is negligible at levels typical of noise in realistic environments. He also noted, however, that some individuals did show some improvement with high-pass filtering under noisy conditions, and that such individual effects need to be studied further to determine if there is a subset of individuals who could benefit from this sort of processing.

5.1.2 Spectral Subtraction

More sophisticated noise-reduction techniques can be implemented in hearing aids once the processing capabilities of DSP aids increase. The INTEL algorithm, developed by Weiss et al. (1974) for the military, is the basis for spectral subtraction techniques (Boll 1979), whereby the noise spectrum is estimated when speech is not present and the spectral magnitude of the noise estimate subtracted from the short-term spectral magnitude of the signal.[1] No processing is performed on the phase of the noisy spectra, but it has been shown that an acceptable result is obtained if the processed amplitude spectra is combined with the unprocessed phase spectra to synthesize the noise-reduced signal (Wang and Lim 1982; Makhoul and McAulay 1989). It should be noted that spectral subtraction assumes a stationary noise signal and it will not work with more dynamic backgrounds such as an interfering talker.

Levitt et al. (1986) evaluated the INTEL algorithm on a prototype digital aid and found no improvement in speech intelligibility, although the SNR was significantly increased. An examination of the processed signal suggested that the algorithm improved formant spectral cues but removed noise-like speech cues germane to fricatives and plosives. Given that consonants carry much of the speech information in continuous discourse, it is a wonder that this processing does not make intelligibility worse. In addi-

[1] The subtraction can also be between the power spectrum of the noise and signal or any power function of the spectral magnitude.

tion, spectral subtraction typically introduces a distortion in the processed signal known as musical noise due to the tone-like quality of the distortion. This is a result of negative magnitude values produced by subtraction, which must be corrected in some manner that usually results in non-Gaussian residual noise. The presence of such distortion may not be acceptable to hearing aid wearers even if it results in increased intelligibility.

5.1.3 Other Techniques

To ascertain what is the best possible performance one can expect from a filter-based noise-reduction algorithm, Levitt et al. (1993) applied Wiener filtering to consonant-vowel (CV) and VC syllables. The Wiener filter (Wiener 1949) maximizes the SNR, given knowledge of the signal and noise spectra. As Levitt et al. point out, this filter assumes stationary signals and is thus not strictly applicable to speech. The authors calculated the Wiener filter for each consonant and applied a separate Wiener filter to each corresponding VC and CV, thereby maximizing the SNR with respect to consonant recognition. Their results show that consonant recognition performance for normal-hearing subjects was worse with the filter than without. However, hearing-impaired subjects did show some benefit from the filtering. The success with the impaired listeners is encouraging, but it should be kept in mind that the Wiener filter requires an estimate of both the noise spectrum and the short-term speech spectrum. While the former can be estimated during pauses in the speech signal, the ability to estimate the spectra of each phoneme in real time is extremely difficult. Thus, this technique provides only an upper bound on what is achievable through filtering techniques.

The disappointing results with most single-microphone techniques has led the hearing aid industry to emphasize the benefit of improved "comfort" rather than intelligibility. Comfort is meant to indicate an improved subjective quality of speech in noise due to the processing. In other words, even though the processing does not improve intelligibility, the listener may prefer to hear the signal with the processing on than off. This quality benefit is a result of the lowering of the overall noise signal presented to the listener and a concomitant reduction in annoyance. While Preminger and Van Tasell (1995) found that intelligibility and listening effort were indistinguishable, meaning that if intelligibility did not improve then the effort required to listen did not improve, the speech intelligibility was reduced by filtering and not by the addition of noise. It is therefore possible that less effort is needed to understand the speech in noise with the noise-reduction processing, and thus an extended session of listening under adverse conditions would be less taxing on the listener. This suggests that an objective measure of the processing's benefit could be obtained with a dual-task experiment, where the listener's attention is divided between understanding speech and some other task that shares the subject's attention. If such noise-reduction techniques truly do reduce the attention necessary to process speech, then a dual-task measure should show improvements.

Other techniques that are candidates for implementation on aids with powerful DSP chips include comb filtering to take advantage of the harmonic structure of speech (Lim et al. 1978), filtering in the modulation frequency domain (Hermansky and Morgan 1994; Hermansky et al. 1997), enhancement of speech cues based on auditory scene analysis (Bregman 1990), sinusoidal modeling (Quatieri and McAuley 1990; Kates 1991), and the application of several different types of speech parameter estimation techniques (Hansen and Clements 1987; Ephraim and Malah 1984). Improvements over current techniques include better estimates of pauses in speech, better use of a priori information about the target signal (e.g., male talker, female talker, music), and identification of different noisy environments for the application of noise-specific processing techniques (Kates 1995).

In general, single-microphone noise reduction techniques are constrained by the need for preserving the quality of the speech, the limited processing capabilities of digital chips, and the need for real-time processing with small processing delays. They have not been shown to improve speech intelligibility for the majority of listeners, impaireds and normals alike. Upward spread of masking does not seem to be severe enough in realistic environments to make frequency-specific gain reduction an effective technique for improving intelligibility. It may be that noise reduction techniques can improve the detection of specific speech phonemes, but there is no improvement in overall word recognition scores. The best one can expect for the time being is that listening effort and the overall subjective quality of the acoustic signal are improved under noisy conditions.

5.2 Multiple Microphone Techniques

Unlike the case with one-microphone noise-reduction techniques, the use of two or more microphones in a hearing aid can provide legitimate potential for significant improvements to speech intelligibility in noisy environments. Array processing techniques can take advantage of source separation between signal and noise, while a second microphone that has a different SNR than the first can be used for noise cancellation.

5.2.1 Directionality

Array processing can filter an interfering signal from a target signal if the sources are physically located at different angles relative to the microphone array, even if they have similar spectral content (Van Veen and Buckley 1988). The microphone array, which can contain as few as two elements, simply passes signals from the direction of the target and attenuates signals from other directions. The difficult situation of improving the SNR when both the target and interfering signal are speech is quite simple with array processing if the target and interferer are widely separated in angle relative to the microphone array.

The microphones used in directional arrays are omnidirectional, meaning that their response in the free field is independent of the direction of the sound. Due to the small distance between the microphones, the signal picked up by each is assumed to be identical except for the direction-dependent time delay caused by the propagation of the wavefront from one microphone to the next. This phase difference between signals is exploited in order to separate signals emanating from different directions.

Ideally, one would prefer the array to span a reasonably long distance but also have a short separation between each microphone in order to have good spatial and frequency resolution and to reduce the directional sensitivity to differences between the microphone transfer functions. Limited user acceptance of head- or body-worn microphone arrays with hardwired connections to the hearing aids and a lack of wireless technology for remote microphone-to-aid communication have resulted in the placement of microphone arrays on the body of the hearing aid. This space constraint has limited the two-microphone arrays in current commercial aids to microphone separations of less than 15 mm.

The microphones are aligned on the aid such that a line connecting the microphones is perpendicular to wavefronts emanating from directly in front of the hearing aid wearer (the direction referred to as 0 degrees). In its simplest implementation, the signal from the back microphone is subtracted from the front, producing a null at 90 and 270 degrees (directly to the left and right of the hearing aid wearer). A time delay is typically added to the back microphone before subtraction in order to move the null to the rear of the listener, thereby causing most of the attenuation to occur for sounds arriving from the rear semisphere of the wearer. This general configuration is shown in Figure 7.15. Figure 7.16 shows two typical polar responses measured in the free field.

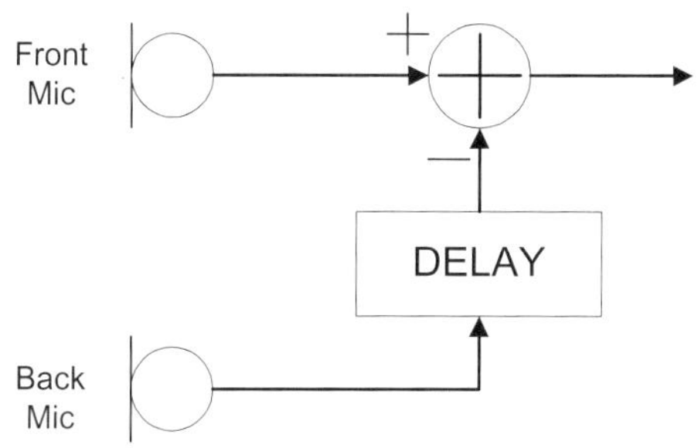

FIGURE 7.15. A typical configuration for a two-microphone (Mic) directional system. The delay to the back microphone determines the angle of the null in the directional pattern.

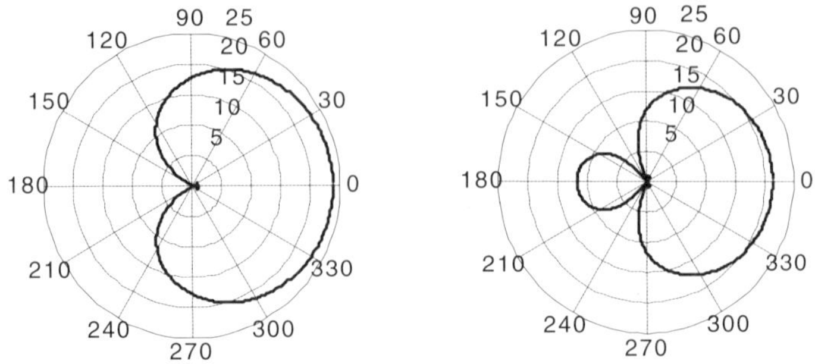

Figure 7.16. Two directional patterns typically associated with hearing aid directional microphones. The angle represents the direction from which the sound is approaching the listener, with 0 degrees representing directly in front of the listener. The distance from the origin at a given angle represents the gain applied for sound from arriving from that direction, ranging here from 0 to 25 dB. The patterns are a cardioid (left) and a hypercardioid (right).

It should be noted that this processing is slightly different from what is known as "beam forming." The simplest implementation of a beam former is to delay the outputs of each microphone by an amount such that the summation of the delayed signals produces a gain in the desired direction of $20 \log N$ dB, where N is the number of microphones (Capon et al. 1967). The target signal from a specified direction is passed unchanged, and the SNR can actually be improved if the noise at each microphone is diffuse and independent due to the coherent summation of the desired signal and the incoherent summation of the interfering noise (such an incoherent noise source in a hearing aid system is microphone noise). Directional patterns change with frequency in standard beam forming, while the subtraction technique shown in Figure 7.15 produces directional patterns that are invariant with frequency. With a two-microphone beam former in an endfire pattern (meaning that a wavefront arriving from the desired direction is perpendicular to a line connecting the microphones, rather than parallel to it as with a broadside array) and a microphone separation of 1.2 cm, effective directional patterns are not achieved for frequencies below 7 kHz, making true beam-forming impractical for hearing aids.

5.2.1.1 Low-Frequency Roll-Off

The directional system shown in Figure 7.15 ideally can produce useful directivity across the whole spectrum of the signal. One effect of the subtraction technique, however, is that a 6 dB/octave low-frequency roll-off is produced, effectively high-pass filtering the signal. As discussed in the previous section on one-microphone techniques, it is unlikely that high-pass filtering, in itself, adds any benefit to speech intelligibility. Indeed, the tinniness produced may be irritating to the listener over prolonged listening periods. Most likely, the only benefit of the roll-off is to emphasize the processing difference between the aid's omni- and directional modes, causing the effect of the directional processing to sound more significant than it actually is.

The roll-off, of course, could be compensated for by a 6-dB/octave boost. One drawback of doing this is that the microphone noise is not affected by this 6-dB/octave roll-off, resulting in a signal-to-microphone-noise ratio that increases with decreasing frequency. Eliminating the tinniness by providing a 6-dB/octave boost will thus increase the microphone noise. While the noise does not mask the speech signal—the noise is typically around 5 dB hearing level (HL)—the greatest level increase in the microphone noise would occur at the lowest frequencies where most hearing aid wearers have relatively near-normal hearing. Since the subtraction of the two microphones as shown in Figure 7.15 already increases the total microphone noise by 3 dB relative to the noise from a single microphone, subjective benefit from compensating for the low-frequency gain reduction has to be weighed against the increased audibility of the device noise. Given that the two-microphone directionality will most likely be used only in the presence of high levels of noise, it is unlikely that the microphone noise would be audible over the noise in the acoustic environment.

5.2.1.2 Measures of Directivity

Directionality in hearing aids can also be achieved with a single microphone using what are known as directional microphones. These are omni-microphones that have two physically separate ports providing two different acoustic paths that lead to opposite sides of the microphone membrane, producing a null when the signals on either side of the membrane are equal. This delay imposed by the separate paths is similar to the delay that occurs between two separate microphones in a two-microphone array, and similar directionality effects are obtained.

Hawkins and Yacullo (1984) investigated the benefit of directional microphones with hearing-impaired subjects and observed that a 3 to 4-dB improvement in SNR was necessary for 50% correct word recognition. Valente et al. (1995) found a 7 to 8-dB improvement in SNR with two-microphone arrays, as did Agnew and Block (1997). Care should be taken when noting the SNR improvement obtained from directionality, since the amount of benefit is dependent on the direction of the noise source relative to the microphone. If a directional aid has a null at 120 degrees, for example, then placing the interfering noise source at 120 degrees will produce maximal attenuation of the noise and maximal SNR improvement. This would not be representative of realistic noise environments, however, where reverberation, multiple interferers, and head movement tend to produce a more dispersed locus of interference that would result in a much lower SNR improvement.

The ratio of the gain applied in the direction of the target (0 degrees) to the average gain from all angles is called the directivity index and is a measure that is standard in quantifying antenna array performance (e.g., Uzkov 1946). It is a measure of the SNR improvement that directionality provides when the target signal is at 0 degrees and the noise source is diffuse, or the difference between the level of a diffuse noise passed by a directional system and the level of the same diffuse noise passed by an omnidirectional system. Free field measures and simple theoretical calculations have shown that the cardioid and hypercardioid patterns shown in Figure 7.16 have directivity indices of 4.8 dB and 6 dB, respectively, the latter being the maximum directionality that can be achieved with a two-microphone array (Kinsler and Frey 1962; Thompson 1997). Unlike SNR improvements with single-microphone techniques, improvements in the SNR presented to the listener using this directional technique translates directly into improvements in SRT measurements with a diffuse noise source. Similar results should be obtained with multiple noise sources in a reverberant environment since Morrow (1971) has shown this to have similar characteristics to a diffuse source for array processing purposes.

Two improvements in the way in which the directivity index is calculated can be made for hearing aid use. Directionality measures in the free field do not take into account head-shadow and other effects caused by the body of the hearing aid wearer. Bachler and Vonlanthen (1997) have shown that the directionality index is generally larger for the unaided listener than for one with an omnidirectional, behind-the-ear hearing aid—the directionality of the pinna is lost because of the placement of the hearing aid microphone. A similar finding was shown by Preves (1997) for in-the-ear aids. Directivity indices, then, should be measured at the very least on a mannequin head since the shape of the polar pattern changes considerably due to head and body effects.

The second modification to the directivity index considers the fact that the directionality of head-worn hearing aids, and of beam formers, are frequency dependent (the head-worn aid due to head-shadow effects, the beam former due to the frequency-dependency of the processing itself). The directivity index, then, varies with frequency. To integrate the frequency-specific directionality indices into a single directionality index, several researchers have weighted the directivity index at each frequency with the frequency importance function of the AI (Peterson 1989; Greenberg and Zurek 1992; Killion 1997; Saunders and Kates 1997). This incorporates the assumption that the target signal at 0 degrees is speech but does not include the effect of the low-frequency roll-off discussed earlier, which may make the lowest frequency region inaudible.

5.2.2 Noise Cancellation

Directionality is not the only benefit that can be obtained with multiple microphones. Two microphones can also be used for noise cancellation tech-

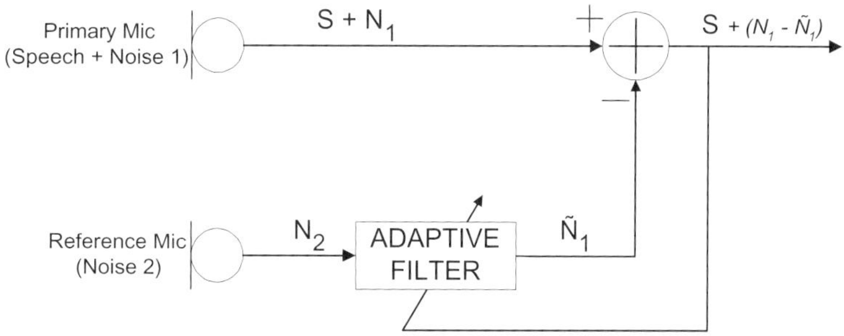

FIGURE 7.17. A typical two-microphone noise cancellation system. Ideally, the primary microphone measures a mixture of the interfering noise and the target speech, and the reference microphone measures only a transformation of the interfering noise.

niques (Fig. 7.17). With such processing, the primary microphone picks up both the target speech and interfering noise. A reference microphone picks up only noise correlated with the noise in the primary microphone. An adaptive filter is adjusted to minimize the power of the primary signal minus the filtered reference signal, and an algorithm such as the Widrow-Hoff least mean square (LMS) algorithm can be used to adapt the filter for optimal noise reduction (Widrow et al. 1975).

It can be shown that if the interfering noise is uncorrelated with the target speech, then this processing results in a maximum SNR at the output. This noise cancellation technique does not introduce the audible distortion that single-microphone spectral subtraction techniques produce (Weiss 1987), and has been shown to produce significant improvements in intelligibility (Chabries et al. 1982).

Weiss (1987) has pointed out that the output SNR for an ideal adaptive noise canceler is equal to the noise-to-signal ratio at the reference microphone. So for noise cancellation to be effective, little or no target signal should be present at the reference microphone and the noise at each microphone needs to be correlated. This could be achieved with, say, a primary microphone on the hearing aid and a reference microphone at a remote location picking up only the interfering noise. If cosmetic constraints require that the microphones be mounted on the hearing aid itself, nearly identical signals will reach each microphone. This eliminates the possibility of using noise cancellation with two omnidirectional microphones on a hearing aid since the target signal is present in both. Weiss has suggested that the reference microphone could be a directional microphone that is directed to the rear of the hearing aid wearer, i.e., passing the noise behind the wearer and attenuating the target speech in front of the wearer. Weiss measured the SNR improvement of this two-microphone noise cancellation system and compared it to the SNR improvement from a single directional microphone. He found that there was little difference between the two, and attributed this lack of improvement with the two-microphone canceler to the presence of speech in the reference microphone, which caused some of the speech to be canceled along with the noise, a common problem in noise cancellation systems. He did find, however, that adapting the filter only when the noise alone was present did produce significant improvements under anechoic conditions. These improvements were reduced as the number of interfering signals increased and when reverberation was introduced, implying that the system will not work when the noise source is diffuse—a static directional system is ideal for this condition.

Schwander and Levitt (1987) investigated the combined effects of head movement and reverberation on a noise cancellation system with an omnimicrophone and a directional microphone described above. It was thought that head movements typical in face-to-face communication might affect the adaptation stage of the processing since the correlation between the noise at the two microphones would be time-varying and affect the adaptation filter. Using normal-hearing subjects, they found significant benefit to intelligibility relative to using a single omnidirectional microphone. Levitt et al. (1993) repeated the experiment in multiple reverberation environments with both impaired and normal listeners. They found improvement in speech intelligibility, as did Chabries et al. (1982) with hearing-impaired listeners.

A trade-off exists in these systems with regard to the adaptation filter. To compensate for reverberation, the filter must have an impulse response duration on the order of the reverberation time, but a longer impulse response results in a slower adaptation rate. This is counter to the requirements necessitated by moving, nonstationary noise sources and moving microphone locations that must have fast adaptation rates to be effective. Any adaptive system will also have to work in conjunction with whatever time-varying hear loss compensation technique is being used, such as multiband compression. Care must be taken that the two dynamic nonlinear systems do not reduce each other's effectiveness or produce undesirable behavior when put in series. Add to this the difficulty in identifying pauses in the target speech that is necessary for successful filter adaptation, and the application of adaptive noise cancellation to hearing aids becomes a nontrivial implementation that no manufacturer has as yet produced in a hearing aid.

5.3 Other Considerations

If microphone placement is not limited to the hearing aid body, greater SNR improvements can be achieved with microphone arrays. Several researchers have shown significant benefit in speech intelligibility with both adaptive and nonadaptive arrays (Soede et al. 1993; Hoffman et al. 1994; Saunders and Kates 1997). Soede et al. investigated an array processor with five microphones placed along an axis either perpendicular (end fire) or parallel (broadside) to the frontal plane. The overall distance spanned by the microphones was 10cm. In a diffuse noise environment, the arrays improved the SNR by 7dB. In general, a large number of microphones can produce better directionality patterns and potentially better noise and echo cancellation if the microphones are placed properly.

It should be remembered that most single-microphone and many multiple-microphone noise reduction techniques that improve SNR do not result in improved speech intelligibility. This indicates a limitation of the ability of the AI to characterize the effect of these noise-reduction techniques and points to the more general failure of SNR as a measure of signal improvement since intelligibility is not a monotonic function of SNR. Clearly, there must exist some internal representation of the stimulus whose SNR more accurately reflects the listener's ability to identify the speech cues necessary for speech intelligibility. The cognitive effects described by auditory scene analysis (Bregman 1990), for example, have been used to describe how different auditory cues are combined to create an auditory image, and how combined spectral and temporal characteristics can cause fusion of dynamic stimuli. These ideas have been applied to improving the performance of automatic speech recognition systems (Ellis 1997), and their application to improving speech intelligibility in noise seems a logical step.

One would like a metric that is monotonic with intelligibility for quantifying the benefit of potential noise reduction systems without having to actually perform speech intelligibility tests on human subjects. It seems likely that the acoustic signal space must be transformed to a perceptually based space before this can be achieved. This technique is ubiquitous in the psychoacoustic field for explaining perceptual phenomenon. For example, Gresham and Collins (1997) used auditory models to successfully apply signal detection theory at the level of the auditory nerve to psychoacoustic phenomenon, better predicting performance in the perceptual signal space than in the physical acoustic signal space. Patterson et al.'s (1995) physiologically derived auditory image model transforms acoustic stimuli to a more perceptually relevent domain for better explanations of many auditory perceptual phenomena. This is also common in the speech field. Turner and Robb (1987) have effectively done this by examining differences in stop consonant spectra after calculating the consonant's excitation patterns. Van Tasell et al. (1987a) correlated vowel discrimination performance with excitation estimates for each vowel measured with probe thresholds. Auditory models have been used to improve data reduction encoding schemes (Jayant et al. 1994) and as front ends for automatic speech recognition systems (Hermansky 1990; Koehler et al. 1994). It seems clear that noise reduction techniques cannot continue to be developed and analyzed exclusively in the acoustic domain without taking into account the human auditory system, which is the last-stage receiver that decodes the processed signal.

6. Further Developments

There have been a number of recent research developments pertaining to hearing-impaired auditory perception that are relevant to hearing aid design. Significant attention has been given to differentiating between outer and inner hair cell damage in characterizing and compensating for hearing loss, thereby dissociating the loss of the compressive nonlinearity from the loss of transduction mechanisms that transit information from the cochlea to the auditory nerve. Moore and Glasberg (1997) have developed a loudness model that accounts for the percentage of hearing loss attributable to outer hair cell damage and inner hair cell damage. Loss of outer hair cells results in a reduction of the cochlea's compression mechanism, while loss of inner hair cells results in a linear reduction in sensitivity. This model has been successfully applied to predicting loudness summation data (Moore et al. 1999b).

Subjects are most likely capable of detecting pure tones in regions with no inner hair cells because of off-frequency excitation, and thus dead regions have been difficult to detect, and the extent of their existence in moderate hearing losses is unknown. Moore et al. (2000) have developed a clinically efficient technique for identifying dead regions using a form of masker known as threshold equalizing noise (TEN). This technique masks off-frequency listening by producing equally masked thresholds at all frequencies. In the presence of a dead region, the masking of off-frequency excitation by the TEN masker elevates thresholds for tone detection well above the expected masked threshold and the measured threshold in quiet.

Vickers et al. (2001) measured dead regions using TEN and then measured the intelligibility of low-pass-filtered speech with increasingly higher cutoff frequencies, increasing the high-frequency content of the speech. They found that speech intelligibility increased until the cutoff frequency was inside of the dead region that had been previously measured. The additional speech energy added within the dead region did not increase intelligibility. In one case, intelligibility actually deteriorated as speech information was added in the dead region. For subjects with no identified dead regions, speech intelligibility improved by increasing the cutoff frequency of the low-pass-filtered speech.

Other researchers (Ching et al. 1998; Hogan and Turner 1998) found that increased high-frequency audibility did not increase speech intelligibility for subjects with severe high-frequency loss, and in some cases the application of gain to speech in the high-frequency region of a subject's steeply sloping loss resulted in a deterioration of speech intelligibility. Hogan and Turner suggested that this was due to the amplification occurring in regions of significant inner hair cell loss, consistent with the later findings of Vickers et al. (2001). These results, along with those of others who found no benefit from amplification in the high-frequency regions of severe steeply sloping losses, suggest that hearing amplification strategies could be improved by taking into account information about dead regions. If hearing aid amplification were eliminated in frequency regions that represent dead regions, battery power consumption could be reduced, speakers/receivers would be less likely to saturate, and the potential for degradation of intelligibility would be diminished. The identification of dead regions in the investigation of speech perception could also lead to alternative signal-processing techniques, as there is evidence that frequency-translation techniques could provide benefit when speech information is moved out of dead regions into regions of audibility (Turner and Hurtig 1999).

Temporal aspects of auditory processing have continued to produce useful results for understanding the deficits of the hearing impaired. Further confirmation of the normal temporal resolving abilities of the hearing impaired was presented with pure-tone TMTFs (Moore and Glasberg 2001). Changes to the compressive nonlinearity can explain differences between normal and hearing-impaired listeners' temporal processing abilities without having to account for changes to central temporal processing (Oxenham and Moore 1997; Wojtczak et al. 2001). Of key importance in this research is evidence that forward-masked thresholds can be well modeled by a compressive nonlinearity followed by a linear temporal integrator (Plack and Oxenham 1998; Oxenham 2001; Wojtczak et al. 2001). If the forward masked thresholds of the hearing impaired are different from normal because of a reduction in the amount of their cochlear compression, then forward masking could be used as a diagnostic technique to measure the amount of residual compression that the hearing impaired subject has, as well as to determine their prescription for hearing aid compression (Nelson et al. 2001; Edwards 2002).

Hicks and Bacon (1999a) extended the forward masking protocol developed by Oxenham and Plack (1997) to show that the amount of compression in a healthy cochlea decreases below 1 kHz, with very little compression measurable at 375 Hz. Similar conclusions have been drawn from results investigating suppression (Lee and Bacon 1998; Hicks and Bacon, 1999b; Dubno and Ahlstrom 2001b). This is consistent with physiological data that show decreasing densities of outer hair cell in the basal end of the cochlea (e.g., Cooper and Rhode 1995). An understanding of how the compressive nonlinearity in a healthy cochlea becomes more linear as frequency decreases can affect how compression in a hearing aid is designed and how the compression parameters are fit since the need to restore compression to normal may decrease as frequency decreases. This may also have implications for estimating the inner versus outer hair cell mix associated with a hearing loss of known magnitude. For example, a 40-dB hearing loss at 250 Hz may be attributable exclusively to inner hair cell loss, while a 40-dB loss at 4 kHz may reflect primarily outer hair cell loss. Further research is needed in this area.

Suppression has been suggested by many as a physiological mechanism for improving speech recognition in noise. That the lack of suppression due to hearing loss can worsen speech recognition in noise was demonstrated with forward masking noise (Dubno and Ahlstrom 2001b). Phonemes that temporally followed a masker became more identifiable as the masker bandwidth increased due to suppressive effects. Subjects with hearing loss were unable to take advantage of bandwidth widening of the forward masker, presumably because of the observed loss of a suppressive effect in the frequency region of the masker.

Physiological evidence also suggests that loss of suppression results in poorer speech intelligibility because of a degraded representation of the speech signal in the auditory nerve. Nerve fibers in a healthy auditory system manifest synchrony capture to vowel formants closest in frequency to the characteristic frequency of the fiber (Young and Sachs 1979). This formant capture could provide a robust method for speech encoding in the auditory nerve to background noise. Miller et al. (1997) have shown that fibers lose the formant capture ability after acoustic trauma has induced moderate sensorineural hearing loss. The loss of formant capture means that the discharge timing of many fibers is no longer encoding formant information, and the fibers are instead entrained to nonformant energy in the frequency region around its best frequency. Miller et al. (1999) have suggested that some form of spectral contrast enhancement could improve the temporal coding of spectral peaks. Speech intelligibility research on spectral enhancement algorithms, however, continues to show negligible overall benefit. Franck et al. (1999) found with one spectral enhancement algorithm that vowel perception was enhanced but consonant perception deteriorated, and the addition of multiband compression to spectral-contrast enhancement resulted in worse results than either alone. Similar effects on vowel and consonant perception were predicted by a physiological model of suppression (Billa and el-Jaroudi 1998).

There is an increasing amount of evidence that damage to the outer hair cells, with attendant reduction (or complete loss) of the compressive nonlinearity, is responsible for more psychoacoustic changes than just loudness recruitment. Moore et al. (1999a) estimated the amount of hearing loss attributable to outer hair cells and to inner hair cells, and then correlated both estimates with reduced frequency resolution measures and forward masking measures. Both frequency resolution and forward masking measures were more highly correlated with the estimate of outer hair cell loss than with the overall level of hearing loss, but neither measure was correlated with the estimate of inner hair cell loss. This result indicates that changes to temporal and spectral resolving capabilities are inherently linked to the compressive nonlinearity and are not independent phenomena. Similar conclusions, including the dependence of suppression on the presence of compression, have been drawn by other researchers using a variety of different experimental methods (e.g., Gregan et al. 1998; Moore and Oxenham 1998; Hicks and Bacon 1999b; Summers 2000). Model predictions of auditory phenomena other than loudness have been improved with the addition of the cochlear nonlinearity, also indicating the dependence of these phenomena on the presence of compression (e.g., Derleth and Dau 2000; Heinz et al. 2001; Wojtczak et al. 2001).

The results of these studies support the application of multiband compression to hearing aids (assuming that listeners would benefit from the restoration of some of these auditory percepts to normal). This conclusion also assumes that the reintroduction of multiband compression in a hearing aid could restore these percepts to normal. This latter point was addressed by Sasaki et al. (2000), who demonstrated that narrowband masking patterns are closer to normal with the application of multiband compression. Similar results were obtained by Moore et al. (2001), who applied multiband compression to gap-detection stimuli. Edwards (2002) also demonstrated that multiband compression can make forward-masking thresholds, loudness summation, and narrowband frequency-masking patterns appear closer to normal. Edwards suggested that such signal-processing-assisted ("aided") psychoacoustic results can be used as a measure of the appropriateness of different hearing aid signal-processing designs. For example, the effect of multiband compression on psychoacoustic thresholds will depend on such compressor specifics as time constants, compression ratio, and filter bank design. Moreover, aided psychoacoustic results can be used to differentiate among the different designs.

Ultimately, though, the most important benefit that a hearing aid can currently provide is improvement in speech understanding. Results from speech perception experiments show mixed benefit from multiband compression; some experiments demonstrate either no benefit or even a negative benefit relative to linear amplification (e.g., Franck et al. 1999; Stone et al. 1999), while others found a positive benefit from multiband compression over linear amplification (Moore et al. 1999a; Souza and Bishop 1999; Souza and Turner 1999). Interestingly, Moore et al. (1999a) were able to demonstrate that the benefit from compression was most significant when the background noise had temporal and spectral dips, presumably because the fast-acting compressor increased audibility of the target speech above the masked threshold since the level-dependent gain could be applied to the target signal separately from the masker.

Auditory models have been applied to directional-processing techniques for enhancing speech perception in a system that uses a binaural coincidence detection model in a beam-forming device (Liu et al. 2000). The coincidence-detector identifies the location of multiple sound sources and then cancels the interfering sounds. An 8- to 10-dB improvement has been obtained with four competing talkers arriving from four separate angles in an anechoic environment. Speech intelligibility improvements also have been shown with other recent array-processing designs (Vanden Berghe and Wouters 1998; Kompis and Dillier 2001a,b; Shields and Campbell 2001).

7. Summary

The nonlinear nature of sensorineural hearing loss makes determining the proper nonlinear compensation difficult. Determining the proper parameter selection of that technique for the hearing loss of the subject can be equally difficult. Added to the problem is difficulty in verifying that one particular technique is optimal, or even that one technique is better than another one. This is partly due to the robustness of speech. Speech can be altered in a tremendous number of ways without significantly affecting its intelligibility as long as the signal is made audible. From the perspective of speech understanding, this is fortuitous since the signal processing of a hearing aid does not have to precisely compensate for the damaged outer hair cells in order to provide significant benefit. This does not address the perception of nonspeech signals, however, such as music, which can have a dynamic range much greater than that of speech and possess much more complex spectral and temporal characteristics, or of other naturally occurring sounds. If the goal of a hearing aid is complete restoration of normal perception, then the quality of the sound can have as much of an impact on the benefit of a hearing aid as the results of more objective measures pertaining to intelligibility.

Obtaining subjective evaluations from hearing aid wearers is made difficult due to the fact that they are hearing sounds to which they haven't been exposed for years. Complaints of hearing aids amplifying too much low-level noise, for example, may be a problem of too much gain but may also be due to their not being used to hear ambient sounds that are heard by normal-hearing people, which simply reflects their newfound audibility.

The assumption that hearing loss is only due to outer hair cell damage is most likely false for many impaired listeners. Dead zones in specific frequency regions may exist due to damaged inner hair cells, even though audiograms measure hearing losses of less than 60 dB HL in these regions because of their ability to detect the test signal through off-frequency listening (Thornton and Abbas 1980). Applying compression in this region would then be an inappropriate strategy. Alternate strategies for such cases will have to be developed and implemented in hearing aids. Even if hearing loss is a result of only outer hair cell damage, no signal-processing strategy may perfectly restore normal hearing. Compression can restore sensitivity to lower signals, but not in a manner that will produce the sharp tip in the basilar membrane tuning curves.

A healthy auditory system produces neural synchrony capture to spectral peaks that does not exist in a damaged auditory system. It is unlikely that hearing aid processing will restore this fine-structure coding in the temporal patterns of auditory nerve fibers. While basilar membrane I/O responses provide evidence that hair cell damage eliminates compression in regions that correspond to the frequency of stimulation, they do not show any effect of hair cell damage on the response to stimuli of distant frequencies (Ruggero and Rich 1991). The amplification applied to a narrowband signal would then have the effect of producing a normal response in the frequency region of the basilar membrane that is most sensitive to the frequency of the stimulus, but would produce an abnormal response in distant regions of the basilar membrane. If off-frequency listening is important for the detection of certain stimuli, as has been suggested for modulation and for speech in noise, then the auditory ability of the listener has not been restored to normal functionality. Wider auditory filters and loss of suppression also confound this difficulty. While a significant literature exists on the physiological coding of sound in the periphery of damaged auditory systems, little has been done with regard to the coding of aided sounds. Since hearing aids hope to produce near-normal responses in the auditory nerve, the effect of various strategies on the neural firing rate and synchrony codes would be of significant benefit.

Hearing aid design and laboratory investigations into different processing strategies need to consider the broad spectrum of psychoacoustic consequences of a hearing deficit. Speech scores, consonant confusions, and limited forms of loudness restoration have been a significant part of validating hearing aid design under aided conditions, but other measures of perception have not. Given the nonlinear nature of multiband compression, perceptual artifacts may be introduced that are not detected with these tests. Two completely different amplification or noise-reduction strategies may produce similar SRTs and gain responses, but could provide completely different perceptions of staccato piano music. Or two different compression systems may make speech signals equally audible, but the dynamics of one might significantly distort grouping cues necessary to provide auditory streaming for the cocktail party effect. The many different ways in which compression can be implemented—rms estimators vs. peak-followers, filter characteristics, dynamic ranges—somewhat confound the ability to compare results across studies. These design choices can have significant impact on the dynamic characteristics of the processing.

Wearable digital hearing aids will allow research to be conducted in settings outside of the typical laboratory environments. This will also allow subjects to acclimatize to new auditory cues, which may be unrecognizable to the wearer at first. This will also ensure exposure to a variety of background environments, listening conditions, and target stimuli that are difficult to provide in a clinic or laboratory. Multiple memories in such devices will also allow subjects to compare competing processing schemes in a way that could expose weaknesses or strengths that may not be captured in more controlled environments.

With these objectives in mind, the continued synthesis of speech, psychoacoustics, physiology, signal processing, and technology should continue to improve the benefit that people with hearing loss can obtain from hearing aids.

List of Abbreviations

AGC	automatic gain control
AI	articulation index
AM	amplitude modulation
ASP	automatic signal processing
BILL	bass increase at low levels
CMTF	compression modulation transfer function
CV	consonant-vowel
CVC	consonant-vowel-consonant
DR	dynamic range
DSP	digital signal processing
ERB	equivalent rectangular bandwidth
ERD	equivalent rectangular duration
HL	hearing level
jnd	just noticeable difference
LMS	least mean square
MTF	modulation transfer function
NAL	National Acoustic Laboratories
rms	root mean square
SL	sensation level
SNR	signal-to-noise ratio
SPL	sound pressure level
SRT	speech reception threshold
STI	speech transmission index
TEN	threshold equalizing noise
TMTF	temporal modulation transfer function
VC	vowel-consonant

References

Agnew J, Block M (1997) HINT threshold for a dual microphone BTE. Hear Rev 4:26–30.

Allen JB (1994) How do humans process speech? IEEE Trans Speech Audio Proc 2:567–577.

Allen JB (1996) Derecuitment by multi-band compression in hearing aids. In: Kollmeier B (ed) Psychoacoustics, Speech, and Hearing Aids. Singapore: World Scientific, pp. 141–152.

Allen JB, Hall JL, Jeng PS (1990) Loudness growth in 1/2-octave bands—a procedure for the assessment of loudness. J Acoust Soc Am 88:745–753.

American National Standards Institute (1987) Specifications of hearing aid characteristics. ANSI S3.22-1987. New York: American National Standards Institute.

American National Standards Institute (1996) Specifications of hearing aid characteristics. ANSI S3.22-1996. New York: American National Standards Institute.

Bachler H, Vonlanthen A (1997) Audio zoom-signal processing for improved communication in noise. Phonak Focus 18.

Bacon SP, Gleitman RM (1992) Modulation detection in subjects with relatively flat hearing losses. J Speech Hear Res 35:642–653.

Bacon SP, Viemeister NF (1985) Temporal modulation transfer functions in normal-hearing and hearing-impaired listeners. Audiology 24:117–134.

Baer T, Moore BCJ (1993) Effects of spectral smearing on the intelligibility of sentences in noise. J Acoust Soc Am 94:1229–1241.

Bakke M, Neuman AC, Levitt H (1974) Loudness matching for compressed speech signals. J Acoust Soc Am 89:1991.

Barfod (1972) Investigations on the optimum corrective frequency response for high-tone hearing loss. Report No. 4, The Acoustic Laboratory, Technical University of Denmark.

Bilger RC, Wang MD (1976) Consonant confusions in patients with sensorineural hearing loss. J Speech Hear Res 19:718–748.

Billa J, el-Jaroudi A (1998) An analysis of the effect of basilar membrane nonlinearities on noise suppression. J Acoust Soc Am 103:2691–2705.

Boll SF (1979) Suppression of acoustic noise in speech using spectral subtraction. IEEE Trans Acoust Speech Signal Proc 27:113–120.

Bonding P (1979) Frequency selectivity and speech discrimination in sensorineural hearing loss. Scand Audiol 8:205–215.

Boothroyd A, Mulhearn B, Gong J, Ostroff J (1996) Effects of spectral smearing on phoneme and word recognition. J Acoust Soc Am 100:1807–1818.

Bosman AJ, Smoorenberg GF (1987) Differences in listening strategies between normal and hearing-impaired listeners. In: Schouten MEH (ed) The Psychoacoustics of Speech Perception. Dordrecht: Martimus Nijhoff.

Breeuwer M, Plomp R (1984) Speechreading supplemented with frequency-selective sound-pressure information. J Acoust Soc Am 76:686–691.

Bregman AS (1990) Auditory Scene Analysis. Cambridge: MIT Press.

Bunnell HT (1990) On enhancement of spectral contrast in speech for hearing-impaired listeners. J Acoust Soc Am 88:2546–1556.

Bustamante DK, Braida LD (1987a) Multiband compression limiting for hearing-impaired listeners. J Rehabil Res Dev 24:149–160.

Bustamante DK, Braida LD (1987b) Principal-component compression for the hearing impaired. J Acoust Soc Am 82:1227–1239.

Byrne D, Dillon H (1986) The National Acoustic Laboratory's (NAL) new procedure for selecting the gain and frequency response of a hearing aid. Ear Hear 7:257–265.

Capon J, Greenfield RJ, Lacoss RT (1967) Design of seismic arrays for efficient on-line beamforming. Lincoln Lab Tech Note 1967–26, June 27.

Caraway BJ, Carhart R (1967) Influence of compression action on speech intelligibility. J Acoust Soc Am 41:1424–1433.

Carlyon RP, Sloan EP (1987) The "overshoot" effect and sensorineural hearing impairment. J Acoust Soc Am 82:1078–1081.

Carney A, Nelson DA (1983) An analysis of psychoacoustic tuning curves in normal and pathological ears. J Acoust Soc Am 73:268–278.

CHABA Working Group on Communication Aids for the Hearing-Impaired (1991) Speech-perception aids for hearing-impaired people: current status and needed research. J Acoust Soc Am 90:637–685.

Chabries DM, Christiansen RW, Brey RH (1982) Application of the LMS adaptive filter to improve speech communication in the presence of noise. IEEE Int Cont Acoust Speech Signal Proc-82 1:148–151.

Ching TY, Dillon H, Byrne D (1998) Speech recognition of hearing-impaired listeners: predictions from audibility and the limited role of high-frequency amplification. J Acoust Soc Am 103:1128–1140.

Coker CH (1974) Speech as an error-resistant digital code. J Acoust Soc Am 55: 476(A).

Cook JA, Bacon SP, Sammeth CA (1997) Effect of low-frequency gain reduction on speech recognition and its relation to upward spread of masking. J Speech Lang Hear Res 40:410–422.

Cooper NP, Rhode WS (1995) Nonlinear mechanics at the apex of the guinea-pig cochlea. Hear Res 82:225–243.

Crain TR, Yund EW (1995) The effect of multichannel compression on vowel and stop-consonant discrimination in normal-hearing and hearing-impaired subjects. Ear Hear 16:529–543.

Danaher EM, Pickett JN (1975) Some masking effects produced by low-frequency vowel formants in persons with sensorineural hearing loss. J Speech Hear Res 18: 261–271.

Darwin CJ (1981) Perceptual grouping of speech components differing in fundametal frequency and onset time. Q J Exp Psychol 33A:185–207.

Darwin CJ (1984) Perceiving vowels in the presence of another sound: constraints on formant perception. J Acoust Soc Am 76:1636–1647.

Davis H, Stevens SS, Nichols RH, et al. (1947) Hearing Aids—An Experimental Study of Design Objectives. Cambridge: Harvard University Press.

Davis SB, Mermelstein P (1980) Comparison of parametric representations for monosyllabic word recognition in continuously spoken sentences. IEEE Trans Acoust Speech Signal Proc 28:357–366.

De Gennaro SV (1982) An analytic study of syllabic compression for severely impaired listeners. S.M. thesis, Department of Electrical Engineering and Computer Science, Massachusetts Institute of Technology, Cambridge.

De Gennaro S, Braida LD, Durlach NI (1986) Multichannel syllabic compression for severely impaired listeners. J Rehabil Res 23:17–24.

Delattre PC, Liberman AM, Cooper FS, Gerstman LJ (1952) An experimental study of the acoustic determinants of vowel colour: observations on one- and two-formant vowel synthesized from spectrographic patterns. Word 8:195–210.

Derleth RP, Dau T (2000) On the role of envelope fluctuation processing in spectral masking. J Acoust Soc Am 108:285–296.

Derleth RP, Dau T, Kollmeier B (1996) Perception of amplitude modulated narowband noise by sensorineural hearing-impaired listeners. In: Kollmeier B (ed) Psychoacoustics, Speech, and Hearing Aids. Singapore: World Scientific, pp. 39–44.

Dillon H (1993) Hearing aid evaluation: predicting speech gain from insertion gain. J Speech Hear Res 36:621–633.

Dillon H (1996) Compression? Yes, but for low or high frequencies, for low or high intensities, and with what response times? Ear Hear 17:267–307.

Dirks DD, Morgan D, Dubno JR (1982) A procedure for quantifying the effects of noise on speech recognition. J Speech Hear Dis 47:114–123.

Dreschler WA (1980) Reduced speech intelligibility and its psychophysical correlates in hearing-impaired listeners. In: Brink G van den, Bilsen FA (eds) Psychophysical, Physiological and Behavioral Studies in Hearing. Alphenaand den Rijn, The Netherlands: Sijthoff and Noordhoff.

Dreschler WA (1986) Phonemic confusions in quiet and noise for the hearing-impaired. Audiology 25:19–28.

Dreschler WA (1988a) The effects of specific compression settings on phoneme identification in hearing-impaired subjects. Scand Audiol 17:35–43.

Dreschler WA (1988b) Dynamic-range reduction by peak clipping or compression and its effects on phoneme perception in hearing-impaired listeners. Scand Audiol 17:45–51.

Dreschler WA (1989) Phoneme perception via hearing aids with and without compression and the role of temporal resolution. Audiology 28:49–60.

Dreschler WA, Leeuw AR (1990) Speech reception in reverberation related to temporal resolution. J Speech Hear Res 33:181–187.

Dreschler WA, Plomp R (1980) Relation between psychophysical data and speech perception for hearing-impaired subjects. I. J Acoust Soc Am 68:1608–1615.

Drullman R (1995) Temporal envelope and fine structure cues for speech intelligibility. J Acoust Soc Am 97:585–592.

Drullman R, Festen JM, Plomp R (1994) Effect of temporal envelope smearing on speech perception. J Acoust Soc Am 95:1053–1064.

Drullman R, Festen JM, Houtgast T (1996) Effect of temporal modulation reduction on spectral contrasts in speech. J Acoust Soc Am 99:2358–2364.

Dubno JR, Ahlstrom JB (1995) Masked thresholds and consonant recognition in low-pass maskers for hearing-impaired and normal-hearing listeners. J Acoust Soc Am 97:2430–2441.

Dubno JR, Ahlstrom JB (2001a) Forward- and simultaneous-masked thresholds in bandlimited maskers in subjects with normal hearing and cochlear hearing loss. J Acoust Soc Am 110:1049–1157.

Dubno JR, Ahlstrom JB (2001b) Psychophysical suppression effects for tonal and speech signals. J Acoust Soc Am 110:2108–2119.

Dubno JR, Dirks DD (1989) Auditory filter characteristics and consonant recognition for hearing-impaired listeners. J Acoust Soc Am 85:1666–1675.

Dubno JR, Dirks DD (1990) Associations among frequency and temporal resolution and consonant recognition for hearing-impaired listeners. Acta Otolaryngol (suppl 469):23–29.

Dubno JR, Schaefer AB (1991) Frequency selectivity for hearing-impaired and broadband-noise-masked normal listeners. Q J Exp Psychol 43:543–564.

Dubno JR, Schaefer AB (1992) Comparison of frequency selectivity and consonant recognition among hearing-impaired and masked normal-hearing listeners. J Acoust Soc Am 91:2110–2121.

Dubno JR, Schaefer AB (1995) Frequency selectivity and consonant recognition for hearing-impaired and normal-hearing listeners with equivalent masked thresholds. J Acoust Soc Am 97:1165–1174.

Duifhuis H (1973) Consequences of peripheral frequency selectivity for nonsimultaneous masking. J Acoust Soc Am 54:1471–1488.

Duquesnoy AJ, Plomp R (1980) Effect of reverberation and noise on the intelligibility of sentences in cases of presbyacusis. J Acoust Soc Am 68:537–544.

Eddins DA (1993) Amplitude modulation detection of narrow-band noise: effects of absolute bandwidth and frequency region. J Acoust Soc Am 93:470–479.

Eddins DA, Hall JW, Grose JH (1992) The detection of temporal gaps as a function of absolute bandwidth and frequency region. J Acoust Soc Am 91: 1069–1077.

Edwards BW (2002) Signal processing, hearing aid design, and the psychoacoustic Turing test. IEEE Proc Int Conf Acoust Speech Signal Proc, Vol. 4, pp. 3996–3999.

Edwards BW, Struck CJ (1996) Device characterization techniques for digital hearing aids. J Acoust Soc Am 100:2741.

Egan JP, Hake HW (1950) On the masking pattern of a simple auditory stimulus. J Acoust Soc Am 22:622–630.

Ellis D (1997) Computational auditory scene analysis exploiting speech-recognition knowledge. IEEE Workshop on Appl Signal Proc Audiol Acoust 1997, New Platz, New York.

Ephraim Y, Malah D (1984) Speech enhancement using a minimum mean-square error short-time spectral amplitude estimator. IEEE Trans Speech Signal Proc 32:1109–1122.

Erber NP (1972) Speech-envelope cues as an acoustic aid to lipreading for profoundly deaf children. J Acoust Soc Am 51:1224–1227.

Erber NP (1979) Speech perception by profoundly hearing-impaired children. J Speech Hear Disord 44:255–270.

Evans EF, Harrison RV (1976) Correlation between outer hair cell damage and deterioration of cochlear nerve tuning properties in the guinea pig. J Physiol 252:43–44.

Fabry DA, Van Tasell DJ (1990) Evaluation of an articulation-index based model for predicting the effects of adative frequency response hearing aids. J Speech Hear Res 33:676–689.

Fabry DA, Leek MR, Walden BE, Cord M (1993) Do adaptive frequency response (AFR) hearing aids reduce "upward spread" of masking? J Rehabil Res Dev 30:318–325.

Farrar CL, Reed CM, Ito Y, et al. (1987) Spectral-shape discrimination. I. Results from normal-hearing listeners for stationary broadband noises. J Acoust Soc Am 81:1085–1092.

Faulkner A, Ball V, Rosen S, Moore BCJ, Fourcin A (1992) Speech pattern hearing aids for the profoundly hearing impaired: speech perception and auditory abilities. J Acoust Soc Am 91:2136–2155.

Fechner G (1933) Elements of Psychophysics [English translation, Howes DW, Boring EC (eds)]. New York: Holt, Rhinehart and Winston.

Festen JM (1996) Temporal resolution and the importance of temporal envelope cues for speech perception. In: Kollmeier B (ed) Psychoacoustics, Speech and Hearing Aids. Singapore: World Scientific.

Festen JM, Plomp R (1983) Relations between auditory functions in impaired hearing. J Acoust Soc Am 73:652–662.

Festen JM, van Dijkhuizen JN, Plomp R (1990) Considerations on adaptive gain and frequency response in hearing aids. Acta Otolaryngol 469:196–201.

Festen JM, van Dijkhuizen JN, Plomp R (1993) The efficacy of a multichannel hearing aid in which the gain is controlled by the minima in the temporal signal envelope. Scand Audiol 38:101–110.

Fitzgibbons PJ, Gordon-Salant S (1987) Minimum stimulus levels for temporal gap resolution in listeners with sensorineural hearing loss. J Acoust Soc Am 81: 1542–1545.

Fitzgibbons PJ, Wightman FL (1982) Gap detection in normal and hearing-impaired listeners. J Acoust Soc Am 72:761–765.

Fletcher H (1953) Speech and Hearing in Communication. New York: Van Nostrand.

Florentine M, Buus S (1984) Temporal gap detection in sensorineural and simulated hearing impairments. J Speech Hear Res 27:449–455.
Florentine M, Buus S, Scharf B, Zwicker E (1980) Freqquency selectivity in normally-hearing and hearing-impaired observers. J Speech Hear Res 23:646–669.
Fowler EP (1936) A method for the early detection of otosclerosis. Arch Otolaryngol 24:731–741.
Franck BA, van Kreveld-Bos CS, Dreschler WA, Verschuure H (1999) Evaluation of spectral enhancement in hearing aids, combined with phonemic compression. J Acoust Soc Am 106:1452–1464.
French NR, Steinberg JC (1947) Factors governing the intelligibility of speech sounds. J Acoust Soc Am 19:90–119.
Gagné JP (1983) Excess masking among listeners with high-frequency sensorineural hearing loss. Doctoral dissertation, Washington University (Central Institute for the Deaf), St. Louis.
Gagné JP (1988) Excess masking among listeners with a sensorineural hearing loss. J Acoust Soc Am 83:2311–2321.
Glasberg BR, Moore BCJ (1986) Auditory filter shapes with unilateral and bilateral cochlear impairments. J Acoust Soc Am 79:1020–1033.
Glasberg BR, Moore BCJ (1992) Effects of envelope fluctuations on gap detection. Hear Res 64:81–92.
Glasberg BR, Moore BCJ, Bacon SP (1987) Gap detection and masking in hearing-impaired and normal-hearing subjects. J Acoust Soc Am 81:1546–1556.
Gordon-Salant S (1984) Effects of acoustic modification on consonant recognition in elderly hearing-impaired subjects. J Acoust Soc Am 81:1199–1202.
Gordon-Salant S, Sherlock LP (1992) Performance with an adaptive frequency response hearing aid in a sample of elderly hearing-impaired listeners. Ear Hear 13:255–262.
Gorga MP, Abbas PJ (1981a) AP measurements of short term adaptation in normal and in acoustically traumatized ears. J Acoust Soc Am 70:1310–1321.
Gorga MP, Abbas PJ (1981b) Forwards-masking AP tuning curves in normal and in acoustically traumatized ears. J Acoust Soc Am 70:1322–1330.
Goshorn EL, Studebaker GA (1994) Effects of intensity on speech recognition in high- and low-frequency bands. Ear Hear 15:454–460.
Graupe D, Grosspietsch JK, Taylor RT (1986) A self-adaptive noise filtering system, part 1: overview and description. Hear Instrum 37:29–34.
Graupe D, Grosspietsch JK, Basseas SP (1987) A single-microphone-based self-adaptive filter of noise from speech and its performance evaluation. J Rehabil Res Dev 24:119–126.
Green DM (1969) Masking with continuous and pulsed sinusoids. J Acoust Soc Am 49:467–477.
Greenberg J, Zurek P (1992) Evaluation of an adaptive beamforming method for hearing aids. J Acoust Soc Am 91:1662–1676.
Greenberg S (1997) On the origins of speech intelligibility in the real world. Proc ESCA Workshop on Robust Speech Recognition for Unknown Communication Channels, pp. 23–32.
Greenberg S, Hollenback J, Ellis D (1996) Insights into spoken language gleaned from phonetic transcription of the switchboard corpus. Proc 4th Int Conf Spoken Lang Proc, pp. S32–35.
Gregan MJ, Bacon SP, Lee J (1998) Masking by sinusoidally amplitude-modulated tonal maskers. J Acoust Soc Am 103:1012–1021.

Gresham LC, Collins LM (1997) Analysis of the performance of a model-based optimal auditory processor on a simultaneous masking task. J Acoust Soc Am 101:3149.

Grose JH, Eddins D, Hall JW (1989) Gap detection as a function of stimulus bandwidth with fixed high-frequency cutoff in normal-hearing and hearing-impaired listeners. J Acoust Soc Am 86:1747–1755.

Gutnick HN (1982) Consonant-feature transmission as a function of presentation level in hearing-impaired listeners. J Acoust Soc Am 72:1124–1130.

Hack Z, Erber N (1982) Auditory, visual, and audiory-visual perception of vowels by hearing-impaired children. J Speech Hear Res 25:100–107.

Hall JW, Fernandes MA (1983) Temporal integration, frequency resolution, and off-frequency listening in normal-hearing and cochlear-impaired listeners. J Acoust Soc Am 74:1172–1177.

Hansen J, Clements M (1987) Iterative speech enhancement with spectral constraints. IEEE Int Conf Acoust Speech Signal Proc, pp. 189–192.

Hawkins DB, Yacullo WS (1984) Signal-to-noise ratio advantage of binaural hearing aids and directional microphones under different levels of reverberation. J Speech Hear Disord 49:278–286.

Heinz MG, Colburn HS, Carney LH (2001) Rate and timing cues associated with the cochlear amplifier: level discrimination based on monaural cross-frequency coincidence detection. J Acoust Soc Am 110:65–2084.

Hermansky H (1990) Perceptual linear predictive (PLP) analysis of speech. J Acoust Soc Am 87:1738–1752.

Hermansky H, Morgan N (1993) RASTA processing of speech. IEEE Trans Speech Audiol Proc 2:578–589.

Hermansky H, Wan EA, Avendano C (1995) Speech enhancement based on temporal processing. Proc Int Cont Acoust Speech Signal Proc-95:405.

Hermansky H, Greenberg S, Avendano C (1997) Enhancement of speech intelligibility via compensatory filtering of the modulation spectrum. 2nd Hear Aid Res Dev Conf, Bethesda, MD.

Hicks ML, Bacon SP (1999a) Psychophysical measures of auditory nonlinearities as a function of frequency in individuals with normal hearing. J Acoust Soc Am 105: 326–338.

Hicks ML, Bacon SP (1999b) Effects of aspirin on psychophysical measures of frequency selectivity, two-tone suppression, and growth of masking. J Acoust Soc Am 106:1436–1451.

Hickson LMH (1994) Compression amplification in hearing aids. Am J Audiol 11: 51–65.

Hickson L, Byrne D (1997) Consonant perception in quiet: effect of increasing the consonant-vowel ratio with compression amplification. J Am Acad Audiol 8:322–332.

Hoffman MW, Trine TD, Buckley KN, Van Tasell DJ (1994) Robust adaptive microhpone array processing for hearing aids: realistic speech enhancement. J Acoust Soc Am 96:759–770.

Hogan CA, Turner CW (1998) High-frequency audibility: benefits for hearing-impaired listeners. J Acoust Soc Am 104:432–441.

Holte L, Margolis RH (1987) The relative loudness of third-octave bands of speech. J Acoust Soc Am 81:186–190.

Horst JW (1987) Frequency discrimination of complex signals, frequency selectivity, and speech perception in hearing-impaired subjects. J Acoust Soc Am 82:874–885.

Hou Z, Pavlovic CV (1994) Effects of temporal smearing on temporal resolution, frequency selectivity, and speech intelligibility. J Acoust Soc Am 96:1325–1340.
Houtgast T, Steeneken HJM (1973) The modulation transfer function in room acoustics as predictor of speech intelligibility. Acustica 28:66–73.
Houtgast T, Steeneken HJM (1985) A review of the MTF concept in room acoustics and its use for estimating speech intelligibility in auditoria. J Acoust Soc Am 77: 1069–1077.
Humes LE (1982) Spectral and temporal resolution by the hearing impaired. In: Studebaker GA, Bess FH (eds) The Vanderbilt Hearing Aid Report: State of the Art—Research Needs. Upper Darby, PA: Monographs in Contemporary Audiology.
Humes LE, Dirks DD, Bell TS, Ahlstrom C, Kincaid GE (1986) Application of the articulation index and the speech transmission index to the recognition of speech by normal-hearing and hearing-impaired listeners. J Speech Hear Res 29:447–462.
Humes LE, Boney S, Loven F (1987) Further validation of the speech transmission index (STI). J Speech Hear Res 30:403–410.
Humes LE, Christensen LA, Bess FH, Hedley-Williams A (1997) A comparison of the benefit provided by well-fit linear hearing aids and instruments with automatic reductions of low-frequency gain. J Speech Lang Hear Res 40:666–685.
Irwin RJ, McAuley SF (1987) Relations among temporal acuity, hearing loss, and the perception of speech distorted by noise and reverberation. J Acoust Soc Am 81:1557–1565.
Jayant NS, Johnston JD, Safranek RJ (1993) Signal compression based on human perception. Proc IEEE 81:1385–1422.
Jerlvall LB, Lindblad AC (1978) The influence of attack time and release time on speech intelligibility. Scand Audiol 6:341–353.
Kates JM (1991) A simplified representation of speech for the hearing impaired. J Acoust Soc Am 89:1961.
Kates JM (1993) Optimal estimation of hearing-aid compression parameters. J Acoust Soc Am 94:1–12.
Kates JM (1995) Classification of background noises for hearing aid applications. J Acoust Soc Am 97:461–470.
Kiang NYS, Liberman MC, Levine RA (1976) Auditory-nerve activity in cats exposed to ototoxic drugs and high-intensity sounds. Ann Atol Rhinol Laryngol 85:752–768.
Killion MC (1996) Talking hair cells: what they have to say about hearing aids. In: Berlin CI (ed) Hair Cells and Hearing Aids. San Diego: Singular.
Killion MC (1997) Hearing aids: past, present, future: moving toward normal conversations in noise. Br J Audiol 31:141–148.
Killion MC, Fikret-Pasa S (1993) The three types of sensorineural hearing loss: loudness and intelligibility considerations. Hear J 46:31–36.
Killion MC, Tillman TW (1982) Evaluation of high-fidelity hearing aids. J Speech Hear Res 25:15–25.
King AB, Martin MC (1984) Is AGC beneficial in hearing aids? Br J Audiol 18:31–38.
Kinsler LE, Frey AR (1962) Fundamentals of Acoustics. New York: John Wiley.
Klatt DH (1980) Software for a cascade/parallel formant synthesizer. J Acoust Soc Am 67:971–995.

Klatt DH (1982) Prediction of perceived phonetic distance from critical-band spectra: a first step. Proc IEEE Int Conf Speech Acoust Signal Proc, pp. 1278–1281.

Klumpp RG, Webster JC (1963) Physical measurements of equally speech-interfering navy noises. J Acoust Soc Am 35:1328–1338.

Kochkin S (1993) MarkeTrack III: why 20 million in US don't use hearing aids for their hearing loss. Hear J 46:20–27.

Koehler J, Morgan N, Hermansky H, Hirsch HG, Tong G (1994) Integrating RASTA-PLP into speech recognition. IEEE Proc Int Conf Acoust Speech Signal Proc, pp. 421–424.

Kompis M, Dillier N (2001a) Performance of an adaptive beamforming noise reduction scheme for hearing aid applications. I. Prediction of the signal-to-noise-ratio improvement. J Acoust Soc Am 109:1123–1133.

Kompis M, Dillier N (2001b) Performance of an adaptive beamforming noise reduction scheme for hearing aid applications. II. Experimental verification of the predictions. J Acoust Soc Am 109:1134–1143.

Kryter KD (1970) The Effects of Noise on Man. New York: Academic Press.

Laurence RF, Moore BCJ, Glasberg BR (1983) A comparison of behind-the-ear high-fidelity linear hearing aids and two-channel compression hearing aids in the laboratory and in everyday life. Br J Audiol 17:31–48.

Lee J, Bacon SP (1998) Psychophysical suppression as a function of signal frequency: noise and tonal maskers. J Acoust Soc Am 104:1013–1022.

Leek MR, Summers V (1993) Auditory filter shapes of normal-hearing and hearing-impaired listeners in continuous broadband noise. J Acoust Soc Am 94:3127–3137.

Leek MR, Summers V (1996) Reduced frequency selectivity and the preservation of spectral contrast in noise. J Acoust Soc Am 100:1796–1806.

Leek MR, Dorfman MF, Summerfield Q (1987) Minimum spectral contrast for vowel identification by normal-hearing and hearing-impaired listeners. J Acoust Soc Am 81:148–154.

Levitt H (1991) Future directions in signal processing hearing aids. Ear Hear 12: 125–130.

Levitt H, Neuman AC (1991) Evaluation of orthogonal polynomial compression. J Acoust Soc Am 90:241–252.

Levitt H, Neuman A, Mills R, Schwander T (1986) A digital master hearing aid. J Rehabil Res Dev 23:79–87.

Levitt H, Bakke M, Kates J, Neuman A, Schwander T, Weiss M (1993) Signal processing for hearing impairment. Scand Audiol 38:7–19.

Liberman MC, Kiang NY (1978) Acoustic trauma in cats: cochlear pathology and auditory-nerve pathology. Acta Otolaryngol Suppl (Stockh) 358:1–63.

Lim JS (1983) Speech Enhancement. Englewood Cliffs, NJ: Prentice Hall.

Lim JS, Oppenheim AV (1979) Enhancement and bandwidth compression of noisy speech. Proc IEEE Int Conf Acoust Speech Signal Proc, pp. 1586–1604.

Lim JS, Oppenheim AV, Braida LD (1978) Evaluation of an adaptive comb filtering method for enhancing speech degraded by white noise addition. IEEE Trans Speech Signal Proc 26:354–358.

Lindemann E (1997) The Continuous Frequency Dynamic Range Compressor. IEEE Workshop on Applications of Signal Processing to Audio and Acoustics, New Paltz, New York.

Lippmann RP, Braida LD, Durlach NI (1981) Study of multichannel amplitude compression and linear amplification for persons with sensorineural hearing loss. J Acoust Soc Am 69:524–534.

Liu C, Wheeler BC, O'Brien WD Jr, Bilger RC, Lansing CR, Feng AS (2000) Localization of multiple sound sources with two microphones. J Acoust Soc Am 108: 1888–1905.
Lunner T, Arlinger S, Hellgren J (1993) 8-channel digital filter bank for hearing aid use: preliminary results in monaural, diotic and dichotic modes. Scand Audiol 38: 75–81.
Lunner T, Hellgren J, Arlinger S, Elberling C (1997) A digital filterbank hearing aid: predicting user preference and performance for two signal processing algorithms. Ear Hear 18:12–25.
Lutman ME, Clark J (1986) Speech identification under simulated hearing-aid frequency response characteristics in relation to sensitivity, frequency resolution and temporal resolution. J Acoust Soc Am 80:1030–1040.
Lybarger SF (1947) Development of a new hearing aid with magnetic microphone. Elect Manufact 1–13.
Makhoul J, McAulay R (1989) Removal of Noise from Noise-Degraded Speech Signals. Washington, DC: National Academy Press.
Miller GA (1951) Language and Communication. New York: McGraw-Hill.
Miller GA, Nicely PE (1955) An analysis of perceptual confusions among some English consonants. J Acoust Soc Am 27:338–352.
Miller RL, Schilling JR, Franck KR, Young ED (1997) Effects of acoustic trauma on the representation of the vowel /ε/ in cat auditory nerve fibers. J Acoust Soc Am 101:3602–3616.
Miller RL, Calhoun BM, Young ED (1999) Contrast enhancement improves the representation of /ε/-like vowels in the hearing-impaired auditory nerve. J Acoust Soc Am 106:2693–2708.
Moore BCJ (1991) Characterization and simulation of impaired hearing: implications for hearing aid design. Ear Hear 12:154–161.
Moore BCJ (1996) Perceptual consequences of chochlear hearing loss and their implications for the design of hearing aids. Ear Hear 17:133–161.
Moore BCJ, Glasberg BR (1988) A comparison of four methods of implementing automatic gain control (AGC) in hearing aids. Br J Audiol 22:93–104.
Moore BCJ, Glasberg BR (1997) A model of loudness perception applied to cochlear hearing loss. Audiol Neurosci 3:289–311.
Moore BC, Glasberg BR (2001) Temporal modulation transfer functions obtained using sinusoidal carriers with normally hearing and hearing-impaired listeners. J Acoust Soc Am 110:1067–1073.
Moore BCJ, Oxenham AJ (1998) Psychoacoustic consequences of compression in the peripheral auditory system. Psychol Rev 105:108–124.
Moore BCJ, Laurence RF, Wright D (1985) Improvements in speech intelligibility in quiet and in noise produced by two-channel compression hearing aids. Br J Audiol 19:175–187.
Moore BCJ, Glasberg BR, Stone MA (1991) Optimization of a slow-acting automatic gain control system for use in hearing aids. Br J Audiol 25:171–182.
Moore BCJ, Lynch C, Stone MA (1992) Effects of the fitting parameters of a two-channel compression system on the intelligibility of speech in quiet and in noise. Br J Audiol 26:369–379.
Moore BCJ, Wojtczak M, Vickers DA (1996) Effects of loudness recruitment on the perception of amplitude modulation. J Acoust Soc Am 100:481–489.
Moore BCJ, Glasberg BR, Baer T (1997) A model for the prediction of thresholds, loudness, and partial loudness. J Audiol Eng Soc 45:224–240.

Moore BCJ, Glasberg BR, Vickers DA (1999a) Further evaluation of a model of loudness perception applied to cochlear hearing loss. J Acoust Soc Am 106: 898–907.

Moore BCJ, Peters RW, Stone MA (1999b) Benefits of linear amplification and multichannel compression for speech comprehension in backgrounds with spectral and temporal dips. J Acoust Soc Am 105:400–411.

Moore BCJ, Vickers DA, Plack CJ, Oxenham AJ (1999c) Inter-relationship between different psychoacoustic measures assumed to be related to the cochlear active mechanism. J Acoust Soc Am 106:2761–2778.

Moore BCJ, Huss M, Vickers DA, Glasberg BR, Alcantara JI (2000) A test for the diagnosis of dead regions in the cochlea. Br J Audiol 34:5–244.

Moore BCJ, Glasberg BR, Alcantara JI, Launer S, Kuehnel V (2001) Effects of slow- and fast-acting compression on the detection of gaps in narrow bands of noise. Br J Audiol 35:365–374.

Morrow CT (1971) Point-to-point correlation of sound pressures in reverberant chambers. J Sound Vib 16:29–42.

Nabelek AK, Robinson PK (1982) Monaural and binaural speech perception in reverberation for listeners of various ages. J Acoust Soc Am 71:1242–1248.

Nabelek IV (1983) Performance of hearing-impaired listeners under various types of amplitude compression. J Acoust Soc Am 74:776–791.

Nabelek IV (1984) Discriminability of the quality of amplitude-compressed speech. J Speech Hear Res 27:571–577.

Nelson DA, Schroder AC, Wojtczak M (2001) A new procedure for measuring peripheral compression in normal-hearing and hearing-impaired listeners. J Acoust Soc Am 110:2045–2064.

Neuman AC, Schwander TJ (1987) The effect of filtering on the intelligibility and quality of speech in noise. J Rehabil Res Dev 24:127–134.

Neuman AC, Bakke MH, Mackersie C, Hellman S, Levitt H (1995) Effect of release time in compression hearing aids: paired-comparison judgements of quality. J Acoust Soc Am 98:3182–3187.

Noordhoek IM, Drullman R (1997) Effect of reducing temporal intensity modulations on sentence intelligibility. J Acoust Soc Am 101:498–502.

Olsen WO, Van Tasell DJ, Speaks CE (1997) Phoneme and word recognition for words in isolation and in sentences. Ear Hear 18:175–188.

Ono H, Kanzaki J, Mizoi K (1983) Clinical results of hearing aid with noise-level-controlled selective amplification. Audiology 22:494–515.

Owens E, Talbott C, Schubert E (1968) Vowel discrimination of hearing-impaired listeners. J Speech Hear Res 11:648–655.

Owens E, Benedict M, Schubert E (1972) Consonant phonemic errors associated with pure-tone configurations and certain kinds of hearing impairment. J Speech Hear Res 15:308–322.

Oxenham AJ (2001) Forward masking: adaptation or integration? J Acoust Soc Am 109:732–741.

Oxenham AJ, Plack CJ (1997) A behavioral measure of basilar-membrane nonlinearity in listeners with normal and impaired hearing. J Acoust Soc Am 101: 3666–3675.

Pascoe DP (1975) Frequency responses of hearing aids and their effects on the speech perception of hearing-impaired subjects. Ann Otol Rhinol Laryngol 84(suppl 23).

Patterson RD, Allerhand MH, Giguere C (1995) Time-domain modeling of peripheral auditory processing: a modular architecture and a software platform. J Acoust Soc Am 98:1890–1894.

Pavlovic CV (1984) Use of articulation index for assessing residual auditory function in listeners with sensorineural hearing impairment. J Acoust Soc Am 75:1253–1258.

Pavlovic CV, Studebaker GA, Sherbecoe RL (1986) An articulation index based procedure for predicting the speech recognition performance of hearing-impaired individuals. J Acoust Soc Am 80:50–57.

Pearsons KS, Bennett RL, Fidell S (1977) Speech levels in various noise environments (EPA-600/1-77-025). Office of Health and Ecological Effects, Office of Research and Development, U.S. Environmental Protection Agency.

Pekkarinen E, Salmivalli A, Suonpaa J (1990) Effect of noise on word discrimination by subjects with impaired hearing, compared with those with normal hearing. Scand Audiol 19:31–36.

Peterson GE, Lehiste I (1960) Duration of syllable nuclei in English. J Acoust Soc Am 30:693–703.

Peterson PM (1989) Adaptive array processing for multiple microphone hearing aids. Ph.D. thesis, Department of Electrical Engineering and Computer Science, Massachusetts Institute of Technology, Cambridge.

Pick GF, Evans EF, Wilson JP (1977) Frequency resolution in patients wit hearing loss of cochlear origin. In: Evans EF, Wilson JP (eds) Psychoacoustics and Physiology of Hearing. London: Academic Press.

Pickett JM (1980) The Sounds of Speech Communication. Baltimore: University Park Press.

Pickett JM, Martin ES, Johnson D, et al. (1970) On patterns of speech feature reception by deaf listeners. In: Fant G (ed) Speech Communication Ability and Profound Deafness. Washington DC: Alexander Graham Bell Association for the Deaf.

Plack CJ, Moore BCJ (1991) Decrement detection in normal and impaired ears. J Acoust Soc Am 90:3069–3076.

Plack CJ, Oxenham AJ (1998) Basilar-membrane nonlinearity and the growth of forward masking. J Acoust Soc Am 103:1598–1608.

Plomp R (1964) The rate of decay of auditory sensation. J Acoust Soc Am 36: 277–282.

Plomp R (1978) Auditory handicap of hearing impairment and the limited benefit of hearing aids. J Acoust Soc Am 63:533–549.

Plomp R (1988) The negative effect of amplitude compression in multichannel hearing aids in the light of the modulation-transfer function. J Acoust Soc Am 83:2322–2327.

Plomp R (1994) Noise, amplification, and compression: considerations of three main issues in hearing aid design. Ear Hear 15:2–12.

Plomp R, Mimpen AM (1979) Speech-reception threshold for sentences as a function of age and noise level. J Acoust Soc Am 66:1333–1342.

Pollack I (1948) Effects of high pass and low pass filtering on the intelligibility of speech in noise. J Acoust Soc Am 20:259–266.

Preminger JE, Van Tasell DJ (1995) Quantifying the relation between speech quality and speech intelligibility. J Speech Hear Res 38:714–725.

Preves D (1997) Directional microphone use in ITE hearing instruments. Hear Rev 4(7): 21–27.

Price PJ, Simon HJ (1984) Perception of temporal differences in speech by "normal-hearing" adults: effects of age and intensity. J Acoust Soc Am 76:405–410.
Punch JL, Beck EL (1980) Low-frequency response of hearing and judgements of aided speech quality. J Speech Hear Dis 45:325–335.
Punch JL, Beck LB (1986) Relative effects of low-frequency amplification on syllable recognition and speech quality. Ear Hear 7:57–62.
Quatieri TF, McAuley RJ (1990) Noise reduction using a soft-decision sine-wave vector quantizer. Proc IEEE Int Conf Acoust Speech Signal Proc, pp. 821–823.
Rankovic CM (1997) Understanding speech understanding. 2nd Hear Aid Res Dev Conf, Bethesda, MD.
Robinson CE, Huntington DA (1973) The intelligibility of speech processed by delayed long-term averaged compression amplification. J Acoust Soc Am 54:314.
Rosen S, Walliker J, Brimacombe JA, Edgerton BJ (1989) Prosodic and segmental aspects of speech perception with the House/3M single-channel implant. J Speech Hear Res 32:93–111.
Rosenthal RD, Lang JK, Levitt H (1975) Speech reception with low-frequency speech energy. J Acoust Soc Am 57:949–955.
Ruggero MA, Rich NC (1991) Furosemide alters organ of Corti mechanics: evidence for feedback of outer hair cells upon basilar membrane. J Neurosci 11:1057–1067.
Sasaki N, Kawase T, Hidaka H, et al. (2000) Apparent change of masking functions with compression-type digital hearing aid. Scand Audiol 29:159–169.
Saunders GH, Kates JM (1997) Speech intelligibility enhancement using hearing-aid array processing. J Acoust Soc Am 102:1827–1837.
Scharf B (1978) Comparison of normal and impaired hearing II. Frequency analysis, speech perception. Scand Audiol Suppl 6:81–106.
Schmidt JC, Rutledge JC (1995) 1st Hear Aid Res Dev Conf, Bethesda, MD.
Schmidt JC, Rutledge JC (1996) Multichannel dynamic range compression for music signals. Proc IEEE Int Conf Acoust Speech Signal Proc 2:1013–1016.
Schroder AC, Viemeister NF, Nelson DA (1994) Intensity discrimination in normal-hearing and hearing-impaired listeners. J Acoust Soc Am 96:2683–2693.
Schwander T, Levitt H (1987) Effect of two-microphone noise reduction on speech recognition by normal-hearing listeners. J Rehabil Res Dev 24:87–92.
Sellick PM, Patuzzi R, Johnstone BM (1982) Measurement of basilar membrane motion in the guinea pig using the Mössbauer technique. J Acoust Soc Am 72: 131–141.
Shailer MJ, Moore BCJ (1983) Gap detection as a function of frequency, bandwidth, and level. J Acoust Soc Am 74:467–473.
Shannon RV, Zeng FG, Kamath V, Wygonski J, Ekelid M (1995) Speech recognition with primarily temporal cues. Science 270:303–304.
Shields PW, Campbell DR (2001) Improvements in intelligibility of noisy reverberant speech using a binaural subband adaptive noise-cancellation processing scheme. J Acoust Soc Am 110:3232–3242.
Sigelman J, Preves DA (1987) Field trials of a new adaptive signal processor hearing aid circuit. Hear J (April):24–29.
Simon HJ, Aleksandrovsky I (1997) Perceived lateral position of narrow-band noise in hearing-impaired and normal-hearing listeners under conditions of equal sensation level and sound pressure level. J Acoust Soc Am 102:1821–1826.

Skinner MW (1976) Speech intelligibility in noise-induced hearing loss: effects of high frequency compensation. Doctoral dissertation, Washington University, St. Louis.

Skinner MW (1980) Speech intelligibility in noise-induced hearing loss: effects of high-frequency compensation. J Acoust Soc Am 67:306–317.

Slaney M, Lyon RF (1993) On the importance of time—a temporal representation of sound. In: Cooke M, Beet S, Crawford M (eds) Visual Representations of Speech Signals. Chichester: John Wiley.

Smoorenburg GF (1990) On the limited transfer of information with noise-induced hearing loss. Acta Otolaryngol 469:38–46.

Snell KB, Ison JR, Frisina DR (1994) The effects of signal frequency and absolute bandwidth on gap detection in noise. J Acoust Soc Am 96:1458–1464.

Soede W, Berhout A, Bilsen F (1993) Assessment of a directional microphone array for hearing-impaired listeners. J Acoust Soc Am 94:799–808.

Souza PE, Bishop RD (1999) Improving speech audibility with wide dynamic range compression in listeners with severe sensorineural loss. Ear Hear 20:461–470.

Souza PE, Turner CW (1999) Quantifying the contribution of audibility to recognition of compression-amplified speech. Ear Hear 20:12–20.

Staab WJ, Nunley J (1987) New development: multiple signal processor (MSP). Hear J August:24–26.

Steeneken HJM, Houtgast T (1980) A physical method for measuring speech transmission quality. J Acoust Soc Am 67:318–326.

Steeneken HJM, Houtgast T (1983) The temporal envelope spectrum of speech and its significance in room acoustics. Proc Int Cong Acoust 7:85–88.

Stein LK, Dempesy-Hart D (1984) Listener-assessed intelligibility of a hearing aid self-adaptive noise filter. Ear Hear 5:199–204.

Steinberg JC, Gardner MB (1937) The dependence of hearing impairment on sound intensity. J Acoust Soc Am 9:11–23.

Stelmachowicz PG, Jesteadt W, Gorga MP, Mott J (1985) Speech perception ability and psychophysical tuning curves in hearing-impaired listeners. J Acoust Soc Am 77:620–627.

Stevens KN, Blumstein SE (1978) Invariant cues for place of articulation in stop consonsants. J Acoust Soc Am 64:1358–1368.

Stillman JA, Zwislocki JJ, Zhang M, Cefaratti LK (1993) Intensity just-noticeable differences at equal-loudness levels in normal and pathological ears. J Acoust Soc Am 93:425–434.

Stone MA, Moore BCJ (1992) Spectral feature enhancement for people with sensorineural hearing impairment: effects on speech intelligibility and quality. J Rehabil Res Dev 29:39–56.

Stone MA, Moore BCJ, Alcantara JI, Glasberg BR (1999) Comparison of different forms of compression using wearable digital hearing aids. J Acoust Soc Am 106: 3603–3619.

Strickland EA, Viemeister NF (1997) The effects of frequency region and bandwidth on the temporal modulation transfer function. J Acoust Soc Am 102:1799–1810.

Stubbs RJ, Summerfield Q (1990) Algorithms for separating the speech of interfering talkers: evaluations with voiced sentences, and normal-hearing and hearing-impaired listeners. J Acoust Soc Am 87:359–372.

Studebaker GA (1980) Fifty years of hearing aid research: an evaluation of progress. Ear Hear 1:57–62.

Studebaker GA (1992) The effect of equating loudness on audibility-based hearing aid selection procedures. J Am Acad Audiol 3:113–118.
Studebaker GA, Taylor R, Sherbecoe RL (1994) The effect of noise spectrum on speech recognition performance-intensity functions. J Speech Hear Res 37: 439–448.
Studebaker GA, Sherbecoe RL, Gwaltney CA (1997) Development of a monosyllabic word intensity importance function. 2nd Hear Aid Res Dev Conf, Bethesda, MD.
Summerfield (1992) Lipreading and audio-visual speech perception. Philos Trans R Soc Lond B 335:71–78.
Summerfield Q, Foster J, Tyler R, Bailey P (1985) Influences of formant bandwidth and auditory frequency selectivity on identification of place of articulation in stop consonants. Speech Commun 4:213–229.
Summers V (2000) Effects of hearing impairment and presentation level on masking period patterns for Schroeder-phase harmonic complexes. J Acoust Soc Am 108:2307–2317.
Summers V, Leek MR (1994) The internal representation of spectral contrast in hearing-impaired listeners. J Acoust Soc Am 95:3518–3528.
Summers V, Leek MR (1995) Frequency glide discrimination in the F2 region by normal-hearing and hearing-impaired listeners. J Acoust Soc Am 97:3825–3832.
Summers V, Leek MR (1997) Intraspeech spread of masking in normal-hearing and hearing-impaired listeners. J Acoust Soc Am 101:2866–2876.
Syrdal AK, Gopal HS (1986) A perceptual model of vowel recognition based on the auditory representation of American English vowels. Lang Speech 29:39–57.
Takahashi GA, Bacon SP (1992) Modulation detection, modulation masking, and speech understanding in noise and in the elderly. J Speech Hear Res 35:1410–1421.
Thibodeau LM, Van Tasell DJ (1987) Tone detection and synthetic speech discrimination in band-reject noise by hearing-impaired listeners. J Acoust Soc Am 82: 864–873.
Thompson SC (1997) Directional patterns obtained from dual microhpones. Knowles Tech Rep, October 13.
Thornton AR, Abbas PJ (1980) Low-frequency hearing loss: perception of filtered speech, psychophysical tuning curves, and masking. J Acoust Soc Am 67:638–643.
Tillman TW, Carhart R, Olsen WO (1970) Hearing aid efficiency in a competing speech situation. J Speech Hear Res 13:789–811.
Trees DE, Turner CW (1986) Spread of masking in normal subjects and in subjects with high-frequency hearing loss. Audiology 25:70–83.
Turner CW, Hurtig RR (1999) Proportional frequency compression of speech for listeners with sensorineural hearing loss. J Acoust Soc Am 106:877–886.
Turner CW, Robb MP (1987) Audibility and recognition of stop consonants in normal and hearing-impaired subjects. J Acoust Soc Am 81:1566–1573.
Turner CW, Smith SJ, Aldridge PL, Stewart SL (1997) Formant transition duration and speech recognition in normal and hearing-impaired listeners. J Acoust Soc Am 101:2822–2825.
Tyler RS (1986) Frequency resolution in hearing impaired listeners. In: Moore BCJM (ed) Frequency Selectivity in Hearing. London: Academic Press, pp. 309–371.
Tyler RS (1988) Signal processing techniques to reduce the effects of impaired frequency resolution. Hear J 9:34–47.

Tyler RS, Kuk FK (1989) The effects of "noise suppression" hearing aids on consonant recognition in speech-babble and low-frequency noise. Ear Hear 10:243–249.

Tyler RS, Baker LJ, Armstrong-Bednall G (1982a) Difficulties experienced by hearing-aid candidates and hearing-aid users. Br J Audiol 17:191–201.

Tyler RS, Summerfield Q, Wood EJ, Fernandes MA (1982b) Psychoacoustic and temporal processing in normal and hearing-impaired listeners. J Acoust Soc Am 72:740–752.

Uzkov AI (1946) An approach to the problem of optimum directive antenna design. C R Acad Sci USSR 35:35.

Valente M, Fabry DA, Potts LG (1995) Recognition of speech in noise with hearing aids using dual-microphones. J Am Acad Audiol 6:440–449.

van Buuren RA, Festen JM, Houtgast T (1996) Peaks in the frequency response of hearing aids: evaluation of the effects on speech intelligibility and sound quality. J Speech Hear Res 39:239–250.

van Dijkhuizen JN, Anema PC, Plomp R (1987) The effect of varying the slope of the amplitude-frequency response on the masked speech-reception threshold of sentences. J Acoust Soc Am 81:465–469.

van Dijkhuizen JN, Festen JM, Plomp R (1989) The effect of varying the amplitude-frequency response on the masked speech-reception threshold of sentences for hearing-impaired listeners. J Acoust Soc Am 86:621–628.

van Dijkhuizen JN, Festen JM, Plomp R (1991) The effect of frequency-selective attenuation on the speech-reception threshold of sentences in conditions of low-frequency noise. J Acoust Soc Am 90:885–894.

van Harten-de Bruijn H, van Kreveld-Bos CSGM, Dreschler WA, Verschuure H (1997) Design of two syllabic nonlinear multichannel signal processors and the results of speech tests in noise. Ear Hear 18:26–33.

Van Rooij JCGM, Plomp R (1990) Auditive and cognitive factors in speech perception by elderly listeners. II: multivariate analyses. J Acoust Soc Am 88: 2611–2624.

Van Tasell DJ (1993) Hearing loss, speech, and hearing aids. J Speech Hear Res 36: 228–244.

Van Tasell DJ, Crain TR (1992) Noise reduction hearing aids: release from masking and release from distortion. Ear Hear 13:114–121.

Van Tasell DJ, Yanz JL (1987) Speech recognition threshold in noise: effects of hearing loss, frequency response, and speech materials. J Speech Hear Res 30: 377–386.

Van Tasell DJ, Fabry DA, Thibodeau LM (1987a) Vowel identification and vowel masking patterns of hearing-impaired subjects. J Acoust Soc Am 81:1586–1597.

Van Tasell DJ, Soli SD, Kirby VM, Widin GP (1987b) Speech waveform envelope cues for consonant recognition. J Acoust Soc Am 82:1152–1161.

Van Tasell DJ, Larsen SY, Fabry DA (1988) Effects of an adaptive filter hearing aid on speech recognition in noise by hearing-impaired subjects. Ear Hear 9:15–21.

Van Tasell DJ, Clement BR, Schroder AC, Nelson DA (1996) Frequency resolution and phoneme recognition by hearing-impaired listeners. J Acoust Soc Am 4: 2631(A).

Van Veen BD, Buckley KM (1988) Beamforming: a versatile approach to spatial filtering. IEEE Acoust Speech Sig Proc Magazine 5:4–24.

van Veen TM, Houtgast T (1985) Spectral sharpness and vowel dissimilarity. J Acoust Soc Am 77:628–634.

Vanden Berghe J, Wouters J (1998) An adaptive noise canceller for hearing aids using two nearby microphones. J Acoust Soc Am 103:3621–3626.
Verschuure J, Dreschler WA, de Haan EH, et al. (1993) Syllabic compression and speech intelligibility in hearing impaired listeners. Scand Audiol 38:92–100.
Verschuure J, Prinsen TT, Dreschler WA (1994) The effects of syllabic compression and frequency shaping on speech intelligibility in hearing impaired people. Ear Hear 15:13–21.
Verschuure J, Maas AJJ, Stikvoort E, de Jong RM, Goedegebure A, Dreschler WA (1996) Compression and its effect on the speech signal. Ear Hear 17:162–175.
Vickers DA, Moore BC, Baer T (2001) Effects of low-pass filtering on the intelligibility of speech in quiet for people with and without dead regions at high frequencies. J Acoust Soc Am 110:1164–1175.
Viemeister NF (1988) Psychophysical aspects of auditory intensity coding. In: Edelman GM, Gall WE, Cowan WM (eds) Auditory Function. New York: John Wiley.
Viemeister NF, Plack CJ (1993) Time analysis. In: Yost W, Popper A, Fay R (eds) Human Psychophysics. New York: Springer-Verlag.
Viemeister NF, Urban J, Van Tasell D (1997) Perceptual effects of anplitude compression. Second Biennial Hearing Aid Research and Development Conference, 41.
Villchur E (1973) Signal processing to improve speech intelligibility in perceptive deafness. J Acoust Soc Am 53:1646–1657.
Villchur E (1974) Simulation of the effect of recruitment on loudness relationships in speech. J Acoust Soc Am 56:1601–1611.
Villchur E (1987) Multichannel compression for profound deafness. J Rehabil Res Dev 24:135–148.
Villchur E (1989) Comments on "The negative effect of amplitude compression in multichannel hearing aids in the light of the modulation transfer function." J Acoust Soc Am 86:425–427.
Villchur E (1996) Multichannel compression in hearing aids. In: Berlin CI (ed) Hair Cells and Hearing Aids. San Diego: Singular, pp. 113–124.
Villchur E (1997) Comments on "Compression? Yes, but for low or high frequencies, for low or high intensities, and with what response times?" Ear Hear 18:172–173.
Wakefield GH, Viemeister NF (1990) Discrimination of modulation depth of sinusoidal amplitude modulation (SAM) noise. J Acoust Soc Am 88:1367–1373.
Walker G, Dillon H (1982) Compression in hearing aids: an analysis, a review and some recommendations. NAL Report No. 90, National Acoustic Laboratories, Chatswood, Australia.
Wang DL, Lim JS (1982) The unimportance of phase in speech enhancement. IEEE Trans Acoust Speech Signal Proc 30:1888–1898.
Wang MD, Reed CM, Bilger RC (1978) A comparison of the effects of filtering and sensorineural hearing loss on patterns of consonant confusions. J Speech Hear Res 21:5–36.
Weiss M (1987) Use of an adaptive noise canceler as an input preprocessor for a hearing aid. J Rehabil Res Dev 24:93–102.
Weiss MR, Aschkenasy E, Parsons TW (1974) Study and development of the INTEL technique for improving speech intelligibility. Nicolet Scientific Corp., final report NSC-FR/4023.
White NW (1986) Compression systems for hearing aids and cochlear prostheses. J Rehabil Dev 23:25–39.

Whitmal NA, Rutledge JC, Cohen J (1996) Reducing correlated noise in digital hearing aids. IEEE Eng Med Biol 5:88–96.

Widrow B, Glover JJ, McCool J, et al. (1975) Adaptive noise canceling: principles and applications. Proc IEEE 63:1692–1716.

Wiener N (1949) Extrapolation, Interpolation and Smoothing of Stationary Time Series, with Engineering Applications. New York: John Wiley.

Wightman F, McGee T, Kramer M (1977) Factors influencing frequency selectivity in normal hearing and hearing-impaired listeners. In Psychophysics and Physiology of Hearing, Evans EF, Wilson JP (eds). London, Academia Press.

Wojtczak M (1996) Perception of intensity and frequency modulation in people with normal and impaired hearing. In: Kollmeier B (ed) Psychoacoustics, Speech, and Hearing Aids. Singapore: World Scientific, pp. 35–38.

Wojtczak M, Viemeister NF (1997) Increment detection and sensitivity to amplitude modulation. J Acoust Soc Am 101:3082.

Wojtczak M, Schroder AC, Kong YY, Nelson DA (2001) The effect of basilar-membrane nonlinearity on the shapes of masking period patterns in normal and impaired hearing. J Acoust Soc Am 109:1571–1586.

Wolinsky S (1986) Clinical assessment of a self-adaptive noise filtering system. Hear J 39:29–32.

Yanick P (1976) Effect of signal processing on intelligibility of speech in noise for persons with sensorineural hearing loss. J Am Audiol Soc 1:229–238.

Yanick P, Drucker H (1976) Signal processing to improve intelligibility in the presence of noise for persons with ski-slope hearing impairment. IEEE Trans Acoust Speech Signal Proc 24:507–512.

Young ED, Sachs MB (1979) Representation of steady-state vowels in the temporal aspects of the discharge patterns of populations of auditory-nerve fibers. J Acoust Soc Am 66:1381–1403.

Yund EW, Buckles KM (1995a) Multichannel compression in hearing aids: effect of number of channels on speech discrimination in noise. J Acoust Soc Am 97:1206–1223.

Yund EW, Buckles KM (1995b) Enhanced speech perception at low signal-to-noise ratios with multichannel compression hearing aids. J Acoust Soc Am 97:1224–1240.

Yund EW, Buckles KM (1995c) Discrimination of mulitchannel-compressed speech in noise: long term learning in hearing-impaired subjects. Ear Hear 16:417–427.

Yund EW, Simon HJ, Efron R (1987) Speech discrimination with an 8-channel compression hearing aid and conventional aids in background of speech-band noise. J Rehabil Res Dev 24:161–180.

Zhang C, Zeng FG (1997) Loudness of dynamic stimuli in acoustic and electric hearing. J Acoust Soc Am 102:2925–2934.

Zurek PM, Delhorne LA (1987) Consonant reception in noise by listeners with mild and moderate sensorineural hearing impairment. J Acoust Soc Am 82:1548–1559.

Zwicker E (1965) Temporal effects in simultaneous masking by white-noise bursts. J Acoust Soc Am 37:653–663.

Zwicker E, Flottorp G, Stevens SS (1957) Critical bandwidth in loudness summation. J Acoust Soc Am 29:548–557.

Zwicker E, Fastl H, Frater H (1990) Psychoacoustics: Facts and Models. Berlin: Springer-Verlag.

8
Cochlear Implants

Graeme Clark

In Memoriam

This chapter is dedicated to the memory of Bill Ainsworth. He was a highly esteemed speech scientist, and was also a warm-hearted and considerate colleague. He inspired me from the time I commenced speech research under his guidance in 1976. He had the ability to see the important questions, and had such enthusiasm for his chosen discipline. For this I owe him a great debt of gratitude, and I will always remember his friendship.

1. Introduction

Over the past two decades there has been remarkable progress in the clinical treatment of profound hearing loss for individuals unable to derive significant benefit from hearing aids. Now many individuals who were unable to communicate effectively prior to receiving a cochlear implant are able to do so, even over the telephone without any supplementary visual cues from lip reading.

The earliest cochlear implant devices used only a single active channel for transmitting acoustic information to the auditory system and were not very effective in providing the sort of spectrotemporal information required for spoken communication. This situation began to change about 20 years ago upon introduction of implant devices with several active stimulation sites. The addition of these extra channels of information has revolutionized the treatment of the profoundly hearing impaired. Many individuals with such implants are capable of nearly normal spoken communication, whereas 20 years ago the prognosis for such persons would have been extremely bleak.

Cochlear implant devices with multiple channels are capable of transmitting considerably greater amounts of information germane to speech and environmental sounds than single-channel implant devices. For profoundly deaf people, amplification alone is inadequate for restoring hearing.

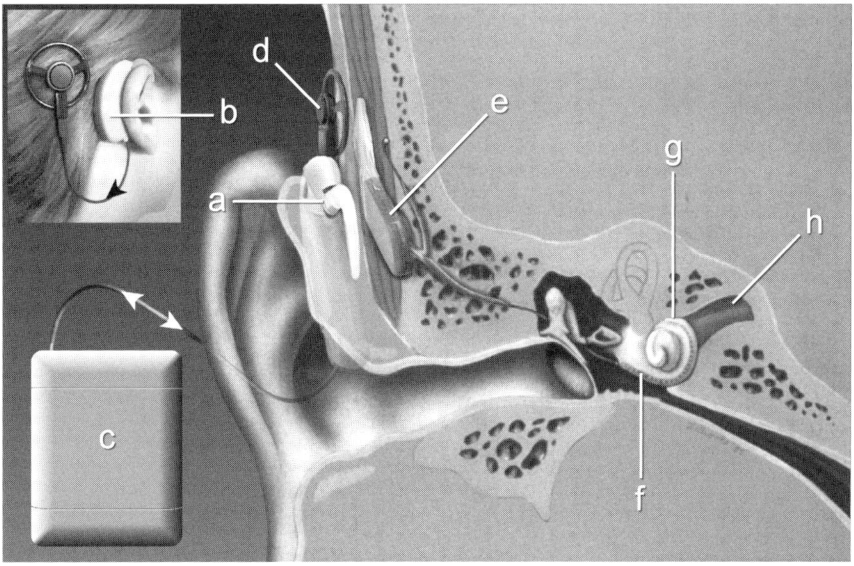

FIGURE 8.1. A diagram of the University of Melbourne/Nucleus multiple-channel cochlear prosthesis manufactured by Cochlear Limited. The components: a, microphone; b, behind-the-ear speech processor; c, body-worn speech processor; d, transmitting aerial; e, receiver-stimulator; f, electrode bundle; g, inner ear (cochlea); h, auditory or cochlear nerve (Clark 2003).

If the organ of Corti is no longer functioning, acoustic stimulation does not produce a sensation of hearing, so it becomes necessary to resort to direct electrical stimulation of the auditory nerve. Sounds are converted into electrical signals, as in a conventional hearing aid, but then, instead of driving a transducer to produce a more intense acoustic signal, they stimulate the auditory nerve directly via a number of electrodes implanted in the cochlea.

This chapter describes the principles involved in the design and implementation of cochlear implants and reviews studies of their effectiveness in restoring speech communication. The University of Melbourne/Nucleus speech processor (Fig. 8.1) and the associated speech-processing strategies are taken as a prime example of a successful cochlear implant, but other processors are reviewed where appropriate. Section 2 outlines the design principles. Sections 3 and 4 introduce the relevant the physiological and psychophysical principles. Speech processing for postlinguistically deaf adults is described in section 5, and that for prelinguistically as well as postlinguistically deaf children in section 6. The main conclusions are briefly summarized in section 7.

The multiple-channel cochlear implant can transmit more information pertaining to speech and environmental sounds than a single-channel implant. However, initial research at the University of Melbourne emphasized that even for multiple-channel electrical stimulation there was an electroneural "bottleneck" restricting the amount of speech and other acoustic information that could be presented to the nervous system (Clark 1987). Nevertheless, improvements in the processing of speech with the University of Melbourne/Nucleus speech processing strategies have now resulted in a mean performance level for postlinguistically deaf adults of 71% to 79% for open sets of Central Institute for the Deaf (CID) sentences when using electrical stimulation alone (Clark 1996b, 1998). Postlinguistically deaf children have also obtained good open-set speech perception results for electrical stimulation alone. Results for prelinguistically deaf children were comparable with those for the postlinguistic group in most tests. However, performance was poorer for open sets of words and words in sentences unless the subjects were implanted at a young age (Clark et al. 1995; Cowan et al. 1995, 1996; Dowell et al. 1995). Now if they receive an implant at a young age, even 6 months, their speech perception, speech production, and language can be comparable to that of age-appropriate peers with normal hearing (Dowell et al. 2002).

The above results for adults are better, on average, than those obtained by severely to profoundly deaf individuals with some residual hearing using an optimally fitted hearing aid (Clark 1996b). This was demonstrated by Brimacombe et al. (1995) on 41 postlinguistically deaf adults who had only marginal benefits from hearing aids as defined by open-set sentence recognition scores less than or equal to 30% in the best aided condition preoperatively. When these patients were converted from the Multipeak to SPEAK strategies (see section 5 for a description of these strategies), the average scores for open sets of CID sentences presented in quiet improved from 68% to 77%. The recognition of open sets of City University of New York (CUNY) sentences presented in background noise also improved significantly from 39% with Multipeak to 58% with SPEAK.

There has been, however, considerable variation in results, and in the case of SPEAK, performance ranged between 5% and 100% correct recognition for open sets of CID sentences via electrical stimulation alone (Skinner et al. 1994). This variation in results may be due to difficulties with "bottom-up" processing, in particular the residual spiral ganglion cell population (and other forms of cochlear pathology) or "top-down" processing, in particular the effects of deafness on phoneme and word recognition.

For a more detailed review the reader is referred to "Cochlear Implants: Fundamentals and Applications" (Clark 2003).

2. Design Concepts

2.1 Speech Processor

The external section of the University of Melbourne/Nucleus multiple-channel cochlear prosthesis (Clark 1996b, 1998), is shown diagrammatically in Figure 8.1. The external section has a directional microphone placed above the pinna to select the sounds coming from in front of the person, and this is particularly beneficial in noisy conditions. The directional microphone sends information to the speech processor. The speech processor can be worn either behind the ear (ESPrit) or on the body (SPrint). The speech processor filters the sound, codes the signal, and transmits the coded data through the intact skin by radio waves to an implanted receiver-stimulator. The code provides instructions to the receiver-stimulator for stimulating the auditory nerve fibers with temporospatial patterns of electrical current patterns that represent speech and other sounds.

Power to operate the receiver-stimulator is transmitted along with the data. The receiver-stimulator decodes the signal and produces a pattern of electrical stimulus currents in an array of electrodes inserted around the scala tympani of the basal turn of the cochlea. These currents in turn induce temporospatial patterns of responses in auditory-nerve fibers, which are transmitted to the higher auditory centers for processing. The behind-the-ear speech processor (ESPrit) used with the Nucleus CI-24M receiver-stimulator presents the SPEAK (McKay et al. 1991), continuous interleaved sampler (CIS) (Wilson et al. 1992), or Advanced Combination Encoder (ACE) strategies (Staller et al. 2002). The body-worn speech processor (SPrint) can implement the above strategies, as well as more advanced ones.

The behind-the-ear speech processor (ESPrit) has a 20-channel filter bank to filter the sounds, and the body-worn speech processor (SPrint) uses a digital signal processor (DSP) to enable a fast Fourier transform (FFT) to provide the filtering (Fig. 8.2). The outputs from the filter bank or FFT are selected, as well as the electrodes to represent them. The output voltages are referred to a "map", where the thresholds and comfortable loudness levels for each electrode are recorded and converted into stimulus current levels. An appropriate digital code for the stimulus is produced and transmitted through the skin by inductive coupling between the transmitter coil worn behind the ear and a receiver coil incorporated in the implanted receiver-stimulator. The transmitting and receiving coils are aligned through magnets in the centers of both coils. The transmitted code is made up of a digital data stream representing the sound at each instant in time, and is transmitted by pulsing a radiofrequency (RF) carrier.

2.2 Receiver-Stimulator

The receiver-stimulator (Figs. 8.1 and 8.2) decodes the transmitted information into instructions for the selection of the electrode, mode of stimulation (i.e., bipolar, common ground, or monopolar) current level, and pulse width. The stimulus current level is controlled via a digital-to-analog converter. Power to operate the receiver-stimulator is also transmitted by the RF carrier. The receiver-stimulator is connected to an array of electrodes incorporated into a carrier that is introduced into the scala tympani of the

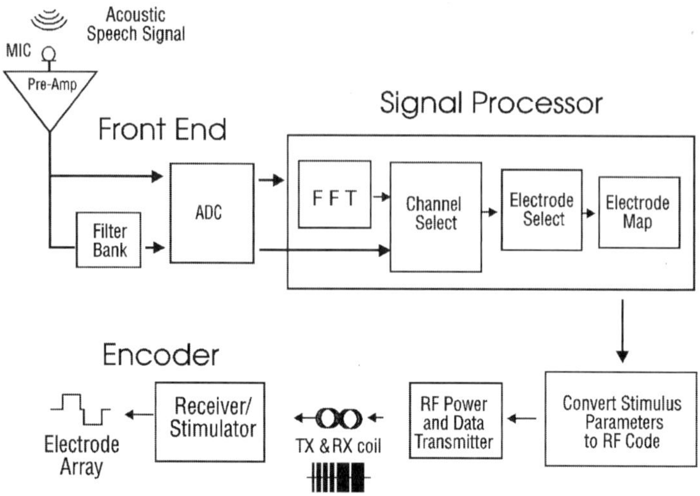

FIGURE 8.2. A diagram of the Spectra-22 and SP-5 speech processors implemented using either: a standard filter bank or fast Fourier transform (FFT) filter bank. The front end sends the signal to a signal-processing chip via either a filter bank or a digital signal processor (DSP) chip, which carries out an FFT. The signal processor selects the filter-bank channels and the appropriate stimulus electrodes and amplitudes. An encoder section converts the stimulus parameters to a code for transmitting to the receiver-stimulator on a radiofrequency (RF) signal, together with power to operate the device (Clark 1998).

basal turn of the cochlea and positioned to lie as close as possible to the residual auditory-nerve fibers.

The receiver-stimulator (CI-24R) used with the Nucleus-24 system can provide stimulus rates of up to 14,250 pulses/s. When distributed across electrodes, this can allow a large number of electrodes to be stimulated at physiologically acceptable rates. It also has telemetry that enables electrode-tissue impedances to be determined, and compound action potentials from the auditory nerve to be measured.

3. Physiological Principles

The implant should provide stimulation for the optimal transmission of information through the electroneural "bottleneck." This would be facilitated by interfacing it to the nervous system so that it can encode the frequencies and intensities of sounds as closely as possible to those codes that occur normally. In the case of frequency, coding is through time/period (rate) and place codes, and for intensity, the population of neurons excited and their mean rate of firing.

3.1 Time/Period (Rate) Coding of Frequency

The time/period coding of frequency (Tasaki 1954; Katsuki et al. 1962; Rupert et al. 1963; Kiang et al. 1965; Rose et al. 1967; Sachs and Young 1979) depends on action potentials being locked to the same phase of the sine wave so that the intervals between the action potentials are an integral multiple of the period. It has been postulated (Rose et al. 1967) that the intervals in a population of neurons and not just individual neurons are important in the decoding of frequency.

3.1.1 Comparison of Unit Responses for Acoustic and Electric Stimulation

Physiological studies in the experimental animal have shown significant limitations in reproducing the time/period coding of frequency by electrical stimulation (Clark 1969; Merzenich 1975). This is illustrated in Figure 8.3, where interval histograms are shown for unit responses from primary-like neurons in the anteroventral cochlear nucleus to acoustic and electrical stimulation.

For electrical stimulation at low rates of 400 pulses/s and below, the distribution of interspike intervals is very different from acoustic stimulation at the same frequency. With an acoustic stimulus of 416 Hz, there is a distribution of intervals around each population mode, referred to as stochastic firing. With electrical stimulation of 400 pulses/s there is a single population of intervals, with a mode in the firing pattern distribution that is the same as the period of the stimulus. There is also very little jitter around the mode, a phenomenon known as "deterministic firing." The jitter increases and the phase-locking decreases, with increasing rates of stimulation, as illustrated in the lower right panel of Figure 8.3 for electrical stimulation of 800 pulses/s.

3.1.2 Behavioural Responses in the Experimental Animal for Electrical Stimulation

The discrimination of acoustic frequency and electrical stimulus rate was found to be significantly different for acoustic and electrical stimulation in experimental animals. Rate discrimination results from three behavioral studies on experimental animals (Clark et al. 1972, 1973; Williams et al. 1976) showed that the rate code, as distinct from the place code, could convey temporal information only for low rates of stimulation up to 600 pulses/s. Similar psychophysical results were also obtained on cochlear implant patients (Tong et al. 1982).

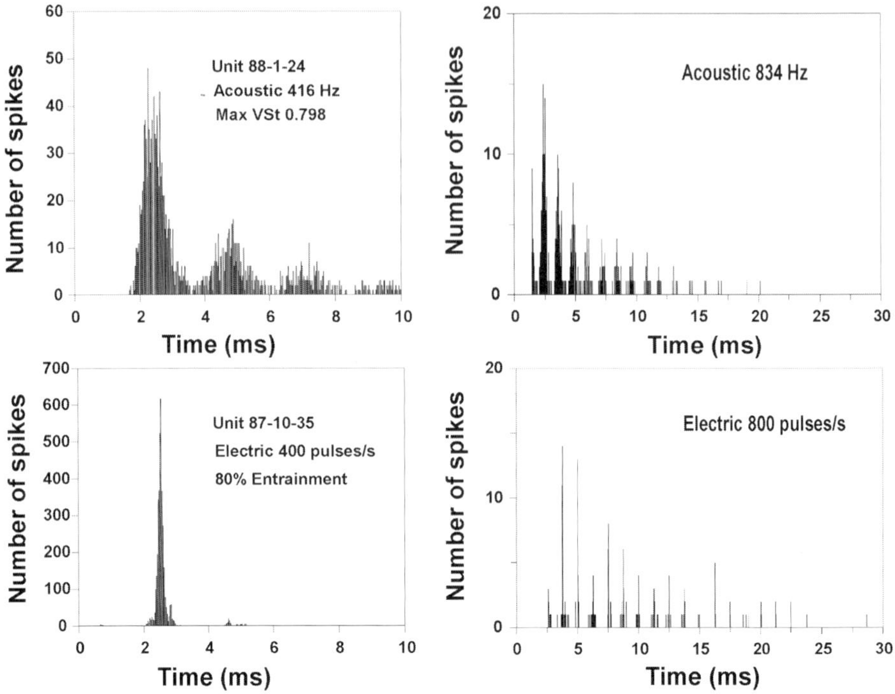

FIGURE 8.3. Interspike interval histograms from primary-like units in the anteroventral cochlear nucleus of the cat. Left top: Acoustic stimulation at 416 Hz. Left bottom: Electrical stimulation at 400 pulses/s (pps). Right top: Acoustic stimulation at 834 Hz. Right bottom: Electrical stimulation at 800 pulses/s (pps).

3.1.3 Simulation of Time/Period Coding of Frequency

Why then is there an apparent contradiction between the above psychophysical and physiological results? Why is rate discrimination more like sound at low stimulus rates, but the interspike interval histograms for electrical stimulation, which reflect temporal coding, not like those for sound? On the other hand, why is rate discrimination poor at high stimulus rates, but the pattern of interspike intervals similar for electrical stimulation and sound? The discrepancy between the physiological and psychophysical results can be explained if we assume that a temporospatial pattern of intervals in a group of fibers is required for the temporal coding of sound, and that the temporospatial pattern is not adequately reproduced by electrical stimulation. A temporospatial pattern of action potentials in a group of nerve fibers is illustrated in Figure 8.4. This figure shows that the individual fibers in a group do not respond with an action potential each

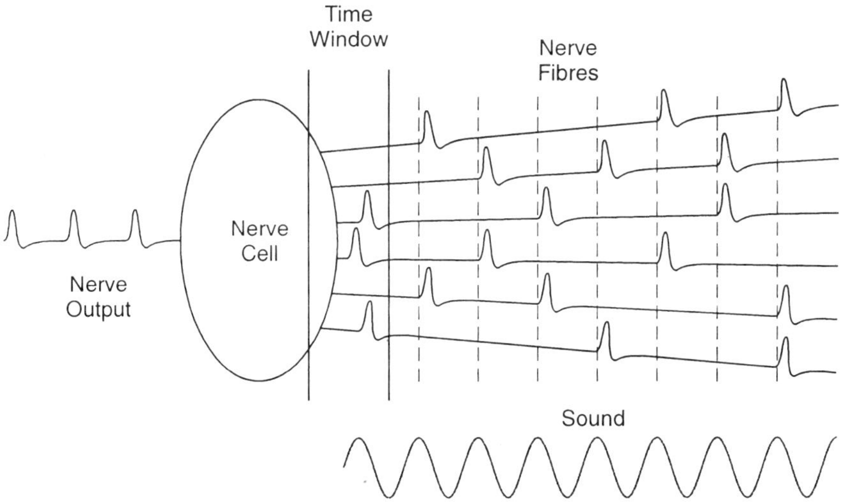

FIGURE 8.4. Temporospatial patterns of action potentials in an ensemble of neurons in response to a low to mid-acoustic frequency. Top: Nerve action potentials in a population of neurons. Bottom: Pure tone acoustic stimulus. This demonstrates the phase locking of neurons to the sound wave, but note that the action potentials do not occur each cycle. The diagram also demonstrates convergent pathways on a cell. The convergent inputs only initiate an action potential in the cell if they arrive within a time window (coincidence detection).

sine wave, but when an action potential occurs it is at the same phase on the sine wave.

Moreover, the data, together with the results of mathematical modeling studies on coincidence detection from our laboratory (Irlicht et al. 1995; Irlicht and Clark 1995), suggest that the probability of neighboring neurons firing is not in fact independent, and that their co-dependence is essential to the temporal coding of frequency. This dependence may be due to phase delays along the basilar membrane, as well as convergent innervation of neurons in the higher auditory centers. A temporospatial pattern of responses for dependent excitation in an ensemble of neurons for acoustic stimulation is illustrated in Figure 8.5.

Further improvements in speech processing for cochlear implants may be possible by better reproduction of the temporospatial patterns of responses in an ensemble of neurons using patterns of electrical stimuli (Clark 1996a). The patterns should be designed so that auditory nerve potentials arrive at the first higher auditory center (the cochlear nucleus) within a defined time window for coincidence detection to occur. There is evidence that coincidence detection is important for the temporal coding of sound frequency (Carney 1994; Paolini et al. 1997) and therefore patterns of electrical stimuli should allow this to occur.

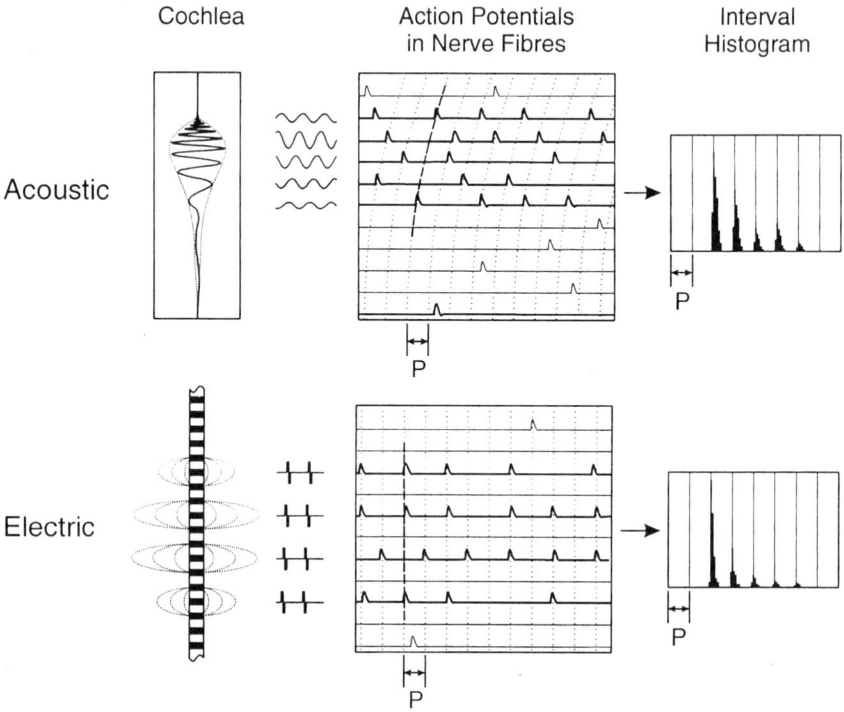

FIGURE 8.5. A diagram of the unit responses or action potentials in an ensemble of auditory neurons for electrical and acoustic stimulation showing the effects of the phase of the basilar membrane traveling were. Top: The probability of firing in an ensemble of neurons to acoustic excitation due to phase delays along the basilar membrane. Bottom: The probability of firing due to electrical stimulation (Clark 2001).

3.2 Place Coding of Frequency

The place coding of frequency (Rose et al. 1959; Kiang 1966; Evans and Wilson 1975; Evans 1981; Aitkin 1986; Irvine 1986) is due to the localized excitation of the cochlea and auditory neurons, which are ordered anatomically so that their frequencies of best response form a frequency scale. Reproducing the place coding of frequency is important for multiple-channel cochlear implant speech processing, particularly for the coding of the speech spectrum above approximately 600 Hz.

3.2.1 Stimulus Mode and Current Spread

Research was required to ascertain how to best localize the electrical current to discrete groups of auditory-nerve fibers in the cochlea for the place coding of frequency. The research showed that bipolar and common-ground stimulation would direct adequate current through the neurons without short-circuiting along the fluid compartments of the cochlea (Merzenich 1975; Black and Clark 1977, 1978, 1980; Black et al. 1981). A resistance model of the cochlea also demonstrated localization of current for monopolar stimulation with electrodes in the scala tympani. With bipolar stimulation the current passes between neighboring electrodes, and with common ground stimulation the current passes between an active electrode and the others on the cochlear array connected together electrically. It has subsequently been shown that if the electrodes are placed close to the neurons, then monopolar stimulation between an active and distant electrode may also allow localized stimulation (Busby et al. 1994). There is thus an interaction among stimulus mode, electrode geometry, and cochlear anatomy for the optimal localization of current for the place coding of speech frequencies.

3.3 Intensity Coding

The coding of sound intensity (reviewed by Irvine 1986) in part may be due to the mean rate of unit firing. For most auditory neurons there is a monotonic increase in discharge rate with intensity, which generally saturates 20 to 50dB above threshold (Kiang et al. 1965; Evans 1975). However, for a small proportion of neurons there is an extended dynamic range of about 60dB. With the limited dynamic range of mean firing rate, and only a 20-dB range in thresholds (Evans 1975), the coding of the greater than 100-dB dynamic range of hearing in the human has not been fully explained. This extended dynamic range may be due to the recruitment of neighboring neurons, as suggested by studies with band-stop noise (Moore and Raab 1974; Viemeister 1974).

3.3.1 Intensity Input/Output Functions for Electrical Stimulation

With electrical stimulation, the dynamic range of auditory-nerve firing was initially shown to be approximately 4dB (Moxon 1971; Kiang and Moxon 1972). Subsequent studies have established the dynamic range for the response rate of auditory-nerve fibers varies between 0.9 and 6.1dB (Javel et al. 1987), and is greater at high stimulus rates (Javel et al. 1987). The narrow dynamic range for unit firing is similar to that obtained from psychophysical studies on implant patients, indicating that mean rate is important in coding intensity. Field potentials, which reflect the electrical activity from a population of neurons, have a dynamic range of 10 to 20dB when recorded over a range of intensities (Simmons and Glattke 1970; Glattke 1976; Clark and Tong 1990). As this range is similar to the psychophysical results in implant patients, it suggests that the population of neurons excited, as well as their mean firing rate, is important in coding intensity.

Experimental animal studies and psychophysical results in humans indicate that the dynamic range for electrical stimulation is much narrower than for sound, and as a result linear compression techniques for encoding speech signals are required.

3.4 Plasticity and Acoustic and Electric Stimulation

It should be borne in mind, when implanting cochlear electrodes into children, that there is a critical period associated with the development of the auditory system; after a certain stage children may not be able to benefit from speech-processing strategies presenting information on the basis of a place or time/period code. It has been demonstrated in psychophysical studies, in particular, that if profoundly deaf children cannot perceive place of electrode stimulation tonotopically, then speech perception will be poor (Busby et al. 1992; Busby and Clark 2000a,b). For this reason studies in experimental animals are very important in determining the plasticity of the central auditory nervous system and the critical periods for the changes. Acute experiments on immature experimental animals have shown that there is a sharpening of frequency tuning postpartum, and that the spike count versus intensity functions are steep in young kittens compared to adults (Brugge et al. 1981). Research on adult animals has also demonstrated that when a lesion is made in the cochlea, tonotopic regions surrounding the affected frequency region are overly represented in the auditory cortex (Robertson and Irvine 1989; Rajan et al. 1990). It has also been shown by Recanzone et al. (1993), with behavioral experiments in the primate, that attention to specific sound frequencies increases the cortical representation of those spectral bands.

Research has also suggested that chronic electrical stimulation may increase cortical representation associated with a stimulus site (Snyder et al. 1990). However, it is unclear if this result is related to current spread or to chronic stimulation per se. For this reason research has been undertaken to examine the uptake of ^{14}C-2-deoxyglucose after electrically stimulating animals of different ages (Seldon et al. 1996). As there was no difference in uptake with age or whether the animal was stimulated or unstimulated, it suggests that other factors such as the position of the stimulating electrode and current spread are important.

These basic studies thus indicate that there is a critical period for the adequate development of the place and time/period codes for frequency, and that implantation should be carried out during this period. Moreover, electrical stimulation during the critical period may cause reorganization of cortical neural responsiveness so that initial global or monopolar stimulation could preclude subsequent improvement of place coding with bipolar or common ground stimulation and, consequently, speech perception.

3.5 Coding of Frequency and Intensity versus the Perception of Pitch and Loudness

Frequency and intensity coding are associated predominantly with the percepts of pitch and loudness, respectively, and these percepts underlie the processing of speech and environmental sounds. For this reason an adequate representation of frequency and intensity coding using electrical stimulation from a cochlear implant is important. The time/period (rate) code for frequency results in temporal pitch, and the place code is associated with place (spectral) pitch. The latter was experienced as timbre from sharp to dull. However, frequency coding may also have an effect on loudness, and loudness coding may affect pitch. It has been shown by Moore (1989) that increasing intensity not only will increase loudness, but also may have a small effect on pitch.

4. Psychophysical Principles

Effective representation of pitch and loudness with electrical stimulation underpins speech processing for the cochlear implant. In the psychophysical studies of pitch perception using electrical stimulation that are discussed below, the intensity of stimuli was balanced to preclude loudness being used as an auxiliary cue.

4.1 Temporal Information

The perception of temporal information has been studied using rate discrimination, pitch ratio, and pitch-matching measures.

4.1.1 Rate Discrimination

Research on the first two implant patients from Melbourne (Tong et al. 1982) showed that the difference limens (DLs) for electric stimulation at 100 and 200 pulses/s ranged from approximately 2% to 6%. The DLs were similar for stimulation at apical and basal electrodes. The rate DLs for up to 200 pulses/s were more comparable with acoustic stimulation than at higher stimulus rates. These results support the use of a time/period code to convey low-frequency information, such as voicing, when stimulating both apical and basal electrodes. They also indicate that temporal pitch perception for low frequencies is at least partly independent of place pitch.

4.1.2 Rate Discrimination versus Duration

Having established that there was satisfactory discrimination of low rates of electrical stimulation, it was necessary to determine if this occurred over the durations required for coding segmental and suprasegmental speech information. Tong et al. (1982) showed that variations in the rate of stimulation from 150 to 240 pulses/s over durations of 25, 50, and 100 ms were well discriminated for the stimuli of longest duration (50 and 100 ms), but not for durations as short as 25 ms (comparable to the duration associated with specific components of certain consonants). These findings indicated that rate of stimulation may not be suitable for the perception of consonants, but could be satisfactory for coding the frequency of longer-duration phenomena such as those associated with suprasegmental speech information, in particular voicing.

4.1.3 Rate of Stimulation and Voicing Categorization

To determine whether a time/period (rate) code was suitable for conveying voicing, it was important to ascertain if the percept for rate of stimulation could be categorized in the speech domain. It was necessary to know whether voicing was transmitted independently of place of stimulation by varying the rate of stimulation on different electrodes. The rate was varied by Tong et al. (1982) from 60 to 160 pulses/s on a single electrode and across electrodes. Patients were asked to categorize each stimulus as a question or a statement according to whether the pitch was rising or falling. The data showed that as the trajectory rose more steeply, the proportions of stimuli labeled as a question reached 100%, while for steeply falling trajectories the number of utterances labeled as questions was close to zero. The data were the same when stimulating apical, middle, and basal electrodes, and when varying repetition rate across four electrodes. The data indicate that rate of stimulation was perceived as voicing, and that voicing was perceived independently from place of stimulation. Moreover, voicing could be perceived by varying the rate of stimulation across different nerve populations.

4.1.4 Pitch Ratio

As rate of stimulation was discriminated at low frequencies, and used to convey voicing, it was of interest to know whether it was perceived as pitch. This was studied by comparing the pitches of test and standard stimuli. Tong et al. (1983) showed that the pitch ratio increased rapidly with stimulus rate up to 300 pulses/s, similar to that for sound. Above 300 pulses/s the pitch estimate did not change appreciably. The pitch ratios were the same for stimulation of apical, middle, and basal electrodes. This study indicated that a low rate of electrical stimulation was perceived as pitch.

4.1.5 Pitch Matching of Acoustic and Electrical Stimuli

The pitch of different rates of electrical stimulation on a single-channel implant was compared with the pitch for acoustic stimulation of the opposite ear with some residual hearing (Bilger et al. 1977). This study found that the pitch of an electrical stimulus below 250 pulses/s could be matched to that of an acoustic signal of the same frequency, but above 250 pulses/s a proportionately higher acoustic signal frequency was required for a match to be made. Subsequently, it was found in a study on eight patients using the Nucleus multiple-electrode implant that a stimulus was matched to within 20% of a signal of the same frequency by five out of eight patients for 250 pulses/s, three out of eight for 500 pulses/s, and one out of eight for 800 pulses/s (Blamey et al. 1995). The pitch-matching findings are consistent with the pitch ratio and frequency DL data showing that electrical stimulation was used for temporal pitch perception at least up to frequencies of about 250 Hz. The fact that some patients in the Blamey et al. (1995) study matched higher frequencies suggested that there were patient variables that were important for temporal pitch perception.

4.1.6 Amplitude Modulation and Pitch Perception

It has been suggested that pitch associated with stimulus rate is similar to that perceived with amplitude-modulated white noise (Evans 1978). With modulated white noise a "weak" pitch was perceived as corresponding to the modulation frequency up to approximately 500 Hz (Burns and Veimeister 1981). In addition, amplitude-modulated electrical stimuli were readily detected up to 100 Hz, above which detectability fell off rapidly (Blamey et al. 1984a,b; Shannon 1992; Busby et al. 1993a), a result that was similar to that for the detectability of amplitude-modulated white noise by normal-hearing and hearing-impaired subjects (Bacon and Gleitman 1992). The pitch perceived with amplitude-modulated electrical stimuli was further studied by varying the modulation depth and comparing the resultant pitch with that of an unmodulated stimulus (McKay et al. 1995). With increasing modulation depth, for modulation frequencies in the range between 80 and 300 pulses/s, the pitch fell exponentially from a value close to the carrier rate to one close to the modulation frequency. The pitch was predicted on the basis of the weighted average of the two neural firing rates in the stimulated population, with the weightings proportional to the respective numbers of neurons firing at each frequency. This model was developed from data reported by Cacace and Margolis (1985), Eddington et al. (1978), and Zeng and Shannon (1992), and predicts the pitch for carrier rates up to 700 Hz. It supports the hypothesis that pitch for amplitude-modulated electrical stimuli depends on a weighting of the interspike intervals in a population of neurons.

4.2 Place Information

4.2.1 Scaling of Place Pitch

The scaling of place pitch needed to be established in cochlear implant patients in order to develop an effective speech-processing strategy, as behavioral research in animals had shown the limitations of using rate of electrical stimulation to convey the mid- and high-frequency spectrum of speech. The typical patient reported that a stimulus presented at a constant rate (using common-ground stimulation) varied in spectral quality or timbre, ranging from sharp to dull for electrodes placed in the higher- or lower-frequency regions of the cochlea, respectively (Tong et al. 1983). The results showed there was a good ranking for place pitch in the sharp-dull domain. This finding supported the multiple-electrode speech-processing strategy, which codes mid- to high frequencies, including the second formant, on a place-frequency basis.

4.2.2 Time-Varying Place Pitch Discrimination and Stimulus Duration

The study described above used steady-state stimuli with a duration of at least 200 to 300 ms, comparable to the length of a sustained vowel. Speech, however, is a dynamic stimulus, and for consonants, frequencies change rapidly over a duration of approximately 20 to 80 ms. For this reason, the discrimination of place pitch for stimulus duration was studied (Tong et al. 1982), and it was found that a shift in the place of stimulation across two or more electrodes could be discriminated 100% of the time for durations of 25, 50, and 100 ms. This finding indicated that place of stimulation was appropriate for coding the formant transitions in consonants.

4.2.3 Two-Component Place Pitch Perception

To improve the initial Melbourne speech-processing strategy, which presented information pertaining to the second formant on a place-coding basis and voicing as rate of stimulation, it was necessary to know if more information could be transmitted by stimulating a second electrode nonconcurrently for frequencies in the low (first formant) or high (third formant) range. Tong et al. (1982) showed that a two-component sensation, as might be expected for two-formant signals, was perceived, and this formed the basis for the first speech-processing improvement that presented the first (F_1) as well as the second formant (F_2) on a place-coding basis.

4.2.4 Dual-Electrode Intermediate Place Pitch Percepts

Concurrent electrical stimulation through two electrodes was first shown to produce a vowel timbre distinct from that produced by two electrodes stimulated separately by Tong et al. (1979), a result originally attributed to an averaging process. This phenomenon was subsequently explored by Townshend et al. (1987), who showed that when two electrodes were stimulated simultaneously, changing the current ratio passing through the electrodes causes the pitch to shift between the two. It has also been shown by McDermott and McKay (1994) that for some patients the electrode separation needs to be increased to 3 mm to produce an intermediate pitch. Too much separation, however, will lead to a two-component pitch sensation (Tong et al. 1983b).

4.2.5 Temporal and Place Pitch Perception

As place of stimulation resulted in pitch percepts varying from sharp to dull, and rate of stimulation percepts from high to low, it was important to determine whether the two percepts could be best described along one or two perceptual dimensions. The data from the study of Tong et al. (1983b) were analyzed by multidimensional scaling. The study demonstrated that a two-dimensional representation provided the best solution. This indicated a low degree of interaction between electrode position and repetition rate. It was concluded that temporal and place information provide two components to the pitch of an electrical stimulus.

4.3 Intensity Information

4.3.1 Loudness Growth Function

Psychophysical studies (Simmons 1966; Eddington et al. 1978; Shannon 1983; Tong et al. 1983a) have shown that a full range of loudness, from threshold to discomfort, can be evoked by varying the current level. However, the loudness growth due to increases in the current level was much steeper than the growth for acoustic stimulation in normal-hearing subjects. It is apparent that to utilize current level for coding acoustic amplitude, an appropriate compression algorithm needs to be used.

4.3.2 Intensity Difference Limens

The just-discriminable difference in electric current varies from 1% to 8% of the dynamic range (Shannon 1983; Hochmair-Desoyer et al. 1983). Electrical current therefore can be used to convey variations in acoustic intensity information in speech.

4.3.3 Loudness versus Mean Stimulus Rate

The loudness of stimuli incorporating either single or multiple pulses per period was compared to study the effect of average electrical-stimulation rate on loudness (Tong et al. 1983b). It was found that if the overall number of pulses over time in the multiple-pulse-per-period stimulus was kept constant, there was little change in loudness as the firing rate of the burst of stimuli increased. On the other hand, with single-pulse-per-period stimuli, there was a significant increase in loudness as the pulse rate increased. Such results suggest that loudness under such conditions is a function of the two physical variables—charge per pulse and overall pulse rate (Busby and Clark 1997).

4.4 Prelinguistic Psychophysical Information

4.4.1 Temporal Information

An early study of one prelinguistically deaf 61-year-old patient revealed a limitation in rate discrimination (Eddington et al. 1978). A more detailed study by Tong et al. (1988) was undertaken on three prelinguistically deaf individuals between the ages of 14 and 24, all of whom had completely lost their hearing at between 18 and 36 months of age. The temporal processing associated with on/off patterns was studied for duration DLs, gap detection, and numerosity. The results of all three tests were worse for two of the three prelinguistically deaf patients than for a control group of two postlinguistically deaf patients. One individual, with better than average speech perception, had results similar to those of the postlinguistic group. This was the youngest patient (14 years old) and the only one receiving an oral/auditory education. With rate identification, the results for all three were initially worse than for the two postlinguistically deaf adults. Furthermore, speech perception scores were poorer than for the average postlinguistically deaf adults using either the F_0/F_2 or $F_0/F_1/F_2$ speech processors (see section 5.3.1). This result applied in particular to the recognition of consonants. It is not clear if this was due to inadequate rate or place identification. However, a multidimensional analysis of the vowel data, using a one-dimensional solution (interpreted as vowel length), indicated that intensity rather than frequency information was responsible. A clustering analysis indicated a high degree of perceptual confusion among consonants, and this suggested that neither electrode place nor pulse repetition rate were being used for identification. It is also of interest that one patient showed an improvement in rate discrimination with training over time, which was accompanied by an improvement in speech perception. These findings were substantiated in a larger study of 10 prelinguistically deaf patients (Busby et al. 1992) (one third were congenitally deaf, one third were postmeningitic, and one third had Usher's syndrome). Their ages at surgery ranged between 5.5 and 23 years.

The ability of prelinguistically deaf patients to discriminate rate when varied over stimulus intervals characteristic of speech was compared with that of postlinguistically deaf patients (Busby et al. 1993b). The study was focused on four prelinguistically deaf children between the ages of 5 and 14 who had lost their hearing between 1 and 3 years of age, and on four postlinguistically deaf adults (ages ranged between 42 and 68) who lost their hearing between the ages of 38 and 47. Stimulus rates were varied over durations ranging between 144 and 240 ms. Duration had no effect on rate discrimination. The postlinguistically deaf adults were more successful in discriminating repetition rate and also had better speech perception scores.

4.4.2 Place Information

An initial finding pertaining to the prelinguistically deaf patient of Eddington et al. (1978) was that this individual experienced more difficulty distinguishing differences in pitch between adjacent electrodes than other postlinguistic patients. In a more detailed study by Tong et al. (1988), it was shown that three prelinguistically deaf patients were poorer in identifying electrode place than the postlinguistic patients, but two of these individuals improved their performance over time. These findings were basically substantiated in a larger study of 10 prelinguistically deaf patients (Busby et al. 1992).

The importance of electrode-place information for the perception of speech by prelinguistically deaf children was demonstrated in further psychophysical studies, where it was shown that if children were not able to perceive place of electrode stimulation tonotopically, it was likely that their speech perception would be poor (Busby and Clark 1996; 2000a,b). On the other hand, it was also shown that children with a limited ability to rank percepts according to electrode site lacked good speech perception. This finding suggests that the critical period for tonotopic discrimination of place of stimulation, together with rate discrimination, is a likely factor in developing speech understanding. Furthermore, it is interesting to observe that some teenagers who have no history of exposure to sound still have reasonable tonotopic ordering of place pitch; these are the older, prelinguistic children who do well with the implant.

4.4.3 Intensity Information

In a study by Tong et al. (1988), electrical-current-level identification for the two prelinguistically deaf patients examined showed discrimination measured as a percentage of the dynamic range that was similar to that obtained by patients with postlinguistic hearing loss. This result was similar to the electrical current DLs reported in a larger study on 10 prelinguistically deaf patients by Busby et al. (1992).

5. Speech Processing for Postlinguistically Deaf Adults

5.1 Single-Channel Strategy

5.1.1 Types of Schemes

5.1.1.1 Minimal Preprocessing of the Signal

The implant system developed in Los Angeles (House et al. 1981) filtered the signal over the frequency range between 200 and 4000 Hz and provided nonlinear modulation of a 16-kHz carrier wave.

5.1.1.2 Preprocessing of the Signal

Some preprocessing of speech was performed by the system developed in Vienna (Hochmair et al. 1979). With their best strategy there was gain compression, followed by frequency equalization from 100 to 4000 Hz and a mapping of the stimulus onto an equal loudness contour at a comfortable level. Speech was preprocessed by the system developed in London (Fourcin et al. 1979). This system stimulated a single extracochlear electrode with a pulsatile current source triggered by a voicing detector.

5.1.2 Speech Feature Recognition and Speech Perception

5.1.2.1 Minimal Preprocessing of the Signal

The Los Angeles cochlear implant presented variations in speech intensity, thus enabling the rapid intensity changes in stop consonants (e.g., /p/, /t/, /k/, /b/, /d/, /g/) to be coded and vowel length to be discriminated. Intensity and temporal cues permitted the discrimination of voiced from unvoiced speech and low first-formant from high first-formant information, but were insufficient to discriminate formant transitions reliably. This was reflected in the fact that no open-set speech recognition was possible with electrical stimulation alone, but closed-set consonant and vowel recognition could be achieved in some of the patients.

5.1.2.2 Preprocessing of the Signal

With the Vienna system some patients were reported to obtain high open-set scores for words and sentences for electrical stimulation alone (Hochmair-Desoyer et al. 1980, 1981), but open-set speech recognition was not found in a controlled study in which this device was compared with the Los Angeles single-channel and the Salt Lake City and Melbourne multiple-channel devices (Gantz et al. 1987). With the London system the signal retained information about the precise timing of glottal closure and fine details of the temporal aspects of phonation. It was found that patients could reliably detect small intonation variations, and when combined with a visual signal the information on voicing improved scores on closed sets of consonants.

5.2 Multiple-Channel Strategies: Fixed Filter Schemes

5.2.1 Types of Schemes

5.2.1.1 Cochlear and Neural Models

Prior to developing the initial Melbourne formant or cue extraction F_0/F_2 strategy in 1978, a fixed-filter strategy was evaluated that modeled the physiology of the cochlea and the neural coding of sound (Laird 1979). This strategy had bandpass filters to approximate the frequency selectivity of auditory neurons, delay mechanisms to mimic basilar-membrane delays, and stochastic pulsing for maintaining the fine time structure of responses and a wide dynamic range.

5.2.1.2 *Minimal Preprocessing of the Signal*

The Salt Lake City (Symbion or Ineraid) device presented the outputs of four fixed filters to the auditory nerve by simultaneous monopolar stimulation. It was used initially as a compressed analog scheme presenting the voltage outputs of four filters as simultaneous analog stimulation (Eddington 1980, 1983). The scheme was also used with the University of California at San Francisco/Storz device but with biphasic pulses (Merzenich et al. 1984). Compression was achieved with a variable gain amplifier operating in compression mode. The compressed analog scheme was subsequently used with eight filters in the Clarion processors (Battmer et al. 1994).

5.2.1.3 *Continuous Interleaved Sampler (CIS)*

A more recent development in the use of fixed filters for cochlear implant speech processing is an electroneural scheme called continuous interleaved sampler (CIS). This type of processing addresses the "bottleneck" by sampling the outputs of six or more filters at a constant high rate in order to reduce channel interactions. The outputs of the bandpass filters are rectified and low-pass filtered, and samples continuously interleaved among the electrodes. In 1992 six filters with low-pass frequency cutoffs between 200 and 800 Hz, and stimulus rates between 500 and 1515 pulses/s, were used (Wilson et al. 1992). In 1993, 200-Hz low-pass filters and stimulus rates up to 2525 pulses/s were used (Wilson et al. 1993). This system was implemented in the Advanced Bionics Clarion processor with eight bandpass channels ranging from 250 to 5500 Hz and a constant stimulus rate between 833 and 1111 pulses/s per channel for bipolar or monopolar stimulus modes (Battmer et al. 1994). However, it is still not clear up to what rates the auditory system can handle the increased information from higher stimulus rates. It has been shown, for example, that there is a decrement in the response of units in the anteroventral cochlear nucleus of the cat to stimulus rates of 800 pulses/s (Buden et al. 1996).

5.2.2 Speech Feature Recognition and Speech Perception

5.2.2.1 *Cochlear and Neural Models*

With this fixed-filter strategy, unsatisfactory results were obtained because simultaneous stimulation of the electrodes resulted in channel interaction (Clark et al. 1987). This negative result led to the important principle in cochlear implant speech processing of presenting electric stimuli nonsimultaneously.

5.2.2.2 Minimal Preprocessing of the Signal

With the Salt Lake City (Symbion or Ineraid) four-fixed-filter strategy, Dorman et al. (1989) reported that the median score in 50 patients for open sets of monosyllable words with electrical stimulation alone was 14% (range 0–60%), and for CID sentences 45% (range 0–100%). When vowel recognition was studied for 12 vowels in a b/V/t context, the median score was 60% (range 49–79%) (Dorman et al. 1989). The errors were mainly limited to vowels with similar formant frequencies. Using a closed set of consonantal stimuli, it was found that the acoustic features of manner of articulation and voicing were well recognized, but that place of articulation for stop and nasal consonants was not well identified. The patients with the highest scores exhibited superior recognition of stop consonant, place of articulation, and improved discrimination between /s/ and /ʃ/, suggesting that more information pertaining to the mid- to high frequencies was effectively transmitted (Dorman 1993).

5.2.2.3 Continuous Interleaved Sampler

An analysis of results for CIS using the Clarion system was undertaken by Schindler et al. (1995) on 91 American patients. The processor used in seven of the patients was of the compressed analog type, while the remaining 84 used the CIS strategy. In a group of 73 patients with CIS, the mean open-set CID sentence score for electrical stimulation alone was 50% at 3 months postoperatively, 58% at 6 months, and 59% at 12 months. A study by Kessler et al. (1995) reported the mean CID sentence score for the first 64 patients implanted with the Clarion device to be 60%. It is not clear to what extent there was overlap in the patients from the two studies. Kessler et al. (1995) also reported a bimodal distribution in results with a significant number of poorer performers. It is also of interest to examine the differences in information transmitted for the compressed analog and CIS strategies. In a study by Dorman (1993), seven patients had better transmission for nasality, frication, place, and envelope using CIS, indicating better resolution of frequency and envelope cues.

5.3 Multiple-Channel Strategies: Speech Feature and Spectral Cue Extraction

5.3.1 Types of Schemes

5.3.1.1 Fundamental and Second Formant Frequencies

With the original formant-extraction strategy developed in 1978 at Melbourne the second formant (F_2) frequency was extracted and presented as place of stimulation, the fundamental or voicing frequency (F_0) as rate of stimulation on individual electrodes, and the amplitude of F_2 as the electrical current level (A_2). Unvoiced sounds were considered present if the energy of the voicing frequency was low in comparison to energy of the second formant, and they were coded by a low random rate, as this was described perceptually as rough and noise-like. The first clue to developing this strategy came when it was observed that electrical stimulation at individual sites within the cochlea produced vowel-like signals, and that these sounds resembled the single-formant vowels heard by a person with normal hearing when corresponding areas in the cochlea were excited (Clark 1995).

5.3.1.2 Fundamental, First, and Second Formant Frequencies

Further research aimed, in particular, at improving the performance of multiple-channel speech processing for electrical stimulation alone, both in quiet and background noise, through better perception of consonants because of their importance for speech intelligibility. Having presented the second formant or spectral energy in the mid-frequency region on a place-coding basis, and having found results for electrical stimulation to be consistent with those for single-formant acoustic stimulation, the next step was to extract additional spectral energy and present this on a place-coded basis. The efficacy of this strategem was supported by a psychophysical study (Tong et al. 1983a), which showed that stimuli presented to two electrodes could be perceived as a two-component sensation. The anticipated improvement expected in providing first-formant information was seen in the acoustic model studies of electrical stimulation on normal-hearing individuals (Blamey et al. 1984a,b, 1985). Patients in these studies showed improved speech perception scores associated with the F_1 information transmitted. To overcome the problems of channel interaction, first demonstrated in the physiological speech-processing strategy used in 1978, nonsimultaneous, sequential pulsatile stimulation at two different sites within the cochlea was used to provide F_1 and F_2 information. The fundamental frequency was coded as rate of stimulation as with the original F_0/F_2 strategy.

5.3.1.3 Fundamental, First, and Second Formant Frequencies and High-Frequency, Fixed-Filter Outputs

The next advance in speech processing was to extract the outputs of filters in the three frequency bands—2.0–2.8 kHz, >2.8–4.0 kHz, and >4.0 kHz—and present these, as well as the first two formants, on a place-coded basis, together with voicing (represented as rate of stimulation). The high-frequency spectral information was used to provide additional high-frequency cues to improve consonant perception and speech understanding in noise. The strategy was called Multipeak, and it was implemented in a speech processor known as the miniature speech processor (MSP). The MSP differed from the WSP III processor, which was used to implement the $F_0/F_1/F_2$ strategy, in a number of ways: (1) making the relative amplitudes (A_1 and A_2) of F_1 and F_2 closer to normal level, rather than boosting the level of A_2; (2) using an alternative peak-picking algorithm for F_0 (Gruenz and Schott 1949); (3) increasing the rate of stimulation for unvoiced sounds from 130 to 260 pulses/s; and (4) making a logarithmic conversion of 256 sound-intensity levels to 32 electrical-stimulation levels rather than using a linear 32-to-32 form of conversion.

5.3.1.4 Spectral Maxima Sound Processor (SPEAK)

In early 1989 Tong et al. (1989) compared the $F_0/F_1/F_2$-WSP III system and a strategy estimating three spectral peaks from 16 bandpass filters that were presented nonsimultaneously to three electrodes on a place-coded basis. The $F_0/F_1/F_2$-WSP III system used filters to separate the speech into two bands, and then estimated the formant frequencies with zero-crossing detectors. The filter bank scheme used a simple algorithm to search for the three highest peaks, and the peak voltages determined the current levels for the three pulses. A plot of the electrode outputs for the filter bank scheme, the "electrodogram," was more similar to the spectrogram of the sound. It was found on an initial subject that the information transmission for vowels was better for the filter bank scheme, the same for consonant features, but the consonant-nucleus-consonant (CNC) word and Bench-Kowal-Bamford (BKB) sentence results were poorer (Clark et al. 1996). With the filter bank scheme, the better result for vowels could have been due to the better representation of the formants, and the worse results for words due to the poorer representation of temporal information. Because the initial results for words with the three peak-picking filter-bank strategy were not as good as the $F_0/F_1/F_2$-WSP III scheme, it was decided to develop schemes that picked more peaks (four and six), and another scheme that selected the six maximal output voltages of the 16 bandpass filters and presented these at constant rates to electrodes on a place-coded basis. The latter strategy was called the spectral maxima sound processor (SMSP). It was considered that the selection of more peaks would provide, in particular, a better place representation of frequency transitions in the speech signal.

In a study by Tong et al. (1990), the $F_0/F_1/F_2$-WSP III system was compared with a strategy where the four largest peaks were coded as place of stimulation, and F_0 coded as rate of stimulation with random stimulation for unvoiced speech. The comparison was made on two research subjects. The $F_0/F_1/F_2$-WSP III system was also compared with a strategy where the four largest spectral peaks were encoded as place of stimulation with the amplitudes of the filters setting the current levels of four electrical pulses presented simultaneously at a constant rate of 125 pulses/s. A constant rate was used to reduce the problem of channel interaction occuring with the introduction of more stimulus channels. The perception of vowels and consonants was significantly better for both filter-bank schemes compared to the $F_0/F_1/F_2$-WSP III strategy. With consonants, an improvement occurred for duration, nasality, and place features. These improvements did not carry over to the tracking of speech, and this could have been due to the small periods of utilization with the filter-bank schemes.

In 1990 the Multipeak-MSP system was compared with a filter-bank strategy that selected the four highest spectral peaks and coded these on a place basis (Tong et al 1989a,b). A constant stimulus rate of 166 pulses/s rather than 125 pulses/s was used to increase the sampling rate and thus the amount of temporal information transmitted. This strategy was implemented using the Motorola DSP56001 digital signal processor. Tong et al. (1990) showed better results for consonants and vowels with the processor extracting four spectral peaks. As the selection of four peaks gave improved scores, it was considered that the selection of six peaks might provide even more useful information, but this was found not to be the case in a pilot study on one patient. It therefore was decided to develop a strategy that extracted six spectral maxima instead of six peaks. In the latter case the voltage outputs of the filters were also presented at a constant rate of stimulation (166 pulses/s), as had been the case with some of the peak-picking strategies described above.

This strategy, the SMSP scheme, was tested on a single patient (P.S.) with an analog implementation of the strategy using an NEC filter-bank chip (D7763), and was found to give substantial benefit. For this reason, in 1990 a pilot study was carried out on two additional patients comparing this SMSP strategy and analog processor with the $F_0/F_1/F_2$-MSP system (McKay et al. 1991). The study showed significantly better scores for closed sets of consonants and open sets of words for electrical stimulation alone using the SMSP system. An initial evaluation of the Multipeak-MSP system was also carried out on one of the patients, and the results for electrical stimulation alone for consonants, CNC words, and CID sentences were better for the SMSP system. The SMSP system was then assessed on four patients who were converted from the Multipeak-MSP system. The average scores for closed sets of vowels (76% to 91%) and consonants (59% to 75%), and open sets of CNC words (40% to 57%) and words in sentences (81% to 92%) improved for the SMSP system (McDermott et al. 1992; McKay et al. 1992). In view of these encouraging results this SMSP strategy was implemented by Cochlear Limited as SPEAK. The SPEAK strategy was implemented (Seligman and McDermott 1995) in a processor referred to as Spectra-22. SPEAK-Spectra-22 differed from SMSP and its analog implementation in being able to select six or more spectral maxima from 20 rather than 16 filters.

5.3.1.5 Advanced Combination Encoder (ACE)

To further improve the transmission of information to the central nervous system, the Advanced Combination Encoder (ACE) presented in SMSP or SPEAK strategy at rates up to 1800 pulses/s, and stimulated on 6 to 20 channels. This allowed individual variations in patients' responses to rate and channel numbers to be optimized.

Table 8.1. Speech features for F_0 and for F_2 speech-processing strategies: electrical stimulation alone—patient MC-1 (Clark et al. 1984)

Features	F_0 (%)	F_0–F_2 (%)
Voicing	26	25
Nasality	5	10
Affrication	11	28
Duration	10	80
Place	4	19
Overall	35	42

5.3.2 Speech Feature Recognition and Speech Perception

5.3.2.1 Fundamental and Second Formant Frequencies

The F_0/F_2 strategy and WSP II speech processor underwent clinical trials supervised by the U.S. Food and Drug Administration (FDA) on 40 postlinguistically deaf adults from nine centers worldwide. Three months postimplantation the patients had obtained a mean CID sentence score of 87% (range 45–100%) for lipreading plus electrical stimulation, compared to a score of 52% (range 15–85%) for lipreading alone. In a subgroup of 23 patients, the mean CID sentence scores for electrical stimulation alone rose from 16% (range 0–58%) at 3 months postimplantation to 40% (range 0–86%) at 12 months (Dowell et al. 1986). The F_0/F_2-WSP II was approved by the FDA in 1985 for use in postlinguistically deaf adults. The transmission of speech information pertaining to 12 consonants utilizing two separate strategies on the University of Melbourne's wearable speech processor was determined for electrical stimulation alone on a single patient, MC-1 (Clark et al. 1984). The results for a strategy extracting only the fundamental frequency, F_0, and presenting this to a single electrode were compared with the F_0/F_2 strategy. The data are shown in Table 8.1, where it can be seen that the addition of F_2 resulted in improved nasality, affrication, duration, and place information.

5.3.2.2 Fundamental, First and Second Formant Frequencies

Before developing the $F_0/F_1/F_2$ strategy for electrical stimulation, it was first implemented as an acoustic model of electrical stimulation, and tested on normal-hearing subjects (Blamey et al. 1984a). The model stimuli were generated from a pseudorandom, white-noise generator with the output fed through seven separate bandpass filters with center frequencies corresponding to the electrode sites. The psychophysical test results were similar to those for multiple-channel electrical stimulation. Information transmission for the F_0/F_2 and $F_0/F_1/F_2$ strategies using the acoustic model is shown in Table 8.2. It should it be noted that the F_0/F_2 acoustic model yielded results comparable to the F_0/F_2 strategy for electrical stimulation except for significantly increased nasality. When the F_0/F_2 strategy was compared with $F_0/F_1/F_2$, there was shown

TABLE 8.2. Speech features for acoustic models of F_0 and F_2 and $F_0/F_1/F_2$ as well as electrical stimulation with $F_0/F_1/F_2$ strategy (Blamey et al. 1985; Clark 1986, 1987; Dowell et al. 1987)

	Acoustic model of electrical stimulation		Electrical stimulation alone
Features	F_0–F_2 (%)	F_0–F_1–F_2 (%)	F_0–F_1–F_2 (%)
Total	43	49	50
Voicing	34	50	56
Nasality	84	98	49
Affrication	32	40	45
Duration	71	81	—
Place	28	28	35
Amplitude envelope	46	61	54
High F_2	68	64	48

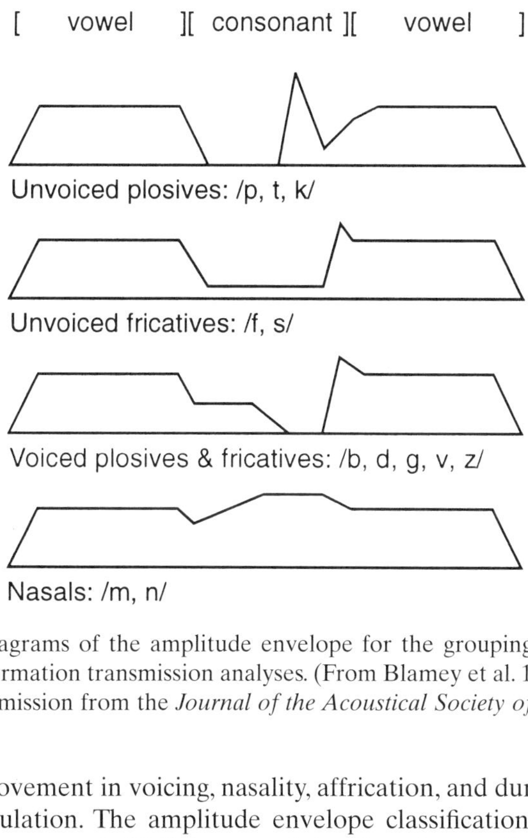

FIGURE 8.6. Diagrams of the amplitude envelope for the grouping of consonants used in the information transmission analyses. (From Blamey et al. 1985, and reproduced with permission from the *Journal of the Acoustical Society of America*.)

to be an improvement in voicing, nasality, affrication, and duration, but not place of articulation. The amplitude envelope classification, as shown in Figure 8.6 (Blamey et al. 1985), improved significantly as well.

A comparison was made in Melbourne of the F_0/F_2-WSP II used on 13 postlinguistically deaf adults and the $F_0/F_1/F_2$-WSP III systems on nine patients (Dowell et al. 1987). The results for electrical stimulation alone were recorded 3 months postoperatively. The average medial vowel score increased from 51% to 58%, the initial and final consonants from 54% to 67%, and the open-set CID sentence score from 16% to 35%. These improvements were the result of adding F_1 information. A comparison was also made of the two speech-processing strategies in background noise on two groups of five patients using each strategy. The results of a four-choice spondee test using multispeaker babble showed the $F_0/F_1/F_2$ was significantly better at a signal-to-noise ratio of 10 dB. The $F_0/F_1/F_2$-WSP III speech processor was approved by the FDA in May 1986 for use in postlinguistically deaf adults. When the $F_0/F_1/F_2$ strategy was implemented as a speech processor for electrical stimulation and evaluated on patients, the information transmission was similar to that of the acoustic model except for nasality, which was significantly less with electrical stimulation as also occurred with F_0/F_2 (Table 8.2) (Clark 1986; Dowell et al. 1987). The place of articulation feature was improved with electrical stimulation with the $F_0/F_1/F_2$.

5.3.2.3 Fundamental, First, and Second Formant Frequencies and High-Frequency Fixed-Filter Outputs (Multipeak Strategy)

An initial study (Dowell et al. 1990) was undertaken to compare a group of four experienced subjects who used the WSP III speech processor with the $F_0/F_1/F_2$ speech-processing strategy, as well as four patients who used the newer MSP speech processor with the Multipeak strategy. The patients were not selected using any special criteria except their availability and their willingness to participate in research studies. The results showed, for quiet listening conditions, a statistically significant difference for vowels using the Multipeak-MSP system; however, this benefit did not extend to consonants. For open-set BKB sentences, there was a statistically significant improvement in quiet and noise. The differences in results became greater with lower signal-to-noise ratios. The information transmitted for vowels and consonants with the $F_0/F_1/F_2$ and Multipeak strategies were compared in four subjects. With vowels the information transmitted for F_1 and F_2 increased with the Multipeak strategy, and the identification scores improved from 80% to 88% (Dowell 1991). With consonants the information increased for place, frication, nasality, and voicing, and the identification scores increased from 48% to 63% (Dowell 1991). The improvement was probably due to additional high-frequency spectral information, but could also have been due to improvements in the speech processor. A study was undertaken at the Washington University School of Medicine to help confirm the findings from the clinic in Melbourne (Skinner et al. 1991). Seven postlinguistically deaf adults who used the $F_0/F_1/F_2$-WSP III underwent clinical trials with $F_0/F_1/F_2$-MSP, and Multipeak-MSP systems. The Multipeak-MSP system yielded significantly higher scores for open-set speech tests in quiet and in noise compared to the $F_0/F_1/F_2$-WSP III system.

The results were similar to those obtained in the Melbourne study. However, there was no significant difference in speech perception scores for the $F_0/F_1/F_2$-WSP III and $F_0/F_1/F_2$-MSP systems, indicating that the improvements with Multipeak-MSP were not due to better engineering of the speech processor. The Multipeak-MSP system was approved by the FDA in 1989 for use in postlinguistically deaf adults.

5.3.2.4 SPEAK

A multicenter comparison of the SPEAK-Spectra-22 and Multipeak-MSP systems was undertaken to establish the benefits of the SPEAK-Spectra-22 system (Skinner et al. 1994). The field trial was on 63 postlinguistically and profoundly deaf adults at eight centers in Australia, North America, and the United Kingdom. A single-subject A/B : A/B design was used. The mean scores for vowels, consonants, CNC words, and words in the CUNY and SIT sentences in quiet were all significantly better for SPEAK at the $p = .0001$ level of significance. The mean score for words in sentences was 76% for SPEAK-Spectra-22 and 67% for Multipeak-MSP. SPEAK performed particularly well in noise. For the 18 subjects who underwent the CUNY and SIT sentence tests at a signal-to-noise ratio of 5 dB, the mean score for words in sentences was 60% for SPEAK and 32% for Multipeak-MSP. SPEAK-Spectra-22 was approved by the FDA for postlinguistically deaf adults in 1994. The speech information transmitted for closed sets of vowels and consonants for the SPEAK-Spectra-22 system (McKay and McDermott 1993) showed an improvement for F_1 and F_2 in vowels, as well as place- and manner-of-articulation distinctions for consonants. The differences in information presented to the auditory nervous system can be seen in the outputs to the electrodes for different words, and are plotted as electrodograms for the word "choice" in Figure 8.7. From this it can be seen there is better representation of transitions and more spectral information presented on a place-coded basis with the SPEAK-Spectra-22 system.

5.3.2.5 ACE

A flexible strategy called ACE was implemented which would allow the presentation of SPEAK at different rates and stimulus channels.

A study on the effects of low (250 pulses/s) and high (800 pulses/s and 1600 pulses/s) rates of stimulation was first carried out for CUNY sentences on five subjects. The mean results for the lowest signal-to-noise ratio (Vandali et al. 2000) show there was a significantly poorer performance for the highest rate. However, the scores varied in the five individuals. Subject #1 performed best at 807 pulses/, subject #4 was poorest at 807 pulses/s, and #5 poorest at 1615 pulses/s. There was thus significant inter- subject variability for SPEAK at different rates. These differences require further investigation.

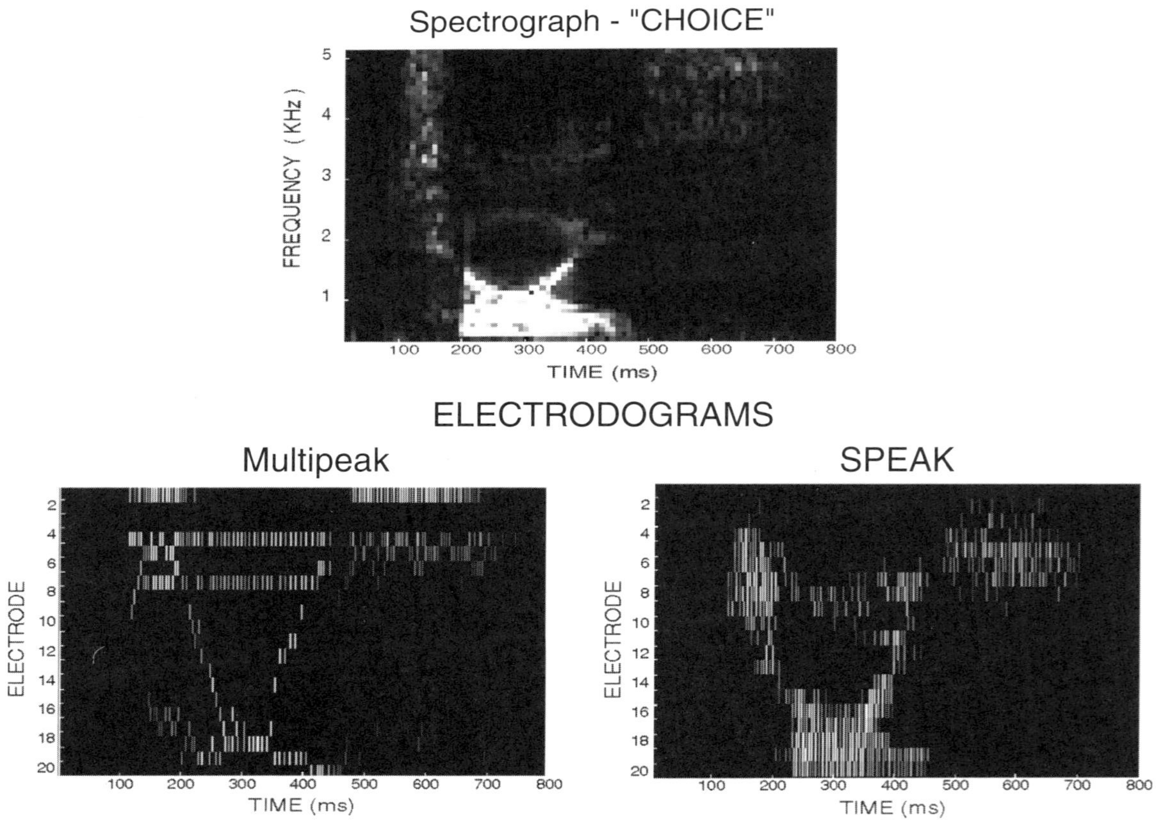

FIGURE 8.7. Spectrogram for the word "choice" and the electrode representations (electrodograms) for this word using the Multipeak, continuous interleaved sampler (CIS), and SPEAK strategies.

5.4 Comparison of Speech-Processing Strategies

5.4.1 $F_0/F_1/F_2$-WSP III and Multipeak-MSP versus the Four-Fixed-Filter Scheme

The Symbion/Ineraid four-fixed-filter system was compared with the Nucleus $F_0/F_1/F_2$-WSP III and Multipeak-MSP systems in a controlled study by Cohen et al. (1993). They tested for prosody, phoneme, spondee, and open-set speech recognition, and found a significant difference between the Multipeak-MSP and Symbion or Ineraid systems, particularly for the perception of open-set speech presented by electrical stimulation alone. There was an increase in the mean speech scores from approximately 42% with the Symbion or Ineraid system to approximately 75% with the Multipeak-MSP system. On the other hand, there was no significant difference between the $F_0/F_1/F_2$-WSP III and Symbion or Ineraid systems. The data suggest that if one looks at the place coding of spectral information alone, the preprocessing of speech into two stimulus channels with the $F_0/F_1/F_2$ strategy gave comparable results to presenting the outputs from four bandpass filters with the Symbion or Ineraid. Intermediate pitch percepts could explain the comparable results with averaging across the filters probably giving a similar representation of the formants.

The advantage of undertaking appropriate preprocessing of speech is suggested from the comparison of the Multipeak-MSP and Symbion or Ineraid systems. Both speech processing strategies presented information along approximately the same number of channels (five for Multipeak and six for Ineraid), but with Multipeak there were significantly better results.

5.4.2 SPEAK-Spectra-22 System versus CIS-Clarion System

Figure 8.8 shows the open-set CID sentence scores for electrical stimulation alone 6 months postoperatively for the CIS-Clarion system on 64 patients (Kessler et al. 1995), as well as the scores associated with the SPEAK-Spectra-22 system on 51 unselected patients tested from 2 weeks to 6 months after the startup. The data for SPEAK-Spectra-22 were presented to the FDA for evaluation in 1996. Both speech-processing systems are similar in that six stimulus channels are stimulated at a constant rate. However, with SPEAK, the stimulus channels were derived from the six spectral maxima, and with CIS from six fixed filters. If it is assumed that the higher stimulus rate of CIS (up to 800 pulses/s), works positively in its favor, then the selection of spectral maxima is an important requirement for cochlear-implant speech processing as the results for SPEAK are at least as good or possibly better.

5.4.3 The ACE versus SPEAK versus CIS strategies

The ACE strategy was also evaluated in a larger study on 62 post-linguistically deaf adults who were users of SPEAK at 21 centers in the US (Arndt et al.

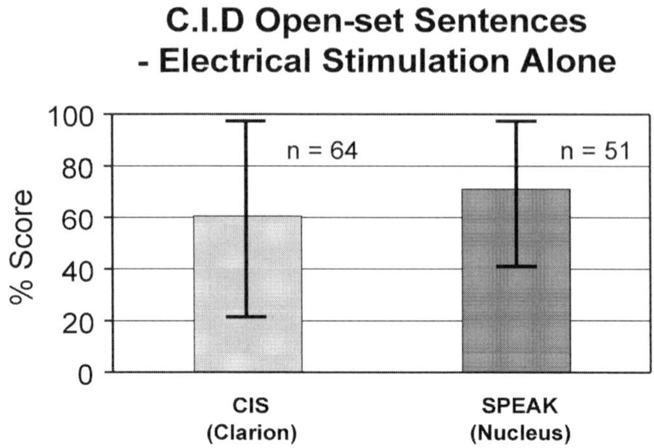

FIGURE 8.8. The mean open-set Central Institute for the Deaf (CID) sentence score of 71% for the SPEAK (University of Melbourne/Nucleus) strategy on 51 patients (data presented to the Food and Drug Administration January 1996) and 60% for the CIS (Clarion) strategy on 64 patients (Kessler et al. 1995).

1999). ACE was compared with SPEAK and CIS. The rate and number of channels were optimised for ACE and CIS. Mean HINT (Nilsson et al. 1994) sentence scores in quiet were 64.2% for SPEAK, 66.0% for CIS, and 72.3% for ACE. The ACE mean was significantly higher than the CIS mean ($p < 0.05$), but not significantly different from SPEAK. The mean CUNY sentence recognition at a signal-to-noise ratio of 10 dB was significantly better for ACE (71.0%) than both CIS (65.3%) and SPEAK (63.1%). Overall, 61% preferred ACE, 23% SPEAK, and 8% CIS. The strategy preference correlated highly with speech recognition. Furthermore, one third of the subjects used different strategies for different listening conditions.

6. Speech Processing for Prelinguistically and Postlinguistically Deaf Children

6.1 Single-Channel Strategies

6.1.1 Speech Feature Recognition and Speech Perception

In the early 1980s the Los Angeles 3M single-channel implant was first implanted in a young patient. The results for this device on 49 children, ranging in age from 2 to 17 years, were reported by Luxford et al. (1987), and showed children could discriminate syllable patterns, but only two patients from this group could be provided with any degree of open-set comprehension using the device. The single-channel device permitted speech and syllable-pattern discrimination, but did not provide sufficient auditory information for most children to identify or comprehend significant amounts of speech information.

6.2 Multiple-Channel Strategies

6.2.1 Speech Feature Recognition and Speech Perception

6.2.1.1 Fundamental, First, and Second Formant Frequencies

The first child to have the $F_0/F_1/F_2$-WSP III system and mini-receiver-stimulator was patient B.D., who was 5 years old when operated on in Melbourne in 1986. When it was shown he was gaining benefit, additional children received similar implants in Melbourne. In 1989 it was reported that five children (aged 6 to 14 years) out of a group of nine were able to achieve substantial open-set speech recognition for monosyllabic words scored as phonemes (range 30% to 72%), and sentences scored as keywords (range 26% to 74%) (Dawson et al. 1989). Four of the five children who achieved open-set scores were implanted before adolescence, and the fifth, who had a progressive loss, was implanted as an adolescent. These children also showed improvement in language communication. The children who were unable to achieve good open-set speech recognition were those implanted during adolescence after a long period of profound deafness. The results of the study were published in more detail in Dawson et al. (1992). After the initial success in Melbourne a clinical trial involving 142 children at 23 centers commenced on February 6, 1987. In this trial at least one speech test was used in the following categories: suprasegmental information, closed-set word identification, and open-set word recognition (Staller et al. 1991). The tests were appropriate for the developmental age of the child, and were administered 12 months postoperatively. The results showed that 51% of the children could achieve significant open-set speech recognition with their cochlear prosthesis compared with 6% preoperatively. Their performance also improved over time, with significant improvement in open- and closed-set speech recognition performance at between 1 and 3 years postoperatively.

When the results on 91 prelinguistically deaf children were examined separately, it was found that improvements were comparable with the postlinguistic group for most tests except for open-sets of words, where the results were poorer. The $F_0/F_1/F_2$-WSP III system was approved by the FDA for use in children in 1990.

6.2.1.2 Multipeak

Ten children with the $F_0/F_1/F_2$-WSP III system were changed over to the Multipeak-MSP system in 1989. Apart from an initial decrement of response in one child, performance continued to improve in five and was comparable for the other children. As a controlled trial was not carried out, it was not clear whether the improvements were due to learning or to the new strategy and processor. The Multipeak-MSP system was also approved by the FDA for use in children in 1990 on the basis of the $F_0/F_1/F_2$-WSP III approval for children and the Multipeak-MSP approval for adults.

6.2.1.3 SPEAK

After it was shown that the results for SPEAK-Spectra-22 were better than Multipeak-MSP for postlinguistically deaf adults, a study was performed to determine if prelinguistically and postlinguistically deaf children could be changed over to the SPEAK-Spectra-22 system and gain comparable benefit. Would children who had effectively "learned to listen" through their cochlear implant using the Multipeak strategy be able to adapt to a "new" signal, and would they in fact benefit from any increase in spectral and temporal information available from the SPEAK system? Furthermore, as children are often in poor signal-to-noise situations in integrated classrooms, it was of great interest to find out if children using the SPEAK processing strategy would show similar perceptual benefits in background noise as those shown for adult patients. To answer these questions, speech perception results for a group of 12 profoundly hearing-impaired children using SPEAK were compared with the benefits these children received using the Multipeak speech-processing strategy. The children were selected on the basis of being able to achieve a score for CNC words using electrical stimulation alone.

Comparison of mean scores for the 12 children on open-set word and sentence scores showed a significant advantage for the SPEAK strategy as compared with Multipeak in both quiet and 15 dB signal-to-noise ratio conditions. The SPEAK-Spectra 22 was approved by the FDA for children in 1994.

6.2.1.4 ACE

The ACE strategy has been evaluated on 256 children for the US FDA (Staller et al. 2002). There were significant improvements for all age appropriate speech perception and language appropriate tests.

7. Summary

During the last 20 years, considerable advances have been made in the development of cochlear implants for the profoundly deaf. It has been shown that multiple-channel devices are superior to single-channel systems. Strategies in which several electrodes (six to eight) correspond to fixed-filter outputs, or the extraction of six to eight spectral maxima for 20 to 22 electrodes offer better speech perception than stimulation with second and first formants at individual sites in the cochlea, provided that nonsimultaneous or interleaved presentation is employed to minimize current leakage between the electrodes. Further refinements such as spectral maxima at rates of approximately 800 to 1600 pulses/s and the extraction of speech transients also give improvements for a number of patients.

Successful speech recognition by many prelinguistically deafened children as well as by postlinguistically deaf children has been achieved.

If children are implanted before 2 years of age and have good language training, they can achieve speech perception, speech production, and expressive and receptive language at levels that are normal for their chronological age. The main restriction on the amount of information that can be presented to the auditory nervous system is the electroneural "bottleneck" caused by the relatively small number of electrodes (presently 22) that can be inserted into the cochlea and the limited dynamic range of effective stimulation. Strategies to overcome this restriction continue to be developed.

List of Abbreviations

ACE	Advanced Combination Encoder
BKB	Bench-Kowal-Bamford (Australian Sentence Test)
CID	Central Institute for the Deaf
CIS	continuous interleaved sampler
CNC	consonant-nucleus-consonant
CUNY	City University of New York
DL	difference limen
DSP	digital signal processor
FDA	United States Food and Drug Administration
FFT	fast Fourier transform
F_0	fundamental frequency
F_1	first formant
F_2	second formant
MSP	miniature speech processor
RF	radiofrequency
SMSP	spectral maxima sound processor

References

Aitkin LM (1986) The Auditory Midbrain: Structure and Function in the Central Auditory Pathway. Clifton, NJ: Humana Press.

Arndt P, Staller S, Arcoroli J, Hines A, Ebinger K (1999) Within-subject comparison of advanced coding strategies in the Nucleus 24 cochlear implant. Cochlear Corporation.

Bacon SP, Gleitman RM (1992) Modulation detection in subjects with relatively flat hearing losses. J Speech Hear Res 35:642–653.

Battmer R-D, Gnadeberg D, Allum-Mecklenburg DJ, Lenarz T (1994) Matched-pair comparisons for adults using the Clarion or Nucleus devices. Ann Oto Rhino Laryngol 104(suppl 166):251–254.

Bilger RC, Black RO, Hopkinson NT (1977) Evaluation of subjects presently fitted with implanted auditory prostheses. Ann Oto Rhino Laryngol 86(suppl 38):1–176.

Black RC, Clark GM (1977) Electrical transmission line properties in the cat cochlea. Proc Austral Physiol Pharm Soc 8:137.

Black RC, Clark GM (1978) Electrical network properties and distribution of potentials in the cat cochlea. Proc Austral Physiol Pharm Soc 9:71.

Black RC, Clark GM (1980) Differential electrical excitation of the auditory nerve. J Acoust Soc Am 67:868–874.

Black RC, Clark GM, Patrick JF (1981) Current distribution measurements within the human cochlea. IEEE Trans Biomed Eng 28:721–724.

Blamey PJ, Dowell RC, Tong YC, Brown AM, Luscombe SM, Clark GM (1984a) Speech processing studies using an acoustic model of a multiple-channel cochlear implant. J Acoust Soc Am 76:104–110.

Blamey PJ, Dowell RC, Tong YC, Clark GM (1984b) An acoustic model of a multiple-channel cochlear implant. J Acoust Soc Am 76:97–103.

Blamey PJ, Martin LFA, Clark GM (1985) A comparison of three speech coding strategies using an acoustic model of a cochlear implant. J Acoust Soc Am 77:209–217.

Blamey PJ, Parisi ES, Clark GM (1995) Pitch matching of electric and acoustic stimuli. In: Clark GM, Cowan RSC (eds) The International Cochlear Implant, Speech and Hearing Symposium, Melbourne, suppl 166, vol 104, no 9, part 2. St. Louis: Annals, pp. 220–222.

Brimacombe JA, Arndt PL, Staller SJ, Menapace CM (1995) Multichannel cochlear implants in adults with residual hearing. NIH Consensus Development Conference on Cochlear Implants in Adults and Children, May 15–16.

Brugge JF, Kitzes L, Javel E (1981) Postnatal development of frequency and intensity sensitivity of neurons in the anteroventral cochlear nucleus of kittens. Hear Res 5:217–229.

Buden SV, Brown M, Paolini G, Clark GM (1996) Temporal and entrainment response properties of cochlear nucleus neurons to intra cochleal electrical stimulation in the cat. Proc 16th Ann Austral Neurosci Mgt 8:104.

Burns EM, Viemeister NG (1981) Played-again SAM: further observations on the pitch of amplitude-modulated noise. J Acoust Soc Am 70:1655–1660.

Busby PA, Clark GM (1996) Spatial resolution in early deafened cochlear implant patients. Proc Third European Symposium Pediatric Cochlear Implantation, Hannover, June 5–8.

Busby PA, Clark GM (1997) Pitch and loudness estimation for single and multiple pulse per period electric pulse rates by cochlear implant patients. J Acoust Soc Am 101:1687–1695.

Busby PA, Clark GM (2000a) Electrode discrimination by early-deafened subjects using the Cochlear Limited multiple-electrode cochlear implant. Ear Hear 21: 291–304.

Busby PA, Clark GM (2000b) Pitch estimation by early-deafened subjects using a multiple-electrode cochlear implant. J Acoust Soc Am 107:547–558.

Busby PA, Tong YC, Clark GM (1992) Psychophysical studies using a multiple-electrode cochlear implant in patients who were deafened early in life. Audiology 31:95–111.

Busby PA, Tong YC, Clark GM (1993a) The perception of temporal modulations by cochlear implant patients. J Acoust Soc Am 94:124–131.

Busby PA, Roberts SA, Tong YC, Clark GM (1993b) Electrode position, repetition rate and speech perception early- and late-deafened cochlear implant patients. J Acoust Soc Am 93:1058–1067.

Busby PA, W[illegible]ford LA, Blamey PJ, Richardson LM, Clark GM (1994) Pitch perception for different modes of stimulation using the Cochlear multiple-electrode prosthesis. J Acoust Soc Am 95:2658–2669.

Clark GM (1969) Responses of cells in the superior olivary complex of the cat to electrical stimulation of the auditory nerve. Exp Neurol 24:124–136.

Clark GM (1986) The University of Melbourne/Cochlear Corporation (Nucleus) Program. In: Balkany T (ed) The Cochlear Implant. Philadephia: Saunders.

Clark GM (1987) The University of Melbourne–Nucleus multi-electrode cochlear implant. Basel: Karger.

Clark GM (1995) Cochlear implants: historical perspectives. In: Plant G, Spens K-E (eds) Profound Deafness and Speech Communication. London: Whurr, pp. 165–218.

Clark GM (1996a) Electrical stimulation of the auditory nerve, the coding of sound frequency, the perception of pitch and the development of cochlear implant speech processing strategies for profoundly deaf people. J Clin Physiol Pharm Res 23:766–776.

Clark GM (1996b) Cochlear implant speech processing for severely-to-profoundly deaf people. Proc ESCA Tutorial and Research Workshop on the Auditory Basis of Speech Perception, Keele University, United Kingdom.

Clark GM (1998) Cochlear implants. In: Wright A, Ludman H (eds) Diseases of the Ear. London: Edward Arnold, pp. 149–163.

Clark GM (2001) Editorial. Cochlear implants: climbing new mountains. The Graham Fraser Memorial Lecture 2001. Cochlear Implants Int 2(2):75–97.

Clark GM (2003) Cochlear Implants: Fundamentals and Applications. New York: Springer-Verlag.

Clark GM, Tong YC (1990) Electrical stimulation, physiological and behavioural studies. In: Clark GM, Tong YC, Patrick JF (eds) Cochlear Prostheses. Edinburgh: Churchill Livingstone.

Clark GM, Nathar JM, Kranz HG, Maritz JSA (1972) Behavioural study on electrical stimulation of the cochlea and central auditory pathways of the cat. Exp Neurol 36:350–361.

Clark GM, Kranz HG, Minas HJ (1973) Behavioural thresholds in the cat to frequency modulated sound and electrical stimulation of the auditory nerve. Exp Neurol 41:190–200.

Clark GM, Tong YC, Dowell RC (1984) Comparison of two cochlear implant speech processing strategies. Ann Oto Rhino Laryngol 93:127–131.

Clark GM, Carter TD, Maffi CL, Shepherd RK (1995) Temporal coding of frequency: neuron firing probabilities for acoustical and electrical stimulation of the auditory nerve. Ann Otol Rhinol Laryngol 104(suppl 166):109–111.

Clark GM, Dowell RC, Cowan RSC, Pyman BC, Webb RL (1996) Multicentre evaluations of speech perception in adults and children with the Nucleus (Cochlear) 22-channel cochlear implant. IIIrd Int Symp Transplants Implants Otol, Bordeaux, June 10–14, 1995.

Cohen NL, Waltzman SB, Fisher SG (1993) A prospective, randomized study of cochlear implants. N Engl J Med 328:233–282.

Cowan RSC, Brown C, Whitford LA, et al. (1995) Speech perception in children using the advanced SPEAK speech processing strategy. Ann Otol Rhinol Laryngol 104(suppl 166):318–321.

Cowan RSC, Brown C, Shaw S, et al. (1996) Comparative evaluation of SPEAK and MPEAK speech processing strategies in children using the Nucleus 22-channel cochlear implant. Ear Hear (submitted).

Dawson PW, Blamey PJ, Clark GM, et al. (1989) Results in children using the 22 electrode cochlear implant. J Acoust Soc Am 86(suppl 1):81.

Dawson PW, Blamey PJ, Rowland LC, et al. (1992) Cochlear implants in children, adolescents and prelinguistically deafened adults: speech perception. J Speech Hear Res 35:401–417.

Dorman MF (1993) Speech perception by adults. In: Tyler RS (ed) Cochlear Implants. Audiological Foundations. San Diego: Singular, pp. 145–190.

Dorman M, Dankowski K, McCandless G (1989) Consonant recognition as a function of the number of channels of stimulation by patients who use the Symbion cochlear implant. Ear Hear 10:288–291.

Dowell, RC (1991) Speech Perception in Noise for Multichannel Cochlear Implant Users. Doctor of philosophy thesis, The University of Melbourne.

Dowell RC, Mecklenburg DJ, Clark GM (1986) Speech recognition for 40 patients receiving multichannel cochlear implants. Arch Otolaryngol 112:1054–1059.

Dowell RC, Seligman PM, Blamey PJ, Clark GM (1987) Speech perception using a two-formant 22-electrode cochlear prosthesis in quiet and in noise. Acta Otolaryngol (Stockh) 104:439–446.

Dowell RC, Whitford LA, Seligman PM, Franz BK, Clark GM (1990) Preliminary results with a miniature speech processor for the 22-electrode Melbourne/Cochlear hearing prosthesis. Otorhinolaryngology, Head and Neck Surgery. Proc XIV Congress Oto-Rhino-Laryngology, Head and Neck Surgery, Madrid, Spain, pp. 1167–1173.

Dowell RC, Blamey PJ, Clark GM (1995) Potential and limitations of cochlear implants in children. Ann Otol Rhinol Laryngol 104(suppl 166):324–327.

Dowell RC, Dettman SJ, Blamey PJ, Barker EJ, Clark GM (2002) Speech perception in children using cochlear implants: prediction of long-term outcomes. Cochlear Implants Int 3:1–18.

Eddington DK (1980) Speech discrimination in deaf subjects with cochlear implants. J Acoust Soc Am 68:886–891.

Eddington DK (1983) Speech recognition in deaf subjects with multichannel intracochlear electrodes. Ann NY Acad Sci 405:241–258.

Eddington DK, Dobelle WH, Brackman EE, Brackman DE, Mladejovsky MG, Parkin JL (1978) Auditory prosthesis research with multiple channel intracochlear stimulation in man. Ann Otol Rhino Laryngol 87(suppl 53):5–39.

Evans EF (1978) Peripheral auditory processing in normal and abnormal ears: physiological considerations for attempts to compensate for auditory deficits by acoustic and electrical prostheses. Scand Audiol Suppl 6:10–46.

Evans EF (1981) The dynamic range problem: place and time coding at the level of the cochlear nerve and nucleus. In: Syka J, Aitkin L (eds) Neuronal Mechanisms of Hearing. New York: Plenum, pp. 69–85.

Evans EF, Wilson JP (1975) Cochlear tuning properties: concurrent basilar membrane and single nerve fiber measurements. Science 190:1218–1221.

Fourcin AJ, Rosen SM, Moore BCJ (1979) External electrical stimulation of the cochlea: clinical, psychophysical, speech-perceptual and histological findings. Br J Audiol 13:85–107.

Gantz BJ, McCabe BF, Tyler RS, Preece JP (1987) Evaluation of four cochlear implant designs. Ann Otol Rhino Laryngol 96:145–147.

Glattke T (1976) Cochlear implants: technical and clinical implications. Laryngoscope 86:1351–1358.

Gruenz OO, Schott LA (1949) Extraction and portrayal of pitch of speech sounds. J Acoust Soc Am 21:5, 487–495.

Hochmair ES, Hochmair-Desoyer IJ, Burian K (1979) Investigations towards an artificial cochlea. Int J Artif Organs 2:255–261.

Hochmair-Desoyer IJ, Hochmair ES, Fischer RE, Burian K (1980) Cochlear prostheses in use: recent speech comprehension results. Arch Otorhinolaryngol 229: 81–98.

Hochmair-Desoyer IJ, Hochmair ES, Burian K (1981) Four years of experience with cochlear prostheses. Med Prog Tech 8:107–119.

House WF, Berliner KI, Eisenberg LS (1981) The cochlear implant: 1980 update. Acta Otolaryngol 91:457–462.

Irlicht L, Clark GM (1995) Control strategies for nerves modeled by self-exciting point processes. In: Clark GM, Cowan RSC (eds) The International Cochlear Implant, Speech & Hearing Symposium, Melbourne 1994. St Louis: Annals, pp. 361–363.

Irlicht L, Au D, Clark GM (1995) A new temporal coding scheme for auditory nerve stimulation. In: Clark GM, Cowan RSC (eds) The International Cochlear Implant, Speech and Hearing Symposium, Melbourne 1994. St Louis: Annals, pp. 358–360.

Irvine DRF (1986) The Auditory Brainstem. A Review of the Structure and Function of Auditory Brainstem Processing Mechanisms. Berlin: Springer-Verlag.

Javel E, Tong YC, Shepherd RK, Clark GM (1987) Responses of cat auditory nerve fibers to biphasic electrical current pulses. Ann Otol Rhinol Laryngol 96(suppl 128):26–30.

Katsuki Y, Suga N, Kanno Y (1962) Neural mechanism of the peripheral and central auditory system in monkeys. J Acoust Soc Am 34:1396–1410.

Kessler DK, Loeb GE, Barker MJ (1995) Distribution of speech recognition results with the Clarion cochlear prosthesis. Ann Otol Rhino Laryngol 104(suppl 166) (9):283–285.

Kiang NYS (1966) Stimulus coding in the auditory nerve and cochlear nucleus. Acta Otolaryngol 59:186–200.

Kiang NYS, Moxon EC (1972) Physiological considerations in artificial stimulation of the inner ear. Ann Otol Rhinol Laryngol 81:714–729.

Kiang NYS, Pfeiffer RF, Warr WB (1965) Stimulus coding in the cochlear nucleus. Ann Otol Rhino Laryngol 74:2–23.

Laird RK (1979) The bioengineering development of a sound encoder for an implantable hearing prosthesis for the profoundly deaf. Master of engineering science thesis, University of Melbourne.

Luxford WM, Berliner KI, Eisenberg MA, House WF (1987) Cochlear implants in children. Ann Otol 94:136–138.

McDermott HJ, McKay CM (1994) Pitch ranking with non-simultaneous dual-electrode electrical stimulation of the cochlea. J Acoust Soc Am 96:155–162.

McDermott HJ, McKay CM, Vandali AE (1992) A new portable sound processor for the University of Melbourne/Nucleus Limited multi-electrode cochlear implant. J Acoust Soc Am 91:3367–3371.

McKay CM, McDermott HJ (1993) Perceptual performance of subjects with cochlear implants using the Spectral Maxima Sound Processor (SMSP) and the Mini Speech Processor (MSP). Ear Hear 14:350–367.

McKay CM, McDermott HJ, Clark GM (1991) Preliminary results with a six spectral maxima speech processor for the University of Melbourne/Nucleus multiple-electrode cochlear implant. J Otolaryngol Soc Aust 6:354–359.

McKay CM, McDermott HJ, Vandali AE, Clark GM (1992) A comparison of speech perception of cochlear implantees using the Spectral Maxima Sound Processor

(SMSP) and the MSP (MULTIPEAK) processor. Acta Otolaryngol (Stockh) 112: 752–761.

McKay CM, McDermott HJ, Clark GM (1995) Pitch matching of amplitude modulated current pulse trains by cochlear implantees: the effect of modulation depth. J Acoust Soc Am 97:1777–1785.

Merzenich MM (1975) Studies on electrical stimulation of the auditory nerve in animals and man: cochlear implants. In: Tower DB (ed) The Nervous System, vol 3, Human Communication and Its Disorders. New York: Raven Press, pp. 537–548.

Merzenich M, Byers C, White M (1984) Scala tympani electrode arrays. Fifth Quarterly Progress Report 1–11.

Moore BCJ (1989) Pitch perception. In: Moore BCJ (ed) An Introduction to the Psychology of Hearing. London: Academic Press, pp. 158–193.

Moore BCJ, Raab DH (1974) Pure-tone intensity discrimination: some experiments relating to the "near-miss" to Weber's Law. J Acoust Soc Am 55:1049–1954.

Moxon EC (1971) Neural and mechanical responses to electrical stimulation of the cat's inner ear. Doctor of philosophy thesis, Massachusetts Institute of Technology.

Nilsson M, Soli SD, Sullivan JA (1994) Development of the Hearing in Noise Test for the measurement of speech reception thresholds in quiet and in noise. Journal of the Acoustical Society of America 95(2):1085–99.

Rajan R, Irvine DRF, Calford MB, Wise LZ (1990) Effect of frequency-specific losses in cochlear neural sensitivity on the processing and representation of frequency in primary auditory cortex. In: Duncan A (ed) Effects of Noise on the Auditory System. New York: Marcel Dekker, pp. 119–129.

Recanzone GH, Schreiner CE, Merzenich MM (1993) Plasticity in the frequency representation of primary auditory cortex following discrimination training in adult owl monkeys. J Neurosci 13:87–103.

Robertson D, Irvine DRF (1989) Plasticity of frequency organization in auditory cortex of guinea pigs with partial unilateral deafness. J Comp Neurol 282:456–471.

Rose JE, Galambos R, Hughes JR (1959) Microelectrode studies of the cochlear nuclei of the cat. Bull Johns Hopkins Hosp 104:211–251.

Rose JE, Brugge JF, Anderson DJ, Hind JE (1967) Phase-locked response to low-frequency tones in single auditory nerve fibers of the squirrel monkey. J Neurophysiol 30:769–793.

Rupert A, Moushegian G, Galambos R (1963) Unit responses to sound from auditory nerve of the cat. J Neurophysiol 26:449–465.

Sachs MB, Young ED (1979) Encoding of steady-state vowels in the auditory nerve: representation in terms of discharge rate. J Acoust Soc Am 66:470–479.

Schindler RA, Kessler DK, Barker MA (1995) Clarion patient performance: an update on the clinical trials. Ann Otol Rhino Laryngol 104(suppl 166):269–272.

Seldon HL, Kawano A, Clark GM (1996) Does age at cochlear implantation affect the distribution of responding neurons in cat inferior colliculus? Hear Res 95: 108–119.

Seligman PM, McDermott HJ (1995) Architecture of the SPECTRA 22 speech processor. Ann Otol Rhinol Laryngol 104(suppl 166):139–141.

Shannon RV (1983) Multichannel electrical stimulation of the auditory nerve in man: I. Basic psychophysics. Hear Res 11:157–189.

Shannon RV (1992) Temporal modulation transfer functions in patients with cochlear implants. J Acoust Soc Am 91:2156–2164.

Simmons FB (1966) Electrical stimulation of the auditory nerve in man. Arch Otolaryngol 84:2–54.

Simmons FB, Glattke TJ (1970) Comparison of electrical and acoustical stimulation of the cat ear. Ann Otol Rhinol Laryngol 81:731–738.

Skinner MW, Holden LK, Holden TA, et al. (1991) Performance of postlinguistically deaf adults with the Wearable Speech Processor (WSP III) and Mini Speech Processor (MSP) of the Nucleus multi-electrode cochlear implant. Ear Hear 12: 3–22.

Skinner MW, Clark GM, Whitford LA, et al. (1994) Evaluation of a new Spectral Peak coding strategy for the Nucleus 22 channel cochlear implant system. Am J Otol 15:15–27.

Snyder RL, Rebscher SJ, Cao KL, Leake PA, Kelly K (1990) Chronic introcochlear electrical stimulation in the neonatally deafened cat. 1: Expansion of central representation. Hear Res 50:7–33.

Staller S, Parkinson J, Arcaroli J, Arndt P (2002) Pediatric outcomes with the Nucleus 24 contour: North American clinical trial. Ann Otol Rhino Laryngol 111(suppl 189):56–61.

Tasaki I (1954) Nerve impulses in individual auditory nerve fibers of the guinea pig. J Neurophysiol 17:7–122.

Tong YC, Black RC, Clark GM, et al. (1979) A preliminary report on a multiple-channel cochlear implant operation. J Laryngol Otol 93:679–695.

Tong YC, Clark GM, Blamey PJ, Busby PA, Dowell RC (1982) Psychophysical studies for two multiple-channel cochlear implant patients. J Acoust Soc Am 7: 153–160.

Tong YC, Blamey PJ, Dowell RC, Clark GM. (1983a) Psychophysical studies evaluating the feasibility of a speech processing strategy for a multiple-channel cochlear implant. J Acoust Soc Am 74:73–80.

Tong YC, Dowell RC, Blamey PJ, Clark GM (1983b) Two component hearing sensations produced by two-electrode stimulation in the cochlea of a totally deaf patient. Science 219:993–994.

Tong YC, Busby PA, Clark GM (1988) Perceptual studies on cochlear implant patients with early onset of profound hearing impairment prior to normal development of auditory, speech, and language skills. J Acoust Soc Am 84:951–962.

Tong YC, Harrison JM, Lim HH, et al. (1989a) Speech Processors for Auditory Prostheses. First Quarterly Progress Report NIH contract No. 1-DC-9-2400. February 1–April 30.

Tong YC, Lim HH, Harrison JM, et al. (1989b) Speech Processors for Auditory Prostheses. First Quarterly Progress Report, NIH contract No. 1-DC-9-2400. February 1–April 30.

Tong YC, van Hoesel R, Lai WK, Vandali A, Harrison JM, Clark GM (1990) Speech Processors for Auditory Prostheses. Sixth Quarterly Progress Report NIH contract No. 1-DC-9-2400. June 1–August 31.

Townshend B, Cotter NE, Van Compernolle D, White RL (1987) Pitch perception by cochlear implant subjects. J Acoust Soc Am 82:106–115.

Vandali AE, Whitford LA, Plant KL, Clark GM (2000) Speech perception as a function of electrical stimulation rate using the Nucleus 24 cochlear implant system. Ear and Hearing 21:608–624.

Viemeister NF (1974) Intensity discrimination of noise in the presence of band-reject noise. J Acoust Soc Am 56:1594–1600.

Williams AJ, Clark GM, Stanley GV (1976) Pitch discrimination in the cat through electrical stimulation of the terminal auditory nerve fibres. Physiol Psychol 4: 23–27.
Wilson BS, Lawson DT, Zerbi M, Finley CC (1992) Twelfth Quarterly Progress Report—Speech Processors for Auditory Prostheses. NIH contract No. 1-DC-9-2401. Research Triangle Institute, April.
Wilson BS, Lawson DT, Zerbi M, Finley CC (1993) Fifth Quarterly Progress Report—Speech Processors for Auditory Protheses. NIH contract No. 1-DC-2-2401. Research Triangle Institute, October.
Zeng FG, Shannon RV (1992) Loudness balance between electric and acoustic stimulation. Hear Res 60:231–235.

Index

SPRINGER HANDBOOK OF AUDITORY RESEARCH

(continued from page ii)

Volume 20: Cochlear Implants
Edited by Fan-Gang Zeng, Arthur N. Popper, and Richard R. Fay

For more information about the series, please visit www.springer-ny.com/shar.